World Soils Book Series

Series Editor

Alfred E. Hartemink
Department of Soil Science, FD Hole Soils Laboratory
University of Wisconsin–Madison
Madison, WI
USA

The World Soils Book Series publishes peer-reviewed books on the soils of a particular country. They include sections on soil research history, climate, geology, geomorphology, major soil types, soil maps, soil properties, soil classification, soil fertility, land use and vegetation, soil management, soils and humans, soils and industry, future soil issues. The books summarize what is known about the soils in a particular country in a concise and highly reader-friendly way. The series contains both single and multi-authored books as well as edited volumes. There is additional scope for regional studies within the series, particularly when covering large land masses (for example, The Soils of Texas, The Soils of California), however, these will be assessed on an individual basis.

More information about this series at http://www.springer.com/series/8915

Allan E. Hewitt • Megan R. Balks •
David J. Lowe

The Soils of Aotearoa New Zealand

Springer

Allan E. Hewitt
Manaaki Whenua - Landcare Research
Lincoln, New Zealand

Megan R. Balks
School of Science
University of Waikato
Hamilton, New Zealand

David J. Lowe
School of Science
University of Waikato
Hamilton, New Zealand

ISSN 2211-1255 ISSN 2211-1263 (electronic)
World Soils Book Series
ISBN 978-3-030-64761-2 ISBN 978-3-030-64763-6 (eBook)
https://doi.org/10.1007/978-3-030-64763-6

This Springer imprint is published by the registered company Springer Nature Switzerland AG
The registered company address is: Gewerbestrasse 11, 6330 Cham, Switzerland

Preface

Aotearoa/New Zealand is a unique archipelago surrounded by the Pacific Ocean and Tasman Sea. Soils in New Zealand are diverse—a reflection of the country's oceanic setting and varied geology, topography, climate, vegetation, and land surface ages. The landscape, ecology, and soils have evolved in isolation with the first humans arriving from eastern Polynesia only about 750 years ago. Following arrival of European settlers in the nineteenth century there was a rapid transformation as agriculture was established with accompanying large-scale changes to the flora and fauna and thus also to soils and ecosystems.

Recognising the unique assemblage of soils in New Zealand, the soil science community saw a need to develop a modern, well defined, New Zealand-specific, soil classification. The resultant *New Zealand Soil Classification* was first published in 1992 with a third edition in 2010. Although the *New Zealand Soil Classification* has been widely used by soil scientists, there has been demand for a more accessible book that gives the wider community a clear understanding of our, now well-established, soil orders as well as an appreciation of the important properties and features of New Zealand soils.

It is our intention that this book will help people understand the diversity, importance, and intrinsic beauty of New Zealand's valuable soil resource in the context of the *New Zealand Soil Classification*. It should be of interest to anyone with an involvement in the whenua/land, including students, farmers and growers, land managers, environmental planners, gardeners, and those who simply want to better appreciate the soils and landscapes. Knowledge of the soil resource and its environment, from the scale of the individual farm paddock to the wider landscape, is essential to support sustainable soil and land management.

The book is the first dealing specifically with New Zealand's soils and landscapes for more than 30 years and hence brings a wealth of new information and understanding. We nevertheless acknowledge earlier national-extent compilations: a three-part book, *Soils of New Zealand*, was published by Soil Bureau of the Department of Scientific and Industrial Research in 1968; a short text entitled *New Zealand Soils—an Introduction*, written by Harry Gibbs, was published in 1980; and the outstanding book, *Soils in the New Zealand Landscape*, written by Les Molloy and illustrated with photographs by Quenton Christie, was published in 1988 (a second printing, which included a new appendix on the *New Zealand Soil Classification*, was published in 1993).

There is a need for all New Zealanders to have a sense of the importance of the whenua/land on which they live and the need for kaitiakitanga/stewardship of our soils so that the soil can sustainably support future generations for millennia to come. We hope this book will contribute to that shared knowledge and understanding.

Lincoln, New Zealand	Allan E. Hewitt
Hamilton, New Zealand	Megan R. Balks
Hamilton, New Zealand	David J. Lowe
September 2020	

Acknowledgements

Many colleagues of Allan Hewitt from Manaaki – Whenua Landcare Research assisted Allan with advice and preparation of diagrams, in particular, Pierre Roudier, James Barringer, and Sam Carrick. Manaaki Whenua – Landcare Research supported Allan Hewitt and allowed access and use of their important soil database, maps, and other resources. The mixed media drawings of soil profiles and related diagrams are by Allan Hewitt. Photos without source acknowledgement in the text are those of the authors.

The University of Waikato provided support for Megan Balks and David Lowe.

We especially thank Marianne Coleman and Max Oulton for help in drafting diagrams, Glen Balks for computer support, and Annette Carshalton for technical assistance.

Special thanks are due to Tanya O'Neill and Malcolm McLeod for their helpful input in reviewing a draft of the entire book.

We acknowledge support from University of Waikato reference librarians Maria McGuire, Cheryl Ward, and John Wort, especially in accessing obscure articles via interloan. Vanessa Clark and Stephen Turner provided technical advice regarding the use of the term 'Aotearoa' in the book title.

Our many talented and dedicated colleagues in science, especially soil scientists, together with farmers, growers, land managers, environmental regulators and planners, and numerous students, have provided us with enthusiasm, ideas, helpful discussions, and a willingness to share with us their expertise and knowledge (and access to their land and soils in many cases) over many years, for which we are most grateful.

Finally, and most importantly, we must recognise and thank our respective spouses: Liz Hewitt, Errol Balks, and Maria Lowe for supporting all of us in our soil science work. Their contributions are often invisible, but without their support works such as this would not be possible.

Contents

About the Authors

Allan E. Hewitt was born in Akaroa, a descendant of the French settlers who arrived in the 1830s. His early years were spent on a hill farm at Pigeon Bay. The family resettled onto the Canterbury plains, firstly to Ashburton where the farm is now a public amenity called 'Trott's Garden', then to Greenpark. Allan completed a B.Sc.(Hons) degree in geology at Canterbury University, then a DipAgSci at Lincoln University. After five years as a pedologist with the DSIR Soil Bureau he was awarded an NRAC scholarship to study at Cornell University, upstate New York, where he completed his Ph.D. in agronomy in 1982. On returning to New Zealand, Allan was given the sole responsibility of developing a soil classification system specific to New Zealand. This took 10 years. The rest of his career as team leader, Senior Pedologist, and Research Leader, in Manaaki Whenua Landcare Research in Otago then Canterbury, used his visionary and innovative skills to help lead soils research. He played a key role in the development of S-map, New Zealand's spatial soil information system. Allan served a term as President of the New Zealand Society of Soil Science (NZSSS) and is a Fellow of the NZSSS. He is also a recipient of the NZSSS Mike Leamy Award and was a Norman Taylor Memorial Lecturer. In 2019, Lincoln University awarded him the Bledisloe Medal, acknowledging his outstanding contributions to soil science and for advancing New Zealand's interests. Allan is a skilled large format landscape painter, sometimes using soil as a medium.

Megan R. Balks was brought up on a hill country farm in the north Wairarapa. She completed a B.Sc.(Hons) in soil science at Massey University and a Ph.D. at the University of Waikato. Megan worked as a soil scientist in DSIR, Dunedin, mainly on soil survey work for irrigation development in Central Otago, for three years. She then taught soil and environmental sciences at the University of Waikato for 30 years, including two terms as Chairperson of the Department of Earth and Ocean Sciences. Megan's research interests have been wide ranging including irrigation of effluent to land, pedology, and applied soil physics. She has been involved in Antarctic Research for over 30 years, including undertaking 19 Antarctic fieldtrips, many as field leader. Megan currently serves on the ITPS (Intergovernmental Technical Panel for Soils) where she represents New Zealand, Australia, and the Southwest Pacific. Megan has served on the Council, and as President, of the New Zealand Society of Soil Science. She has also served two terms on the Waikato Conservation Board and two terms on the Board of the QEII National Trust. Megan is a recipient of the NZSSS Norman Taylor Memorial Lecture Award and a Fellow of the NZSSS. Megan, with her husband Errol, has a small sheep farm on the slopes of Mt. Pirongia where she gains hands-on experience in endeavouring to sustainably manage a small piece of New Zealand's soil resource.

David J. Lowe although born in Waipukurau, the son of a veterinarian, grew up in Tauranga from the age of one. He undertook his B.Sc., M.Sc., and Ph.D. degrees at the University of Waikato in Hamilton, being an early graduate in the new integrative discipline of Earth sciences developed by foundation staff John McCraw, Michael Selby, and Harry Gibbs. During his formative student years David transformed carbon-bearing samples into benzene in the fledgling radiocarbon dating laboratory at Waikato University (1975–1976), undertook soil description work for Soil Bureau DSIR (Hamilton) in the Matamata County survey (1977–1978), and sledged and mapped in the Britannia Range in Antarctica (1978–1979). David's masterate thesis was in pedology, the study of soils in the landscape, and his Ph.D. was in tephrochronology, the characterisation or 'fingerprinting' of volcanic-ash layers—tephras—and their use in dating geological, climatic, and archaeological events. He has combined pedology and tephrochronology ('tephra' providing the nexus) in his teaching and wide-ranging research interests at Waikato University for ~40 years. David's research topics include the origins and unique properties of tephra-derived soils and paleosols, dating tephras and evaluating volcanic eruption histories and hazards, explaining why altered-tephra deposits may collapse as landslides, climatic change over millennia, and the timing of Polynesian settlement of New Zealand. His team was the first to develop a method to extract and purify DNA from ancient buried paleosols on tephras, and he is currently leading research using liquefied ash layers in lakes to evaluate prehistoric earthquake history in the Hamilton lowlands. David, Team Leader of Earth sciences in the School of Science until 2020, served for several years as Chair of Earth and Ocean Sciences, and has co-led tephra studies globally as an officer of the International Focus Group on Tephrochronology and Volcanism for 16 years, including four as President. He represented New Zealand on the International Council, International Union for Quaternary Research (INQUA), for 12 years. David is a recipient of the NZSSS Norman Taylor Memorial Lecture Award and the Mackay Hammer Award of the Geoscience Society of New Zealand. He is a Fellow of NZSSS, INQUA, and the Royal Society of New Zealand Te Apārangi and a life member of NZSSS and INTV.

1 Introduction

Toitu he whenua, whatungarongaro he tangata.

The land is permanent, people come and go

Abstract

New Zealand's soils are diverse, reflecting the varied and dynamic landscape. Located on the boundary between the Pacific and Australian tectonic plates, in the southern mid-latitudes, and surrounded by oceans, New Zealand consequently has: active tectonism, high mountains, and generally fast rates of erosion; active volcanism; a varying but generally temperate, mainly moist climate; and unique flora and fauna. The soils of New Zealand were first cultivated about 750 years ago by Māori with more intensive agricultural development commencing around 1840 following the arrival of Europeans. About two-thirds of New Zealand soils have now been developed for food production or other human use including plantation forestry, with much of the remainder in native vegetation, often in protected areas. Soils have an essential role in the New Zealand economy, underpinning food, wool, and forestry production as well as providing a substrate for urban development, filtering of water, and recycling of nutrients and wastes. Without agricultural inputs New Zealand's soils are generally of moderate or low fertility. The use of fertilisers has greatly increased soil productive capacity. The *New Zealand Soil Classification* identifies 15 Soil orders: Allophanic, Anthropic, Brown, Gley, Granular, Melanic, Organic, Oxidic, Pallic, Podzol, Pumice, Raw, Recent, Semiarid, and Ultic Soils.

1.1 Importance, Diversity, and Development, of New Zealand Soils

The story of Aotearoa/New Zealand's soils is particularly rich and inspiring. The diverse character of the land and the activity of people within the short (≤ 750 year) history of human settlement all contribute to a unique soil assemblage. Māori people traditionally regard the whenua (land) and all that derives from it as sacred, interlinked to all life, and an integral part of human ancestry. In Māori tradition Papatuanuku, the 'Earth Mother', along with Ranginui, the 'Sky Father', are the source of all life on Earth. These deities encapsulate the earth, air, and water that are the critical elements that interact to form the soils on which most life depends.

Soils are essential to the New Zealand economy as they support the food production industries, notably dairy, meat, wine, apples, stone fruit, kiwifruit, vegetables, and other horticultural products, along with wool and forestry, which together accounted for about 45% of the country's foreign exchange income prior to 2020. Soils are vital to many other

A. E. Hewitt et al., *The Soils of Aotearoa New Zealand*, World Soils Book Series,
https://doi.org/10.1007/978-3-030-64763-6_1

aspects of life including providing the substrate for roads and buildings, recreational fields, filtering of water, recycling of nutrients and wastes, and preserving archaeological and historical features. Soils and their associated landscapes provide the visually outstanding and remarkably variable landforms throughout the country that inspire New Zealanders and visitors alike.

The soils of New Zealand are diverse, reflecting the wide range of environmental conditions. The major groupings of soils (orders) that are identified in global soil classification systems can all be found in New Zealand. However, like snowflakes, no two soils are identical and so the study of soils gives an appreciation of the complexity, as well as the beauty, of soil.

Humans have impacted on New Zealand soils since the arrival of Polynesians in about 1280 AD from tropical eastern Polynesia. The indigenous descendants of those first settlers, Māori, first cultivated soils using techniques imported from the Pacific to grow the tropical kumara (sweet potato), and other crops. The first Polynesian settlers were confronted with a landscape largely clothed in extensive forest. About 96% of the North Island and 72% of the South Island were under forest (~82% forest cover in total). The early Māori cleared large tracts of forest, using fire, so that by ~1800 AD about half of the South Island and a third of the North Island forest had been replaced by bracken fern-land, grassland, and tall scrub.

The arrival of Europeans from the early-mid 1800s brought considerable changes with further deforestation as pastoral agriculture was established and new plants and animals were introduced. The development of exotic grasslands for pasture accelerated from the 1870s when landowners were required by law to 'improve' the land. Native forest cover today amounts to about one-quarter of New Zealand. Over the twentieth century, in common with other parts of the world, there was a greatly increased scale and rate of change in the New Zealand soil environment due to rapid changes in technology and marked increase in population. Key activities, along with pasture establishment, included use of drainage and fertilisers. About 85% of natural wetlands were drained to develop flat land for highly productive agriculture. From the late 1940s-early 1950s, the advent of aerial topdressing of fertiliser helped New Zealand to greatly increase food production from its soils, which generally have relatively low natural fertility. Improved pasture also helped to stem the rate of loss of soils in erosion-prone hill country in the eastern North Island and elsewhere. The widespread use of fertilisers has led to issues with water quality in the twenty-first century because of the increase in nutrients, especially nitrogen and phosphorus, in the soil-water system. The rate of transformation of the New Zealand landscape following (relatively recent) human arrival is among the fastest known in the world.

Within this chapter we introduce the environmental factors (called the soil-forming factors)—geology (parent material), topography, climate, organisms, and time—that interact to form the diversity of New Zealand's soils. We discuss some key attributes of the soil, including study of the soil profile that represents a soil 'individual', provide an overview of the *New Zealand Soil Classification*, and introduce the 15 New Zealand soil orders and their distribution. In following chapters, each soil order is described in a systematic format that documents its main features, genesis, relationships with landscapes, key properties, classification, and use and management. The book then concludes with a chapter on soils in the Ross Sea region of Antarctica and a final chapter that considers New Zealand soils in a global context, soil formation pathways and soil and land evaluation. A series of 'ratings tables' that define terms such as 'low', 'medium', and 'high' for a range of soil chemical and physical properties are included in Appendix A, Tables A.1 and A.2. The terms in the ratings tables are followed throughout this book.

1.2 The Environmental (Soil-Forming) Factors that Influence New Zealand Soil Development

1.2.1 Geology, Soil Parent Materials, and Topography of New Zealand

A varied range of geological materials and topographic features contribute to the diversity of New Zealand soils. Although New Zealand appears as a series of islands, geologists have come to recognise the land mass of New Zealand as the emergent part of a recently identified eighth continent, named Zealandia, most (94%) of which is currently beneath the sea (Fig. 1.1). New Zealand lies on the collisional plate boundary between the Pacific and Australian tectonic plates which are moving obliquely towards each other at a rate of about 40 mm/year. In the North Island the Pacific Plate is being subducted beneath the Australian plate resulting in current volcanic activity in the Taupo Volcanic Zone in central North Island and the Taranaki region (with the Auckland Volcanic Field also regarded as active). To the south of the South Island the subduction is in the opposite direction with the Australian Plate being subducted beneath the Pacific Plate.

Through the central part of New Zealand the boundary of the two plates is marked by the Alpine Fault, a transform or strike-slip fault which passes mainly along the western margin of the Southern Alps, splitting into a number of sub-faults in the northern South Island that continue north-eastwards through, and east of, North Island. There has been substantial uplift of sediments on the Pacific Plate,

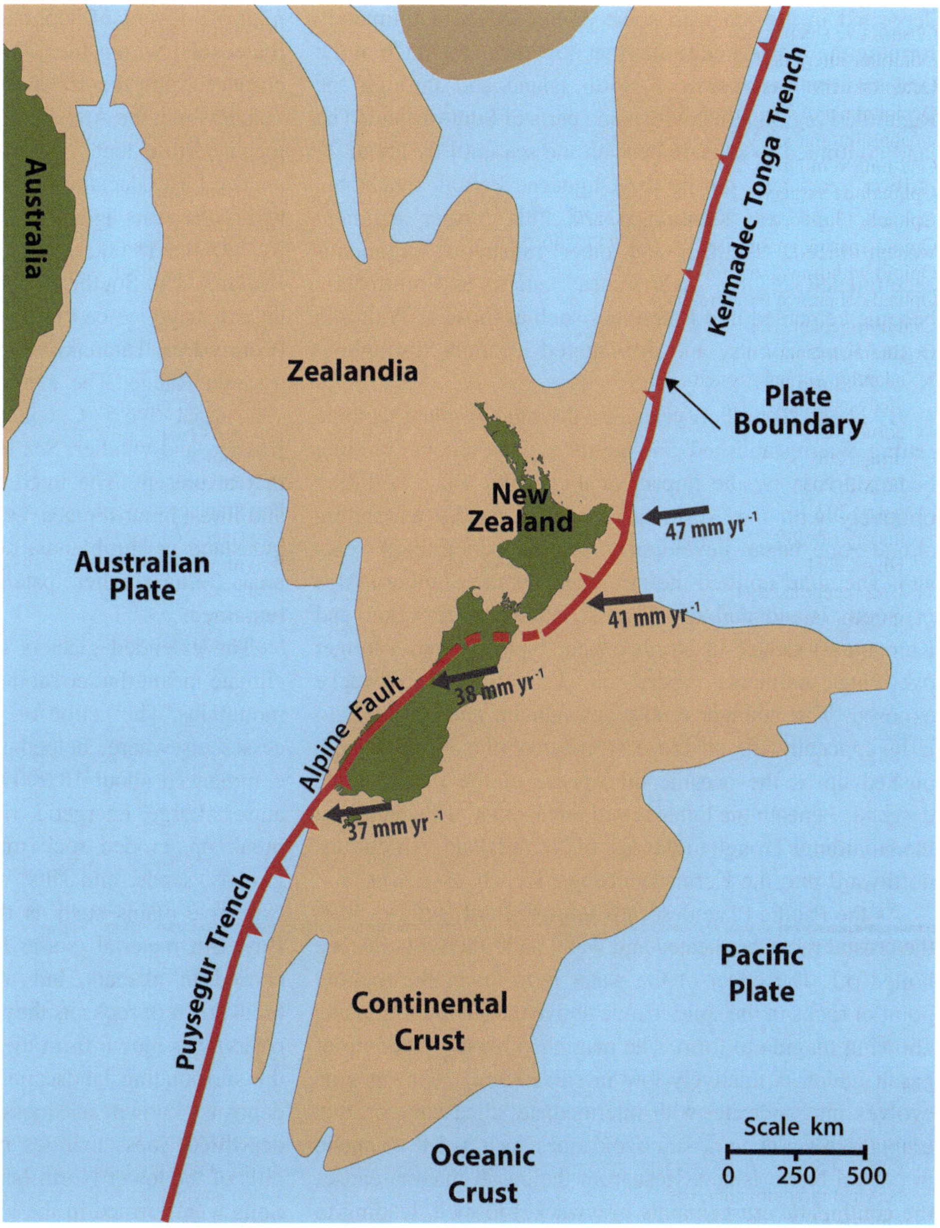

Fig. 1.1 The Zealandia continent (largely submerged) and modern plate tectonic setting of New Zealand. The 'teeth' on the red (plate boundary) line mark subduction trenches; the black arrows show the direction and rate of convergence of the Pacific Plate relative to the Australian Plate (adapted from Q-map, GNS Science)

as the plates have crumpled upwards to form the Southern Alps, which are among the fastest-rising mountains of the world. Maximum rock uplift rates are about 10 mm/year.

Lateral movement, where the plates grind past each other, has resulted in the separation of rocks on opposite sides of the Alpine Fault by about 480 km over the past 23 million years. The crumpling, uplift, and lateral movement are marked by hills, mountain ranges, and multiple faults, in both the North and South Islands.

The oldest rocks in New Zealand are the granites, schists, and marble rocks of Fiordland and trilobite-fossil-bearing sedimentary rocks of northwest Nelson (dated at 510 million years old in Cobb Valley) which trace their origins back to ancient Gondwana. Much of New Zealand is underlain by hard, but highly fractured, sandstone (greywacke), and siltstone (argillite) rocks here collectively referred to as greywacke and as 'basement' rocks. The greywacke rocks were formed from silt- and sand-layered sediments that accumulated in the sea off the coast of Gondwana mainly during the Jurassic Period (about 200–145 million years ago). During the Cretaceous Period (at around 85 million years ago) a plate tectonic spreading zone started to operate in what is now the Tasman Sea, pushing the land that now makes up New Zealand away from what is now Australia. As a result, the greywacke sediments were lithified and uplifted out of the sea and then gradually eroded into a fairly low-lying, flat, landmass.

Over time, the land gradually subsided relative to sea level. Initially a lot of sediment was eroded off the land and deposited in the sea with some of the sediment ultimately forming the 'papa' mudstones that now make up much of the land on the east coast of North Island and through the Rangitikei region along with some parts of South Island. The land continued to subside beneath the sea until by about 25 million years ago, during the Oligocene Period, much, but not all, land was submerged and little further sediment was produced. Shellfish and other calcareous organisms accumulated in the shallow sea waters, and ultimately become cemented into limestones such as those at Waitomo in the King Country and in scattered locations throughout New Zealand.

By about 20 million years ago the modern plate tectonic setting was established and uplift and mountain building began, driven by the impact of the Pacific Plate colliding obliquely with the Australian Plate. A rapidly expanding alpine zone began developing from about 5 million years ago. The areas uplifted the most, such as the Southern Alps in South Island and the Tararua, Ruahine, Kaweka, and Kaimanawa ranges in North Island, typically had younger overlying sediments eroded off, leaving the greywacke exposed. The younger mudstones remain, and are particularly concentrated, on the east coast of the North Island, pushed up as the oceanic lithosphere of the Pacific Plate descends beneath the lighter continental crust of Zealandia at the Hikurangi Trough to the east of North Island (which runs northward into the Kermadec Tonga Trench, Fig. 1.1).

As the Pacific Plate descends into the Earth's hot mantle, the crustal rocks are heated and water and other volatiles are boiled off. The effect of the water is to lower the melting point of rocks in the solid mantle above the subducting plate, allowing magma to form. The magma is mostly made up of basalt which is relatively low in silica (SiO_2). The magma evolves into andesite, with intermediate silica content, that erupts to form cone or stratovolcanoes such as Mt Ruapehu in central North Island. Heat from the basalt magma causes the continental crust (mainly greywacke) to melt, leading to huge rhyolite eruptions, with high silica content, that can result in collapse of the land to form calderas such as those occupied by lakes Taupo and Rotorua.

The recently active volcanoes, other volcanic landforms, and associated geothermal activity, are concentrated in a linear zone, stretching from Ruapehu in the south to White Island (Whakaari) in the north, the Taupo Volcanic Zone (TVZ). Magma erupted from Taranaki Maunga (Mt Taranaki) in the western North Island derives from a much deeper source than that beneath the TVZ. Large volumes of pyroclastic (fragmental) material or tephra (volcanic ash) have been explosively erupted from volcanoes in the TVZ, and from Taranaki, and deposited over the landscape of the central North Island and beyond. Hence many soils in the central North Island are formed from tephra deposits which, because of their intermittent deposition, often comprise multiple layers of tephra beds with buried soil horizons (paleosols) within them. Both lavas and tephras have been erupted from the most recently active intraplate basaltic volcanism in the Auckland Volcanic Field, but they are more localised in extent.

The Last Glaciation, which extended from about 115,000 to 11,700 years ago, with a glacial maximum between about 31,000 and 18,000 years ago, had marked impacts on New Zealand. The Southern Alps were sufficiently high to allow an extensive ice cap to develop on them and small glaciers occurred on Taranaki, Tongariro and Ruapehu, and in the Tararua Range. The average temperature in New Zealand was about 6–6.5 °C colder, and it was about 25% drier, frostier, and windier. Sea level was about 135 m lower than that at present. The treeline was lowered by about 800 m, and forest in most places (except Northland) was replaced by grassland or shrublands, or both, except in small sheltered areas (refugia) where patches of beech and/or conifer forest remained.

The extended glaciers and the cold, drier, and windier, climate meant that erosion increased in the high country and mountains. The action of glacier ice, and freeze-thaw processes elsewhere, helped shatter rocks. The glacial retreat commenced about 18,000 years ago with warming temperatures. Large, energetic, rivers were able to transport, and break up, eroded rock materials, depositing the resulting gravels, sands, and silts, on the lowlands to build up the extensive plains such as those in Canterbury and Hawkes Bay. Silt material (some formed by the grinding of rocks carried in glaciers, but most derived from abrasion and breakdown of rocks as they were transported by the vigorous rivers) was blown from the wide riverbeds and deposited on the surrounding landscape, at relatively slow rates, to form deposits known as loess. The resulting aeolian (wind deposited) loess mantles many of the terraces and rolling hills of the lower North Island and the eastern South Island. Soils were formed in the loess at the same time as it slowly accumulated (i.e., loess deposition and soil formation occurred concurrently).

From about 18,000 years ago, when the glaciation started to transition towards the warmer Holocene period that began 11,700 years ago, sea level rose and many former river valleys were drowned to form extensive harbours and fjords. As climate warmed, forest vegetation was re-established, initially in the north from 17,500 years ago and then towards the south by the start of the Holocene, or soon after. The rate of erosion, and reworking, declined in many regions as the landscape was stabilised. However, ongoing volcanic activity in the Taupo Volcanic Zone and at Taranaki, and river floodplain deposition, has provided some new substrates for soil formation.

Thus our active geological history has led to the wide range of geological materials that underpin the soils of New

Zealand (Fig. 1.2). Hence the New Zealand land surface comprises a wide variety of loose or *unconsolidated* deposits, such as the thick mantles of tephra beds in central North Island along with extensive loess, alluvium, landslide debris, and peat, as well as *consolidated* rocks, both 'hard' and 'soft'. The 'soft' rocks, such as the papa mudstones of the east coast and Rangitikei regions, are less indurated (not as hard) and more readily broken down or eroded than the so-called 'hard' rocks.

The topography of New Zealand is characterised by complexity, ranging from flat plains, rolling and hilly land, to steep land and alpine environments, as a consequence of the uplift (and subsidence in places such as the Rangitaiki Plains in Bay of Plenty) caused by New Zealand's active plate tectonic setting. New Zealand comprises about 39% flat to undulating land ($\leq 7°$ slopes), 36% rolling to hilly land (7–25°), and about 25% steep to very steep land (>25°). The steeper hilly and mountain lands are prone to erosion by mass movement especially land-sliding. In the eastern North Island, the uplifted calcareous (soft rock) mudstones are especially vulnerable to landslide erosion.

Important regional geomorphic features include the Southern Alps; the hill country of the east coast and lower North Island; the volcanoes and associated landforms of the central North Island and Taranaki, along with other intraplate volcanoes scattered throughout New Zealand such as the Auckland Volcanic Field; the alluvial plains and terraces, especially the widespread gravelly plains of Hawke's Bay, Wairarapa, Canterbury, and Otago; and the glacially gouged valleys, lakes, and fjords of South Westland and Otago.

In hill and mountainous country, slope and aspect (the downhill direction of the slope) are important factors influencing soil processes and site microclimates. The interactions between slope and aspect, and the resultant microclimate and vegetation, impact many soil processes, and thus the soils that form. For example, steeper slopes are more prone to rainfall runoff and erosion than flatter areas which may be sites of accumulation of material derived from further upslope. Slopes with a northerly aspect tend to be warmer and drier than those with a southerly aspect because north-facing slopes receive more incoming solar radiation. Frost and snow may prevail for longer on cooler southerly slopes. In many regions slopes with a westerly aspect are more exposed to prevailing north-westerly winds, which enhance evapotranspiration rates and directly impact on vegetation establishment and survival, and thus the development of the associated soil.

1.2.2 New Zealand's Climate

The climate of New Zealand is moderated by the surrounding ocean and the mid-latitudinal location, with the three main islands extending from 34° to 47° south. Thus compared with many other regions, New Zealand generally has a 'goldilocks' climate, that is, not too hot, not too cold, not too wet, and not too dry. However, marked climatic contrasts and extremes occur over short distances and the regional variations have a marked influence on the soil pattern. Mean annual temperature near sea level ranges from about 15 °C in Northland to 9 °C in Southland (Fig. 1.3) with cooler temperatures at higher altitudes. Basins in the Central Otago region, situated inland, and in the rain shadow of the Southern Alps, and basins in the inland Kaikoura Range, are the nearest any part of New Zealand gets to a continental climate. New Zealand's warmest recorded temperature of 42.4 °C was measured simultaneously, on 7 February 1973, in both Rangiora (Canterbury) and Jordan (Marlborough). New Zealand's record coldest temperature of −25.6 °C was recorded in Ranfurly in Central Otago on 17 July 1903. Nelson, Blenheim, and Whakatane are the sunniest regions of New Zealand with over 2300 annual sunshine hours. Rain-prone Westland has about 1800 sunshine hours per year while lowest annual sunshine hours (about 1700 h) are recorded in Southland and coastal Otago.

The prevailing wind in New Zealand is from the west. Cold fronts can bring strong southerly winds to the central part of the country. Much of New Zealand is located within the 'roaring forties', the latitudinal zone of 40° S–50° S that is particularly prone to strong north-westerly winds, especially in the lower North Island, Marlborough, and Canterbury regions. (During the Last Glaciation, the zone of strong westerly dominated winds shifted northwards to about 30° S.) The hot dry nor'wester sucks moisture from plants and soils, and can readily erode soils that are not protected by vegetation.

The surrounding oceans, in combination with prevailing westerly winds, mean that much of New Zealand has a relatively high and consistent rainfall in many regions (Fig. 1.3). However, the main hill and mountain ranges, especially the Southern Alps, cause a strong rainfall gradient across New Zealand from west to east as wet ocean air coming in from the west is forced up over the mountain ranges. With uplift, the air cools and the water vapour condenses releasing rain on the west and in the mountains. As the air continues down the eastern sides of the ranges it warms and becomes drier, bringing dry conditions (a rain shadow effect), especially in Central Otago and the McKenzie Basin, and also across the eastern areas of coastal Otago, Canterbury, Wairarapa, Hawke's Bay, and the Gisborne region.

The Northland, Coromandel, and Bay of Plenty regions are sometimes subjected to the effects of subtropical cyclones that occasionally track southwards, bringing intense rainfalls and strong winds. Most regions of New Zealand receive between 600 and 1600 mm of rain per year, with

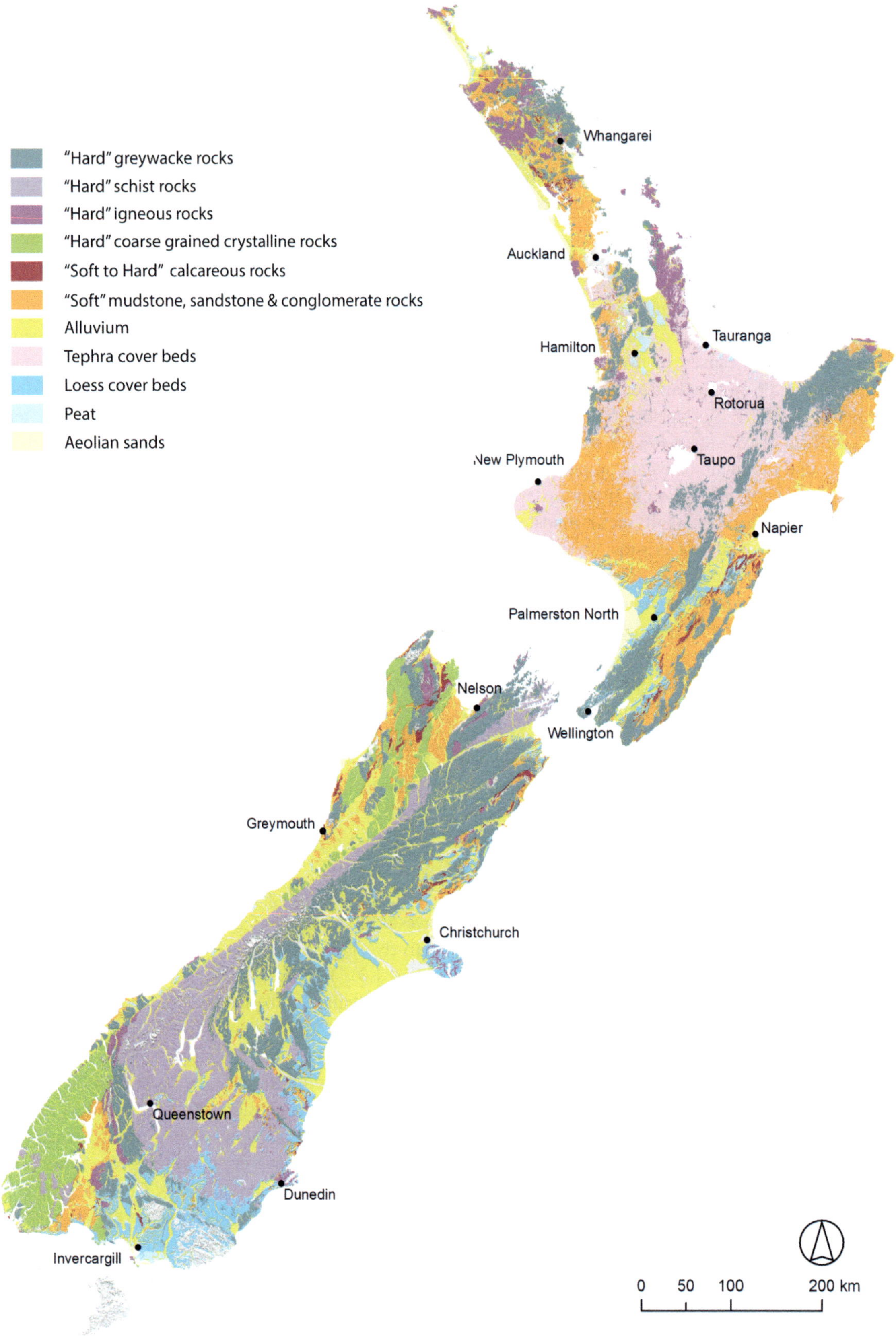

Fig. 1.2 The distribution of the main geological materials on which the majority of New Zealand soils are formed. Map from Manaaki Whenua – Landcare Research

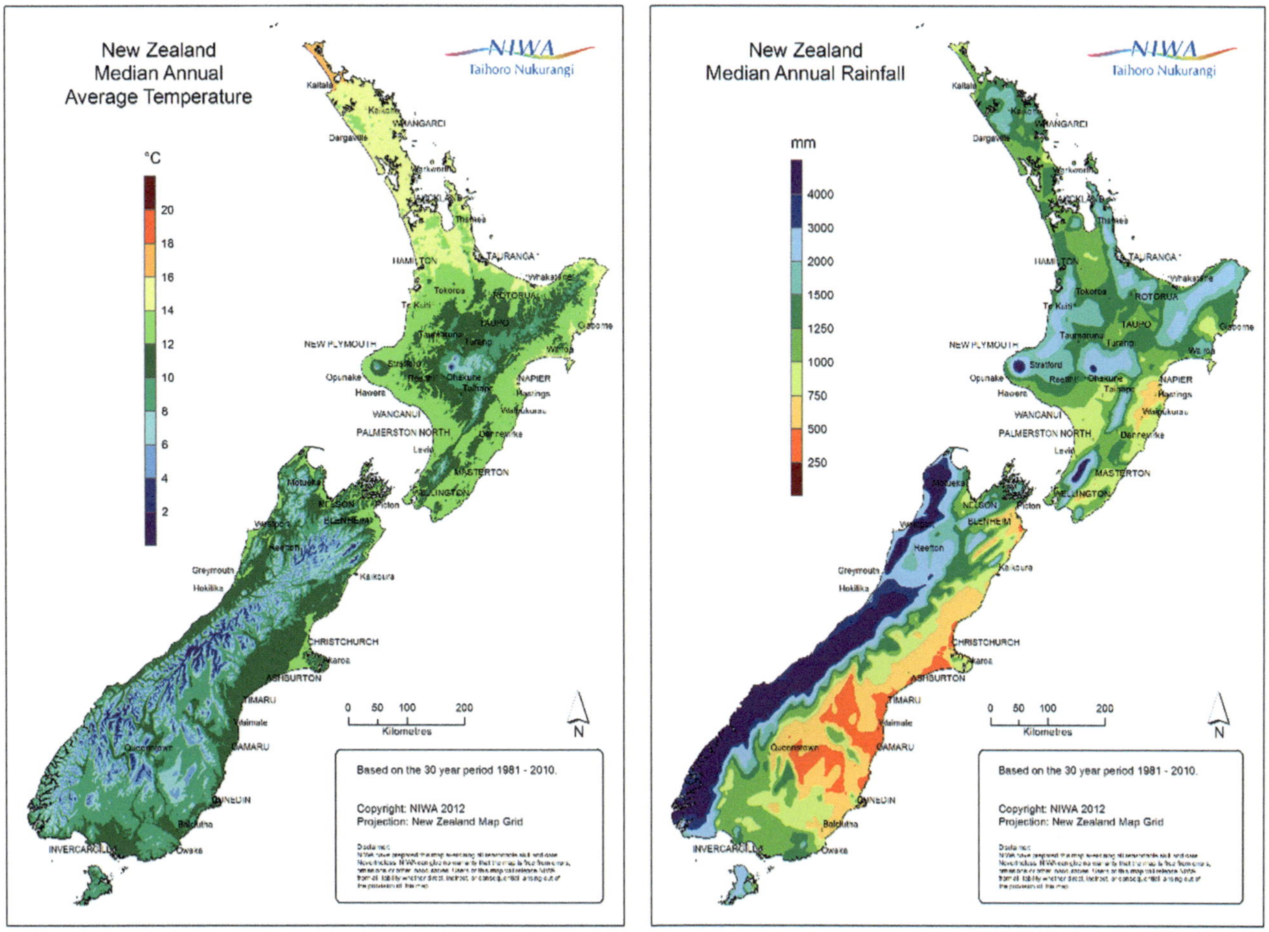

Fig. 1.3 Mean annual temperature (left) and rainfall (right) of New Zealand. *Source* NIWA

more in the high mountains where some precipitation falls as snow. Mean annual rainfall ranges from about 300 mm in Central Otago to over 6000 mm at Milford Sound on the west coast of the South Island. Taking New Zealand as a whole, the mean annual rainfall is >2000 mm. The highest recorded annual rainfall was 16 617 mm at Cropp River in the Hokitika catchment in the Southern Alps in 1998.

Thus New Zealand's generally warm (temperate) and moist climate provides excellent conditions for soil development and plant growth. Many of the processes that operate in soil depend on the presence of water. Temperature strongly influences the rate of weathering processes as well as the growth of organisms. Examples of the balance of rainfall and potential evapotranspiration (Fig. 1.4) show the monthly surplus rainfall typical of much of the west coast and some parts of Northland as well as the moisture deficits experienced in Central Otago and on the east coast of the South Island and lower North Island.

1.2.3 Organisms

Soil potentially forms the most complex ecosystem on Earth and contains the greatest biodiversity due to the countless different microbes that live in soil. For instance, Jackie Aislabie and Julie Deslippe reported that estimates of the number of species of bacteria in a gram of soil range from 2000 to 18,000. Plants, animals (including humans), insects, spiders, worms, and microbes are all important to soil formation and development. Plants extract the nutrients and moisture they need from soil. In return, following death, plant roots, stems, and leaves, along with animal remains, are all incorporated back into the soil, providing the organic matter that is an integral part of a fertile soil. The work of soil organisms is essential to life on Earth through the recycling of nutrients, carbon, and oxygen.

Some scientists have reported that the total fresh weight of organisms below temperate grassland may exceed 45

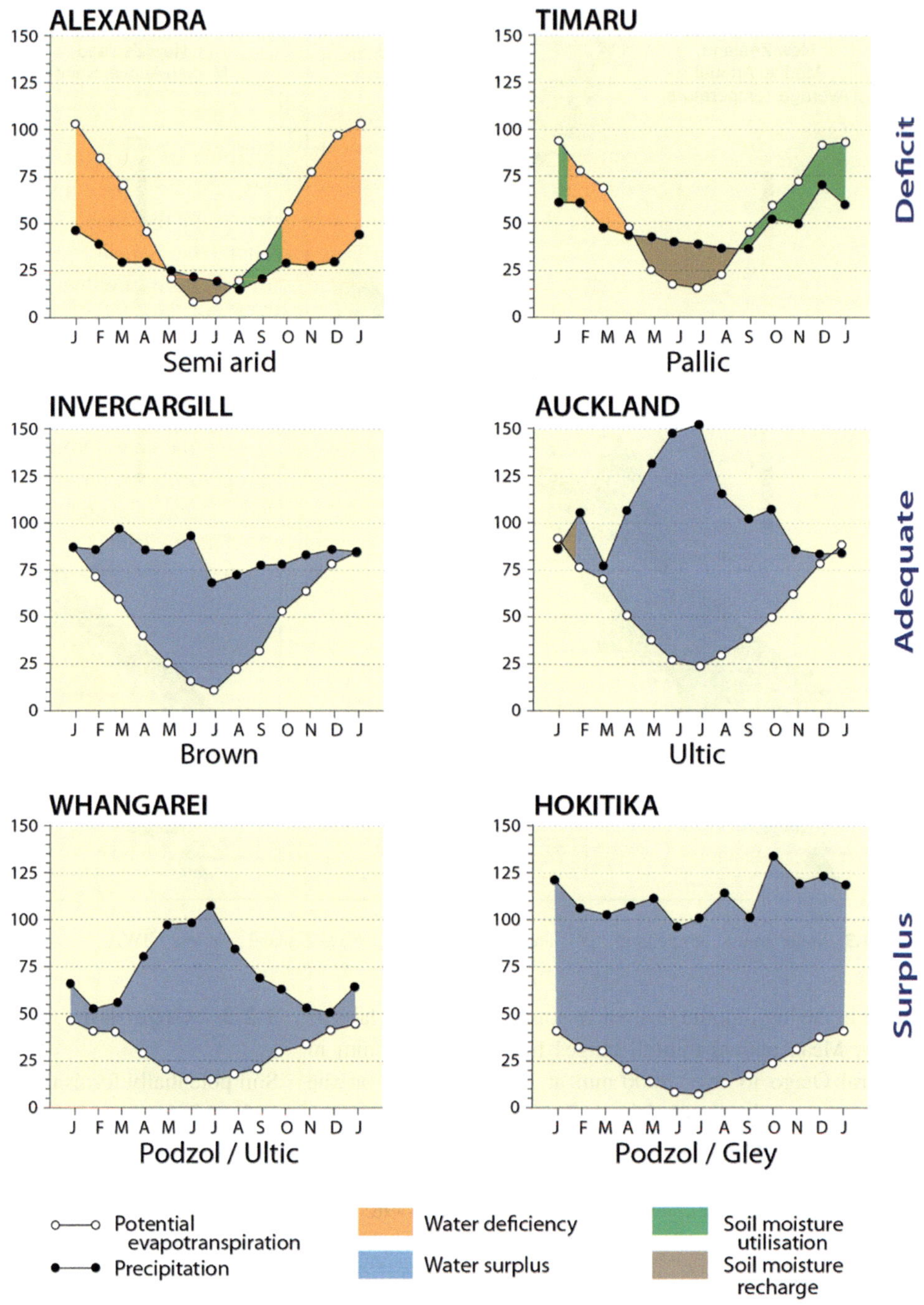

Fig. 1.4 Examples of soil moisture balance for selected areas and soil orders of New Zealand (adapted from New Zealand Soil Bureau 1968)

tonnes per hectare. A teaspoon of New Zealand topsoil will likely have over 100 billion bacteria and more than 15 km of fungal hyphae threads (Fig. 1.5). The myriads of small insects, worms, and microbes that live in a soil provide nature's vital recycling service, breaking down organic matter and re-incorporating the remaining constituents into soil where the nutrients are stored and released to be available for the next generation of life. The organic matter in soil also helps bind soil aggregates together, improving the soil's ability to hold and transmit water, and enabling plant roots to readily extend down through the soil.

Animals interact with, and influence, the soil in many ways. They remove and move plant material; some make their homes within the soil, moving soil materials by their burrowing activities. Animal dung and urine form concentrated bursts of nutrients which are incorporated into the soil.

Humans, for millennia, and more particularly within the last century, have had profound impacts on soil as they harvest food, fuel, and other products to support growing populations and improving standards of living. Many land management activities impact on soil organisms. Prior to human arrival many of New Zealand's soils developed, after

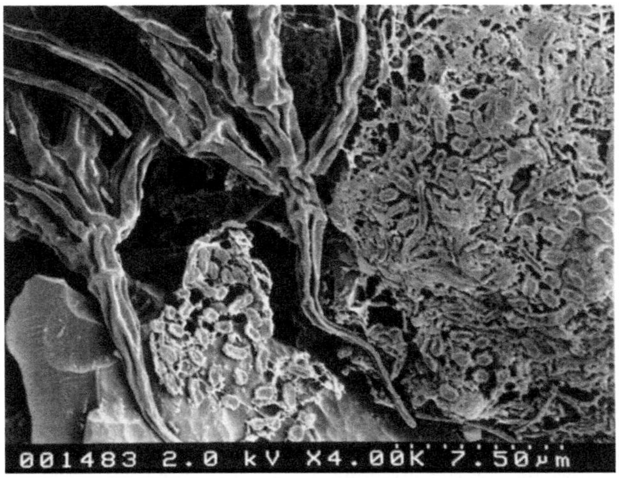

Fig. 1.5 Scanning electron microscope photo showing soil bacteria (the small oval shapes) and fungal hyphae growing in response to irrigation of effluent on a Waikato topsoil. The picture is of an area about 0.03 mm wide

the end of the last glaciation (during the Holocene, last 11,700 years or so), under forest vegetation, mainly mixed conifer (podocarps, kauri, cedar) and broadleaf hardwoods together with southern beeches. Under conifer forests the organic matter tends to build up as a litter horizon (sometimes called mor-humus) on the soil surface as the acid conditions and lack of nitrogen inhibit biodegradation of organic materials. Under broadleaf forest, and grasslands, the pH of litter materials is higher and more nutrients are available, thus the organic matter tends to be more rapidly broken down and incorporated into the soil (forming mull-humus).

Changing vegetation from native forests to pasture has occurred over about two-thirds of New Zealand. Soil carbon in topsoils tends to be higher under pasture than under forests as the grass roots grow and die quickly and contribute to a buildup in soil organic matter. Removal of vegetation can lead to increases in soil erosion and loss of soil materials. Addition of fertiliser stimulates plant growth, and also microbial activity.

Recent advances in DNA sequencing are giving a much greater appreciation of the biodiversity of the soil microbial population. There is still a lot of work to do to apply the new tools to understanding the myriad microbes in the soil, and their role in nutrient recycling and other soil processes.

1.2.4 Time

The longer a land surface has been stable the more time weathering, leaching, and other soil-forming processes have had to operate. Very young soils show minimal alteration of the parent materials from which they are formed. By contrast soils that have had long periods (at least tens of thousands of years) of time in which to weather and develop are likely to be dominated by clays and have low fertility as nutrients have been leached from them. By global standards many (but not all) New Zealand soils are relatively young with landforms and associated soils less than about 15,000 years old. The youngest soils in New Zealand are scattered throughout the country on newly deposited materials including alluvium on river flood plains, landslide debris, colluvium, and recent volcanic deposits. Young soils also occur on sites where erosion has removed material, such as the steep lands and mountains of the North Island axial ranges and the Southern Alps. Thus even though a parent material, such as greywacke, can be millions of years old, frequent erosion in steep terrains means that any soil formed on it is lost and fresh underlying rock is exhumed as a new parent material.

Where soils are formed on tephra deposits that have been dated, the age of the soils (defined here for the purpose of age estimation as approximately the uppermost 1 m of the soil profile) can sometimes be estimated. For example, Allophanic Soils developed on thinly bedded composite tephras in the eastern Waikato date back about 25,000 years; those in western and southwestern Waikato date back about 50,000 years. Soils on older strongly weathered tephra (Hamilton Ash beds) are between 50,000 and 340,000 years old depending on the depth of constituent beds.

Some of New Zealand's most stable land surfaces are in the Marlborough, northern Waikato, and Northland regions where a long time, combined with a warm, moist, climate, and relatively low tectonic activity, have produced New Zealand's most strongly weathered soils, some of which are likely to be of the order of several hundred thousand years old. Some surprisingly old soils, of a similar age, have also been identified in Central Otago.

1.3 The Soil Profile and Key Soil Properties

1.3.1 The Definition of Soil

Most people think they know what 'soil' is. However, it is not necessarily a simple matter to accurately define what is, or is not, soil. One widely accepted definition is that of the United States Department of Agriculture, who in the soil classification system, *Soil Taxonomy*, have (since 1999) defined soil as 'a natural body comprised of solids (minerals and organic matter), liquid, and gases that occurs on the land surface, occupies space, and is characterised by one or both of the following: horizons, or layers, that are distinguishable from the initial material as a result of additions, losses, transfers and transformations of energy and matter *or* the ability to support rooted plants in a natural environment'.

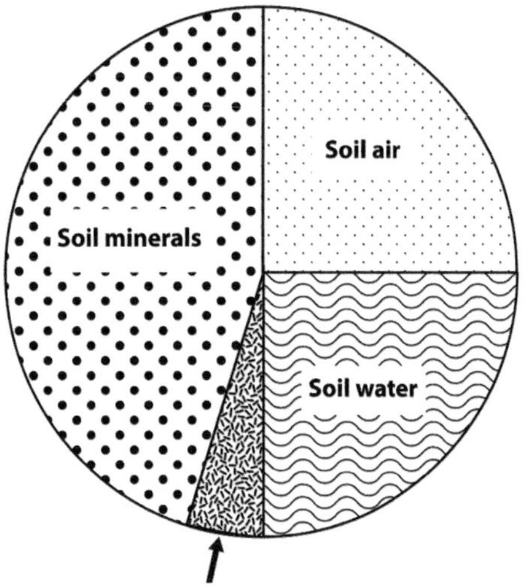

Fig. 1.6 Generalised relative proportions of soil constituents

A simpler definition is that soil is 'a natural, evolving, body, formed on the land surface, which is the product of its environment and contains mineral and organic constituents'. This definition captures some of the key features of soil: they are naturally occurring products of the environment and landscape in which they form. Soils are constantly changing as materials are added (via inputs of plant matter, animal dung or remains, or deposits of dust, tephra, or other mineral material), removed (by leaching, erosion, biodegradation, or removal in crop harvests), or altered (through physical or chemical weathering or biodegradation). Soil can be thought of as occurring in four dimensions, the *x*, *y*, *z*, co-ordinates of three-dimensional space, plus time.

While we tend to think of soil as 'terra firma' in fact about 50% of the volume of topsoil is pore-space that may contain, interchangeably, air, or water (Fig. 1.6). In New Zealand, many of our topsoils contain about 5% soil organic matter.

1.3.2 The Soil Profile

The soil profile is the top metre or two of soil that is exposed when a pit is dug in the ground. Most soil descriptions focus on a soil profile. When a soil profile is excavated, the different horizons (sometimes also geological layers) in the soil are immediately evident (Fig. 1.7).

Above the soil surface plant litter from trees or other vegetation may accumulate. Within the litter three horizons, L, F, and H, may be recognised. The 'L'-horizon comprises undecomposed plant materials with botanic structures easily distinguished. In some sites the L-horizon is underlain by partly decomposed organic material in which some of the original plant structures are visible (fibric), denoted F. In transition from the L- or F-horizon to the topsoil (A-horizon) an H-horizon may be recognised which comprises decomposed (humified) organic material where plant structures are no longer recognisable. The L-, F-, and H-horizons occur on soils under forests in New Zealand but are not usually present in agricultural soils or where other surface disturbance has occurred. Where soil is formed predominantly from organic material in a permanently wet environment, such as a peat bog, an O-horizon is recognised.

The dark coloured topsoil, where organic matter accumulates, is termed an 'A'-horizon. The A-horizon is the uppermost part of the root zone where most of the soil microbial activity is concentrated. The subsoil material that is altered so that the parent material is largely no longer readily recognised is termed the 'B'-horizon. Often a soil can be divided into a number of B-horizons or sub-horizons as the soil properties vary with depth. Beneath the B-horizon weathered parent materials are normally termed 'C'-horizons. C-horizons are essentially unmodified geological materials or deposits, usually only weakly- or un-consolidated and little affected by soil-forming processes. Transitional horizons can bridge the gaps where either two different materials are physically intermixed (such as an A/B-horizon) or where a soil horizon has the properties intermediate between one and another (such as an AB-horizon, or a barely modified parent material denoted BC).

In some soils, such as soils formed in strongly weathered tephras, there are no C-horizons because all the original geological materials have been sufficiently transformed by pedogenesis to become A-, E-, or B-horizons. In soils where a thick layer of new material has been added on top of a profile (such as fresh alluvium from a flood event, or deposition of thick tephra-fall material), the pre-existing surface soil horizons become buried, forming a 'paleosol'. Such buried horizons are shown with a 'b' in front of the horizon name, e.g. bA for a buried A-horizon. Where buried paleosols become buried deeply (perhaps several metres below the contemporary land surface), they are effectively cut off from the surface of the modern soil and thus are generally no longer impacted by near-surface soil-forming processes. Some buried paleosols (e.g. in the Waikato region, developed on clay-rich, strongly altered tephras) are extreme in their properties and age, dating to a former life at the land surface that took place around a million years ago.

Where the parent material is rock or strongly consolidated geological material that is too hard to dig practicably with a spade, the pragmatic soil scientist refers to this as an 'R'-horizon (where R stands for 'rock'). In some soils, beneath the A-horizon but above the B-horizons is a pale coloured, strongly leached, horizon which is referred to as an 'E'-horizon, where E stands for 'eluviation', the process of mobilising and removing material down through the soil either in solution or in suspension, or both. 'E'-horizons are usually underlain by an 'illuvial' B-horizon, which has received material moved into it from the horizon above. The

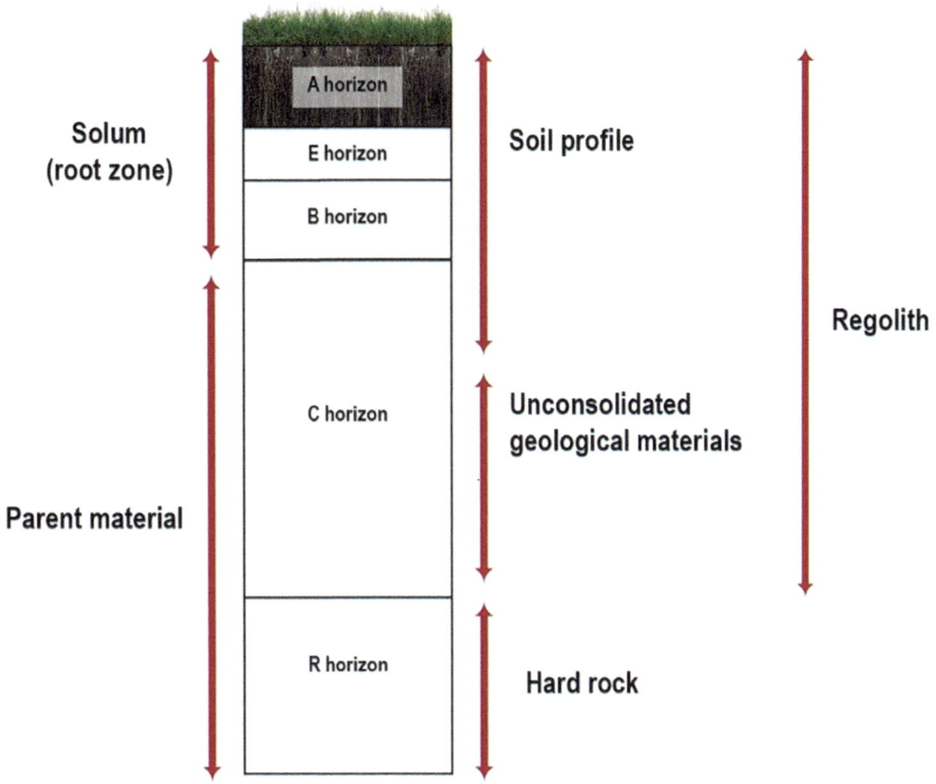

Fig. 1.7 Names and typical arrangement of master mineral soil horizons. Many soils will differ from this generalised scheme and the *New Zealand Soil Description Handbook* should be consulted for detailed information on all soil horizon, sub-horizon, prefix, and suffix names and notations

formal New Zealand soil horizon names, definitions, and further details, are included in the *New Zealand Soil Description Handbook.*

1.3.3 Key Soil Properties

a. Colour

The most obvious property of a soil is its colour. While most soils appear as shades of brown, the colour of a soil can reveal a great deal about the soil's properties and origins. In order to distinguish the various shades and tones of soil colour, scientists use specialised colour books that define the many hues (e.g. red, blue, orange), values (how pale or dark a colour is), and chromas (the range from grey to bright saturated colour) (Fig. 1.8). The darkness of a topsoil (A-horizon) provides clues as to the amount of organic matter that may be in the soil. Topsoil colour may also provide clues about the history of the site. For instance, soils formed under bracken fern (*Pteridium aquilinum*) have a dark black colour, whereas those formed under manuka (*Leptospermum scoparium*) tend to have brown topsoil. Very pale, blueish grey colours are usually indicative of anaerobic soils that form in saturated conditions. Pale coloured soils with bright brown or orange spots (called mottles) or black to reddish precipitates (called concretions) often indicate a soil with a periodically high water table (i.e. fluctuating conditions of oxidation and reduction, Chap. 5).

b. Soil texture

Soil texture describes the relative proportions of sand, silt, and clay in a soil sample. The soil particle size classes used by the New Zealand soil science community are described in Table 1.1. The combinations of different particle sizes determine the soil texture (Fig. 1.9). The 'best' soils for supporting plant growth are generally loams which contain a reasonably even mixture of sand, silt, and clay so that they are able to hold and release nutrients and also store water while remaining well drained.

c. Soil structure (pedality)

Soil materials often cluster together to form aggregates (also sometimes called 'peds', Fig. 1.10). The shape or pattern of aggregates, and the pore spaces between them, is referred to as soil structure or 'pedality'. The shape, size, and strength of soil aggregates strongly influence the soil's physical properties. Large, closely fitting aggregates, such as prismatic and block structures, can limit the passage of water and plant roots through a soil. Aggregates that are readily broken up form friable soils in which it is easy to establish a

Fig. 1.8 Soil scientist, Les Basher, using a soil colour chart. Soil colours are described using their hue, value, and chroma. For example, a dull orange colour (top right-hand corner of the chart) might be reported more precisely as 2.5YR 8/6. Photo: Ian Lynn, reproduced courtesy of Manaaki Whenua – Landcare Research

Table 1.1 Particle sizes as defined in the *New Zealand Soil Description Handbook*. The examples under 'relative size' give a feel for the orders of magnitude of difference in size between (large) boulders and (tiny) clay particles

Particle size fraction	Diameter (mm)	Relative size
Boulders	>200	Rangitoto Island
Very coarse gravel	200–60	Large irrigator circle
Coarse gravel	60–20	Road roundabout
Medium gravel	20–6	Spa pool
Fine gravel	6–2	Truck tyre
Coarse sand	2–0.6	Basket ball
Medium sand	0.6–0.2	Orange
Fine sand	0.2–0.06	Plum
Silt	0.06–0.002	Tack head
Clay	<0.002 (<2 μm)	Pin point

good seed-bed. However, such soils may also be prone to erosion, especially if plant cover is lacking.

Soil horizons that do not comprise aggregates, or that have aggregates that are <6 mm in diameter, are termed 'apedal'. Apedal horizons often comprise loose individual sand grains (single grained), crumb-like aggregates of <6 mm diameter (termed 'earthy'), or a massive structure with no clear natural breaks in the soil material.

d. Soil chemical properties

A wide range of soil chemical properties may be measured to determine the fertility status of a soil and its potential for a variety of crops and uses. The soil pH (a measure of acidity or alkalinity) impacts on nutrient availability and on the suitability of a soil for particular crops. Soil nutrient contents, and in particular P, N, K, Ca, Mg, and Na are commonly measured. Soil carbon contents give an indication of soil fertility because organic matter is important to both plant nutrient availability and soil physical properties (such as aiding water holding capacity). Soil carbon is also of interest in determining global carbon budgets as soils contain about the same amount of carbon as the plants plus the atmosphere. Many other elements, including Al, Cd, Pb, and As are measured in soils to give indications of potential toxicity or contamination issues. In New Zealand, and in this book, we often refer to the levels of soil chemical components as low, medium, high, etc. following a ratings table (reproduced in Appendix A) first published by Les Blakemore and others in 1987.

1.4 *New Zealand Soil Classification*

1.4.1 Introduction

The first workers in New Zealand who set out to map soils from the 1920s to 1940s gradually developed a soil classification system based on ideas developed mainly by V.V. Dokuchiaev (1848–1903) in Russia on soil genesis and the

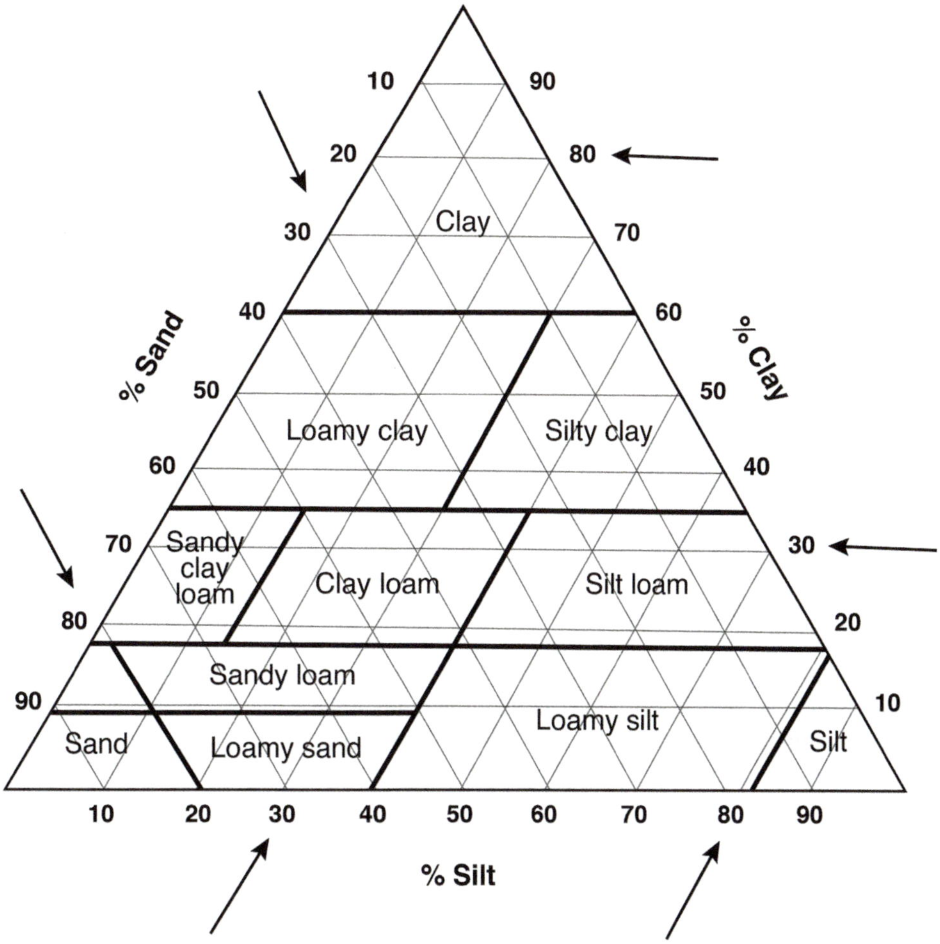

Fig. 1.9 New Zealand soil textural triangle. For example, a soil with 50% sand, 20% clay, and 30% silt would be defined as a clay loam (adapted from the *New Zealand Soil Description Handbook*)

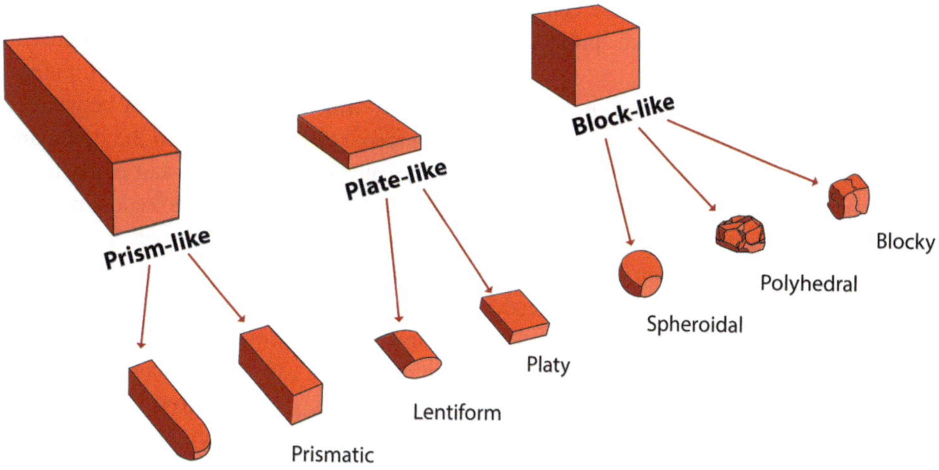

Fig. 1.10 Soil aggregate shapes (adapted from the *New Zealand Soil Description Handbook*)

zonal, climate related, pattern of soils. Norman H. Taylor, following on from Leslie I. Grange (who developed the first functioning soil classification in New Zealand in 1944) captured the essence of what came to be known as the 'New Zealand genetic soil classification' which was first published in a book entitled 'Soil Survey Method' in 1962, and then in more detail in 1968 in 'Soils of New Zealand, Part 1'. However, the various classes of soils were not defined in detail and thus were difficult to apply consistently. In the late 1970s and early 1980s *Soil Taxonomy* (which had been recently developed by the United States Department of Agriculture) was trialled in New Zealand, as a possible alternative, with varying success. However, in 1983, it was decided to develop a modern soil classification system for

New Zealand that incorporated the best parts of the New Zealand 'genetic classification' alongside some of the new ideas and principles established in *Soil Taxonomy*. This effort culminated in the publication of the '*New Zealand Soil Classification*' in 1992, by Allan Hewitt, which has now been widely adopted for use in New Zealand. Here we will follow the 3rd edition published in 2010.

The *New Zealand Soil Classification* is a hierarchical system with 15 soil orders that are in turn divided into 60 groups, and 250 subgroups (Fig. 1.11). Subgroups are divided into families based on four criteria: the nature of the profile material, the parent rock, the dominant texture class, and the permeability of the slowest horizon. Families are given geographic names with a small italicised '*f*' at the end of the name to distinguish them from the widely used, older, soil series names. Families are further subdivided into siblings on the basis of soil depth, topsoil stoniness, soil texture, soil drainage, and soil functional horizons. Each sibling within a family has the family name (abbreviated) followed by a number. Older soil maps comprise soil series and soil types, which also had geographic names. Within the family/sibling system some soil series may fall into more than one family or sibling. Correlation between established soil series and families and/or siblings is included in the fact-sheet part of the S-map database https://smap.landcareresearch.co.nz/.

The soil order names used in the *New Zealand Soil Classification* departed from the naming systems used in many other soil classification systems, which are based mainly on Greek or Latin. The New Zealand system was designed for use by New Zealanders and, where possible, common English words that were connotative, or used terms that were well known from previous New Zealand or international usage, are used.

All of the subgroups depart in some specified way from the central concept of the soil group. The central concept soil is identified as the typic subgroup. 'Typic' does not necessarily mean 'typical'; rather, it means 'no aberrant properties'. This book gives only an overview of the *New Zealand Soil Classification*. For more detailed technical information, or to accurately classify a soil, the reader should refer to the '*New Zealand Soil Classification*' by Allan Hewitt. Detailed information relating to specific soils is available in the National Soil Data Repository at https://soils.landcareresearch.co.nz/soil-data/national-soils-data-repository-and-the-national-soils-database/

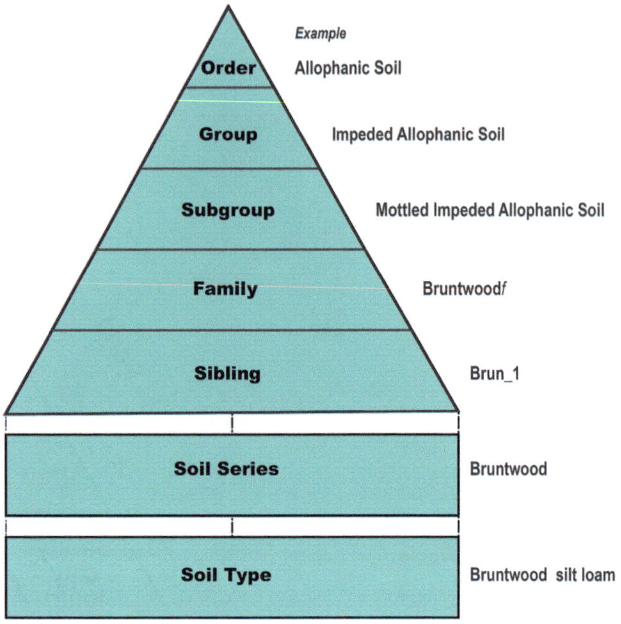

Fig. 1.11 The *New Zealand Soil Classification* hierarchy with an example of the naming system

1.4.2 Overview of the New Zealand Soil Orders

a. Introduction

In this section the 15 soil orders of the *New Zealand Soil Classification* are briefly reviewed. Each of the soil orders is described in detail in the following chapters. A map of the distribution of the New Zealand soil orders (Fig. 1.12) shows their broad pattern of occurrence.

A useful way of understanding the patterns of occurrence, and differences between, the soil orders is to arrange them according to the dominant factors that have influenced their formation and character: age, climate, wetness, and type of rock or geological deposit (Table 1.2).

The Brown Soil order is the most widely occurring soil order, making up just over 40% of New Zealand, mainly on the hill country throughout the lower North Island and South Island (Figs. 1.12 and 1.13). Podzol, Pallic, Pumice, and Allophanic Soils are the next largest areas, each ranging from 5 to 13% of New Zealand. The other ten soil orders cumulatively account for the remaining 20% of the country.

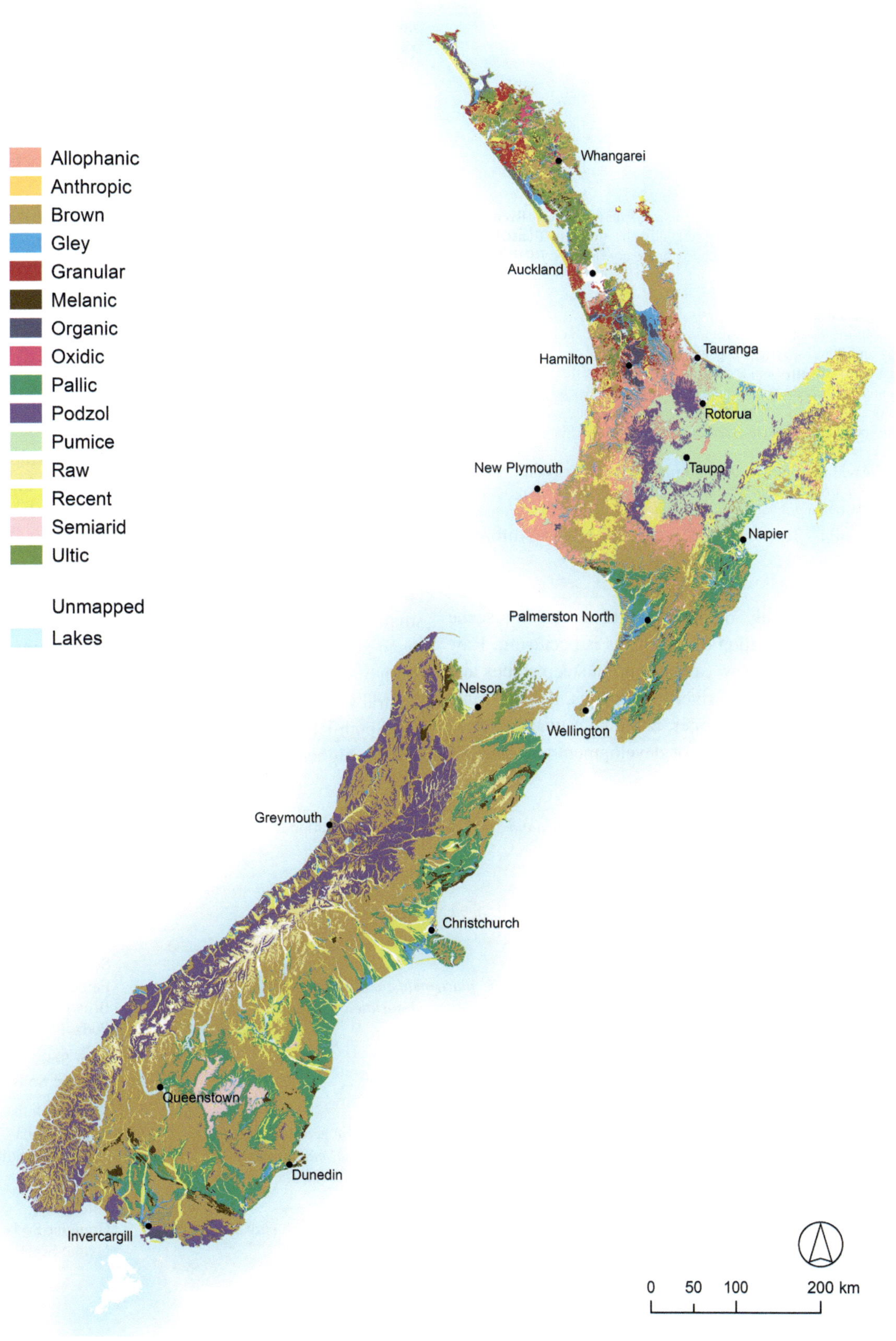

Fig. 1.12 Distribution of New Zealand soil orders. From Manaaki – Whenua Landcare Research

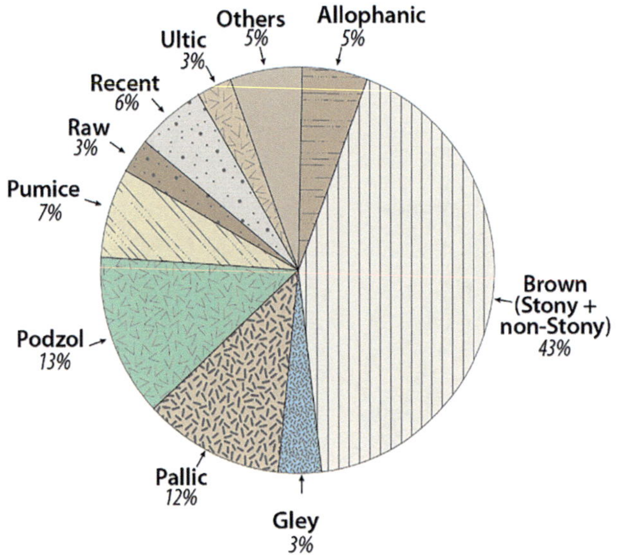

Fig. 1.13 Areal distribution of soil orders in New Zealand

b. Young soils—Raw Soils, Recent Soils, and Anthropic Soils

Where time is limiting, on active parts of the landscape where erosion or deposition of sediments is occurring, Raw and Recent Soils form. Raw Soils occur on steep slopes and rocky outcrops, sand dunes, river flood plains, new volcanic deposits close to active volcanoes, and tidal estuaries, where soils show very little evidence of development and topsoils have not yet formed. Recent Soils form on relatively young parent materials such as river flood plain deposits; however, Recent Soils do have some incipient soil weathering and an A-horizon (a distinct topsoil).

Anthropic Soils occur where humans have had a strong impact, greatly altering or rebuilding a soil by truncating, burying, or deep mixing. Minor human impacts, such as a disturbed A-horizon, are insufficient for a soil to qualify in the Anthropic order. Anthropic Soils occur, for example, in areas within cities that were recontoured during subdivision, areas that have been restored or rebuilt following activities such as open cast mining, Māori midden or gardening sites, and some forest harvest sites.

c. Climate-dominated soils—Semiarid Soils, Pallic Soils, Brown Soils, and Podzol Soils

Where soils are formed predominantly from silicic (quartzo-feldspathic) rocks such as greywacke, schist, and mudstone (or on materials derived from them such as loess and colluvium), climate is a major determinant of the soil properties. Thus a soil-climate sequence is recognised, commencing with the Semiarid Soils in Central Otago, and small areas of the central Waitaki River catchment, where mean annual rainfall is less than about 500 mm. In summer drought-prone areas, particularly around the margins of Central Otago and along the eastern lands of both the North and South Islands, Pallic Soils form, many on loess deposits. On the higher hill country and areas where the rainfall is

Table 1.2 Organisation of the soil orders of the *New Zealand Soil Classification* in relation to age, climate, soil wetness and geological material

Dominant influence		Soil order
Young soils *Soils on recently deposited or exposed materials with minimal, or no, horizon development and usually low clay content*		Raw Soils Recent Soils Anthropic Soils
Mature soils *Soils that have well developed topsoil and subsoil horizons*	***Climate*** *Soils formed mainly in quartzofeldspathic-rich materials that show the effects of climate and associated vegetation*	Semiarid Soils Pallic Soils Brown Soils Podzol Soils
	Wetness *Soils with prolonged high water tables*	Gley Soils Organic Soils
	Geological material *Soil parent materials formed from rocks or geological deposits that dominate the soil character: limestone, mafic material, and tephra (including pumice and other pyroclastic materials)*	Melanic Soils Pumice Soils Allophanic Soils
Old soils *On land surfaces with materials that have attributes of advanced weathering including high clay content*		Ultic Soils Granular Soils Oxidic Soils

higher, and more evenly distributed through the year, Brown Soils occur.

Podzol Soils form where there is a high rainfall, along with good drainage, to facilitate leaching, and particular plants that provide a source of acidic organic material to the soil. Rimu *(Dacrydium cupressinum)*, kauri (*Agathis australis*), and beech (*Nothofagus* sp.) trees are all common sources of acid litter associated with Podzol Soils in New Zealand.

Collectively, the Semiarid Soils, Pallic Soils, Brown Soils, and Podzol Soils extend over 73% of New Zealand (Figs. 1.12 and 1.13).

d. Soils dominated by wetness—Organic Soils and Gley Soils

In topographically low-lying areas, or in soils with subsoil impediments to drainage, water tends to accumulate on or near the soil surface. Where plants grow in a saturated environment, dead plant material tends to accumulate in the water and, rather than biodegrading, forms peat. Soils with thick peat, described as 'organic soil materials', are classified as Organic Soils, as are soils under forest with thick ($\geq$ 40 cm) surface layers of organic matter, including decomposed leaf litter. Climate and in some cases altitude play roles additional to that of topography for maintaining peatlands because the number of rain-days per year (not just amount of rain) can become critical to the ongoing development of peat and thus Organic Soils.

The Gley Soils are saturated with water for at least part of the year, so that all the oxygen present is used by microbes, resulting in anaerobic conditions. Saturated conditions can strongly affect the growth and types of plants by limiting the availability of oxygen to roots. The duration and depth to wetness is indicated by accumulation of organic matter and soil colours. Dominant pale grey or blueish colours indicate prolonged wetness (reducing conditions), and rust-coloured mottles, or concretions, in a paler matrix indicate intermittent wetness (redox conditions). Gley Soils form in low-lying areas with high water tables, or where water saturates the soil surface for extended periods because of low permeability or impeded drainage.

Large areas of Gley and Organic Soils are now used for productive agriculture or horticulture but the wetness indicators persist as evidence of the extent of wetlands prior to farm development and drainage.

e. Soils dominated by particular geological materials—Pumice Soils, Allophanic Soils, and Melanic Soils

In the Pumice Soils, Allophanic Soils, and Melanic Soils the parent material has a strong influence on the nature and development of the resulting soil. Through central North Island and across Taranaki, the influence of tephra (unconsolidated, fragmental, explosively erupted, pyroclastic material) dominates the soil properties. In a large area centred on Lake Taupo are soils, called Pumice Soils, formed mainly in glassy, pumice-dominated tephra (vitric materials) erupted from the Taupo volcano about 1800 years ago. Beyond the Pumice Soils, reaching into the Bay of Plenty, King Country, parts of the Waikato, and across to Taranaki, are Allophanic Soils that have formed mainly in finer textured, multiple tephra-fall deposits. Where limestone dominates the soil parent material, Melanic Soils occur which have distinct, deep, dark blackish, topsoils over weathering rock and relatively high pH. Melanic Soils also form on dark-coloured basic or mafic (magnesium- and iron-rich) rocks (such as basalt) or ultramafic rocks (such as dunite).

f. Ancient soils—Ultic Soils, Granular Soils, and Oxidic Soils

Among the rolling land, hills, and terraces, from Northland to the northern Waikato, extensive areas have generally escaped erosion or significant deposition for more than about 50,000 years. Small areas of stable land surfaces also occur in the Wellington, Marlborough, Nelson, Otago, and Southland regions. The comparative stability has enabled soil development to proceed to more advanced weathering and leaching than in other parts of New Zealand, resulting in soils with high clay contents, acidic pHs, and low natural soil fertility.

On some older basalt lava flows and especially on older tephra deposits, particularly in the South Auckland and northern Waikato region, Granular Soils are formed. These tend to have clay-rich subsoils, well-developed pedality, and can be very productive. On the more silicic parent materials, such as greywacke, Ultic Soils form with yellow, sticky clays in subsoils, high acidity, and low soil fertility. Small areas of Oxidic Soils, similar to soils of tropical regions, form north of Auckland on sites that have been stable for long periods with a warm wet climate but more iron-rich parent materials than the sedimentary rocks that weather to form the Ultic Soils. Upper parts of Oxidic Soils tend to be more friable and less sticky than Granular or Ultic Soils.

1.4.3 Key to New Zealand Soil Orders

Follow the key in Fig. 1.14 to determine the soil order(s) that occur at any site of interest. Once a soil order is identified the appropriate chapter should be consulted for more detailed information. Correlations with international soil classifications (*Soil Taxonomy* and *World Reference Base*)

are included in Appendix B, with details for each New Zealand Soil Order in the relevant chapter.

1.5 Soils in the Landscape

An understanding of the relationship between various aspects of the landscape and the soils that form is a key skill of pedologists (scientists who study the soil within the landscape). Traditionally, as a pedologist worked across the landscape observing the subtle variations within the soil and relating them to the soil-forming factors that influence those differences s/he developed a mental predictive 'model' of the soil-landscape relationships that could then be used to predict the properties of soils on sites not yet investigated. Often air photographs were used to give a clear overview of the topography and as a basis for delineating the distribution of soils based on an understanding of the soil-landscape relationships. Modern tools such as GIS allow systematic development and recording of predictive models.

While it is usually only practical to describe individual soil profiles the soil forms a continuum across the landscape. In some places quite different soils gradually merge from one to the other, with a range of 'intergrades' between them. In other places there may be distinct boundaries between soils with contrasting properties that relate to landscape units or landforms such as an alluvial terrace versus an adjacent hillslope.

Across landscapes and regions, geological, topographic, climate, and vegetation differences can be observed which provide clues to the underlying soil. For example, consider a transect across the lower half of the North Island from Foxton Beach (west coast) through Shannon and Eketahuna to Castle Point (east coast). Raw Soils occur on the coastal sand dunes of the west coast. Further inland Recent Soils are formed on the older, vegetated, sand dunes and the flood plains of the Manawatu River. In low-lying wetland areas, where flax swamps used to hold sway, Organic Soils occur that, following drainage and fertiliser addition, are often favoured nowadays for potato growing. Towards Shannon above the terrace scarp from the current river flood plain is an older, uplifted, terrace surface which was coated in loess (wind-blown dust) during the Last Glaciation. The summer drought and fine silt-rich parent materials mean a Pallic Soil will predominate. In the foothills of the Tararua Ranges, as altitude and rainfall increase, Brown Soils form. High in the Tararua Ranges in the cloud forest, where beech and rimu trees grow, pockets of Podzol Soils form within a mosaic of Brown Soils and Recent and Raw Soils near the mountain tops in areas that have been subject to erosion. Down the other side of the range, the mountains give way to the high rainfall and green grass of sheep and dairy farms on the rolling hills (Brown Soils) and river flats (Recent Soils) near Eketahuna. Further on, towards Castle Point on the east coast, the rainfall rapidly drops away and Pallic Soils dominate.

Soil maps seek to capture the distribution and variability of soils across a landscape in two dimensions. The detail that can be included on a map depends on the scale of the map. With the increasing move to digital soil mapping, and the ease of 'zooming in' on a map to a large scale (called the visualisation scale), caution needs to be exercised in interpreting such maps as it is important to know the scale, and thus the level of detail, of the original data that were used to compile the map.

In order to capture the variability of soils in the landscape, soil maps may include soil associations and soil complexes. A soil association can be mapped where two or more soils form a predictable pattern in the landscape but occur at a scale too fine to show at the map scale being used. Soil complexes can be mapped where two or more soils occur within a map unit but the pattern is not readily predictable using microtopography or other visual clues.

Digital soil mapping may be used to make soil surveys by replicating the decision-making process of conventional soil survey using the field observations made by a pedologist and relating them to the soil-forming factors (defined in a GIS) that control soil distribution. For example, attributes of the climate, landform, vegetation, geology, and soil age. The relationships can be expressed as a statistical soil-landscape model to predict soils at locations that have not been sampled. Gradual soil changes across the landscape and measures of map confidence, as well as links to more detailed information, may be included with digital soil maps. A digital soil map of New Zealand (S-map), developed by Manaaki Whenua – Landcare Research, is accessible online at https://smap.landcareresearch.co.nz/.

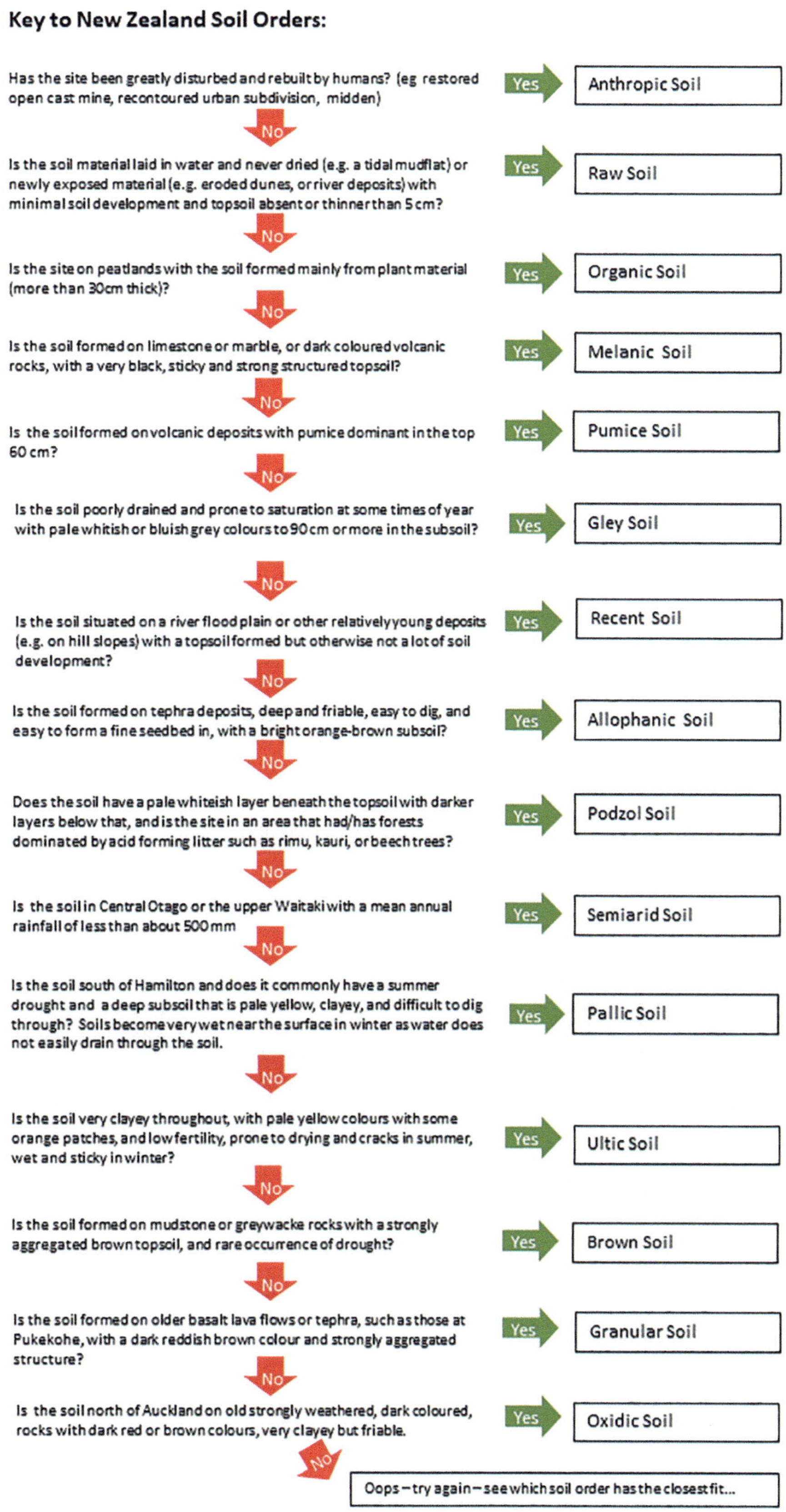

Fig. 1.14 Simplified key to New Zealand soil orders. This is a general guide. For a more detailed and accurate approach see the *New Zealand Soil Classification* (Hewitt 2010)

Further Reading

Aislabie J, Deslippe JR (2013) Soil microbes and their contribution to soil services. In: Dymond JR (ed) Ecosystem services in New Zealand—conditions and trends. Manaaki Whenua Press, Lincoln, New Zealand, pp 143–161

Anderson A (2015) Speaking of migration AD 1150–1450. In: Anderson A, Harris A, Williams B (eds) Tangata Whenua—an illustrated history. Bridget Williams Books, Auckland, pp 42–69

Barrell DJA (2011) Quaternary glaciers of New Zealand. In Ehlers J, Gibbard PL, Hughes PD (eds) Developments in Quaternary science, vol 15. Amsterdam, pp 1047–1064, ISBN: 978-0-444-53447-7

Blakemore L, Searle PL, Daly BK (1987) Methods of soil chemical analysis. Soil Bureau Scientific Report 80. DSIR, Lower Hutt, 103 p

Brooking T, Wood V (2013) The grasslands revolution reconsidered. In: Pawson E, Brooking T (eds) Making a new land—environmental histories of New Zealand, new edn. Otago University Press, Dunedin, pp 193–208

Clayden B, Hewitt AE (1993) A new system of horizon notation for New Zealand soils. Catena 20:405–410

Eaves SR, Winckler G, Mackintosh AN et al (2019) Late-glacial and Holocene glacier fluctuations in North Island, New Zealand. Quat Sci Rev 223:105914

Edbrooke SW (2017) The geological map of New Zealand (accompanying geological map of New Zealand 1:1,000,000). GNS Science Geological Map 2. GNS Science, Lower Hutt

Gibbs HS (1980) New Zealand Soils—an introduction. Oxford University Press, Wellington, 115p

Graham I (ed) (2015) A continent on the move: New Zealand geoscience revealed, 2nd edn. Geoscience Society of New Zealand with GNS Science, Wellington, ISBN 978-1-877480-47-8, 397 p

Grange LI (1944) A basic scheme for land classification. N Z J Sci Technol 26A:136–141

Grange LI (1945) Farming in New Zealand, North Island soils. N Z J Agric 70:387–397

Grange LI (1946) Farming in New Zealand, South Island soils. N Z J Agricu 72:583–591

Hewitt AE (1992) Soil classification in New Zealand—legacy and lesson. Aust J Soil Res 30:843–854

Hewitt AE (2010) New Zealand Soil Classification, 3rd edn. Landcare Research science series No. 1. Manaaki Whenua Press, Lincoln, ISBN 978-0-478-34710-4, 136 p

Jenny H (1941) Factors of soil formation. A system of quantitative pedology. Republished in 1994 with foreword by Ronald Amundsen. Dover Publishers, New York. ISBN 0-486-68128-9, 281 p

Leathwick J, Wilson G, Rutledge D et al (2003) Land environments of New Zealand. Bateman, Auckland. 184 pp + 2 maps

Molloy L (1993) Soils in the New Zealand landscape—the living mantle, revised edition. New Zealand Society of Soil Science. ISBN 0-908606-55-9, 239 p. http://nzsss.science.org.nz/publications.html

McBratney AB, Santos MLM, Minasny B (2003) On digital soil mapping. Geoderma 117:3–52

McGlone MS, Wilmshurst JM (1999) Dating initial Maori environmental impact in New Zealand. Quatern Int 59:5–16

McLaren RG, Cameron KC (1996) Soil science, sustainable production and environmental protection. Revised edition. Oxford Univeristy Press, Auckland, 304 p

McSaveney E, Nathan S (2020) "Geology—overview—Quaternary mountains and glaciers", Te Ara—the encyclopedia of New Zealand. http://www.TeAra.govt.nz/en/geology-overview/page-8. Accessed 17 Sept 2020

McWethy DB, Whitlock C, Wilmshurst JM et al (2009) Rapid deforestation of South Island, New Zealand, by early Polynesian fires. The Holocene 19(6):883–897

Milne JDG, Clayden B, Singleton PL, Wilson AD (1995) Soil description handbook, Revised Edition. Manaaki Whenua Press, Lincoln, New Zealand. ISBN 0-478-04549-2, 157 p

Mortimer N, Campbell HJ (2014) Zealandia: our continent revealed. Penguin, Auckland, New Zealand, p 272 p

Mortimer N, Campbell HJ, Tulloch AJ et al (2017) Zealandia: Earth's hidden content. GSA Today 27. https://doi.org/10.1130/GSATG321A.1

Newnham RM, Lowe DJ, Williams PW (1999) Quaternary environmental change in New Zealand: a review. Prog Phys Geogr 23 (4):567–610

Newnham RM, McGlone MS, Moar NT et al (2013) The vegetation cover of New Zealand at the Last Glacial Maximum. Quatern Sci Rev 74:202–214

New Zealand Soil Bureau (1968) Soils of New Zealand Part 1. Soil Bureau Bulletin 26(1). DSIR, Wellington, New Zealand, 142 p

New Zealand Soil Bureau (1973) Map of parent rocks of New Zealand soils 1:1 000 000. 2 Sheets. N.Z. Soil survey report 5. DSIR, Wellington

Pawson E, Brooking T (eds) (2013) Making a new land—environmental histories of New Zealand, new edn. Otago University Press, Dunedin

Perry GLW, Wilmshurst JM, McGlone MS et al (2012) Explaining fire-driven landscape transformation during the Initial Burning Period of New Zealand's prehistory. Glob Change Biol 18:1609–1621

Schulmeister J (ed) (2017) Landscape and Quaternary environmental change in New Zealand. Atlantis Press (Springer), The Netherlands

Soil Survey Staff (2014) Keys to Soil Taxonomy, 12th edn. USDA Natural Resources Conservation Service, Washington, DC, 362 p

Soil Survey Staff (2015) Illustrated guide to Soil Taxonomy, version 2. U.S. Department of Agriculture, Natural Resources Conservation Service, National Soil Survey Center, Lincoln, Nebraska. https://www.nrcs.usda.gov/wps/portal/nrcs/detail/soils/survey/class/taxonomy/?cid=nrcs142p2_053580. Accessed 2 Mar 2020

Taylor NH, Pohlen IJ (1962) (Reprinted in 1979 with amendments) Soil survey method. Soil Bureau Bulletin 25. DSIR, Lower Hutt, 242 p

Webb TH, LR Lilburne (2011) Criteria for defining the soil family and soil sibling. The fourth and fifth categories of the New Zealand Soil Classification, 2nd edn. Landcare Research science series No. 3. Manaaki Whenua Press, Lincoln, 38 p

Williams PW (2017) New Zealand Landscape—Behind the Scene. Elsevier, Amsterdam, 470 p

World Reference Base (2015 update); http://www.fao.org/3/i3794en/I3794en.pdf. Accessed 2 Mar 2020

2 Allophanic Soils

In the North Island where tephras fall
Allophanic Soils have it all,
Plants will thrive without much toil
Roots reach down to moist subsoil

Abstract

Allophanic Soils cover 5% of New Zealand, predominantly in the Taranaki, Waikato, and western Bay of Plenty regions. Allophanic Soils are usually formed in deep accumulations of tephra (volcanic ash) and other volcanic debris, or occasionally on rapidly weathering rocks, usually with good drainage and relatively high rainfall. Thus they occur in the North Island around Mt Taranaki and in areas peripheral to the Pumice Soils where tephras from the central North Island volcanoes and Mt Taranaki have accumulated, typically millimetre by millimetre, over millennia. Most Allophanic Soils originated by developmental and/or retardant upbuilding pedogenesis. Many are $\geq$20,000 years old. Allophanic Soils provide an excellent medium for plant growth because of their friability, low bulk density, fine structure, high water-storage capacity, and ability to support deep-rooting plants. Allophanic Soils contain significant amounts of a globally uncommon clay mineral, allophane, which is formed mainly from the dissolution products of volcanic glass. Allophane, an Al-rich aluminosilicate, comprises tiny spherules with a huge surface area which strongly adsorbs anions, notably of phosphorus and sulphur. Allophane will adsorb a proportion of the phosphorus applied as fertiliser, making it unavailable for plant uptake, but can be used to advantage to adsorb phosphorus from irrigated effluent.

2.1 Important Features of Allophanic Soils

2.1.1 Concept and Key Features of the Soil Order

Allophanic Soils are deep, very friable, bright brown to yellowish brown or, less commonly, reddish in colour (Fig. 2.1). Allophanic soil material has a distinct ‘greasy’ feel when moist and rubbed between the thumb and fingers. Most Allophanic Soils have formed as a result of gradual accumulation of glass-rich tephra erupted from the major active volcanoes in central and western North Island. Both rhyolitic and andesitic volcanoes contribute, including Okataina, Taupo, Ruapehu, Tongariro, Ngauruhoe, and Taranaki Maunga (Mt Taranaki).

As the name suggests, Allophanic Soils have properties that are strongly influenced by the presence of an aluminosilicate clay mineral called allophane. Allophane is common in soils formed on explosively erupted pyroclastic (fragmental) deposits associated with recently active vol-

A. E. Hewitt et al., *The Soils of Aotearoa New Zealand*, World Soils Book Series,
https://doi.org/10.1007/978-3-030-64763-6_2

canism, such as in the circum-Pacific 'Ring of Fire'. However, elsewhere allophane is relatively uncommon (soils dominated by allophane—the Andisols of '*Soil Taxonomy*'—make up only about 1% of the world's soils). Allophanic Soils, especially on mafic (scoriaceous) materials, may also contain small to moderate quantities of the iron oxide clay mineral called ferrihydrite, in effect an iron-rich mineral equivalent to allophane. Both allophane and ferrihydrite are now recognised as being nanocrystalline clays (or nanominerals)—that is, they comprise tiny structural units in the nanoscale range (1–100 nm), as described in Sect. 2.4.

Allophanic Soils have an unusually low soil dry bulk density, high porosity, and high to very high soil-water holding capacity, and may also be sensitive and/or thixotropic (i.e. they lose strength when remoulded). They are formed mainly in glassy pyroclastic material, also referred to collectively as tephra (volcanic ash). Allophanic Soils form mainly in andesitic or rhyolitic materials, very commonly both together. They can also form in coarse pumice, basaltic scoria, and tephra-derived materials reworked and deposited by rivers, wind, or gravity (described as volcaniclastic). Occasionally soils containing allophane, but with insufficient amounts or thicknesses to qualify for Allophanic Soils per se, are formed on tuffaceous greywacke (fine sandstone) and in strongly leached loess and other regolith materials in high country situations where strong leaching and desilication have occurred.

A key concept is that Allophanic Soil profiles, in contrast to those of many other soil orders, typically form by upbuilding pedogenesis as younger tephras are deposited on top of older ones so that the resulting profile character is a reflection of the interplay, or 'competition', between the rate at which tephras have accumulated on the land surface (by geological processes) and their modification by topdown soil-forming or pedogenic processes that transform the material into soil horizons. Thus soil stratigraphy is at the heart of understanding the genesis of many Allophanic Soils.

2.1.2 Areas of Occurrence

The Allophanic Soils cover 5% of New Zealand (Fig. 2.2). They occur only in the North Island, with concentrations in Taranaki, Waikato, and western Bay of Plenty, and to the south and southeast of the Tongariro Volcanic Centre in the central North Island. Location is mainly controlled by the distribution of andesitic and rhyolitic tephra-fall deposits, tephric loess, or volcanogenic alluvium derived mainly from reworked tephra deposits. Allophanic Soils also form on the weathering products of andesite lahar mounds on the Taranaki and Ruapehu ring plains, and from some mafic pyroclastic deposits including scoria, such as in Auckland and Northland on basaltic scoria cones.

Fig. 2.1 Allophanic Soil and associated landscape. Left: a Typic Orthic Allophanic Soil in the Waikato formed in incrementally accumulating thin tephras derived from central North Island and Taranaki volcanic eruptions. The 1 m deep soil is one of the oldest Andisols in the world (containing Rotoehu Ash, c. 45,000 years old at ~1 m depth and Kawakawa/Oruanui Tephra at ~55 cm). Every part of the profile has been an A-horizon at the surface at some time. Right: friable Allophanic Soils are readily ploughed to form an exceptionally fine seed-bed

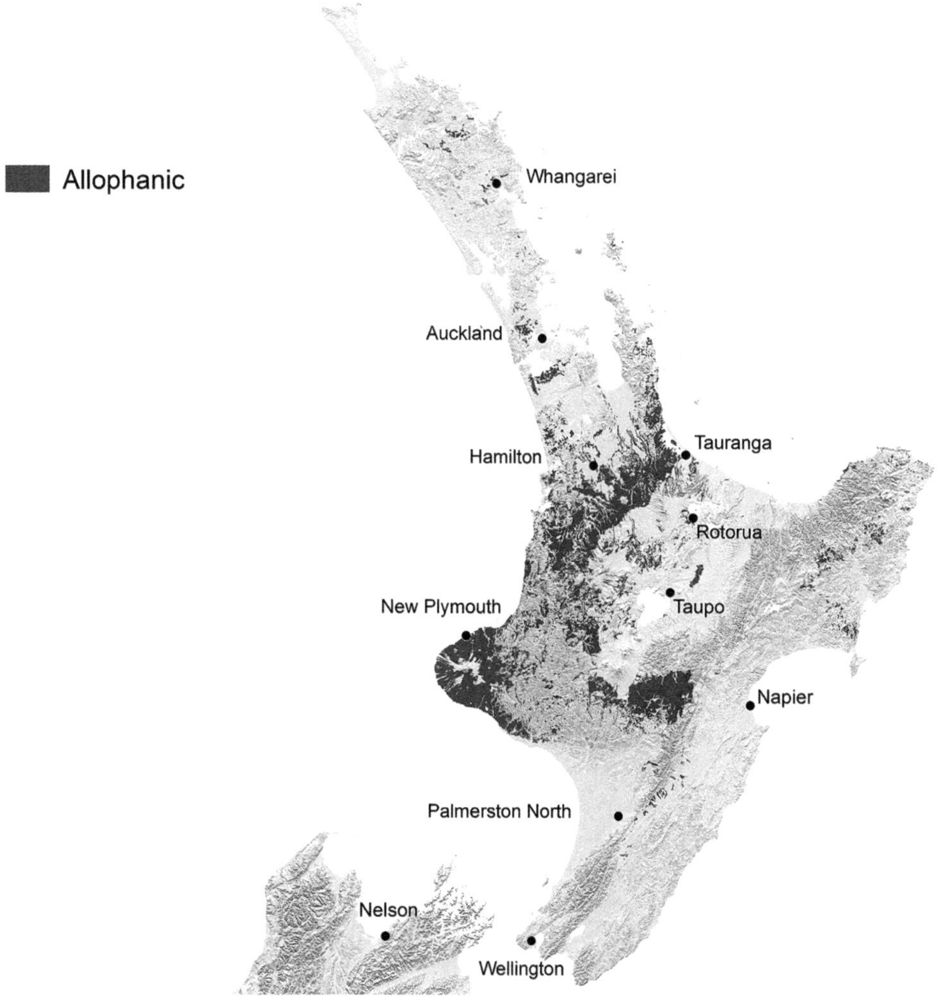

Fig. 2.2 Distribution of Allophanic Soils in New Zealand

Some allophane-bearing soils occur in the South Island associated with Allophanic Brown Soils, and many Podzol Soils, contain allophane formed by translocation of proto-imogolite sols in the podzolization process involving acid leaching (see Chap. 11). Most South Island soils have no (or only tiny amounts) of tephra (from North Island volcanoes), and the allophanic soil materials are formed instead through the weathering of mainly quartzofeldspathic minerals (such as in loess or greywacke) and strong leaching of the soluble products under high rainfall, the loss of silicon (desilication) and associated concentration of aluminium being the key process favouring the formation of allophane. Neverthless, the distal, Taupo-volcano-derived Kawakawa/Oruanui Tephra, aged c. 25,400 years old, is an important and widespread but diminutive deposit in South Island. The Kawakawa/Oruanui Tephra is present as a thin visible layer in peat deposits in northern South Island but elsewhere it occurs mainly as a cryptotephra ('hidden tephra', a sparse glass-shard concentration not visible to the naked eye) including as far south as Otago and Southland.

2.1.3 Variation Within the Allophanic Soil Order

The following soil groups are defined in the *New Zealand Soil Classification* (Fig. 2.3):

Perch-gley Allophanic Soils—periodic wetness (usually marked by redox mottling or low chroma colours in the upper subsoil only) caused by water perching on a slowly permeable horizon (such as an ironstone-pan),

Gley Allophanic Soils—periodic wetness (usually marked by redox mottling or low chroma colours) caused by a groundwater table,

Impeded Allophanic Soils—have a slowly impermeable layer that may impede downwards movement of plant roots and water (within 90 cm of soil surface),

Orthic Allophanic Soils—Other Allophanic Soils.

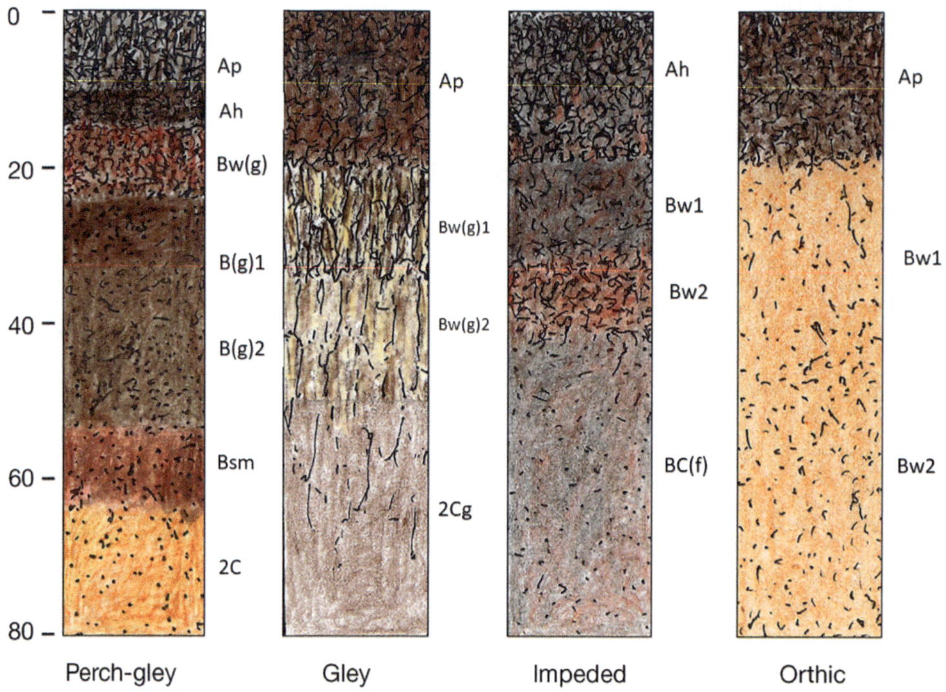

Fig. 2.3 Groups within the Allophanic Soil order (vertical axis is depth in cm)

Gley Allophanic Soils are not common in the landscape because usually halloysite (a crystalline aluminosilicate clay relatively enriched in silicon) is the stable clay mineral formed under slow drainage or wet reducing conditions. In comparison allophane requires good drainage to form, therefore, in poorly drained soils allophanic soil material does not form (see Fig. 2.6).

Three subgroups recognise transitions from Allophanic Soils to other related soil orders or soil groups: Mottled (intergrade to Gley Soils), Peaty (to Organic Soils), and Vitric (to Pumice Soils). Two subgroups, which depart from the soil group, are recognised, namely Ironstone and Acidic. The ironstone subgroup occurs where a distinct iron pan has formed within the subsoil. Such soils occur most commonly in Taranaki. The acidic subgroup comprises soils with pH < 5.5 in the B-horizon. The subgroup closest to the central concept of the group is identified as Typic.

2.1.4 Origin of the Soil Order Name

'Allophanic' refers to the dominance of the mineral allophane or other materials (ferrihydrite, or aluminium- or iron-humus complexes) that impart similar properties. The name 'allophane' was originally coined by mineralogists from the Greek *allos,* meaning 'other', and *phanos*, meaning 'to appear', because of its unusual reaction to the blowpipe flame used in early mineralogical testing.

2.2 Soil Profile Genesis

2.2.1 Soil Profile Formation

The mode of formation of most Allophanic Soils is different from that of many other soil orders because it often comprises two processes (one geological, and one pedological) operating at the same time:

(1) the gradual accumulation of tephra deposits, and
(2) the concurrent weathering and modification of the accumulating tephra deposits by topdown soil-forming processes to form soil horizons containing organic matter, allophane, and associated minerals.

The overall process is referred to as 'upbuilding pedogenesis', meaning the soils are forming *at the same time* as new tephra materials continue to be added to the land surface, building it upwards. The resultant soils may show distinctive layering and buried horizons (often referred to as paleosols), forming distinct multi-layered profiles where successive tephra deposits are relatively thick (Fig. 2.4), or homogenous-looking apedal earthy-fabrics where the tephras are thin (Fig. 2.1).

In contrast, classical soil formation occurs *sequentially in two steps*:

(1) a new parent material is exposed and then,

Fig. 2.4 Complex soil profile near Mt Tarawera chosen to illustrate retardant upbuilding pedogenesis. The sequence shows a series of five tephra layers dating back to 9500 years ago (as indicated on the labels) with soil horizons on each of them developed when at the land surface. The deposition of each new tephra layer has buried the pre-existing soil horizons, forming a paleosol, and isolating them from the land surface, for example, between the Taupo and Kaharoa tephras. The uppermost soil horizons are formed in material erupted from Mt Tarawera on 10 June 1886 (mainly the Rotomahana Mud). The modern multi-layered soil profile (to ~1 m depth) includes the soil on Rotomahana Mud and the paleosols on the Kaharoa and Taupo tephra layers. The soil is classified as Buried-pumice Tephric Recent Soil (which does contain some allophanic soil materials)

(2) soil formation (pedogenesis) subsequently transforms the (stable) parent material through topdown processes, the horizons developing in a downward moving front according to site conditions.

This model, well entrenched in almost all text books, applies to stable landscapes and 'set' parent materials where there are effectively no ongoing geological processes in play, such as the deposition of loess, tephra, colluvium, or alluvium.

The frequency and thickness of tephra accumulation determines, in part, how much impact topdown pedogenic processes have on the resulting soil profile, and if 'developmental upbuilding' or 'retardant upbuilding', or both, take place. Developmental upbuilding occurs when the rate of addition of tephra is sufficiently slow (in thin increments) that topdown soil formation effectively keeps pace, as the soil surface gradually rises, and the separate thin tephra layers are modified to form soil horizons. Because the land surface is slowly rising, there is not much time for horizon features to develop and soils tend to be reasonably uniform (Fig. 2.1). A key point is that every part of the profile has been an A-horizon at one time and so will have some imprint of pedogenesis.

The soil profile in Fig. 2.1 (an example of a soil formed as a result of developmental upbuilding, to ~1 m depth) has a composite age of ≤c. 45,000 years, i.e. it has an age *range* rather than a specific 'single' age—because the profile has taken 45,000 years to accumulate thin distal tephra layers while being simultaneously modified by pedogenesis. The profiles of many Allophanic Soils (formed by developmental upbuilding) in New Zealand have composite ages of 20,000 years or more. By way of contrast, the soil formed on the pumiceous Taupo ignimbrite, which was every where emplaced within about 10 min (Chap. 12), has a single, specific age of ~1788 years (± 10 years) before 2020.

Retardant upbuilding (as exemplified in Fig. 2.4) occurs when a relatively thick layer of tephra is 'instantaneously' added to the surface, or where the rate of accumulation of thinner additions is fast, so that the original soil is rapidly buried. Thus the buried topsoil (now a paleosol) is isolated from the new land surface, and its development is effectively retarded. At the ground surface, soil formation begins anew with a Raw Soil. Over time, if conditions remain stable, the Raw Soil will develop into a Recent Soil and then to a Pumice or an Allophanic Soil. Retardant upbuilding is more likely to occur close (proximal) to the source volcano, whereas developmental upbuilding occurs further away (at medial to distal sites) where the layers of tephra that fall are thinner and/or less frequent. Both forms of upbuilding can be recognised as taking place in some Allophanic Soils.

2.2.2 Formation of Allophane

Allophane, the presence of which imparts the key properties of an Allophanic Soil, is a clay mineral that comprises tiny hollow spherules which are 3.5–5 nm in diameter (which is much finer than many other clay minerals which are measured in the micro- rather than nanoscale). The most common ingredient for the formation of allophane is volcanic glass (similar to window glass) which occurs in tephra (often 50–90%), along with crystals of feldspar, quartz, various mafic (or heavy) minerals, and rock fragments. The volcanic glass in tephra weathers readily because the glass is in the form of tiny shards (fragments) and is often vesicular (with holes), so the particles have a large surface area that is open to rapid attack by acids.

To form allophane the glass shards must first be weathered to release Si and Al species (along with interstitial ions in the glass such as those of Ca, Mg, Na, K, and Fe) into solution. Such weathering occurs primarily by the dissolution of the glass via hydrolysis, forming monosilicic acid ($Si(OH)_4$) and gibbsitic compounds ($Al_2(OH)_3$). Then, within the soil solution, these two ingredients are precipitated as allophane spheres. The formation of allophane is thus dependent ultimately on the concentration of silicic acid and the availability of aluminium species in soil solution. Such concentrations are influenced by various environmental conditions including annual rainfall, soil drainage, the nature of the soil organic matter, soil acidity, tephra composition (rhyolitic versus andesitic versus basaltic) and accumulation rate, depth of burial, and vegetation type. These factors in turn dictate the rate of silicon removal from a soil and form the basis of the 'silica-leaching model' for the formation of allophane (Fig. 2.5).

In the silicon-leaching model, if silicon concentrations in soil solution are $< \sim 10$ ppm, allophane will form, but if concentrations are $> \sim 10$ ppm then halloysite forms. Where the silicon concentration is close to 10 ppm either or both minerals may form. Studies involving Allophanic Soils have shown that if about 250 mm of drainage water passes through the profile annually, typically under an annual rainfall of $> \sim 1600$ mm, then the decreased silicon concentration (due to more leaching) will favour the formation of allophane. If total drainage is $< \sim 250$ mm, typically where annual rainfall is $< \sim 1200$ mm, then halloysite will normally form. The formation of halloysite, rather than allophane, can be the result therefore of either low rainfall or poor drainage, or both. Poor drainage may be caused by a high water table or layers of fine material, a clayey paleosol, or a pan, within the soil profile that supports a perched water table. Thus Allophanic Soils do not form at sites where soils are saturated for extended periods. Although rhyolitic and/or andesitic tephras are the most common parent material for formation of allophane, other materials may also weather and the weathering products can form allophane or related clays (such as imogolite, ferrihydrite, metal-humus complexes, and hydroxyl-interlayered vermiculite). Materials that may weather to allophane, or related minerals, include greywacke, schist, basaltic, tephras (scoria and ash) and basalt, and the widely varying quartz- and feldspar-rich parent materials of many Podzol Soils.

2.3 Soil-Landscape Relationships

2.3.1 Overview

Because of the relationship between tephra and the formation of allophane, the most common soil-landscape patterns of Allophanic Soils relate to the accumulation of tephras to sufficient depth to dominate the soil parent material. The soil also needs to be free draining and under a moderately high rainfall to provide conditions that promote the desilication that favours the formation of common Al-rich allophane. On flat to gently rolling slopes, within regions of regular tephra additions, the consistent blanketing of the landscape results in relatively uniform Allophanic Soils over wide areas. However, in areas more distant from the volcanoes, and in areas with steeper slopes or redistributed tephra materials, then a variety of soil-landscape patterns occur. The particular situations where Allophanic Soils occur in the vicinity of andesitic volcanoes, and at greater distances from them, are discussed along with an example of the pattern of Allophanic Soils that occurs on partly tephra-mantled alluvial deposits in the central Waikato basin.

2.3.2 Allophanic Soils in Proximity to Andesitic Source Volcanoes: The Taranaki Ring Plain and Areas South of Tongariro

On the lower slopes of Mt Taranaki, and on the southern side of Ruapehu (where the rhyolitic pumice, erupted from Taupo in c. 232 AD, did not reach), Allophanic Soils are formed in tephras derived predominantly from the frequent explosive eruptions of the adjacent volcano.

Around Mounts Taranaki and Ruapehu, ring plains are formed from repeated lahar (mud and debris) flows from partial collapses of the volcano edifice or (in the case of Ruapehu) from emptying the crater lake from time to time. The lahar material is, in many places, overlain by tephra—the older the lahar, and the closer to the mountain, the thicker are the overlying tephra layers. The generally deep mantle of tephras that covers the landscape provides a relatively uniform, stable, parent material that supports Allophanic Soil on many surfaces. In some places streams have cut down to underlying lava, or lahar deposits, and so Allophanic Soils form on the stable wide ridges but are intersected by Recent and Raw Soils in the steep gullies.

The soils on the hummocky, undulating, lower parts of the Taranaki ring plain are often mapped as a soil association

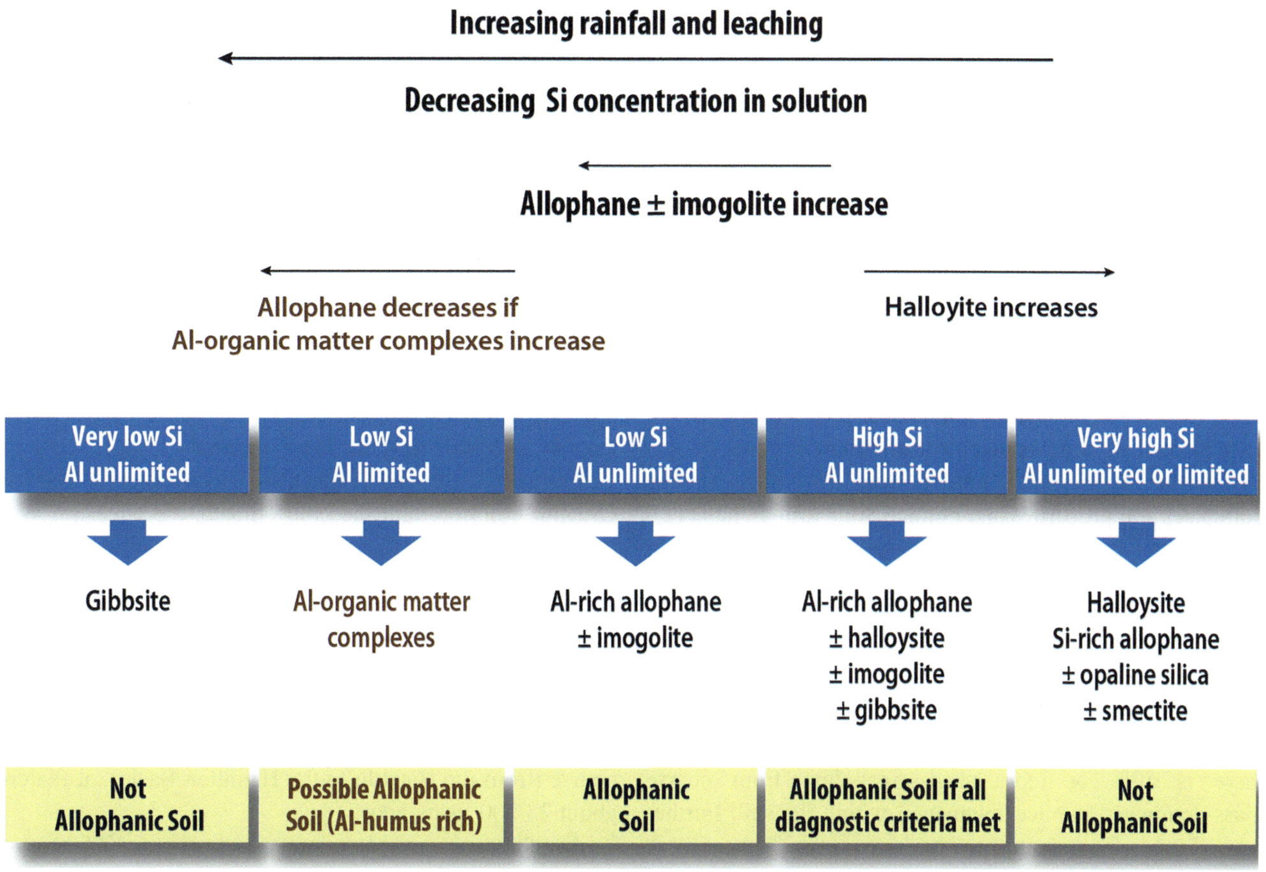

Fig. 2.5 Relationship between environmental conditions and the formation of allophane and other minerals according to the silicon-leaching model and the availability of aluminium

where Orthic Allophanic Soils form on the higher parts of the landscape with Gley Soils in hollows and Impeded Allophanic Soils in the intermediate positions between. Where steeper slopes occur, such as on lahar mounds, much of the tephra may have been eroded off and soils tend to be bouldery, stony, and gravelly. Some are still sufficiently Allophanic to qualify as Allophanic Soils (Opua soils). Rainfall increases with altitude and hence more strongly leached Allophanic Soils are recognised at higher altitudes (250–1150 m). Above about 1150 m, on Mt Taranaki, younger volcanic deposits (Burrell Formation, erupted about 1655 AD, and the Tahurangi Formation, erupted about 1755 AD) have Recent Soils developed on them that grade into Raw Soils in alpine areas where rock is exposed by erosion and minimal soil development has occurred. In the Taranaki region, where rainfall is high, iron has often been moved down through the profile and accumulated to form distinct, hard, iron pans in the subsoil made up mainly of the iron oxide minerals goethite and ferrihydrite. East of Mt Taranaki, soils become progressively more gravelly with inputs of the young rhyolitic pumice from the Taupo volcano.

On the steep mudstone hill country between Mt Taranaki and Mt Ruapehu, tephra mantles less-steep slopes on flattish ridge tops, terraces that are above the modern flood plain, and the rolling or gently sloping foot slopes. Where a tephra mantle is preserved, Allophanic Soils have formed. Eruption of the pumiceous Taupo Tephra from Tauop volcano resulted in the infilling of many lower lying areas with pumice, forming lower terraces on which Pumice Soils have formed (Chap. 12). On steeper slopes, any tephras may have been eroded off, leaving soils formed in the exhumed, underlying, mudstone parent materials. Brown Soils, Recent Soils, or Raw Soils form on the exposed mudstone parent materials, depending on how recently material has been eroded or deposited.

2.3.3 The Waikato-Bay of Plenty Region

Much of the central Waikato area, along with the Bay of Plenty, is mantled by tephra-fall deposits, predominantly derived from the central North Island volcanoes, but also with some input from Taranaki Maunga and Mayor Island. The tephra cover generally thins with increasing distance away from the central North Island source volcanoes. David Lowe and David Palmer published a map in 2005 that shows the dominance of andesitic versus rhyolitic tephra components, making up soils, in relation to the main volcanic centres. In the Waikato there is a gradient in tephra components from rhyolite dominated in the east to andesite dominated towards the southwest and King Country. All the Waikato soils contain both andesitic and rhyolitc tephras, just in different proportions. David Lowe's study of lake cores, where numerous fine tephra layers have been preserved within the lake sediment, has provided insight into the compositions of the tephras that have merged to form the soils in the Waikato.

Near to Taupo a thick surface layer of rhyolitic pumice dominates soil properties and Pumice Soils form (Chap. 12). As the pumice deposit thins with increasing distance from Taupo, gradually the finer underlying tephra and allophanic soil properties, become nearer to the modern land surface, and eventually dominant within the soil profile. Buried-allophanic Orthic Pumice Soils marry the two orders together nicely in many places. On the margins of the Pumice Soils, Vitric Orthic Allophanic Soils occur where there is 50% or more sand (including from coarse glass-dominated pumice fragments) within the soil. Further from Taupo, for example in the Tirau area, Typic Orthic Allophanic Soils are formed mainly in incrementally deposited thin tephra layers that form a fine, weakly developed, deep soil profile (similar to that shown in Fig. 2.1). On alluvial plains (in the central Waikato Basin, Hauraki Plains, and the Rangitaiki Plains in the Bay of Plenty) many of the soils are formed in tephric alluvium where tephra and small amounts of other materials have been reworked by rivers, carried farther from the source volcanoes, and deposited on the pre-modern river floodplains. Further north the tephra layers thin and are preserved only intermittently, mainly on flat or gently sloping surfaces, with Granular and Ultic Soils on steeper slopes and in areas where younger tephras are not present. Some of the younger tephra layers have been transformed into halloysitic, not allophanic, clays (see Chaps. 6 and 16).

2.3.4 Allophanic Soil Pattern on the Alluvial Fans and Plains of the Waikato Basin and Hauraki Area

The alluvial plains in the central Waikato (Hamilton) Basin, such as the near-flat areas between Cambridge and Taupiri, and in the Hinuera Valley and Matamata-Hauraki Plains, were built up by material carried and deposited by the ancestral Waikato River. During the last cold period in the last glaciation, between about 32,000 and 18,000 years ago, the ancestral Waikato River carried a high sediment load. Such a load was partially due to increased erosion in the headwaters as a result of less vegetation cover in the colder and windier climate. Also contributing to the high sediment load was a huge volume of loose volcanic material, in the catchment, derived from repeated rhyolitic eruptions, particularly the caldera-forming Oruanui eruption of Taupo volcano about 25,400 years ago. The Oruanui eruption (generating the Kawakawa/Oruanui tephra) was the largest eruption on Earth in the last 70,000 years and is classified as a 'super-eruption' because of the huge volume of pyroclastic material generated, around 530 km^3 (as dense rock equivalent) or 1170 km^3 as bulk loose material.

Because of the high bedload of loose material from the Oruanui eruption, and the cold climate, the ancestral Waikato River was a high energy braided river system, somewhat like those of Canterbury today. As the river deposited material and filled its bed it repeatedly overflowed and moved to new channels, gradually filling the Matamata-Hauraki Basin with volcanogenic alluvium ranging from coarse gravelly sands to silts. The river then began infilling the Hamilton Basin with alluvium after shifting its course (from the Hauraki to the Hamilton Basin), at Piarere, about 23,500 years ago.

As climate warmed, commencing about 17,500 years ago, forest vegetation became quickly re-established, erosion rates declined markedly as the landscape became stabilised, and the bedload in the river dropped. As a result the river started to cut down through the earlier sediment deposits. The downcutting process resulted in 'degradational' terraces being formed, such as those evident in the Karapiro area. Multiple abandoned channels are evident in the modern landscape surface of the Hamlton Basin as shallow, wide paleochannels. Eventually around, or soon after, about 17,000 years ago, the river became entrenched in the path we see today, about 20 m lower than the now-abandoned wider alluvial plain. The plain surface, such as that between Hamilton and Cambridge (called the Hinuera Surface), was left with the pattern of a braided river bed with many slightly raised deposits of gravelly sands, and sands, on the former levees (low ridges) and finer material, typically silts, in the low-lying former back-swamp areas (swales).

Ongoing volcanic eruptions, since c. 17,000 years ago, led to the gradual accumulation of thin tephra deposits on the land surface, forming a composite, homogenous, tephra mantle about 0.5–0.8 m thick, in which the upper soil has formed through developmental upbuilding. The equivalent thin tephras are preserved as separate layers in organic sediments in lakes formed about 20,000 years ago in the Hamilton Basin, and in peat bogs formed about 15,000 years ago. Thus the land surface comprises a fine network of tephra-mantled, sandier, well drained, soils on mounds or low ridges with poorly drained, finer textured, silty soils in the hollows or swales.

In the Hamilton Basin, and in parts of the Matamata-Hauraki area, on alluvial plains and very low-angle fans, with low ridge and swale topography, a distinct soil pattern occurs with Orthic Allophanic Soils, Impeded Allophanic Soils, and Tephric Gley Soils all formed in close proximity, often within the same paddock (Fig. 2.6). The soil pattern has sometimes been described and mapped as a 'soil complex', suggesting that it is not easy to predict which soil is where within the landscape. However, a soil association is a better description as there is a clear, predictable, relationship between the three soils at the micro-topographic scale. The Allophanic Soils are formed mainly on the mounds in an incrementally-accumulating tephra mantle over well drained coarse volcanogenic alluvium (i.e. the soils have two parent materials). In the adjacent, low-lying, swales, Gley Soils are formed in poorly drained fine-grained volcanogenic alluvium. In between the low ridge tops and the swale bottoms is an intermediate soil, an Impeded Allophanic Soil, with a modified tephra mantle on poorly drained volcanogenic alluvium.

The properties of the soils are dominated by the thickness of the tephra mantle on the alluvium, the small topographic differences of the landsurface (mound vs. swale), and changes in drainage, that together dictate the relative loss of silicon through leaching and hence the types of clay formed (Fig. 2.5). Similar soils occur in the Hinuera Valley and Matamata-Hauraki Plains.

The tephra mantle is obvious on the low ridges but, puzzlingly, equivalent tephra deposits are not evident in the hollows, in which the silts seem entirely alluvial in origin, suggesting either a different propensity for preservation spatially or that some sort of reworking has taken place after deposition. The mineral weathering sequence (Sect. 2.4.1) led to formation of Typic Orthic Allophanic Soils (locally known as Horotiu series in the Hamilton Basin and Waihou series in the Matamata-Hauraki area) on the tephra-mantled, sandy, well drained, low ridges with Typic Tephric Gley Soil (Te Kowhai series in the Hamilton Basin and Waitoa series in Matamata-Hauraki area) in the hollows or swales. The two endpoints of the weathering sequence tend to get the most attention, but the majority of the landscape comprises the intergrade between the two endpoints. The intergrade (known locally as Bruntwood series in the Hamilton Basin, Piarere series around Matamata, and Te Puninga series in the Hauraki area) has allophanic soil material at the surface and a slowly permeable layer at depth and thus is classified as a Typic Impeded Allophanic Soil (Fig. 2.6).

2.3.5 Southland Occurrences of Allophanic Soil Materials

A few small, isolated, pockets of allophane-bearing soils occur, particularly in northern Southland, in the Hokonui Range, formed in areas where ancient andesitic tephra has been preserved within the greywacke/argillite sequences. The modern acid leaching regime has resulted in the loss of silicon in soil solution and the formation of allophane on the greywacke and andesitic material.

2.4 Key Soil Properties

2.4.1 Soil Composition

Allophane, the dominant clay in Allophanic Soils, is one of the most reactive materials found in soil. Because allophane has a huge influence on the properties of Allophanic Soils, it is worth learning a little more about it. The structures of clay minerals were mainly determined by measuring the way that X-rays are diffracted by structural sheets in crystalline clay minerals. However, allophane does not normally diffract X-rays and so it was described as 'amorphous to X-rays'.

Later it was discovered that allophane does in fact have structural order in the form of very small hollow spherules (3.5–5 nm in diameter). The structure has therefore been termed 'short-range order' or, most recently, 'nanocrystalline' because the tiny spherules are structured within the nanoscale range (1–100 nm). Individual spherules are about the same size as a virus. Strictly, allophane qualifies as a mineral because it has demonstrable crystallinity, albeit in the nanoscale range (on the other hand, its common 'parent', volcanic glass, is not a mineral because glass is amorphous and therefore defined as a mineraloid). The chemical formula of allophane is $(1–2)SiO_2{\cdot}Al_2O_3{\cdot}(2–3)H_2O$. The molar ratio of silicon to aluminium varies but is generally about 1:2, meaning that allophane contains approximately twice as many atoms of Al as Si. Thus allophanes can be described as either Al- or Si-rich.

In the North Island, most Allophanic Soils contain the Al-rich variety of allophane, but Si-rich allophanes occur in eastern North Island, including Hawke's Bay where the rainfall is lower and more seasonal in distribution. Allophane spherules comprise a defective outer gibbsitic Al-octahedral sheet and an inner Si-tetrahedral sheet, a structure described as 'proto-imogolite'. The 'broken bond defects' in the gibbsitic sheet wall provide perforations where variable surface-charge characteristics arise because of reactive hydroxyl groups $[(OH)Al(OH_2)]$ at these sites. The gain of a proton generates a positive charge:

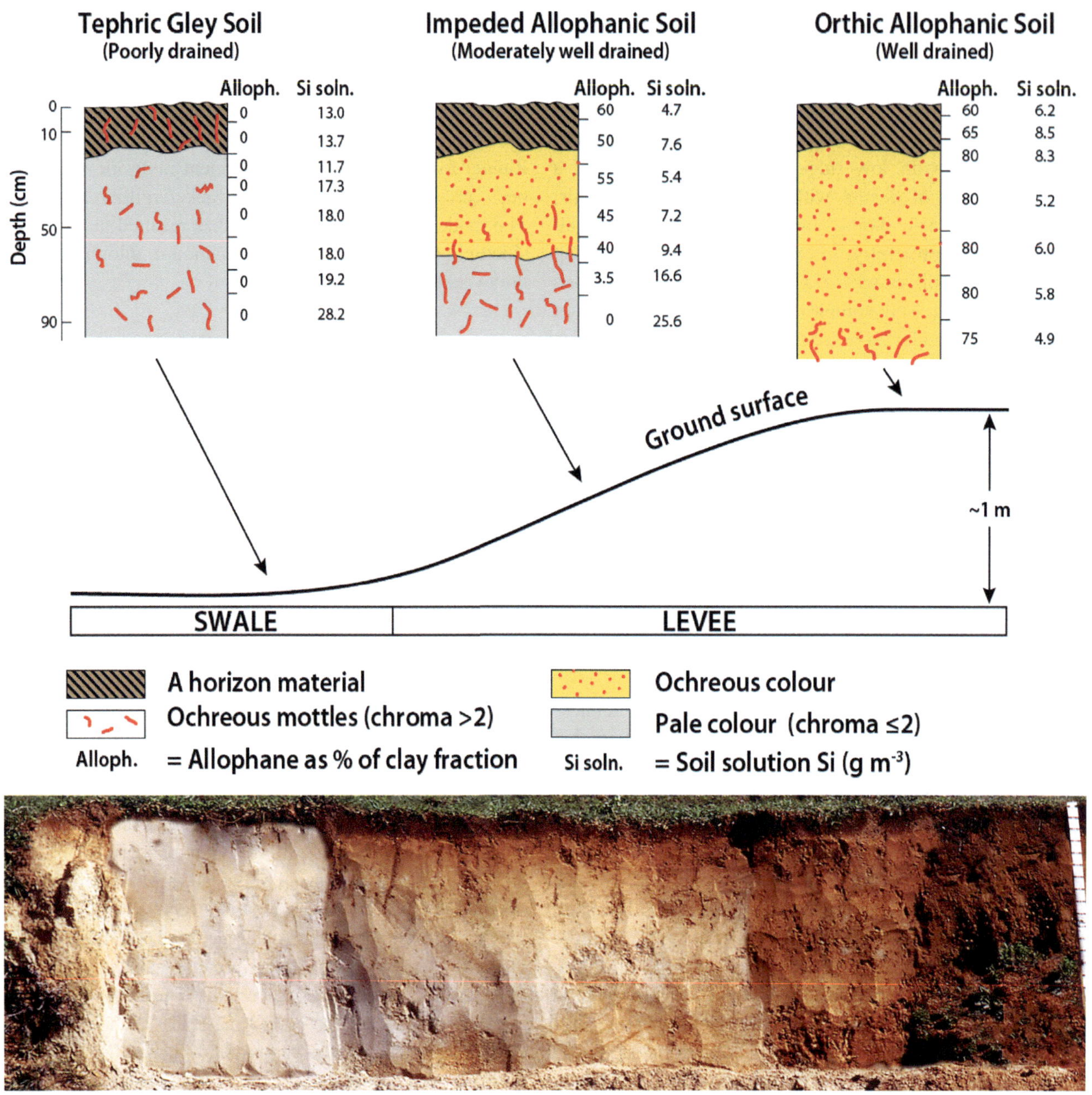

Fig. 2.6 Soil-landscape relationship between Tephric Gley (e.g. Te Kowhai), Impeded Allophanic (e.g. Bruntwood), and Orthic Allophanic (e.g. Horotiu) Soils that occur on the oldest alluvial surface (Hinuera Surface) in the Hamilton Basin. The section is about 7 m across and 2 m high, illustrating changes in soil over short distances. Diagram is modified after Singleton et al. (1989). Photo courtesy of NZ Society of Soil Science

$(OH_2)Al(OH_2)^+$; the loss of a proton generates a negative charge: $(OH)Al(OH)^-$. Allophane thus has strong affinity for water, metal cations, anions, organic molecules, and DNA. The perforations also allow water to pass into the spherule interior.

The amount of allophane in a soil with allophanic soil properties can be as low as 2–5% of the soil by weight. However, due to the extremely small size of the clay particles they have a correspondingly extremely large surface area, ranging from about 250 m^2 g^{-1} to as much as 1125 m^2 g^{-1} (see grey box). Thus, a small amount of allophane can have a potent effect on the properties of the soil mass as it can coat the soil matrix and pore surfaces.

Allophane is often associated with other nanocrystalline (short-range order) minerals including ferrihydrite and aluminium- or iron-humus complexes, and with small amounts of imogolite (which as a long thin tube has long-range order in one direction). The crystalline clay mineral halloysite, a member of the kaolin subgroup, may be found together with allophane or commonly instead of allophane in soils formed on tephras. This because the glass-rich tephra parent materials that weather to allophane may also weather to halloysite in less well drained soils (Sect. 2.2.2).

The parent materials of Allophanic Soils are mostly andesitic and/or rhyolitic tephras, with limited occurrences on basaltic tephras (chiefly scoria). Allophanic Soil materials may also form from indurated greywacke or schist rocks, or their derivatives such as loess or alluvium, but the thickness of the resulting allophanic soil material is usually too thin to enable the soils to qualify as Allophanic Soils. Such soils invariably have good drainage and are formed under precipitation sufficiently high to cause leaching and removal of silicon. They are generally recognised as Allophanic Brown Soils. The rock minerals most likely to weather to allophane are volcanic glass and feldspars, and sometimes minerals with a composition similar to kaolinite with a molar ratio of Al:Si = 2. Rapidly weatherable volcanic glass and feldspars dominate the sand fractions of soils in igneous (including volcanic) parent materials and are the primary source of the nanocrystalline minerals. Plagioclase feldspars are the most likely primary source of allophanic soil material in non-igneous parent materials.

2.4.2 Physical Properties

The physical properties of Allophanic Soils (Fig. 2.7, Table 2.1) are excellent for supporting plant growth. Allophanic soil materials have exceptionally low soil dry bulk densities (around 0.7 t m^{-3} on average, potentially as low as $\sim$0.5 t m^{-3}) and generally high to very high macro-porosities. Thus most Allophanic Soils are well drained as water passes readily through the highly permeable soil. However, where there are slowly permeable subsoils, mottled Bw(f) or B(g) horizons may occur lower in the soil, in which case the soils would be regarded as moderately well or imperfectly drained. In such cases, allophane predominates in upper profiles but may be subservient to halloysite, or absent, lower in the profiles.

The clay contents of allophanic soil materials are generally reported as being $\sim$10–25%. However, particle size measurement is difficult because the clay particles are tightly aggregated (hence clay-sized material is underestimated and sand-sized material overestimated) and so the 'true' or 'primary clay content' may be considerably higher than the reported measurement. Also, irreversible changes in allophane clay aggregation may occur when soils are air dried, hence allophanic soil materials should **not** be dried prior to measurement of soil particle size. However, Brent Alloway and colleagues reported a method for soils in Taranaki that circumvented these problems of measurement. Allophanic soil materials were dissolved using acidified ammonium oxalate (enabling total clay content to be estimated gravimetrically) and the particle-size distribution of the inorganic residuum (sand and silt) was then able to be measured free of clay aggregates.

Although Allophanic Soils are well drained they also have particularly high soil-water retention. Plant roots move readily through the friable, low penetration resistance, soil and, for some plants such as kiwifruit, roots may extend to depths of several metres. The high water-holding capacity and deep potential rooting depths mean Allophanic Soils have some of the greatest plant-available moisture storage of any mineral soil.

Allophanic Soils have stable topsoils that resist puddling or pugging under the impact of machinery or grazing animals in wet weather. The soil is readily dug and samples crumble easily when crushed in the hand and are very friable. Drying may lead to changes in soil structure so rewetting may not achieve the original volume.

Allophane or allophane-like soil clay minerals are strongly expressed in the field-observed properties. The minerals coat the sand and silt grains and maintain a porous, low-density, soil structure with weak strength. The most tactile property is the distinctly greasy feel when moistened and rubbed firmly between the fingers. Sensitivity, an important physical property in Allophanic Soils, is the loss of soil strength that occurs when a soil sample is remoulded by applying heavy pressure, shear, or mixing (a related property is thixotopy). The sensitivity is the origin of the distinctive 'greasy' feel the soil material has when moistened and rubbed between the fingers, sometimes exuding droplets of water in the process.

2.4.3 Chemical Properties

The most outstanding soil chemical property of Allophanic Soils is their ability to adsorb anions (particularly phosphate and sulphide) as evidenced by their very high anion storage capacity or P retention (Fig. 2.8, Table 2.2) which may be as high as 99%. Thus Allophanic Soils will retain a significant proportion of applied fertiliser phosphorus in a form unavailable to plants. Roger Parfitt and collaborators estimated that in 2001/02 about 11 kg ha^{-1} of phosphorus was gained in Waikato and Bay of Plenty topsoils that have high P retention. The ongoing accumulation means that in areas where fertiliser has been regularly used, some tonnes of P are now 'locked away' in Allophanic Soils in a form not easily available to plants.

Allophanic Soils provide excellent plant growing conditions due to their high water holding capacity and good physical properties, thus supporting production of organic matter. Organic matter is readily produced, and predominantly stored, in the topsoil. Moderate amounts are also stored in buried (former) topsoil horizons (paleosols) within

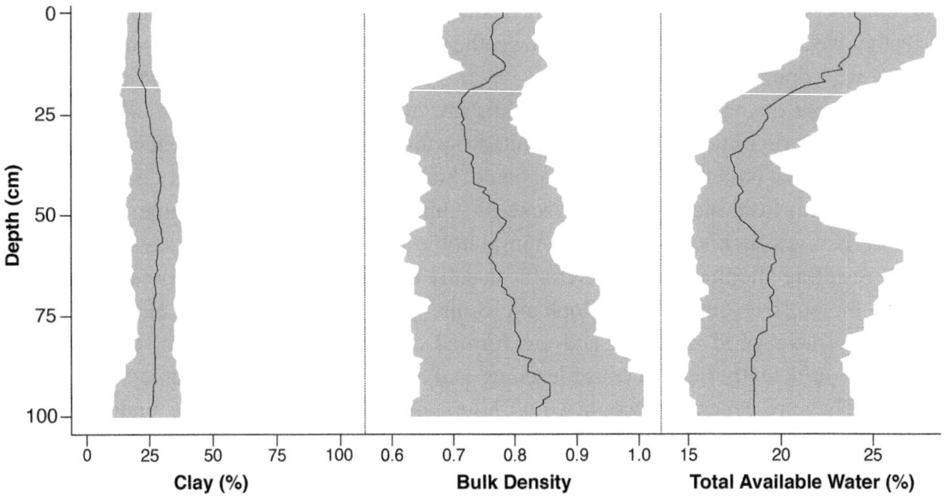

Fig. 2.7 Median and upper and lower quartiles of clay content, soil dry bulk density (t m^{-3}), and soil total available water capacity for Allophanic Soils in the NZ Soil Data Repository

Table 2.1 Typical example of physical properties of an Allophanic Soil: Horotiu series (Typic Orthic Allophanic Soil) SB09944 (SWAMP)

Horizon	Depth cm	Sand %	Silt %	Clay %	Dry bulk density (t m^{-3})	RAW[a] %	TAW[a] %
Ap1	0–6	33	48	17	0.86	13	30
Ap2	6–17	33	49	16	0.82	14	24
AB	17–31	35	47	16	0.80	15	25
Bw1	31–55	46	24	29	0.83	9	15
Bw2	55–73	53	28	18	0.82	7	10
Bw3	73–91	45	21	32	0.81	5	10
2Cu(g)	91–107	63	26	9	0.99	10	12
2Cu	107–130	93	5	1	1.46	2	5

[a]RAW = readily available water, TAW = total available water, both in % by volume

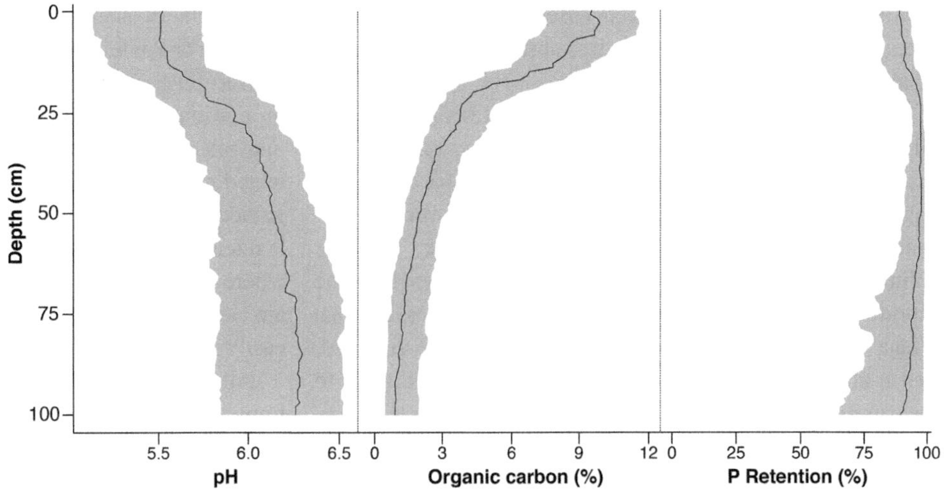

Fig. 2.8 Median and upper and lower quartiles of soil pH, organic carbon and P retention for Allophanic Soils in the NZ Soil Data Repository

Table 2.2 Typical example of chemical properties of an Allophanic Soil: Horotiu series (Typic Orthic Allophanic Soil) SB09944 (SWAMP)

Horizon	Depth cm	pH (in H_2O)	Carbon %	Nitrogen %	CEC[a] $cmol^{(+)}\ kg^{-1}$	Sum bases $cmol^{(+)}\ kg^{-1}$	P retention %
Ap1	0–6	5.7	8.2	0.77	28.2	13.9	85
Ap2	6–17	5.3	5.5	0.55	21	5.5	91
AB	17–31	5.9	3.3	0.32	17.3	7.1	96
Bw1	31–55	6.6	1.7	0.15	12	7.5	96
Bw2	55–73	6.7	1.7	0.14	12.9	8.5	96
Bw3	73–91	6.7	1.2	0.12	11.6	7.5	94
2Cu(g)	91–107	6.7	0.7	0.07	6.5	4.5	76
2Cu	107–130	6.7	0.1	0.01	1.6	1.2	18

[a]CEC = Cation exchange capacity

multi-layered profiles (such as in Fig. 2.4). Complexing between allophane and organic matter means that organic carbon is captured and stored in the soil, much in micro- or nanoaggregates. Allophanic Soil, therefore, has a high capacity to hold carbon in a form that is not readily broken down and so provides a carbon sink. Thus relatively deep, dark, topsoils are formed. Soil carbon contents are medium to high (up to ~8–12% organic carbon, Fig. 2.8) and organic/mineral complexes are stable.

In subsoils and unfertilized topsoils the sum of bases are low to very low and range from less than 1 to about 10 $cmol^{(+)}\ kg^{-1}$. Allophanic Soils, although having excellent physical properties, need careful management of nutrients (especially phosphorus) to maintain high productivity. Sulphate reserves are held in B-horizons.

Allophane, and allophanic soil material, has pH-dependant charge. This means that measured soil cation exchange capacity, which is a measure of the negative charge on the surfaces of clay and organic matter within the soil, increases as the pH of the soil increases.

2.4.4 Biological Properties

The Allophanic Soils are possibly New Zealand's most favourable biological substrate, having low drought risk, generally moderate acidity, high soil organic matter, good rooting depth, high porosity, and moderate soil temperature. The soils have the capability of hosting large and diverse populations of soil organisms, and microbial biomass is generally high. They form in regions where rainfall is generally >1000 mm $year^{-1}$ and soil-water deficits are uncommon.

It has been recognised by Yu-Tuan (Doreen) Huang, David Lowe, and colleagues that DNA can be entrapped and preserved by allophane. They determined that the formation of stable allophane nanoaggregates and microaggregates enabled large amounts of DNA to be adsorbed. In an experiment, about 80% of added DNA was physically adsorbed within tiny spaces (nanopores) between allophane spherules and allophane nanoaggregates. The rest of the added DNA (~20%) was adsorbed chemically on the surfaces of the allophane spherules. The stability of the allophane-DNA aggregates prevents encapsulated DNA from being exposed to oxidants, and DNA within small pores between allophane spherules and nanoaggregates is not accessible to enzymes or microbes, hence enabling its protection and preservation. Thus, substantial organic carbon is likely to be sequestered and protected, in this way, in Allophanic Soils.

The extraction of miniscule amounts of DNA from the soil is difficult as the DNA has strong bonds with soil organic matter and with the allophane spherules (bound tightly to it through the DNA's phosphate group). To release the trapped DNA, Huang and colleagues took a novel approach and dissolved all the allophane using acidified ammonium oxalate (Tamm's reagent), and then purified the resultant DNA. They obtained amplifiable DNA, mainly of native New Zealand plants, from the buried Ah-horizon on Rotoma Tephra, about 9500 years old, from a site near Mt Tarawera (Fig. 2.4). The extraction of non-modern DNA from allophanic materials, especially paleosols, has potential for reconstruction and improved understanding of past environments including their biodiversity, the effects of volcanic activity, and even for use in forensic science.

Fitting ~1000 m^2 of surface area into a gram of soil

Surprisingly, because of the extremely small size of the particles (between 3.5 and 5.0 nm (or three-and-a-half to five millionths of a millimetre) in diameter, the surface area of one gram of allophane can extend to about 1000 m^2. Imagine if you were to take a sheet of very thin paper 40 m long and 20 m wide and scrunch it up to make a tight ball, or perhaps rip the paper into very small pieces and compress them together very tightly. Could you reduce it to the size of a gram of soil? With paper it is clearly impossible but it can be imagined at a molecular scale. The enormous surface area extends over the many tiny spherules of allophane that invariably cluster together to form nanoaggregates (small clusters $\leq$ 100 nm across of spherules) or microaggregates (small groups of nanoaggregates).

2.5 Distinguishing Between Allophanic Soils and Related Soil Orders

Allophanic Soils are identified by the *New Zealand Soil Classification* as containing 'allophanic soil material'. The minerals and compounds responsible for such properties are allophane, imogolite, ferrihydrite, and metal-humus compounds (of Fe and/or Al). The presence of allophanic soil materials can be estimated analytically (phosphate retention of 85% or more, and dry bulk density of $\leq$ 0.9 t m^{-3}). However, in keeping with the *New Zealand Soil Classification* principle that where possible class criteria should use field observable properties, the criteria include greasy sensitivity, weak or very weak soil strength, non-sticky or slightly sticky consistence, and a strong or very strong reactive-aluminium field test for allophane (sometimes also referred to as the Fieldes-Perrott or NaF test).

The presence of allophane in soils can be confirmed by a simple field test, developed by Maurice Fieldes and Ken Perrott, where a small sample of soil is placed on a filter paper that has been soaked in phenolphthalein (a pH indicator) solution. A few drops of NaF are added and if allophane is present a resulting increase in pH (due to dissociated hydroxyl groups) will cause the filter paper to turn a dark red colour. The amount of allophane may be indicated by the rate and strength of the reaction.

Granular Soils occur marginally to Allophanic Soils in north and west Waikato, Auckland, and Northland, where allophanic character is recognised in Allophanic subgroups of the Melanic, Oxidic, and Orthic groups of the Granular Soils.

Pumice Soils occur marginally to Allophanic Soils where allophanic character is recognised in Allophanic and Buried-allophanic subgroups of the Orthic Pumice Soils. Allophanic Soils may occur in pumice parent materials where the pumice has weathered to the extent that the volcanic glass component of the pumice has weathered to allophane to meet thickness and other criteria.

Podzols are distinguished from Allophanic Soils by the presence of a dark humus- or sesquioxide-rich horizon (podzolic-B horizon, Bh and/or Bs) in the Podzol Soil, or the presence of a pale coloured E-horizon, and other features including the site environment and pedological context.

2.6 Correlation with Other Classification Systems

New Zealand, along with Japan and other Pacific Rim countries, that have frequent volcanic eruptions of andesitic and/or rhyolitic tephra, have large areas of allophane-rich soils. However, such soils are rare in continental Europe (though they do occur, for example, in Iceland and Italy) and the contiguous USA where international soil classifications were developed. Thus Allophanic Soils were not classed as an order in early editions of *Soil Taxonomy* (an international soil classification developed by the United States Department of Agriculture). Soils that contained allophane were clearly recognised in the USA but many were under forest in the Pacific Northwest, in very remote parts of Alaska, or on the Hawaiian islands, and so Guy Smith, the architect for *Soil Taxonomy*, took the view that the low abundance or remoteness, and limited agricultural application, meant they were best classed at a suborder level (Andepts). However Mike Leamy, initially with Smith in 1978, proposed and chaired a group who successfully developed the Andisol order which was added to *Soil Taxonomy* in 1990 (supplanting the Andept suborder) (Table 2.3).

2.7 Use and Management of Allophanic Soils

2.7.1 Productive Use

Allophanic Soils underpin some of New Zealand's most productive agriculture and horticulture (Fig. 2.9). The combination of high water-holding capacity, free drainage, characteristically flat to gently sloping topography, low soil dry bulk density, friability, good tilth, and thus ease of cultivation and root penetration, ensure that Allophanic Soils are excellent for agriculture and horticulture.

Table 2.3 Correlation between Allophanic Soils and equivalent classes[a] of *Soil Taxonomy*, *World Reference Base*, and the earlier NZ genetic classification

New Zealand Soil Classification	*Soil Taxonomy*	*World Reference Base*	NZ genetic soil classification
Perch-gley Allophanic Soils	Epiaquands	Gleyic Andosols	Yellow-brown loams
Gley Allophanic Soils	Endoaquands	Gleyic Andosols	Yellow-brown loams
Impeded Allophanic Soils	Hapludands	Andosols	Yellow-brown loams
Orthic Allophanic Soils	Hapludands	Andosols	Yellow-brown loams

[a]The correlations given here are a guide only and for accurate classifications the relevant soil classification documents should be consulted. The two major international soil classification systems are *Soil Taxonomy*, which was developed in the USA, and *World Reference Base*, which was developed primarily in Europe. The NZ genetic soil classification was used in NZ prior to 1992

Fig. 2.9 Typic Orthic Allophanic and Impeded Allophanic Soils in the Hamilton area (Horotiu and Bruntwood soils) are highly suitable for intensive horticulture including specialty vegetables, gourmet leaf salads, and fresh herbs. Maize (harvested in foreground) is another widely grown crop associated with Allophanic Soils in the Waikato region

In the Waikato and Taranaki regions with generally reliable rainfall, dairy production thrives on Allophanic Soils. In the Bay of Plenty where there are few frosts, and good rainfall, kiwifruit, avocados, citrus, and other horticulture crops are all intensively cultivated on Allophanic Soils. A practice of 'recontouring' land supporting Allophanic Soils in the 1970s and early 1980s so as to increase the area perceived to be suitable for horticultural development, especially kiwifruit vines, in the Te Puke-Tauranga area, was demonstrated to be deleterious, and subsequently stopped. As well as creating numerous problems including altering drainage and compacting soils, the loss of Allophanic Soils, ~25,000 years or more in the making, was regarded as unacceptable (reported by Gary Orbell in 1985).

South of Ruapehu, frost-tolerant root crops such as carrots, potatoes, parsnips, and swedes are grown, all benefitting from the friable, low bulk density, well-drained soil. Allophanic Soils are also an excellent receiving environment for irrigation of effluent, particularly because of their ability to adsorb phosphorus and other anions. An example is the Rotorua Land Treatment System, developed in part on Vitric Orthic Allophanic Soils in Whakarewarewa Forest near Rotorua.

To maintain production there are usually nutrient requirements for phosphorus, potassium, and magnesium on dairy farms. Generally, there are no marked trace element deficiencies except in more strongly leached Allophanic Soils where cobalt may be required to prevent 'bush sickness' (vitamin B12 complex deficiency, see Chap. 12 Pumice Soils). Pasture may respond to lime where the pH is less than ~5.5.

The high capacity to adsorb anions is a challenge for soil fertility management because of allophane's potential to adsorb significant proportions of fertiliser phosphate making it unavailable to plants. Frequent additions of smaller amounts of P fertiliser will help ensure that plants have some phosphorus available. A newly recognised issue is the accumulation of fluorine, derived from fertilisers, which has been comparatively rapid and widespread in productive Allophanic Soils which have a history of strong fertiliser use. Nick Kim and colleagues reported that over 50 years, average total fluorine concentrations in some surface soils have doubled from ~220 to 440 mg kg^{-1}. They suggest that chronic fluorosis in grazing animals could develop. The addition of phosphate-containing fertiliser has also been identified as enhancing the dissolution of volcanic glass, consistent with phosphoric acid and fluorine-induced removal of aluminium and silica. Thus, according to Matthew Taylor and others, soil weathering has likely been accelerated because of the increases in fluorine.

The very high anion retention capacity in Allophanic Soils can be useful to prevent loss of phosphate and nitrate to surface or groundwater and a wide range of other contaminants may be adsorbed including heavy metals and pesticides. When the soil storage capacity holds water-soluble nitrates, the nitrate may be detained for capture by root systems for plant consumption. If the soil is poorly or imperfectly drained, biochemical reduction can promote denitrification which transforms nitrate to benign nitrogen gas which returns to the atmosphere.

Topsoils are stable and can resist moderate impact of machinery or grazing animals in wet weather. However, soil quality monitoring shows that, like most soils, compaction can result from heavy grazing when wet. Orthic Allophanic Soils are free draining and water moves rapidly through the soil after rainfall and the soil rarely becomes boggy or pugged. However, Impeded Allophanic Soils may become saturated for some time following heavy rain and hence, in wet weather the use of short grazing periods and stand-off pads for cattle, to prevent pugging and compaction, can be an advantage.

Perch-gley and Gley Allophanic Soils have a slowly permeable horizon which may perch water and provide a barrier to root penetration. Where the slowly permeable horizon is deeper in the soil (it may occur as deep as 90 cm) it may not be a hindrance for most uses. However, where the slowly permeable layer is less than about 50 cm deep, the soil may benefit from installation of drainage.

Allophanic Soil materials with their high water-holding capacity, especially Orthic Allophanic Soils, where plant roots can penetrate to great depths, provide high profile water-holding capacity. Most of the Allophanic Soils occur in the Taranaki, Waikato, and western Bay of Plenty regions where rainfall is moderate to high and so, generally, good productivity can be achieved without need for irrigation. Nevertheless, summer droughts do occur and for high value, or shallow rooting crops, irrigation can underpin confidence in successful production. Some of the sparse Allophanic Soils on basaltic scoria cones in Northland are highly prized in a region where many far less versatile soils, such as Ultic Soils, prevail.

2.7.2 Erosion Risk

Allophanic Soils are generally characterised as having 'slight to insignificant risk of erosion under pasture' and the soil will likely remain stable if rainfall is able to be absorbed into, or rapidly drained through, the soil. However, if soil on slopes does become saturated, then there is some risk of hillslope movement. Where vegetation has been removed, or soils have been cultivated, topsoils may be at risk of wind erosion (Fig. 2.10). On slopes, cultivated soil is also vulnerable to surface sheet and rill erosion. Because of the low bulk density, a rapid runoff event, or strong wind on a dry soil, can quickly cause major loss of soil.

Work led by Les Basher and Craig Ross on the intensely cultivated land used primarily for carrot production near Ohakune showed that water erosion of cultivated soil had caused extensive redistribution of soil within paddocks with some areas eroded by up to 100 t ha^{-1} $year^{-1}$ while other areas had received deposition of 300–500 t ha^{-1} $year^{-1}$. They observed that compacted wheel tracks were the source of much of the runoff and subsequent water erosion as the wheel tracks had much lower infiltration rates (4 mm h^{-1}) than the cultivated beds (800 mm h^{-1}, which can easily allow rainfall to infiltrate).

On central and eastern North Island hills and steeplands, tephra, and thus potentially Allophanic Soil, is frequently found on flatter ridge summits but not on associated slopes >~25° where it has apparently been eroded preferentially. It is likely that in some locations the erosion has occurred by natural processes, including redistribution in the landscape soon after tephra deposition. Overseas studies have shown that differences in vegetation cover can affect the thickness of tephra preserved, although it is evident that native forest covered most, if not all, of the region during the Holocene and until human arrival around 750 years ago. Work on the

Fig. 2.10 Ploughed Allophanic Soils are vulnerable to erosion. Top: wind erosion occurring on a bare soil near Ohakune. Bottom: reduced crop production on an eroded slope

Whatawhata hill country research farm (to the west of Hamilton) in the early 1970s, by Robin Palmer, showed that the tephra mantle on weathered greywacke had been lost from steeper slopes by rotational land-sliding (slumping) or soil creep. He found that on slopes of about 20°, only the upper horizons of soils are derived from tephra, their lower horizons being derived from weathered greywacke. Where slope angles exceed about 26°, the soils are formed largely from weathered greywacke alone.

In managed land, erosion risk may be reduced by maintaining good vegetation cover. The low bulk density, apedal earthy, aggregates are vulnerable to erosion if exposed on slopes following animal disturbance, vegetation clearance, or other management activities.

2.7.3 Geotechnical Engineering

Soils containing allophane and halloysite carry a 'user-beware' warning for geotechnical engineers. Allophanic soil materials lose soil strength after remoulding, for example, when loads are placed on the soil by machinery, heavy animals, earth-moving equipment, or earthquakes. Manipulation (such as by bulldozing) of allophanic soil material, followed by soil saturation, may lead to catastrophic loss of bearing strength, a property described as sensitivity. Recent studies by Max Kluger, Vicki Moon, Jock Churchman, and others, have shown that some forms of halloysite can also lose soil strength after remoulding. It turns out that a newly discovered 'mushroom-cap' shaped halloysite morphology, essentially spheroidal but with an opening on one side, was responsible for the catastrophic landslides on Omokoroa Peninsula in the Tauranga region of the western Bay of Plenty, the most infamous being the Bramley Drive slide dating from the very wet winter of 1979.

Lessons were learned the hard way in 1981 when a canal, with sides formed from allophanic and halloysitic Soil materials, built to convey water to the Ruahihi hydropower station in the Tauranga area, collapsed the day after the official opening of the power station. About 1.5 million cubic metres of soil, which had formed a wall of the canal, flowed, with the consistency of porridge, across the adjacent farm and state highway. A 600 m stretch of the canal was destroyed and a section of the road was washed out. A chasm about 500 m long, 100 m wide, and 40 m deep was gouged out in about an hour by the flowing debris. When allophanic (or halloysitic) soil materials are saturated with water (in this case following the filling of the canal, or in the case of many landslides in the region, following heavy rainfall), unsupported sensitive soils may fail suddenly, and dramatically, with the material behaving as a liquid capable of carrying debris over large run-out distances. It is now suspected, following the revelation of the identical role played by defective spheroidal halloysite in a highly sensitive layer at Omokoroa, and elsewhere, that the Ruahihi collapse may have been caused by such halloysite as well as, or instead of, allophane. Halloysite is the predominant clay mineral in the pyroclastic sequences underlying much of the Tauranga region, with allophane generally limited to the uppermost metre or so of the modern land surface (supporting the extensive Allophanic Soils).

Further Reading

Allbrook RF (1983) Some physical properties of allophane soils from the North Island, New Zealand. N Z J Sci 26:481–492

Alloway BV, Neall VE, Vucetich CG (1992) Particle size analyses of late Quaternary allophane-dominated andesitic deposits from New Zealand. Quat Int 13–14:167–174

Basher LR, Ross CW, Dando J (2004) Effects of carrot growing on volcanic ash soils in the Ohakune area, New Zealand. Aust J Soil Res 42(3):259–273

Bruce JG (1978) Soils of part Raglan County, South Auckland. New Zealand Soil Bureau Bulletin 41. DSIR, Wellington, New Zealand, 102p

Childs CW (1992) Ferrihydrite: a review of structure, properties and occurrence in relation to soils. Zeitschrift für Pflanzenernährung und Bodenkunde 155:441–448

Childs CW, Palmer RWP, Ross CW (1990) Thick iron oxide pans in soils of Taranaki. Aust J Soil Res 28:245–257

Churchman GJ, Pasbakhsh P, Lowe DJ, Theng BKG (2016) Unique but diverse: some observations on the formation, structure, and morphology of halloysite. Clay Miner 51:395–416

Degens BP, Schipper LA, Clayden JJ et al (2002) Irrigation of an allophanic soil with dairy factory effluent for 22 years: responses of nutrient storage and soil biota. Aust J Soil Res 38(1):25–35

Fieldes M, Perrott KW (1966) The nature of allophane in soils. Part 3—rapid field and laboratory test for allophane. N Z J Sci 9(3):623–9

Franks AM, Neall VE, Pollok JA (1991) Soils of part Eltham County, North Island. New Zealand, NZ DSIR Land Res Sci Rep 14, 131p

Gibbs HS (1968) Volcanic-ash soils in New Zealand. N.Z. DSIR Information Series 65, 30p

Grange LI, Taylor NH, Sutherland CF et al (1939) Soils and agriculture of part of Waipa County. New Zealand DSIR Research Bulletin 76

Hewitt AE (2010) New Zealand soil classification, 3rd edn. Manaaki Whenua Press, Lincoln

Huang Y-T, Lowe DJ, Zhang H et al (2016) A new method to extract and purify DNA from allophanic soils and paleosols, and potential for paleoenvironmental reconstruction and other applications. Geoderma 247:114–125

Huang Y-T, Lowe DJ, Churchman GJ et al (2016b) DNA adsorption by nanocrystalline allophane spherules and nanoaggregates, and implications for carbon sequestration in Andisols. Appl Clay Sci 120:40–50

Kim ND, Taylor MD, Drewry JJ (2016) Anthropogenic fluorine accumulation in the Waikato and Bay of Plenty regions of New Zealand: comparison of field data with projections. Environ Earth Sci 75:147. https://doi.org/10.1007/s12665-015-4897-2

Kluger MO, Moon VG, Kreiter S et al (2017) A new attraction-detachment model for explaining flow sliding in clay-rich tephras. Geology 45:131–134

Lowe DJ (1986) Controls on the rates of weathering and clay mineral genesis in airfall tephras: a review and New Zealand case study. In: Coleman SM, Dethier DP (eds) Rates of chemical weathering of rocks and minerals. Academic Press, Orlando, pp 265–330

Lowe DJ, Palmer DJ (2005) Andisols of New Zealand and Australia. J Integr Field Sci 2:39–65

Lowe DJ, Tonkin PJ (2010) Unravelling upbuilding pedogenesis in tephra and loess sequences in New Zealand using tephrochronology. In: Gilkes RJ, Prakongkep N (eds) Proceedings of the 19th world congress of soil science "Soil Solutions for a Changing World" (1–6 Aug., 2010, Brisbane), symposium 1.3.2 geochronological techniques and soil formation, pp. 34–37. Published at http://www.iuss.org

McCraw, JD (2011) The wandering river—landforms and geological history of the Hamilton Basin. Geoscience Society of New Zealand Guidebook No. 16, 88p

McDaniel PA, Lowe DJ, Arnalds O et al (2012) Andisols. In: Huang PM, Li Y, Sumner ME (eds) Handbook of soil sciences. 2nd edn. Vol. 1: Properties and Processes. CRC Press, Boca Raton, FL, pp 33.29–33.48

McLeod M (1984) Soils of the proposed Te Kawa irrigation scheme. New Zealand Soil Bureau District Office Report HN 10, 57p

McLeod (M) (1984) Soils of the Waikato lowlands. New Zealand Soil Bureau District Office Report HN 11, 32p

McLeod M (1992) Soils of part northern Matamata County, North Island, New Zealand. DSIR Land Resources Scientific Report No. 18, 96 p

Milne D, Clayden B, Singleton P et al (1995) Soil description handbook. Revised edition. Manaaki Whenua Press, 157p

Moon VG (2016) Halloysite behaving badly: geomechanics and slope behaviour of halloysite-rich soils. Clay Miner 52:517–528

Neall VE (1977) Genesis and weathering of andosols in Taranaki. New Zealand Soil Sci 123:400–408

Neall VE (ed) 1982. Soil Groups of New Zealand. Part 6 Yellow-brown loams. New Zealand Society of Soil Science. Wellington, New Zealand, 114p

Neall VE (2006) Volcanic soils. In: Verheye W (ed.) Land use and land cover. Encyclopaedia of life support systems (EOLSS). EOLSS Publishers with UNESCO, Oxford, U.K. p 1–24. http://www.eolss.net

Orbell GE (1982) Yellow-brown loams of the Waikato area. In: Neall VE (ed) Soil Groups of New Zealand. Part 6 Yellow-brown loams. New Zealand Society of Soil Science, Wellington, New Zealand, 25–29

Orbell GE (compiler) (1985) Recontouring guidelines for Tauranga County. New Zealand Soil Bureau Scientific Report 73, 33p

Orbell GW (1983) Report on the soils of the proposed irrigation scheme area at Kihikihi. New Zealand Soil Bureau District Office Report HN 9, 33p

Palmer RWP, Neall VE, Pollok JA (1981) Soils of Egmont and part Taranaki counties, North Island, New Zealand. NZ Soil Survey Report 64. DSIR, Wellington, 68p. ISSN 0110-2079

Parfitt RL (1990) Allophane in New Zealand—a review. Aust J Soil Res 28(3):343–360

Parfitt RL (2009) Allophane and imogolite: role in soil biogeochemical processes. Clay Miner 44:135–155

Parfitt RL, Clayden B (1991) Andisols—the development of a new order in Soil Taxonomy. Geoderma 49:181–198

Parfitt RL, Henmi T (1980) Structure of some allophanes from New Zealand. Clays Clay Miner 28:285–295

Parfitt RL, Webb TW (1984) Allophane in some South Island yellow-brown shallow and stony soils and high country yellow-brown earths. N Z J Sci 27:37–40

Parfitt RL, Wilson AD (1985) Estimation of allophane and halloysite in three sequences of volcanic soils, New Zealand. Catena Supplement 7:1–8

Parfitt RL, Pollok JA, Furkert RJ (1980) Guide Book for Tour I, North Island. In: Soils with variable charge conference, Palmerston North (Feb 1981)

Parfitt RL, Robertson SM, Orbell GE (1982) Soil stratigraphy of Kereone, Ohaupo and Mairoa soils in the Waikato and Te Kuiti districts. In: Neall VE (ed) Soil groups of New Zealand. Part 6. Yellow-brown loams. New Zealand Society of Soil Science. Wellington, pp 13–16

Parfitt RL, Russell M, Orbell GE (1983) Weathering sequence of soils from volcanic ash involving allophane and halloysite. Geoderma 29:41–57

Parfitt RL, Saigusa M, Cowie JD (1984) Allophane and halloysite formation in a volcanic ash bed under different moisture conditions. Soil Sci 138:360–364

Parfitt RL, Childs CW, Clayden B et al (1988) International committee on the classification of Andisols (ICOMAND). Circular Letter No. 10 (final), 80p

Pullar WA (1967) Volcanic ash beds in the Waikato district. Earth Sci J 1(1):17–30

Pullar WA, Birrell KS (1973) Parent materials of Tirau silt loam. N Z Jl Geol Geophys 26:677–686

Rijkse WC, Cotching WE (1995) Soils and land use of part Tauranga County, North Island, New Zealand. Landcare Research Technical Record, 111 p

Rijkse WC, Guinto DF (2010) Soils of Bay of Plenty volume 1: western Bay of Plenty. Environment Bay of Plenty Publication 2010/11-1. Bay of Plenty Regional Council. Whakatane. 172p

Schaetzl R, Thompson ML (2015) Soils—genesis and geomorphology, 2nd edn. Cambridge Univ. Press. 778p

Singleton PL (1991) Soils of Ruakura—a window on the Waikato. DSIR Land Resources Scientific Report No. 5. DSIR Lower Hutt, New Zealand, 127p

Singleton PL, McLeod M, Percival HJ (1989) Allophane and halloysite content and soil solution silicon in soils from rhyolitic volcanic material, New Zealand. Aust J Soil Res 27:67–77

Smalley IJ (1979) A spherical structure for allophane. Nature 281:339

Smith GD (1978) The Andisol proposal: a preliminary proposal for the classification of Andepts and some andic subgroups. NZ Soil Bureau, 22p

Soil Survey Staff (2014) Keys to Soil Taxonomy, 12th edn. USDA Natural Resources Conservation Service, Washington, DC 362 p http://www.nrcs.usda.gov/wps/PA_NRCSConsumption/download?cid=stelprdb1252094&ext=pdf. Accessed 4 Mar 2020

Soil Survey Staff (2015) Illustrated guide to Soil Taxonomy, version 2. U.S. Department of Agriculture, Natural Resources Conservation Service, National Soil Survey Center, Lincoln, Nebraska. https://www.nrcs.usda.gov/wps/portal/nrcs/detail/soils/survey/class/taxonomy/?cid=nrcs142p2_053580. Accessed 2 Mar 2020

Taylor NH (1933) Soil processes in volcanic ash-beds. N Z J Sci Technol 14:193–202, 338–352

Taylor MD, Lowe DJ, Hardi P et al (2016) Comparing volcanic glass shards in unfertilised and fertilised Andisols derived from rhyolitic tephras, New Zealand: evidence for accelerated weathering and implications for land management. Geoderma 271:91–98

Theng BKG, Yuan G (2008) Nanoparticles in the soil environment. Elements 4:395–399

Theng BKG, Russell M, Churchman GJ, Parfitt RL (1982) Surface properties of allophane, halloysite, and imogolite. Clays Clay Miner 30(2):143–149

Vucetich CG, Pullar WA (1969) Stratigraphy and chronology of late Pleistocene volcanic ash beds in central North Island, New Zealand. N Z J Geol Geophys 12:784–837

Wesley L (2009) Behaviour and geotechnical properties of residual soils and allophane clays. Obras y Proyectos 6:5–10

Wilson (AD) 1980 Soils of Piako County, North Island, New Zealand. New Zealand Soil Survey Report 39, 174p

Yuan G, Wada SI (2012) Allophane and imogolite nanoparticles in soil and their environmental applications. In: Barnard AS, Guo H (eds) Nature's Nanostructures. Pan Stanford, Singapore, pp 494–515

World Reference Base (2015 update); http://www.fao.org/3/i3794en/I3794en.pdf. Accessed 2 Mar 2020

Anthropic Soils

3

On land recontoured for new cities
on land after dump activities
on land where miners once did toil
there you'll find Anthropic Soil

Abstract

Anthropic Soils comprise soils constructed by people to achieve engineering, recreational, agricultural, land-rehabilitation, or waste-management goals. The original soil may be recontoured, stripped (leaving a truncated soil) and relocated, severely mixed, have inorganic or organic spoil or refuse added (≥ 30 cm in thickness), reformed into sports fields, or utilised in urban amenity landscaping, rain gardens, or roof gardens. The original character of the soil may be considerably altered, thus there is no typical Anthropic Soil. Some are formed as a result of careful planning and management for a particular purpose and so may be an 'improvement' on the original soil at the site. However, other Anthropic Soils are an inadvertent side effect of human activities and represent land that has been altered sufficiently for much, or all, of the original soil character to be lost. Relatively minor modifications (such as mixing an A-horizon by ploughing) do not normally qualify the soil as an Anthropic Soil. Four groups are recognised within Anthropic Soils: Truncated, Refuse, Mixed, and Fill, reflecting the mechanisms by which the soils were constructed.

3.1 Important Features of Anthropic Soils

3.1.1 Concept and Key Features of the Soil Order

Anthropic Soils are constructed, or severely disturbed, by people (Fig. 3.1). Anthropic Soils may include materials formed by (i) removing, or truncating, the natural soil, (ii) deposition of inorganic or organic soil materials or refuse that have thicknesses ≥ 30 cm, or (iii) by soil mixing. Much, or all, of the character of the original soil may be lost.

New areas of Anthropic Soil are constantly being created as a result of human activities such as urban development and road construction. Some Anthropic Soils are formed as a result of careful planning, including those on rehabilitated mine land or urban development sites and some agricultural soils, such as those recontoured to "improve" drainage or topography. Well-managed constructed soils may have improved productivity or versatility compared to the original soil.

Other Anthropic Soils have been formed as an inadvertent side effect of past human activities. For example sites where; Māori discarded shells after dining on shellfish (middens),

A. E. Hewitt et al., *The Soils of Aotearoa New Zealand*, World Soils Book Series,
https://doi.org/10.1007/978-3-030-64763-6_3

Fig. 3.1 A soil mixed by 'flipping' near Murupara in the central North Island. Left: the original Immature Orthic Pumice Soil profile with two coarse pumice layers (Kahroa tephra overlying Taupo tephra) near the surface. Centre: soil 'flipping' using a digger to bring underlying buried soil horizons to the surface. Right: the resulting Mixed Anthropic Soil ([formerly] Immature Orthic Pumice Soil)—in this case the topsoil was put to one side and replaced at the soil surface

gum-diggers sought kauri gum, miners sluiced or dredged for gold in the nineteenth and early twentieth centuries, and foresters have modified soils during tree planting and harvesting.

Given the wide variety of activities that lead to development of Anthropic Soils there is no one set of typical features. Anthropic Soil properties depend on the nature of manufactured or natural materials and the nature of the soil manipulation. Thus many Anthropic Soils may be recognised by their lack of the usual natural soil properties such as distinct, naturally formed, horizons. In other cases the soil features and landforms or landscapes do not conform with natural patterns, and land surfaces can be artificial and drainage modified. Some Anthropic Soils may be identifiable by odour (e.g. methane at landfill sites).

Conceptually, Anthropic Soils are soils in which the original plant rooting environment and soil moisture regime are changed to the degree that the new soil is distinctly different from the original soil. The character of Anthropic Soil lies in the contrast between the natural soil and human-made soil.

3.1.2 Areas of Occurrence

Areas of Anthropic Soils are scattered throughout New Zealand (Fig. 3.2). They are extensive in urban areas, land that has been mined, areas of agricultural development involving flipping or humping and hollowing activities, the addition of material to the land surface, and highly disturbed areas in forest plantations. In the South Island, areas of Mixed Anthropic Soils occur on sites of former gold dredging. For example, 1700 ha of Stony-tailings Fill Anthropic Soils formed on tailings, left behind following gold dredging, are mapped on the flood plains of the Waikaka Stream in Southland. Anthropic Soils are estimated to cover <1% of New Zealand, but accurate identification of Anthropic Soils can be difficult. For example, a restored soil may become indistinguishable from undisturbed land once the new soil surface is developed and new topsoil becomes established. The extent of Anthropic Soils is continuing to increase, particularly as a result of ongoing urban expansion.

3.1.3 Variation Within the Anthropic Soil Order

Four soil groups are recognised within the Anthropic Soil order (Fig. 3.3):

Truncated Anthropic Soils—most of the pre-existing soil profile has been removed,
Refuse Anthropic Soils—waste material that contains significant organic material has been added to the soil, for instance in a landfill,
Mixed Anthropic Soils—severe disturbance and loss of original character by mixing at depth. Currently no subgroups are recognised in this group but if the original soils are known and evident then its classification may be appended in parentheses (e.g. Fig. 3.1),
Fill Anthropic Soils—formed from deposited materials, including reconstructed soils such as those formed following urban development or mine rehabilitation.

Fig. 3.2 Distribution of Anthropic Soils. Numerous small areas are scattered throughout New Zealand that cannot be shown on a map at this scale

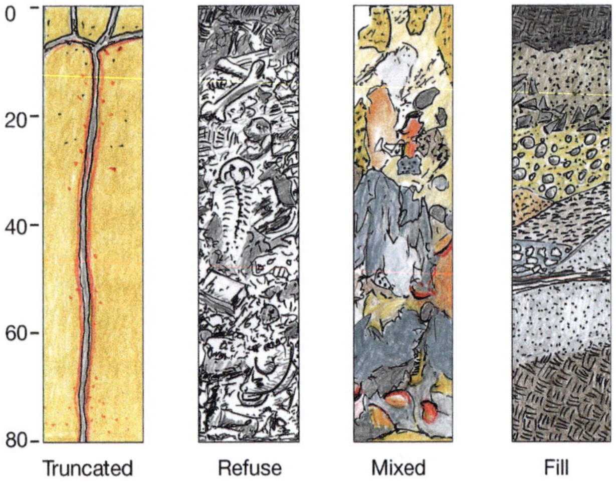

Fig. 3.3 Groups within the Anthropic Soil order (vertical axis is depth in cm)

The soil groups and subgroups recognised in New Zealand Anthropic Soils reflect the dominance of the primary production industries in the New Zealand landscape. In many other countries there is greater emphasis on anthropic soils associated with land occupied by urban, industrial, traffic, and mining use.

3.1.4 Origin of the Soil Order Name

'Anthropic' is derived from the Greek word 'anthropos' meaning 'human', so these are soils formed, or substantially modified, as a result of human activity. The name is widely used internationally—*Soil Taxonomy* uses anthropic diagnostic features and subgroups, and the *World Reference Base* has an Anthrosol Soil Reference group.

3.2 Soil Profile Genesis and Landscape Relationships

Because Anthropic Soils are the result of human actions, rather than natural soil processes, the soil-landscape relationships have often been greatly altered and the resulting soil properties may not relate to their position in a natural landscape. Thus the relationships between soils and landforms do not have the predictable natural relationships of undisturbed soils. For instance, surface topography and soil drainage are often markedly changed from the original state. The disconnect with usual soil-landscape relationships, along with site history, provide important evidence in diagnosing Anthropic Soils.

3.3 Key Soil Properties

3.3.1 Introduction

Anthropic Soils can have a diverse range of soil properties depending on the cause of soil disruption, the nature of any materials added to or removed from the soil, and the means by which the changes were undertaken. Anthropic Soils not only result as inadvertent by-products of human activities but may also be planned and constructed to achieve particular desired soil properties. Anthropic Soils can be designed to take best advantage of existing soil and landscape properties. A number of examples are discussed below.

3.3.2 Humping and Hollowing

Since about 1980 substantial areas of the wet Gley Soils and Perch-gley Podzols in the Westland region of the South Island, and Gley Soils and Organic Soils in the Waikato region, have been recontoured by 'humping and hollowing' to increase the runoff of excess water. Hollows were formed with central open drains, and/or underlying tile drains, that feed runoff, and/or drainage water, into natural streams or community drainage networks. Soil removed from the hollows has been relocated to build intervening 'humps' in the form of broad gently sloping mounds of sufficient height to provide a rooting zone free of continuous water saturation.

Many of the Westland soils that have been altered were Perch-gley Podzols that had strongly cemented, often iron-rich, horizons (placic horizons, ortstein pans, or humus pans, Chap. 11). The pans were suspected to be the main cause of the slow drainage but Craig Ross showed that massive, silty, gleyed, E-horizons, formed in loess above the pans, were the major problem. The very slow permeability within the E-horizon prevented water from draining. Simply breaking the pans would not effectively correct the poor drainage. However, humping and hollowing not only disrupted pans, and the E-horizon, but also provided the gentle slope needed to promote runoff direct to surface drains.

3.3.3 Flipping

Flipping involves the use of diggers to move material up from lower in the profile to, or near, the soil surface (Fig. 3.1). The bucket is forced downward to cut and extract a whole profile slice. The digger lifts the soil and dumps it

Fig. 3.4 An experiment demonstrating the benefits of soil mixing on summer pasture growth near Murupara. The green squares have Mixed Anthropic Soils, the balance of the area (pale brownish colour) is undisturbed control on the natural Pumice Soil

behind the machine, causing mixing and disruption of firm soil material as it falls. Diggers can work systematically, and surprisingly quickly, across a landscape.

Flipping is practised in a range of New Zealand soils particularly to transform marginal land into profitable dairy production in Westland, and to improve summer pasture production in parts of the central North Island. In the Gley Soils and Perch-gley Podzols of the Westland region flipping has been used to disrupt profile drainage impediments, and/or to bury infertile surface horizons.

In the central North Island, on dairy farms near Murupara, flipping has been used to raise fine-grained, clay- and organic-bearing buried soil horizons on mid- to late-Holocene tephra deposits closer to the surface and to relocate the near-surface, coarse-grained, pumice-rich, tephra layers deeper into the profile. The resulting 'inverted' soil profile may comprise chunks of materials from throughout the original soil all mixed together (Fig. 3.1). At some sites desirable clayey buried soil horizons (especially that formed on the Whakatane tephra, c. 5500 years old) are kept separate and placed at the soil surface, with the less desirable coarse soil materials buried below. Either way, the resultant soil has a higher-water holding capacity and greater potential root depth than the original droughty, pumice-rich soil. The outcome of the soil mixing is a marked increase in pasture growth, especially during dry summers (Fig. 3.4).

The original topsoil, along with the improvements made by pasture development, may be lost by burial, though at some sites the topsoil is kept to one side and replaced at the soil surface. At most sites, fertiliser inputs are needed to bring the new surface soil up to appropriate levels to support pasture growth and development of new topsoils. The large areas of land subjected to flipping in Westland attests to the agricultural and economic benefits, and research from Murupara has demonstrated that increased pasture production may recover the costs of flipping within 3–5 years with the benefit of improved water-holding capacity remaining for the long term.

3.3.4 Soil as a Mineable Resource

Soil is sometimes used as a resource to be 'mined' for its bulk properties or constituents. Mining inevitably involves some combination of mixing, processing, transport and removal of soil materials, resulting in formation of an

Anthropic Soil. In most instances requirements under the Resource Management Act (1991) to 'avoid, remedy, or mitigate, adverse effects on the environment' ensure that land restoration, and reconstruction of a viable soil, and revegetation are undertaken.

There has been a long history of peat mining in the Waikato with peat mined, for potting mix and mushroom growing, on the Hauraki Plains. For example, one company had a 'coal mining' licence for an area of about 60 ha of peat and has been required to pay tens of thousands of dollars in bonds to ensure appropriate rehabilitation works are completed following the mining activity.

Growing 'instant lawn' is also a form of soil mining and export. The removal of a thin topsoil layer would not meet the Anthropic Soils requirement of a 'drastic' change from the original soil. However, repeated removal, to the degree that the soil resource is depleted (and its diagnostic horizons and features are lost, or eventually fail the necessary thickness criteria), may lead to a remaining soil that should be classified as a Truncated Anthropic Soil.

Clay materials are mined at a number of sites in New Zealand including halloysite at Matauri Bay in Northland for fine china clay, 'fire clay' at Huntly for brick making, Waikari and Patumahoe clays for cricket pitch construction, and bentonite on the Hawke's Bay coast at Porangahau for drill lubricant, 'kitty litter', and an increasing number of health and beauty products.

3.3.5 Soil Relocation from Cromwell Gorge

The Clyde hydroelectric dam in Central Otago was built as part of the government development 'Think Big' strategy in the early 1980s. The dam flooded the valley of the Cromwell Gorge and part of the upper Clutha valley upstream of Cromwell town. The Cromwell Gorge was a particularly important area for stone fruit production. The area had deep, high water-holding capacity, silty, soils and was subject to down-river cold air drainage that reduced the risk of frost damage to developing fruit. Amid much controversy, a suggestion was made that the silty water-retentive soil material should be relocated onto river terraces that would not be flooded by the dam. At the relocation site, the soils were stony with poor water-holding capacity and would clearly benefit from addition of the relocated soil. A crop trial was initiated by Gary Beecroft and Don Brash to estimate the expected gain in fruit production. However, soil relocation was never undertaken due to the perceived costs. Perhaps in the twenty-first century, with stronger legislation requiring people to avoid, remedy, or mitigate adverse effects of development, soil relocation could become a viable option.

3.3.6 Urban Anthropic Soils

Urban soils were first recognised internationally in large, long established, cities that have been repeatedly rebuilt on top of previous habitation. Archaeologists have used deposits of human artefacts for historical interpretation. In New Zealand where soil disturbance is being undertaken for urban construction it is important to be alert to any artefacts that may be uncovered and protocols need to be in place at construction sites to consult local iwi and archaeological authorities should any artefacts be identified.

Since implementation of the Resource Management Act in 1991, any development that involves earth moving requires resource consent. As a result, a lot of effort and detailed engineering design goes into planning all major earthworks to ensure a sustainable outcome and to prevent adverse effects on the environment. Effects are monitored and managed both during the construction process, and following completion of the development. For example, all storm-water runoff must be contained and treated before discharge to prevent sediment runoff during construction. Designs must also provide for ongoing treatment and attenuation of storm-water runoff once a development is completed.

Urban Anthropic Soils may be designed, and natural soils may be included in development plans, to achieve multiple landscape aesthetic, productivity, hydrological, environmental, and engineering goals. Features of urban designs may include:

- use of permeable surfaces, rather than sealed surfaces, to slow and reduce the amount of runoff,
- 'rain gardens' where water can pond then slowly infiltrate to decrease runoff,
- light-weight soils for living roofs,
- provision of areas with sufficient soil root volume to support large trees within paved areas such as carparks and footpaths,
- re-use of soil displaced by buildings and road construction, and
- designed sites for placement of 'waste materials', both inert fill materials and modern landfills.

Where urban development is designed to utilise the natural land surface (without major recontouring), it is useful to understand the original soils and landscape so that the development can be designed to best fit the site and soil properties. For example, the existence of naturally deep permeable, soils can save the construction of costly drainage, runoff, and irrigation works. Urban development without recontouring is mainly used in lower density developments.

Fig. 3.5 Shell midden, deposited over a Podzol Soil, then buried by a later sand deposit (the sand was likely to have been mobilised and moved as a result of human activities) at Cook's Beach on the Coromandel Peninsula. The soil is classified as an Artifact Anthropic Soil. There is a distinct buried Podzol Soil beneath. (Photo: Andrew Hoffman)

Increasingly in New Zealand, urban subdivision and road development are undertaken with major recontouring of land surfaces. Topsoils are generally removed and stockpiled, the land is recontoured to provide raised sections for building construction and the lower lying streets thus become a conduit for removing storm-water to prevent flooding of buildings. Topsoil is then re-spread on the new land surface.

As land use becomes more intensive it is harder to find good matches between land use requirements and available soil. Techniques of soil design and construction are likely to become increasingly valued. Costs include the preparation, blending, and processing of soil materials, and the energy costs of soil movement and emplacement.

3.3.7 Māori Impacts on Soils: Middens, Gardens, Excavations, and Fortified Pā Sites

Even prior to the advent of large-scale earthmoving equipment, humans have impacted soils in New Zealand in a number of ways. Small pockets of Anthropic Soils occur at sites of previous Māori habitation (Fig. 3.5). Activities that impacted the soil included manipulation of the soil for kumara growing, accumulation of shell materials in middens, and excavations to build whare (houses), kumara storage pits, and fortified pā sites.

The Māori people became particularly adept at manipulating soil to grow kumara (sweet potato, *Ipomoea batatas*) a subtropical crop that they brought with them to New Zealand (originally from South America). Kumara are frost tender and good drainage and warmth were needed to maximise production in the temperate New Zealand environment. Kumara were grown on small mounds, sometimes called puke (Fig. 3.6), formed from gravels and sands excavated from small quarries, on land adjacent to the garden sites, referred to as 'borrow pits' (Fig. 3.7). Burnt material, including charcoal, was also added to the mounds.

Based on techniques needed for growing yams in tropical Polynesia, the sandy mounds had multiple functions. The mounds generated an interface with the buried loamy topsoil material, which resulted in larger and more robust tuber formation. Planting on mounds ensured good drainage (or improved drainage compared with that of the surrounding natural soil), thus preventing tubers from rotting in wet soil. The lower moisture content in the mounds would have also lowered the overall soil heat capacity, allowing the soil to warm more quickly in the spring. The gravel would have helped to improve drainage and may have helped prevent

Fig. 3.6 Traditional Māori kumara growing. Top left: kumara grown on traditional mounds at Te Parapara garden, Hamilton Gardens. Top right: nineteenth-century image of ko in use (from Best 1924). Bottom left: the preserved quincunx pattern of mounds used for kumara growing near Ngaruawahia in pre-European times. Poles are 1 m and 2 m long (photo: Warren Gumbley). Bottom right: a traditional ko (digging stick) used by Wiremu Puke at Te Parapara garden

frost damage by absorbing heat during the day and re-radiating it at night and the mounds themselves would have helped to prevent frost damage through cold air flow. Some plant nutritional benefits (e.g. K^+) likely arose from the addition of burnt materials (although studies in central New Zealand and elsewhere have shown the kumara to be well adapted to low-nutrient sandy soil). The additions of fresh sandy materials helped reduce disease and may have also been part of a purification ritual. Certainly planting and harvesting were accompanied by essential rituals and chants.

The mounds were often constructed in neat rows in a regular quincunx pattern (the pattern used on dice for number five). Such patterns were described by Joseph Banks on Cook's first voyage, and by William Colenso, and other early missionaries who travelled through the Waikato region. Such a pattern was discovered during construction excavation in Hamilton and near Ngaruawahia (Fig. 3.6), and intact mounds have been found preserved under colluvium draped over the edges of terraces. The traditional method of kumara growing has been recreated at the Te Parapara garden within the Hamilton Gardens (Fig. 3.6).

The gravelly sands used for building the puke (mounds) near the Waikato and Waipa rivers in the Hamilton Basin were usually excavated from the high (older) terraces underlain by volcanogenic alluvium dating to c. 20,000 years old (Hinuera Formation) and deployed nearby. In some places sands and gravels from the low terrace adjacent to the Waikato River, that was formed in part from pumice carried down the river in the 20–30 years after the AD 232 ± 10 eruption of Taupo, were also used. The Māori name for the area of Hamilton is Kirikiriroa which means a long strip of gravelly soil and likely refers to the modified soils used for growing kumara along the terraces above and adjacent to the Waikato River. In places along the river there

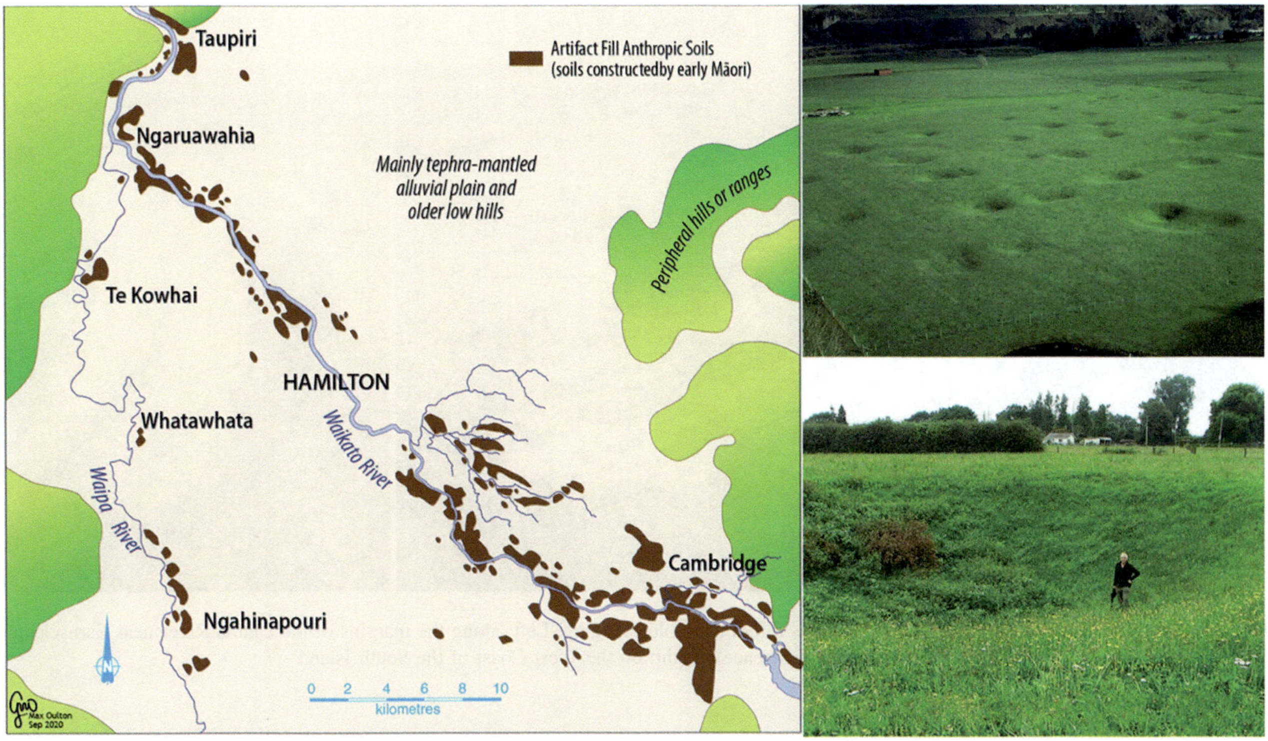

Fig. 3.7 Anthropic Soils in the central Waikato Basin that were modified by pre-European Māori mainly for growing kumara. Left: extent of the modified soils on, and adjacent to, the floodplains and terraces of the Waipa and Waikato Rivers and tributaries (map adapted from Singleton, 1988). Top right: a low terrace adjacent to the Waikato River near Arapuni with borrow pits (small quarries) from which gravel was removed to form mounds for kumara growing on a higher, less flood- and frost-prone, terrace. Bottom right: a large borrow pit at Hooker Road near Cambridge

is still evidence of the 'borrow pits' formed where the gravel was removed as well as the extensive areas nearby where the kumara growing was undertaken (Fig. 3.7).

Around the coastline of New Zealand, and sometimes at quite long distances inland, midden sites can be seen where large piles of shells have been added to the soil. A result of a long period of habitation and the importance of shellfish in the Māori diet (Fig. 3.5).

Evidence of pre-European Māori activity which altered and moved soil materials is still evident in the landscape. The terraces of fortified pā sites are visible on many hilltops. Pits were excavated for a range of purposes, including kumara storage and for whare which sometimes comprised a low roof over a pit dug in the ground.

3.3.8 Gold Dredging Sites

There have been extensive areas of gold dredging on alluvial deposits in the South Island, particularly in Otago, Westland, and Southland. A floating dredge would dig up the gravels, separate out the gold, and discharge the remaining sand and gravel behind it, gradually working its way across the landscape. A hummocky deposit of gravels was left behind that can best be described as a Stony-tailings Fill Anthropic Soil.

In Southland John Bruce mapped 1700 ha (4% of the Gore-Waikaka district), on the floodplain of the Waikaka Stream, as Anthropic Soils. Much of the area had been dredged several times, between the 1890s and 1940s, to depths of up to 19 m. The original soil morphology was destroyed. The resulting Anthropic Soils were variable but generally stony. Many ponds were scattered over the area, forming in the hollows of the hummocky topography. Bruce reported that the majority of the area had been developed for pastoral farming and that cereal and feed crops were grown in some places. However, some areas remained undeveloped and others were used as a source of gravel.

An interesting comparison of the effect of climate can be made by comparing the gold dredging deposits near Earnscleugh, in semiarid Central Otago, with similar deposits on the wet South Island West Coast (Fig. 3.8). At Earnscleugh, vegetation has been slow to become established and much of the area is now managed for recreation/historic interest.

Fig. 3.8 Stony-tailings Fill Anthropic soils formed as a result of gold dredging. Left: along the margins of the Clutha River near Earnscleugh, Central Otago (photo: courtesy of NZ Society of Soil Science). Right: on the West Coast of the South Island

Some gravel extraction has also been undertaken. On the West Coast, vegetation provides a basis for extensive grazing.

3.3.9 Sports Fields

Because of intensive use under varying climatic conditions sports fields are increasingly being constructed to meet specific requirements. Cricket pitches have long been subject to careful construction and import of particular clay materials. The state of 'the pitch' and its impact on bowlers and batspersons, and how it evolves over the course of a game of cricket, is the subject of endless discussion and debate in the cricket world.

Ideally a cricket pitch provides a resilient, even, surface which allows the ball to bounce strongly and not to slow greatly when it hits the ground. The pitch also has to survive up to five days of intensive use by players in a test match without breaking up. Some cracking of the surface as it dries over the course of a match is deemed acceptable as cracking adds some character to the game, but breaking to an uneven surface, which can send the ball bouncing off in unexpected directions, is not appreciated. To get the preferred level of surface hardness and smoothness requires a high soil bulk density which is achieved by compaction using a roller. The process of compaction is sensitive to the moisture content of the clay. Thus there is a need to get water to move deeply into the clay to provide an even moisture content to facilitate survival of the grass cover as well as compaction. Managing the balance between grass growth, water content, and surface compaction in cricket pitches is often considered as much an art as a science.

In New Zealand four clay sources have been widely used for cricket pitches in recent decades—the 'Patumahoe clay' and 'Naike clay' of the Waikato, the 'Waikari clay' of north Canterbury, and the 'Kakanui clay' of North Otago. In 2017, Seddon Park in Hamilton had nine cricket pitches, five constructed from Patumahoe clay (previously a similar soil, Naike clay, was used), and four from Waikari clay. The chocolate-brown coloured Patumahoe clay is derived from the subsoil of a Granular Soil and is described as a 'strong', but weakly swelling, clay (the dominant clay type in the soil is halloysite, a 1:1 clay which has a low, but significant, shrink-swell range). The Patumahoe clay has the ability to dry faster, after wetting, than some other clays (especially montmorillonite) meaning less time is needed for pitch preparation and presumably its use is less 'risky' in the early summer when showers and higher humidity tend to be more likely. The Patumahoe clay is favoured for the longer cricket test matches. Cricket player and coach Grant Bradburn commented in the *NZ Herald* newspaper that 'I'm a huge fan of Patumahoe [clay] which is what they've got at the Basin Reserve. I think that is the best track in the country for pace and bounce—Colin Maiden Park and Eden Park use the same soil too. Batsmen love it because they get consistent bounce. If it lands halfway down the track they know they can play cross-bat shots with freedom. That's critical for exciting cricket. … Patumahoe soil also forces batsmen to play off the back foot better. Waikari [clay] produces a front foot track because it tends to result in more placid pitches' (*NZ Herald* 14/3/2010).

The Waikari and Kakanui clays are both from dark blackish coloured Melanic Soils. The Melanic Soils are commonly derived from limestone or marble parent materials and contain smectite clays (notably montmorillonite, which has a large shrink-swell range) that shrink and swell strongly on wetting and drying. The advantage is that, once dry, water can be moved quickly into the soil through the cracks that form allowing rewetting from depth. The Waikari clay is preferred for one-day games but it will tend to crack more during drying in longer games.

The soils in the major rugby stadiums, such as the Waikato Stadium and the 'Cake Tin' in Wellington, are constructed and designed for purpose. The days of the All Blacks and opponents wading through a 'sea of mud' are long gone and the pressure is on to have a perfect looking green turf, for the TV cameras, that will stand the heavy trampling of rugby players, coatings of paint from sponsors messages, and not lead to any condition of surface roughness or slipperiness that could lead to the injury of an expensive professional player.

For example, the Waikato stadium 'soil' consists of sand that has been mixed with a plastic binding material with a subsurface drainage system and a 'pop-up' irrigation system. This provides a free draining medium for grass growth. The sand has minimal water or nutrient holding capacity and so the grass is grown with regular additions of water and fertiliser, including generous use of nitrogen, a system akin to hydroponics.

Cricket pitch construction—why bother with the grass?

New Zealand Cricket suggests that a good cricket pitch should produce consistent 'pace and bounce' of the ball and to do this needs minimal soil organic matter, no soil layering, good compaction, and even grass cover. The even grass cover should comprise very short grass, pressed onto the ground surface with the clay visible through it. There should be no weeds and the bare soil should not exceed 15% of the surface with no bare spot greater than 2 cm in diameter. These requirements mean that the grass is 'decimated' during the final pitch preparation. This begs the question—why bother with grass at all? Grass is important to cricket pitch preparations as a healthy plant root system helps achieve even drying of the pitch to depth. Without grass, removal of water via surface evaporation can lead to steep moisture gradients and inconsistent surface cracking. The plant roots, and rolled in surface turf, help hold the pitch together. However, a buildup of organic material in the soil is not wanted. Grass cover helps create 'pace' and consistency in how the ball behaves and can also protect the ball from wear. It is a tribute to the resilience of grass that it can withstand the conditions in which it is asked to grow on cricket pitches—where moisture contents are allowed to go extremely low, soil dry bulk densities are exceedingly high, and the grass is cut to 12–15 mm throughout the playing season and as low as 3 mm on playing day.

3.4 Distinguishing Between Anthropic Soils and Related Soil Orders

Land use practices can have strong effects on natural soils. Over the last 50 years or more, land development operations, such as deep ploughing, have caused deep soil mixing. The construction of mid-twentieth century border-dyke irrigation systems resulted in scalping of topsoils and soil relocation over short distances. Historically, there have also been changes made to soil by cutting and burning of native bush for pasture establishment, fertilisation, and seeding, especially by the aerial topdressing industry, leading to widespread changes in soil chemical properties. The question may be asked—have the changes wrought by these management interventions produced Anthropic Soils? It has not been the practice to consider such altered soils as Anthropic Soils. Where then do we draw the line between modified soils and Anthropic Soils?

The boundary between Anthropic and other soil orders depends of the nature of the changes. For a soil to be considered an Anthropic Soil, it must be so severely changed from its original state that its prior order cannot be positively identified using the *New Zealand Soil Classification* key to soil orders. Thus the Anthropic Soils are placed third-to-last in the key (ahead of only Recent and Raw Soils).

For all of the soil orders other than Anthropic Soils identification relies on objective soil properties (presence or absence of diagnostic horizons and features of designated thickness and placement in the profile, the 'control section'), but the identification of the diverse range of Anthropic Soils inevitably involves historical research and inferences from apparent changes in landforms, as well as direct soil observations. The critical observations include the presence or absence of the normal natural soil profiles and soil horizons expected in the area, and the interpretation of any non-normal soil features.

Raw Soils and Recent Soils can potentially be confused with Anthropic Soils, especially where there is evidence of thick burial of a soil by young sediment. An interesting example of this dilemma is the 'German' soil series

Table 3.1 Correlation between Anthropic Soils and *Soil Taxonomy*, *World Reference Base*, and the New Zealand genetic soil classification[a]

New Zealand Soil Classification	*Soil Taxonomy*	*World Reference Base*	NZ genetic soil classification
Truncated Anthropic Soils	See note below[b]	Anthrosols	Not recognised
Refuse Anthropic Soils	See note below[b]	Technosols	Not recognised
Mixed Anthropic Soils	See note below[b]	Anthrosols	Not recognised
Fill Anthropic Soils	See note below[b]	Anthrosols	Not recognised

[a]The correlations given here are a guide only and for accurate classifications the relevant soil classification documents should be consulted. The two major international soil classification systems are *Soil Taxonomy*, which was developed in the USA, and *World Reference Base*, which was developed primarily in Europe. The NZ genetic soil classification was used in New Zealand prior to 1992

[b]In *Soil Taxonomy*, Anthropic Soils are recognised at lower levels in the classification, either (1) as diagnostic horizons such as anthropic (see below) and plaggen epipedons; or through (2) the human-altered and human-transported material class at family level, including for example, artifactic; or (3) as subgroups, such as Anthraquic, Anthropic, or Anthroportic. For example, in the Keys to Soil Taxonomy 2014, the 'anthropic epipedon consists of mineral soil material that shows evidence of the purposeful alteration of soil properties or of earth-surface features by human activity. The field evidence of alteration is significant and excludes agricultural practices such as shallow ploughing or addition of amendments, such as lime or fertiliser'

identified in areas near Bannockburn where material washed from gold sluicing was deposited. The German soil series has been mapped as a 'Recent Soil'. However, it could potentially be described as a Stony-tailings Fill Anthropic Soil. Classifying a soil as an Anthropic Soil, rather than as a Recent or Raw Soil, requires some understanding of the site history and mode of emplacement or removal of soil materials.

3.5 Correlation with Other Classification Systems

There is only limited recognition of Anthropic Soils in other soil classification systems (Table 3.1). Thus the *New Zealand Soil Classification* is perhaps unique in recognising such soils at the highest (order) level. Given the ever increasing impacts of human activities on the environment, the recognition and study of Anthropic Soils is likely to gain in importance in future.

3.6 Key Soil Data

Soil data on Anthropic Soils are limited and highly variable, depending on the origin of the soil or the manner of the original soil's modification. Thus it is not practical to give any indicative soil data.

3.7 Use and Management of Anthropic Soils

3.7.1 Introduction

Well-designed, constructed, Anthropic Soils can be successfully used for a wide range of purposes. However, some of the co-incidental Anthropic Soils have little potential for productive use without some remediation effort. For example, Anthropic Soils at abandoned nineteenth-century gold sluicing sites in Otago comprise piles of gravel and, over a century later, little soil development or plant colonisation has occurred. However, over time Anthropic Soils, like all others, continue to be subject to the pedogenic and geomorphic processes normally expected in the environment and will gradually undergo soil development and ecological succession.

The soil properties of Mixed Anthropic Soils may be highly variable thus little general information on land use can be provided. Management will need to be tailored to the particular soil and site conditions and proposed activities.

In Truncated Anthropic Soils, where most of the pre-existing soil profile has been removed, use of the remaining soil may be problematic because of such factors as high soil bulk density, poor drainage, lack of soil organic matter, potentially limited water-holding capacity, and low soil fertility. However, there is potential to restore such soils. A range of improvements may be required, such as addition of topsoil or organic matter, installation of drainage,

Fig. 3.9 Rehabilitated land (foreground) on Fill Anthropic Soils that was mined for coal at Rotowaro near Huntly. The original site comprised steep hill country. The restored land has a gentler contour, making it improved for both pastural and forestry land use

ploughing, or ripping to break up compacted materials, and addition of fertiliser and/or lime.

3.7.2 Land Restoration Following Mining or Landfilling

Consent for mining activities now requires that disturbed land is rehabilitated. Where land is mined the soil cover can be initially removed and stockpiled and can later be re-spread over a restored land surface. Land restoration is relatively straight forward as there is ample experience available. If the land is sloping, and especially when working with erodible materials, then runoff and erosion risks during reestablishment must be well mitigated particularly by establishing vegetation cover as soon as possible.

Whether the original disturbance was a mine or a landfill site (and many sites are now first mined, then landfilled with urban waste), the basic considerations for land restoration are similar. Planning for the final site contour and elevation, along with end land uses, has to be undertaken at the start of the project. Topsoil is likely to have to be removed and stored in safely vegetated sites with the total depth of topsoil piles controlled (to no more than 1.5 m) to prevent over-compaction or saturated conditions from degrading the soil material. At the completion of mining, or filling, the land surface needs to be restored and recontoured. Often mine overburden or other subsoil cut-to-waste materials are used. At some sites, such as the Rotowaro open cast coal mine near Huntly (Fig. 3.9), the final contour may provide for a better range of land use options than the pre-existing steeplands. On landfill sites the subsoil layer needs to provide sealing to prevent water moving in and gas from moving out except via gas collection systems that must be installed. To this end highly compacted clays or liners may be used. The upper subsoil layers, however, need to be carefully emplaced to prevent over-compaction so they can be a source of soil moisture and anchorage for any proposed vegetation cover.

To complete land restoration, topsoil must be emplaced, without over-compaction, and vegetation cover established to prevent surface erosion. Until continuous vegetation cover is established, storm-water runoff must be managed and treated to prevent sediment discharge from the site. Mine and landfill sites can both be restored for pastoral farming or in some cases forestry has been established. Amenities such as parks and golf courses are also good uses for restored land. Hamilton Gardens, an internationally recognised garden destination, is the site of an old Hamilton city landfill that was used up until the 1970s. A site in Auckland, near Mt Wellington, that was New Zealand's largest aggregate quarry, has been restored for urban housing. The 1100 ha suburb of Stonefields is expected to house about 6000 people. Meanwhile at the nearby former Greenmount Landfill site, which finally closed in 2016, it is proposed to form a 54 ha park that will include areas of native forest, dog exercise areas, children's play grounds, and walking and cycling tracks.

Refuse Anthropic Soils are likely to contain significant volumes of waste material which, as it biodegrades, may subside or emit methane (CH_4) gas. The potential for subsidence limits the use of such sites for buildings or other intensive developments. However, former landfills are often used to provide welcome parkland or greenspace as cities over-take their former rubbish dumps.

Methane emissions are a potential hazard associated with Refuse Anthropic Soil. Being heavier than the ambient air, methane may accumulate in nearby low-lying areas such as house basements where it becomes an explosion hazard. Methane is also a potent greenhouse gas so allowing it to escape to the atmosphere is not desirable. Thus it is best to engineer the site to instal an impermeable cap and a methane collection system so the methane can be collected. Where there is sufficient volume generated, the methane can be used as a valuable source of natural gas for heating or electricity generation. Where volumes are low the methane is often 'flared off' (burned) as it is less damaging to emit the combustion products (CO_2 and H_2O) to the atmosphere than the methane (which has a higher, though shorter lived, warming potential than CO_2).

At many Refuse Anthropic Soil sites there is also potential for leachate to contain toxic substances. At modern engineered landfills, the base of the site will have been sealed prior to infilling and leachate collection and treatment systems must be included. However, at older sites of former rubbish dumps it is likely that no such provisions are in place. In some sites where leachate has been recognised as a problem, it has been decided to excavate the whole site and remove contaminated materials to a contained and managed landfill. If potential for leachate is a concern, monitoring wells around the margins of the site can provide information on the extent and form of any contamination. Pumping wells, to remove the leachate and draw surrounding groundwater back towards the site, may be effective. For example, research at a former rubbish dump near the Gisborne coast showed that pumping of shallow groundwater adjacent to the 'dump', to prevent leachate escape, had drawn sea water back into the site.

3.7.3 Moving Soil Materials and Ecosystems

Fill Anthropic Soils can be designed to provide appropriate soil properties for future uses, for example, by designing appropriate plant rooting depth, water-holding capacity, drainage, and fertility. Where good design and implementation have been employed, such as at well restored mine sites, the resulting soils may have no limitations for future use and could be used for intensive food production or urban development. The relocation of soil material must be carefully managed to prevent adverse impacts on the soil due to inappropriate soil mixing, storage conditions, or mode of placement. If Fill Anthropic Soils are not well designed or installed, problems may occur. A common problem is over-compaction where clay materials have been emplaced with heavy machinery, at high water contents. The resulting dense, massive, material is often impermeable to water and inaccessible to plant roots. Thus increased runoff will occur and the soil will provide a poor substrate for establishment of vegetation.

Where fill soil materials are used (for example, for land restoration, urban amenity, or ecosystem design), unless the fill is derived from surplus material from another site, it has to be mined from somewhere. The management of Anthropic Soils should weigh the effects of mining at the source as well as the use at the site of relocation. A potentially powerful means of analysing the net consequences of relocation is to estimate and compare soil ecosystem functions before and after relocation and balance the loss of functions at the mined site against the gain of functions at the relocation site.

The relocation of soil material may affect soil properties that are considered valuable. Original soil profiles are often mixed during excavation, transport, and relocation. For example, soil-water movement is altered by mixing, bulking, compaction, and by changes in soil biological, and soil organic matter, processes. Relocation may also lead to the incorporation of foreign materials, such as bricks, concrete or other human-made materials, weeds, and pests.

Where the original area hosts a native ecosystem then care must be taken to restore not just the soils but also the native biota. Some restoration may seek to relocate, not just the soils but whole below-and-above-ground ecosystems. Direct transfer of soils with the overlying vegetation can be an important option for preserving habitats for rare and endangered species and can save some costs and risks of providing alternative habitats for endangered species. Soil, soil animals, and vegetation are salvaged by lifting soil and plants intact, with minimum disturbance. Transportation can then be undertaken, by large specialist machinery, to a new site. The process may also involve temporary storage. Relocation is costly but the major benefits are rapid ecosystem recovery and minimised loss of ecosystem values.

When soil is relocated, there are risks that contaminants, weeds, or pests may be transported in the soil. Weeds and pests can easily be transferred in soil and require eradication, thus creating additional costs. For instance, roadside soils have been used to rehabilitate new areas of disturbance with consequent spread of gorse, broom, and pasture grasses. In many natural areas there is a tension between soil fertility and competitiveness of desired native plants against pasture species or gorse. Placement of relocated fertile soil material into a native area of lower fertility can give undesirable weeds a competitive advantage to establish a foothold and invade native areas.

The large cost involved in constructing Anthropic Soils means it is important to manage the process for best possible outcomes as such work is generally a long-term investment.

Further Reading

Beecroft FG, Brash DW (1983) Field trial investigation of horticultural response on relocated soil in a semi-arid environment, Central Otago, New Zealand. DSIR Soil Bureau District Office Report No. DN 14

Best E (1924) The Maori as he was: a brief account of Maori life as it was in pre-European days. NZ Board of Science and Art No. 4. Dominion Museum, Wellington. 272 p

Best E (1925) Māori agriculture: The cultivated food plants of the natives of New Zealand, with some account of native methods of agriculture, its ritual and origin myths. Dominion Museum Bulletin No. 9. 172 p

Bruce JG (1979) Soils of Hamilton City, North Island, NZ. NZ Soil Survey report 31, 65 p + map

Bruce JG (1984) Soils of the Gore-Waikaka District, South Island. New Zealand. NZ Soil Survey Report 74, 86p

Buller Community Development Co. Ltd, undated. Land development by flipping and humping and hollowing, 40 p. http://maxa.maf.govt.nz/sff/about-projects/search/01-231/land-development-booklet.pdf. Accessed April 2018

Clarke A (1977) Maori modified soils of the upper Waikato. N Z Archaeol Assoc Newslett 20:204–222

Denham T (2013) Ancient and historic dispersals of sweet potato in Oceania. Proc Nat Acad Sci USA 110:1982–1983

Ell G (1985) Shadows on the land: signs from the Māori Past. N.Z. Bush Press, Auckland, New Zealand, 112p

Grange LI, Taylor NH, Sutherland CF et al (1939) Soils. In: Soils and agriculture of part of Waipa County. DSIR Bulletin No. 76. pp 30–63

Gumbley W, Laumea M (2019) Archaeological investigation report: pre-European Maori horticultural site S15/465. Heritage New Zealand Authority 2018/133, 84p

Gumbley W, Higham TFG, Lowe DJ (2004) Prehistoric horticultural adaptation of soils in the middle Waikato Basin: review and evidence from S14/201 and S14/185, Hamilton. NZ J Archaeol 25:5–30

Hewitt AE (2012) New Zealand Soil Classification. 3rd edn, Landcare Research Science Series No. 1. Manaaki Whenua Press, Lincoln, New Zealand

Higham TFG, Gumbley WJ (2001) Early preserved Polynesian kumara cultivations in New Zealand. Antiquity 75:511–512

Joblin MH (2010) Environmental assessment of the effects of leachate irrigation and seepage from the Paokahu Landfill, Gisborne, New Zealand. MSc thesis, University of Waikato, New Zealand, 222 p

Laubscher N (2014) Improvement in soil water availability in pastures by excavating and mixing buried soil horizons from multi-layered Pumice Soils (Vitrands) at Galatea, central North Island, New Zealand. MSc thesis, University of Waikato, New Zealand, 152p

McCraw JD (2011) The wandering river—landforms and geological history of the Hamilton Basin. Geoscience Society of New Zealand Guidebook 16. 88 p

McFadgen BG (1980) Maori plaggen soils in New Zealand, their origin and properties. J Royal Soc NZ 10:3–18

New Zealand Cricket (2017) Pitch preparation—the basic fundamentals. V1.0. 82p. https://www.nzc.nz/media/10212/nzc_pitch-prep_pages_digital_1608.pdf on 2 March 2020

Puke WT (2021) Conception, construction, and cultural significance of Te Parapara garden in Hamilton, Aotearoa New Zealand. J NZ Pac Stud 9(1) (in press)

Rossiter DG (2007) Classification of urban and industrial soils in the World Reference Base for Soil Resources. J Soils Sediments 7 (2):96–100

Rogers D, Bartlet R, Simcock R et al (2001) Benefits of vegetation direct transfer as an innovative mine rehabilitation tool. In: Nichils O, Viluclcis N (eds) Proceedings of the Australian mine rehabilitation workshop—achieving rehabilitation success. Adelaide, 17–19 Aug. 2011. JK Tech SMI Technology Transfer, Indooroopilly, Queensland, pp 285–302

Singleton PL (1988) Cultivation and soil modification by the early Maori in the Waikato. N Z Soil News 36:49–57

Soil Survey Staff (2014) Keys to Soil Taxonomy, 12th edn. USDA Natural Resources Conservation Service, Washington, DC, 362 p

Soil Survey Staff (2015) Illustrated guide to Soil Taxonomy, version 2. U.S. Department of Agriculture, Natural Resources Conservation Service, National Soil Survey Center, Lincoln, Nebraska. https://www.nrcs.usda.gov/wps/portal/nrcs/detail/soils/survey/class/taxonomy/?cid=nrcs142p2_053580. Accessed 2 Mar 2020

Taylor NH (1958) Soil science and prehistory. N Z Sci Rev 16:71–79

Waterhouse BR, Adair KL, Boyer S et al (2014) Advanced mine restoration protocols facilitate early recovery of soil microbial biomass, activity and functional diversity. Basic Appl Ecol 15 (7):599–606

World Reference Base (2015 update); http://www.fao.org/3/i3794en/I3794en.pdf. Accessed 2 Mar 2020

Yen D (1963) The New Zealand kumara or sweet potato. Econ Bot 17:31–41

4 Brown Soils

From the Wairarapa hills to the Southland plains

Brown Soils occur in places with good rains

The hard-working Brown Soils support our pastoral farms,

polyhedral structures are just one of their charms.

Abstract

Brown Soils are New Zealand's most extensive soil order covering about 43% of the country. Brown Soils occur on the moister hill and mountain lands throughout New Zealand except where tephra deposits predominate. Brown Soils also occur on flat to rolling lands in wetter regions such as the Southland lowlands. Most Brown Soils occur in areas with mean annual rainfall greater than about 800 mm in the South Island, and 1000 mm in the North Island. Sandy Brown Soils and stony Brown Soils may form in areas with lower rainfalls in materials with low (<75 mm) water holding capacity. Brown Soils have grey-brown topsoils, brown to yellowish brown subsoils, and profiles with A–B–C horizonation. They form from a wide range of parent materials, predominantly siliceous quartzo-feldspathic rocks (e.g. greywacke, mudstone, schist, rhyolite) and their derivatives (e.g. sand dunes, colluvium, alluvium, and loess). Brown Soils are moderately weathered with moderate to strong leaching. The distinctive brown colours are due to thin coatings of iron and aluminium hydrous oxides, or aluminosilicates, potentially including allophane and ferrihydrite, on particle surfaces. The Brown Soils underlie much of New Zealand's sheep and beef production along with dairying, cropping, plantation forestry, and extensive areas of public conservation land.

4.1 Important Features of Brown Soils

4.1.1 Concept and Key Features of the Soil Order

Brown Soils are the work horse of New Zealand's hill country and the Southland plains (Fig. 4.1). Brown Soils have a brown or yellow-brown subsoil below a dark grey-brown topsoil. The Brown Soils are part of the zonal sequence of soils that reflect increasing rainfall from Semiarid Soils (dry spring to autumn), through Pallic Soils (dry summer), to Brown Soils (usually moist), Ultic Soils (moist and warmer), and Podzol Soils (highly leached and rarely dry). Because Brown Soils are our predominant soil, they tend to exemplify some of New Zealand's common national soil characteristics, namely, moist cool temperate climate, generally moderate weathering, acidity, and well-stocked soil organic matter.

The characteristic brown colouring of Brown Soils is attributable to iron and aluminium oxides dispersed throughout the soil mass. Some Allophanic Brown Soils occur in areas with limited tephra inputs that intergrade towards the Allophanic Soils of the central and western North Island. However, Allophanic Brown Soils also occur in the absence of tephra in the South Island high country as a result of strong leaching (especially desilication) that enables small to moderate amounts of allophane and ferrihydrite to form.

A. E. Hewitt et al., *The Soils of Aotearoa New Zealand*, World Soils Book Series,
https://doi.org/10.1007/978-3-030-64763-6_4

Fig. 4.1 Landscape at the northern end of the Tararua Ranges, looking towards the Pahiatua Track, and associated Brown Soil profile (depth about 70 cm), formed on weathered greywacke. Greywacke saprolite is visible at the base of the profile

4.1.2 Areas of Occurrence

Covering about 43% of the country, Brown Soils are New Zealand's most extensive soils, occurring on large areas of the moist mountains, hills, and plains (Fig. 4.2). Brown Soils form mainly on quartz- and feldspar-dominated parent materials in places where summer drought is uncommon, but which are not waterlogged in winter. Most Brown Soils occur in areas with mean annual rainfall greater than about 800 mm in the South Island and more than about 1000 mm year^{-1} in the North Island. Brown Soils may also occur in lower rainfall areas where stony and sandy materials occur. In stony or sandy soils, that have low water-holding capacity, leaching (solutes carried by water flow through the profile) may be sufficient for Brown Soil formation under a relatively low rainfall (600–800 mm year^{-1}). Brown Soils may also occur in lower rainfall areas on sites with low evapotranspiration where leaching is thus relatively high.

4.1.3 Variation Within the Brown Soil Order

Brown Soils are extensive and comprise a wide variety of soils encompassed by seven soil groups (Fig. 4.3) and 48 subgroups. The seven soil groups are as follows:

Allophanic Brown Soils—contain a relatively thin layer of Allophanic soil material or allophanic material coating grains,
Sandy Brown Soils—dominated by sand or loamy sand to depth,
Oxidic Brown Soils—similar to Oxidic Soils but with significant weatherable minerals remaining,
Mafic Brown Soils—in materials derived from dark-coloured (Fe- and Mg-rich) igneous (including volcanic) rocks,
Acid Brown Soils—strongly or extremely acid,
Firm Brown Soils—slightly firm or stronger B-horizon soil strength, apedal subsurface horizon, and
Orthic Brown Soils—other Brown Soils.

In some cases the allophane in Allophanic Brown Soils is the result of thin layers of tephra where the soils are intergrading towards the Allophanic Soils of the central North Island or Taranaki. However, in many Allophanic Brown Soils, the allophanic material is not derived from weathering of tephra, but from weathering of greywacke or schist (or deposits derived from these), mainly in mountain environments where rainfall is relatively high and drainage is good.

Because of the prominence of the Brown Soils there are many (48) subgroups. Subgroups recognise transitions from the Brown Soils to other related soil orders or soil groups

Fig. 4.2 Distribution of Brown Soils

(termed 'intergrading subgroups'). Brown Soil intergrades include Mottled (to Gley Soils), Peaty (to Organic Soils), and Immature (to Recent Soils), Firm (to Pallic Soils), Pallic (to Pallic Soils), Placic (to Podzol Soils), and Allophanic (to Allophanic Soils). Other subgroups depart in some specific way from the soil group or soil order (Acidic, Aged, Calcareous, Firm, Humose, Mafic, Magnesic, Pedal, Sodic, or Weathered).

4.1.4 Origin of the Soil Order Name

The Brown Soils correlate most closely with the 'Yellow brown earths' of the former New Zealand genetic soil classification and the name was derived from that. 'Brown soil' also has international precedence. The term relates to the *Sols Bruns* of France, *Braunerden* of Germany, *Burozems* of Russia, and *Cambissolos* of Brazil.

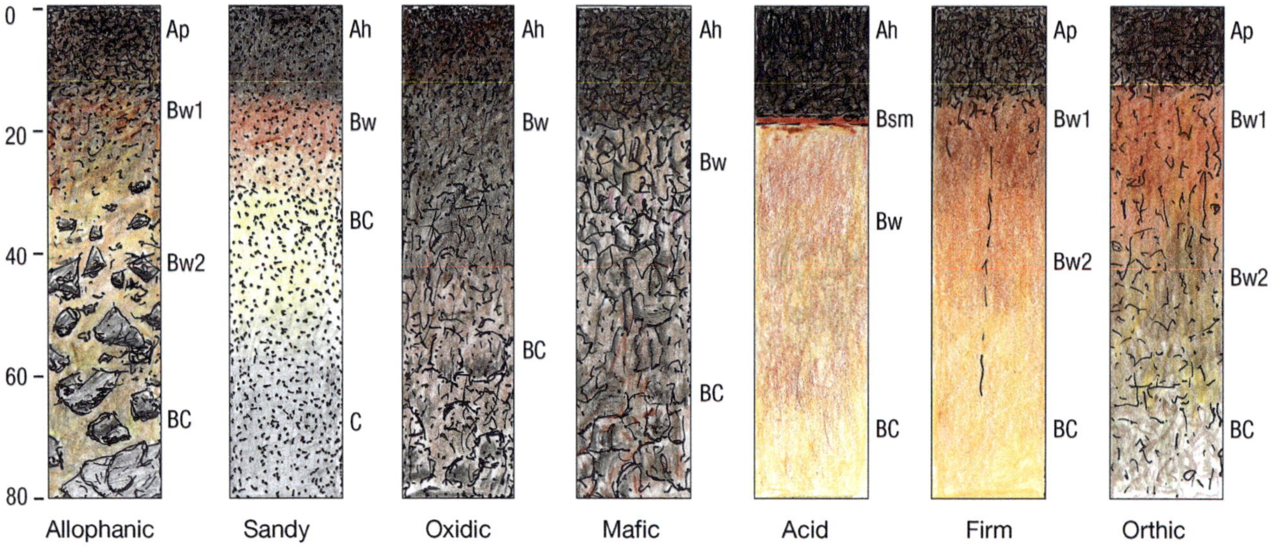

Fig. 4.3 Groups within the Brown Soil order. Vertical axis is depth in cm

4.2 Soil Profile Genesis

An essential character of Brown Soils, and indeed all orders other than Raw Soils, is the topsoil, developed from what would have originally been a Raw Soil, by the process of melanisation (or topsoil darkening).

Another important characteristic of Brown Soils is the weathered B-horizon (usually a Bw or associated subhorizon, Fig. 4.3) formed by weathering and leaching of the soil parent material. The weathering processes that lead to the development of the distinctive brown colour of the subsoil have been termed brunification.

To form a Brown Soil a site must experience sufficient rainfall to leach and lower the quantities of exchangeable base cations (calcium, magnesium, potassium, and sodium) to the extent that they are replaced by acid cations (hydrogen and aluminium). The soil becomes acidified and brunification is activated with the formation of (nanocrystalline) ferrihydrite and related crystalline aluminium and iron oxide minerals dispersed in the subsoil. The soil parent materials are non-calcareous so, unlike soil formation in many parts of the globe, the soil does not need to first undergo a decalcification stage.

The amount of water moving through the soil to achieve brunification has been defined as the 'leaching threshold'. The leaching threshold is a function of the mean annual rainfall less evapotranspiration, and the profile available water capacity. In soils where the soil profile available water capacity is very low (for example, in stony or sandy soils), it takes little rain to fill the soil water storage capacity to the point that drainage, and consequent leaching, occurs. Thus the leaching threshold is reached relatively quickly and brunification will proceed, just as it would in sites with plentiful rainfall. Therefore in marginally dry areas, the limit of brunification depends on soil water storage capacity. In semiarid environments, Brown Soils will not form because the leaching threshold cannot be reached. Further research is needed to elucidate the precise value of the leaching threshold for the formation of Brown Soils in New Zealand.

4.3 Soil-Landscape Relationships

4.3.1 Introduction

The Brown Soils occur predominantly on quartz- and feldspar-rich (sometimes referred to as quartzo-feldspathic) sedimentary rocks, and on reworked materials derived from these such as alluvium, colluvium, fan deposits, sand dunes, and loess. They cover a variety of terrains ranging from the steep greywacke-dominated land of the major hill and mountain ranges, through the highly erodible soft mudstones of the east coast and Rangitikei-Whanganui hill country, to the schist-dominated Otago hill country, and the lowlands in moister regions such as Southland. On each of the parent rocks or deposits, distinct landscapes can be recognised with characteristic geomorphology, slope stability, soil patterns, land use potential, and management requirements.

4.3.2 The Greywacke Hill and Mountain Country

a. The broad soil-landscape patterns of the South Island high country

The Mesozoic (c. 252 million to 66 million year old) greywacke basement rocks of New Zealand are exposed in our rapidly uplifting and eroding main mountain ranges, including the Southern Alps, forming the main soil parent materials. Greywacke is hard, resistant sandstone comprising roughly equal amounts of quartz, feldspar, and fragments of pre-existing rocks cemented together in a mud-rich matrix. The resulting landscapes tend to be cut through with faults and comprise steep hills with sharp, linear ridges, planar side slopes, and well drained hollows.

Generally, there is an east to west trend in the South Island greywacke ranges approaching the Alpine Fault. Nearer the Alpine Fault there is higher rainfall, greater rates of tectonic uplift, steeper slopes, a more rugged landscape, weaker expression of soil weathering, stronger expression of glacially carved topography, and fewer pockets of weathered rock, than in the more distant areas. Thus the broad scale soil-landscape of the central spine of the South Island can be divided into three regions: the eastern hills and mountains, the main divide, and the western alpine region.

In the eastern hill and mountain region the soils are dominated by Orthic and Allophanic Brown Soils. The main divide region has a higher rainfall and the land is steeper and more rugged than the eastern area. The dominant soils in the main divide region are Acid Brown Soils, Allophanic Brown Soils, Recent Soils, and Orthic Podzol Soils. The active uplift and erosion mean that there is often soil disturbance in the main divide area and hence soils of varying ages are common. Soil profiles frequently include buried features and composite soil profiles—for example, a Brown Soil buried by younger sediment in which a Recent Soil is formed. In the western alpine region, where rainfall is considerable, the soils are dominantly Gley Soils, Gley Podzol Soils, Recent Soils, and Raw Soils.

b. The Gibbs-Raeside line between Allophanic Brown Soils and Orthic Brown Soils

In the 1940s Harry Gibbs and James Raeside embarked on the challenging task of the first general soil survey of the South Island high country, partly to map accelerated erosion. They identified a significant boundary, now referred to as the Gibbs-Raeside line. The Gibbs-Raeside line marked the eastern limit of the Allophanic Brown Soils (then known as the High country yellow brown earths) which are characterised by strong brown colours, low soil dry bulk density, and a friable crumb (apedal earthy) structure. Orthic Brown Soils to the east and south (then known as Southern yellow-brown earths) had paler colours, and pedal or massive structure. In their 1942 publication, Gibbs and Raeside and others suggested that the boundary (between what were then called Southern yellow-brown loams (Allophanic Brown Soils, which they termed Kaikoura series), and what they called Omarama series (Orthic Brown Soils) in drier areas, occurred at about 1000–1200 m altitude in the dry inland basins, but at about 450 m in wetter areas, and roughly coincided with the lower limit of the winter snowline and the boundary between tall and low tussock grasslands at the time of European settlement. They also noted that the Allophanic Brown Soils coincided with areas of beech forest at lower altitudes.

In the twenty-first century, Andre Eger and Peter Almond studied contrasting soils, similar to those recognised by Gibbs and Raeside, in the vicinity of Porters Pass. They identified Allophanic Brown Soils on southern and western aspects with the contrasting non-allophanic Orthic Brown Soils on northern and eastern slopes.

The conclusion reached by Eger and others was that the Allophanic Brown Soils were likely to have developed beneath acidifying beech forest (Nothofagus sp.) which generated soil pHs sufficiently strong to mobilise the aluminium (especially) and silicon ions, the precursors of allophane formation (which involves loss of silicon through leaching and retention of high amounts of aluminium). It was proposed that the more eastern Orthic Brown Soils were developed under a shrubby vegetation. If the boundary between the two soils was in fact related to a former Nothofagus forest boundary then it may represent a significant biogeographic, natural ecosystem boundary.

c. Loess in the South Island high country

Some of the high country Brown Soils are formed on loess materials. Loess is often thought of as being associated with flat to rolling land of the lowlands, usually seen as smooth, undulating 'downland' landscapes, but it is also important in hill and mountain landscapes. For loess to accumulate there needs to be a source of fine (silt or fine sand), loose, dry, sediment that is susceptible to wind erosion. Australian insights suggest that clay-sized aeolian 'parna' may also behave as loess if it is flocculated into silt, or fine-sand, sized particles. The bulk of loess in New Zealand is derived from riverbeds, the 'dust engines' in which most of the fine material that ends up as loess is derived from abrasion and breakdown of rocks during river transport, especially during glaciations. Smaller amounts of fines are contributed by glacial grinding and by freeze-thaw processes. Aeolian loess transport from the river beds is driven by the belt of strong westerly/north-westerly winds in the mid-latitude regions.

For loess accumulation there needs to be vegetation to trap the airborne material otherwise it may be rapidly removed by water runoff. At the site of deposition the ground surface needs to be stable to accumulate a significant loess deposit. If the receiving surface is a highly porous gravel the loess may be washed into the gravel matrix. Eventually the pores will be plugged, preventing further loess infiltration.

Loess usually accumulates very slowly—even a dusting of loess on leaves will, in time, accumulate on the soil surface to form substantial deposits many metres thick. In general, the fastest rates of loess accretion were in the cold glacial periods, when the mean rate was c. 3–10 mm per century. The fastest rates were 15–25 mm per century, the slowest < 1 mm per century. Although most loess was deposited before the Holocene, loess is still being deposited today albeit at a much reduced scale, especially along the margins of South Island braided river beds. Wind gusts may reach high velocities (> 100 $km.hr^{-1}$). Such wind is erosive and contributes to transport of loess in the uplands, predominantly from slopes exposed to northerly and westerly winds, to the relatively sheltered south or east facing slopes. Thus soils on slopes with southerly or easterly aspects may have deeper, finer textured, horizons than soils with northerly and westerly aspects. Such loess deposits are always modified to some degree by pedogenic processes that continue while the loess is accumulating, slowly and with weak impact during cold periods but faster and more effectively during warm periods (i.e. by developmental upbuilding pedogenesis, Chap. 10).

d. Soil variation in the high country

Spatial soil variability in hill and mountain land is often high and falls into two broad types: soil patterns where soil age (time) is the dominant determinant of differences (temporally arrayed), and soils of similar age where topography is the dominant influence (spatially arrayed).

The soils of alluvial terraces, fans, or glacial moraines can serve as important time markers in mountainous regions. Where land surfaces can be dated, a relationship may be discerned between estimated ages and a number of soil and land development indicators, including soil properties, land surface morphology, rock weathering rinds, and lichens. Soil classification may provide a useful indicator of soil development but use of specific chemical, physical, or other morphological properties is likely to provide more precise time/soil development correlations. Although tephra deposits in much of the central and western North Island can provide marker beds of known age only one tephra, the Taupo-volcano-derived Kawakawa/Oruanui Tephra (aged c. 25,400 years) provides an important age datum in the South Island. This tephra has been identified as a thin layer or as a concentration of tiny glass shards (a cryptotephra) in peats, loess, and alluvial and glacial deposits in Canterbury, Otago, Westland, and the north-west Nelson region.

Correlations between time and soil development, or other indicators, can then be applied in cautious reconstruction of chronologies of soil patterns in a local terrain. Trevor Chinn and Ian Whitehouse used rock weathering rinds and lichen growth correlations with time to estimate the ages of screes in the Canterbury mountains, and thus the timing of scree activity. They found evidence that the prominent mountain screes in the landscape were not caused by human management but had formed long before grazing animals were introduced. Thus the screes are recognised as geologic features that pre-existed human settlement.

Temporally arrayed soil patterns are generally formed by erosion-deposition processes. Younger soil profiles result from relatively recent deposition of eroded soil, or recent truncation by erosion. Older soil profiles result from comparatively longer periods of stability when the soil has escaped losses by erosion or burial by sediment. The pattern of soil maturity and the evidence of erosion or sedimentation enable inferences to be made about the history, processes, and nature and severity of erosion, at a site. Erosion and deposition can be triggered by several causes including heavy and prolonged rainfall, earthquakes, and heightened susceptibility of soil materials to erosion due to advanced weathering or changes in protective vegetation cover. Instability may occur as a periodic phenomenon that, once triggered, can lead to extensive erosion within a catchment. An example is the periodic destruction of patches of forest by big storm events, leading to a mound-pit microtopography and soil disruption/mixing as a result of tree overturn.

Spatially arrayed soils have similarly mature profile development, but systematic variations in the soil profile are related to topography. Phil Tonkin developed a model of the relative stage of soil development in relation to aspect and altitude in the South Island high country. At lower altitudes, in the drier (mean annual rainfall 600–700 mm) environment, there were no differences between soils with northerly and southerly aspects and vegetation was equally limited by dryness.

At intermediate altitudes where the mean annual rainfall is somewhat higher (700–1200 mm) the relative stage of soil development was more advanced on the wetter, more leached, south aspect compared to soils with a north aspect. As a result of the higher leaching, the wetter southerly aspect soils had lower soil fertility but, because of higher water availability, better vegetation growth than soils with northerly aspects.

In contrast, at the highest altitudes with a mean annual rainfall of 1200–1400 mm, sufficient rain falls to promote leaching on both aspects. However, the northerly aspect was favoured by higher temperature with consequently greater soil development and lower fertility.

The degree of leaching can be estimated from the soil pH, total phosphorus, and exchangeable (sulphuric-acid-soluble) phosphorus, whereby more-leached soils have lower pHs and lower total and exchangeable P.

4.3.3 The Mudstone Hill Country

The mudstones, interspersed with some sandstones and limestones, are mainly weak and poorly indurated, thus weather readily to form soil. In calcareous mudstone the main cementing agent is calcium carbonate which is easily dissolved and leached in high rainfall environments. Sandstone and limestone layers are generally more indurated, and resistant to erosion, than the mudstones and so tend to stand out as steeper cliff faces within hill slopes. The softer mudstones are readily eroded and soils are in a continual state of development as drainage lines incise, steepening side slopes, thus causing continued slope instability. Such areas have potential for severe landslide erosion and deposition in storm or earthquake events.

The less steep areas, such as rounded ridge crests and rolling land, tend to be stable, thereby allowing Brown Soils to develop. However, in more strongly dissected, actively eroding, land, such as much of the land between Whanganui and Raetihi, and between Gisborne and Wairoa, many areas have narrow ridge summits with steep valley sides and river flood plains on which mainly Recent or Raw Soils will occur.

In the North Island (some distance from the active volcanoes) where shallow ($< \sim 35$ cm) blanketing of tephra occurs, the soil pattern often comprises Allophanic Brown Soils on stable ridge tops, Orthic Brown Soils on more stable side (back) slopes that lack tephra accumulation, and Raw or Recent Soils on eroded slopes as well as on valley floors where eroded material has been redeposited. A slope angle of about 26°, along with vegetation cover, seems to be critical in determining if tephra deposits remain in place or not, generally being lost (eroded) from slopes $> \sim 26°$. Gley Soils form in hollows where drainage is poor and on valley floors where the water table is high. Where tephra deposits are absent the stable, flatter, areas generally support Orthic Brown Soils.

A feature of the soft rock hill country is a general lack of catenary development (a regular, predictable, pattern of soil development down a slope). Because of the relatively frequent erosion events, which lead to soil material moving down-hill, leaving scars from which material was removed and areas of redeposited soil downslope, the soil pattern relates more strongly to the history of erosion events than to position on the slope.

4.3.4 Otago Schist Landscapes

In Otago, Brown Soils often occur at higher altitudes where evapotranspiration is lower, and rainfall is higher, than in the lowland semiarid basins. The hard schist rock tends to form landscapes with broad rounded crests and convex to concave side (back) slopes. The drainage patterns and slopes are often related to the strike and dip of the underlying schist rock with steeper slopes tending to form where the schistosity is steeply dipping (as can be observed in Skippers Canyon and the Cromwell Gorge).

On stable moister slopes, where leaching is sufficient, Brown Soils form alongside Pallic Soils on drier slopes and at lower altitudes, and with Gley Soils in poorly drained shady hollows. The interaction between slope and aspect provides an environment of high soil variability, as well as high biodiversity. Northburn Station on the west side of the Dunstan Range provides an example of the landscape relationship between the Semiarid, Pallic, and Brown Soils (Fig. 4.4) resulting from more effective precipitation at higher altitudes. Also evident is a strong topographic effect with sparse vegetation, enhanced erosion, and shallow soils, on the drier, windward, north and north-westerly facing, slopes, and deeper, moister, more strongly vegetated soils on the cooler, leeward, south-facing, slopes.

Particularly at higher altitude on the Central Otago ranges, such as the Old Man Range and the Pisa Range, there is some slope movement. Periglacial activity (still continuing slowly today) is evident with solifluction lobes and ploughing boulders visible on some higher slopes together with rock glaciers and permafrost at some alpine sites. Large-scale earth-slide features are also visible in the landscape. Where the boundary between the debris mantle and the underlying bedrock is 'rough' then it is likely that the materials are locked together and surfaces have been stable for long periods with subsequent stable soil development. However, it is likely that many of the landslides were triggered by earthquakes (potentially also associated with intense rainfall events) and could again be remobilised by such events. A series of large, long-term, landslides on the margins of the Cromwell Gorge became the subject of much study and work to stabilise them while the Clyde dam was being constructed in the 1980s. The fear was that as the dam filled, the rising groundwaters could mobilise some of the large landslides with potential for huge volumes of material to fall into the

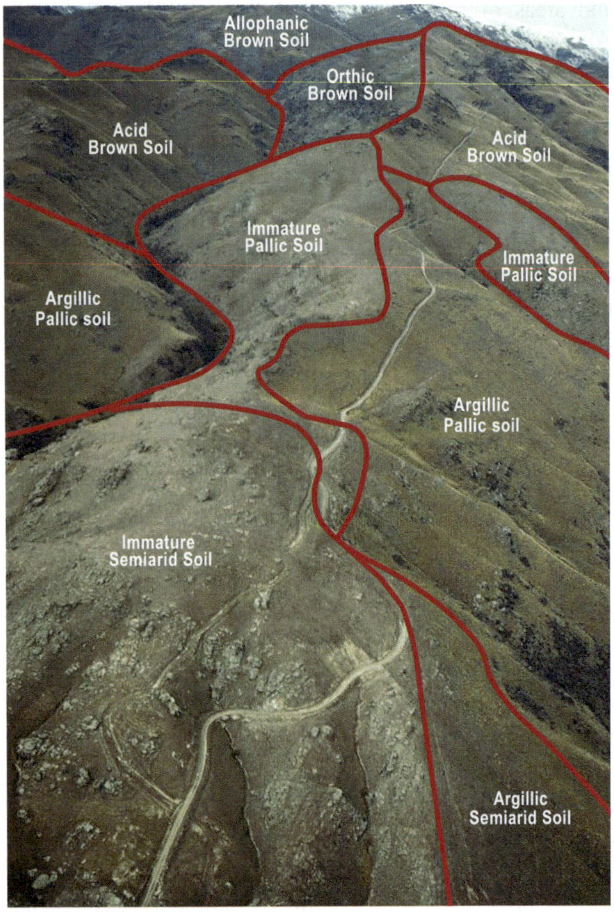

Fig. 4.4 View towards the east on Northburn Station on the west side of the Dunstan Range in Central Otago showing the soil-landscape relationship. The drier north-west facing slopes have much shallower soils than the more strongly vegetated south-facing slopes. Widespread rock outcrops occur, especially on the north-west facing slopes, and so although the diagram is labelled for the dominant soil in each area, Raw and Recent soils are also evident within each map unit

dam leading to catastrophic flooding downstream and potential for tsunami-like waves in Lake Dunstan.

4.3.5 Contrasting Terrains in Canterbury and Otago

There is an interesting contrast between the greywacke terrains of Canterbury and the schist terrains of Otago. A notable boundary lies between the greywacke terrain on the northern side of the Waitaki River, and southern schistose terrain. The shattered greywacke is more readily eroded (forming small, cubic blocks) than schist (larger tabular blocks) and, as a result, the greywacke landscape is dominated by narrow ridge lines, with planar steep slopes that terminate in V-shaped valleys. On the steep slopes a larger proportion of rainfall runs off, infiltration is lower, and hence streams may be dry for long periods between rainfall events. In the greywacke terrain the soils tend to be fairly uniform with relatively lower habitat- and biodiversity. In contrast, the schist terrain tends to have rounded crests, concave slopes, and meandering streams, with stream flow maintained for longer during dry spells. The schist terrain supports a wider range of soil materials and consequent higher habitat- and biodiversity.

4.3.6 Southland Plains, Terraces, Downlands, and Rolling Hills

Brown Soils are the predominant soil order in Southland, occurring on the productive lowland plains, terraces, rolling downlands, and the lower hills, as well as on some steeplands. A diverse range of Brown Soils has been recognised, particularly in the soil surveys of John Bruce and Peter McIntosh. The soil diversity reflects the range of topography and soil parent materials that include loess, greywacke, tuffaceous (containing volcanic material) greywacke, sandstone, siltstone, and schist, along with glacial moraine and outwash alluvium on intermediate terraces. Allophanic Brown Soils form on tuffaceous greywackes and may be associated with Melanic Soils where both limestone and base-rich tuffaceous greywacke overlie the greywacke basement rocks.

Recent Soils from alluvium form on river flood plains, sometimes with adjacent Gley or Organic Soils in poorly drained hollows. The Brown Soils merge into Pallic soils in warmer and drier areas. In the drier areas where Pallic Soils predominate, Brown Soils are found on sites with shallower stony soils where leaching is more effective.

4.4 Key Soil Properties

4.4.1 Soil Composition

Most of the Brown Soil groups are derived from quartzo-feldspathic sedimentary rocks (schist, greywacke, mudstones and sandstones) or acid igneous rocks (e.g. rhyolites and granites), or derivatives including alluvium, colluvium, and loess. However, Mafic Brown Soils are recognised on soils derived from weakly weathered intermediate or basic igneous rocks (e.g. phonolite and basalt) or sediments derived from such rocks. Brown Soil profiles tend to be mineralogically uniform with depth.

Many Brown Soils are formed on calcareous mudstone parent materials yet the calcium carbonate content in most Brown Soils is generally low as a result of the leaching that has occurred during soil formation. However, in some places calcium carbonates do form a calcareous horizon and are recognised as a Calcareous subgroup of the Orthic Brown Soils.

Brown Soils may contain a wide range of clay minerals including mica/illite, and vermiculite.

Ferrihydrite is the main iron oxide mineral in Brown Soils and has been described by Roger Parfitt and Cyril Childs. Ferrihydrite has a net positive charge in acid situations, and so has anion exchange properties, and provides the moderate to high phosphate retention of many Brown Soils. Ferrihydrite is related to allophane in being a nanocrystalline clay (with structure in the nanoscale range 1–100 nm). Ferrihydrite is one of the nanominerals comprising the diagnostic 'allophanic soil material' in the *New Zealand Soil Classification*.

Volcanic materials, which are the most common source of allophanic soil materials, are rare among greywacke rocks. However, in greywacke rocks, allophane or associated weathering products form by the rapid dissolution (by hydrolysis) of plagioclase or other feldspars, especially if they have been pre-weathered within the rock. Weathering to form allophanic materials is aided by strong leaching of silicon (desilication) under moderately high rainfall and good drainage to promote the synthesis of aluminium-rich allophane (i.e. aluminium content > silicon content) and also of ferrihydrite. The complexes of iron and aluminium hydrous oxides, and organic matter, help to stabilise the soil structure and thus generate the high soil porosity.

The alteration status of gravel or hard rock substrates is usually fresh to moderately weathered, but highly weathered rock (saprolite) occurs in some places. Some Brown Soils in the Wellington region, and their extension into Marlborough, have a loess component overlying deeply weathered greywacke with red colouration and significant kaolinite clay (Chap. 16, Ultic Soils). The soil survey of the Taita catchment, on the margin of the Hutt Valley near Wellington, provides an example of Brown Soils on deeply weathered greywacke substrate and on subordinate colluvium and loess.

4.4.2 Physical Properties

Brown Soils are mostly weakly to moderately weathered and clay contents range from 5 to 40% of the soil and are typically around 20% (Fig. 4.5, Table 4.1). Soil dry bulk densities are moderate to high, and water-holding capacity is high in topsoils and moderate to high at depth except on sandy or stony profiles.

No poorly drained, or very poorly drained, soils are included in the Brown Soils. Imperfectly drained soils are included in mottled subgroups. Macroporosity is generally

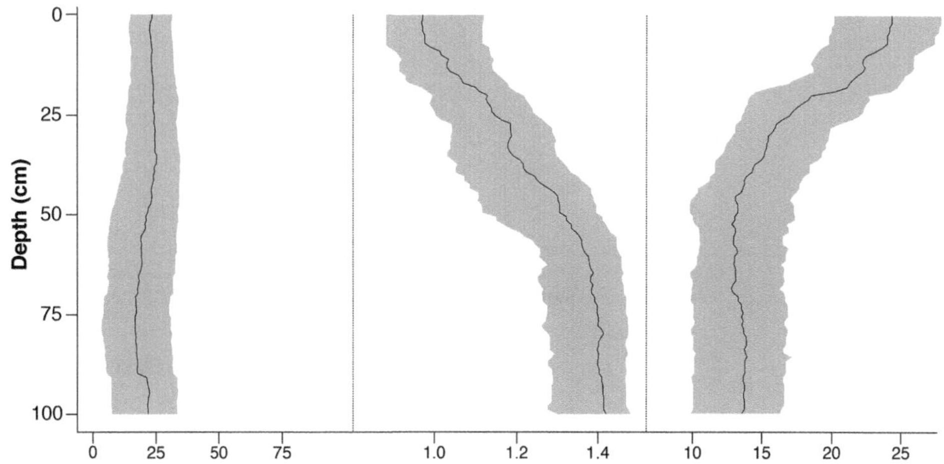

Fig. 4.5 Median and upper and lower quartiles of clay content (percentage in <2 mm fraction), soil dry bulk density (t m^{-3}), and soil total available water-holding capacity for Brown Soils in the New Zealand National Soil Data Repository

Table 4.1 Soil Physical properties of a Typic Orthic Brown Soil (Waikiwi, SB09215)

Horizon	Depth cm	Sand %	Silt %	Clay %	Dry bulk density t m^{-3}	RAW[a]	TAW[a] %
Ap	0–20	7	68	24	1.1	17.4	26.4
A/Bw	20–26						
Bw	26–39	6	81	12			
BC	39–55	8	73	18	1.5	7.4	13.1
C1	55–92	4	74	21	1.4	7.9	14.0
C2	92–120	4	67	28			

[a]RAW = readily available water holding capacity, TAW = total available water holding capacity

moderate (10–14%) except in subsurface horizons of Firm Brown Soils. Brown Soils have good infiltration capacity provided they are not damaged by stock trampling or vehicle compaction, which may occur, particularly when soils are wet. Profile permeability ranges widely depending mainly on the nature of the parent material. For example, soils on stony colluvium on hill slopes are frequently moderately or rapidly permeable, but soils on silty loess of firm strength, or soils with pans, are likely to be slowly permeable. The Brown Soils do not include soils that have prolonged perched water tables, but mottled subgroups (with mottled profile form) are included.

Compared with Pallic Soils, Brown Soils generally have a favourable topsoil structure and aggregates are not readily dispersed, as indicated by their oxide-dominated secondary mineralogy and typically moderate phosphate retention. Bypass flow is generally medium in Brown Soils. Root exploration of the soil is generally favourable but may be somewhat limited in soils with firm subsoil horizons, and prevented by cemented pans or shallow coherent rock.

4.4.3 Chemical Properties

The chemical properties of Brown Soils, especially pH, base saturation, nutrient cations, acidity, KCl-extractable aluminium, and anion exchange capacity (Fig. 4.6, Table 4.2), express the leaching environment in which Brown Soils are formed. Without fertiliser they have low to moderate base saturation (usually < 50%). Aluminium associated with acidity can be high, especially in the Acid Brown Soils and the phosphate retention (or anion exchange capacity) is usually moderate to high.

Topsoil CEC levels are medium or high, and subsoils medium. CEC varies with depth in relation to change in soil organic matter levels. As for many soils, base saturation levels of topsoils vary widely—even as much as 100% if they have had high fertiliser inputs, particularly elevation of calcium by liming. Subsoil base saturations range widely from very low to medium with a mean of 46% indicating that a high proportion of the CEC is occupied by acid cations (Al^{3+} and H^+).

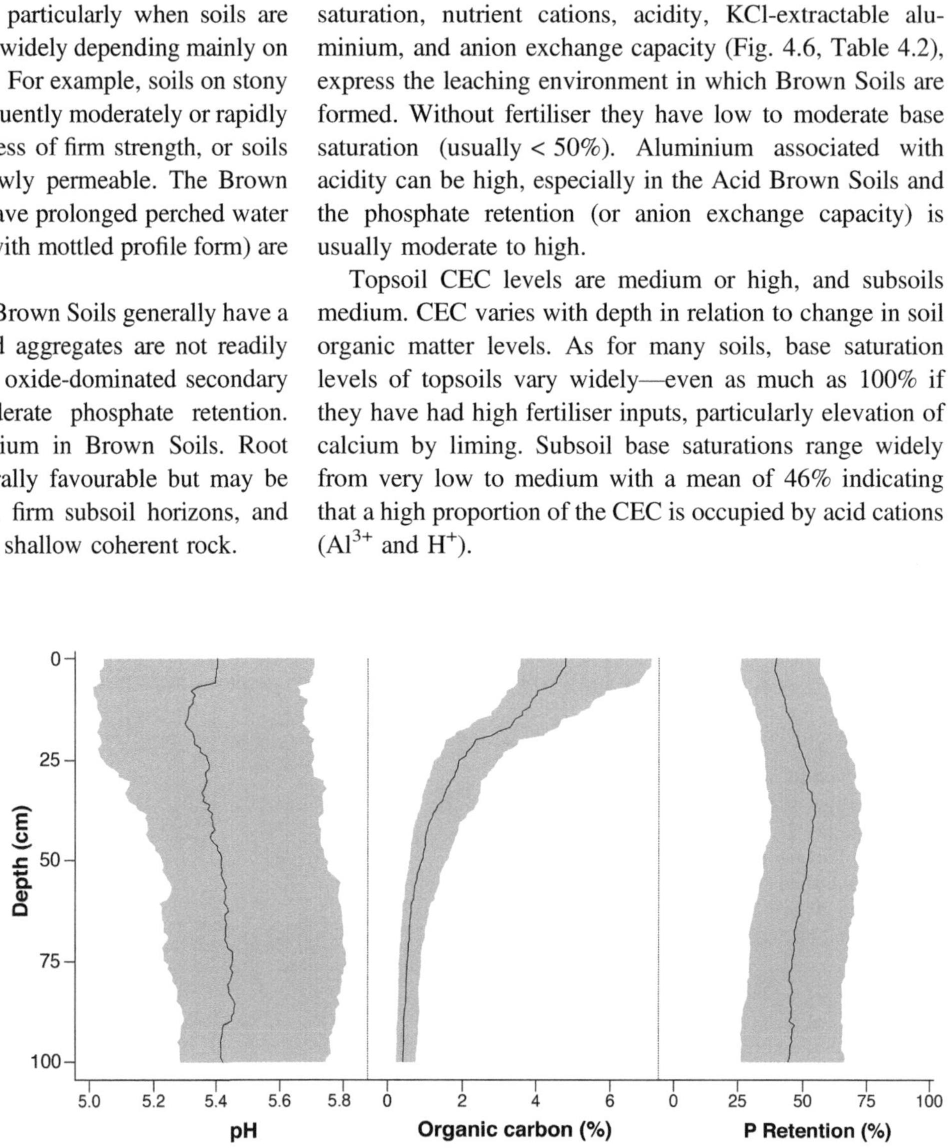

Fig. 4.6 Median and upper and lower quartiles of soil pH, organic carbon, and P retention for Brown Soils in the New Zealand Soil Data Repository

Table 4.2 Soil chemical properties of a Typic Orthic Brown Soil (Waikiwi, SB09215)

Horizon	Depth cm	pH (in H_2O)	Carbon %	Nitrogen %	CEC[a] $cmol^{(+)}kg^{-1}$	Sum bases $cmol^{(+)}kg^{-1}$	P retention %
Ap	0–20	5.7	4.1	0.32	15.4	7.68	59
A/Bw	20–26	6	2.7	0.21	12.1	5.26	65
Bw	26–39	6.1	1.6	0.12	9.1	2.64	76
BC	39–55	5.8	0.8	0.06	7.1	1.06	72
C1	55–92	5.5	0.4	0.03	7.5	1.59	54
C2	92–120	5.3	0.3	0.02	8.8	2.56	46

[a]CEC = cation exchange capacity

The pH of topsoil is, like base saturation, strongly affected by fertiliser inputs. Subsoil pH values are generally moderately acid and in many places are extremely acid, especially in Acid Brown Soils. Where pH values are less than a threshold of about 4.8, soils are likely to have high levels of potassium-chloride-extractable labile (mobile) aluminium (KCl–Al). KCl–Al values are usually more than 1.5 $cmol^{(+)}kg^{-1}$ except where clay contents are relatively low.

Brown Soil total P values are frequently very high in topsoils, again influenced by any fertiliser inputs. In soils with low fertiliser inputs, Olsen P (a measure of plant - available P) is likely to be low. Subsoil total P values range from very low to very high, with only a few sites with low total P levels indicating high soil weathering. P retention is a property of the soil mineralogy and less affected by fertility management. Both topsoil and subsoil P retentions are generally medium. However, Allophanic Brown Soils have very high P retentions.

4.4.4 Biological Properties

Brown Soils generally contain active populations of soil organisms. Except in soils limited by coldness or acidity, worm casts are common in topsoils, and carbon to nitrogen ratios are moderate to low. The roots of plants penetrate deeply. However, the low pH of some subsoils means aluminium becomes soluble and thus there is a risk of aluminium toxicity in many plant species. The rooting depth of some pasture plants may be limited by aluminium toxicity in Brown Soils with low subsoil pH.

4.5 Distinguishing Between Brown Soils and Related Soil Orders

In common with the Raw and Recent Soils, the Brown Soils are keyed out near the end of the *New Zealand Soil Classification* key to orders. This means that the soils are relatively simple and the key for identification filters out soils with specific notable features, leaving Brown Soils in the remainder. Thus the typifying Brown Soil is distinctive because of the specific features that it lacks, for example, the absence of gley profile features, thick allophanic soil material, thick organic soil material, thick pumice, podzolic features, and so on.

By definition, Brown Soils have a distinct topsoil, and a weathered B-horizon or cutanic (clay rich) horizon ≥ 10 cm in thickness, specified brown colours, and phosphate retention of 30% or more. They do not have perched-gley features but may be mottled.

Because they are so extensive, wherever Brown Soils are recognised they can merge laterally into soils of many other orders. Brown Soils are more highly leached and weathered than Semiarid Soils and Pallic Soils, as expressed in lower natural fertility, lower soil pH, and brownish colours.

Boundaries between Pallic Soils and Brown Soils are common throughout New Zealand in soils of similar high water storage capacity. The brunification (thus Brown Soil) boundary is simply controlled by effective rainfall with Brown Soils consistently recognised under higher rainfall.

Brown Soils are ordinarily formed in areas of humid climate where droughts are rare. It is perhaps surprising then to find them formed in sites that commonly experience drought, as on the sandy gravels of the Canterbury Plains, in association with Pallic Soils. Detailed soil surveys by Ted Cox and later Trevor Webb have demonstrated the nature of the Pallic Soil—Brown Soil boundary in the Canterbury Plains. Areas with varying thicknesses of silty loess overlying sandy gravel demonstrated that Brown Soils were confined to low water storage capacity sites where loess was absent or thinner than about 45 cm, and Pallic Soils (Immature or Argillic soil groups) were confined to high water storage capacity sites with loess thicker than $\sim$45 cm over gravel. The slower permeability of the loess materials probably inhibits leaching, and thus supports the development of Pallic, rather than Brown, Soils. Nevertheless, the mantle of loess does make both soils more versatile than if formed directly on coarse alluvium.

4.6 Correlation with Other Classification Systems

The Brown Soil groups mainly correlate with the Inceptisols of *Soil Taxonomy* and Cambisols in the *World Reference Base* and many were formerly classified as Yellow-brown earths in New Zealand (Table 4.3).

The *World Reference Base* correlations are problematic in that Brown Soils can fit both Cambisols and Umbrisols. The Umbrisol concept is of soils with weakly developed profiles with well-structured dark-coloured acidic topsoils. The emphasis is on the topsoil, and the typifying environment is cool humid, mountain lands. They are the acidic variant of Soil Taxonomy's Mollisols. The Cambisols have moderately developed profiles and are recognised globally in young landscapes in both wet humid and dry areas. The cambic horizon is the equivalent concept to the New Zealand-defined weathered B-horizon (Bw). The emphasis is on the subsoil. For New Zealand it is recommended that

Table 4.3 Correlation[a] between Brown Soils and equivalent classes of *Soil Taxonomy*, *World Reference Base*, and the earlier New Zealand genetic soil classification

New Zealand Soil Classification	*Soil Taxonomy*	*World Reference Base*	NZ genetic soil classification
Allophanic Brown Soils	Andic Dystrudepts	Andic Cambisols	High country yellow-brown earths
Sandy Brown Soils	Sandy families of Typic Dystrustepts or Typic Haplustepts	Dystric Arenosols	Yellow-brown sands
Oxidic Brown Soils	Oxic Dystrudepts	Ferralic Cambisols	Yellow-brown earths
Mafic Brown Soils	Dystrudepts	Haplic Cambisols	Brown granular loams
Acid Brown Soils	Dystrudepts	Cambisols (Dystric)	Podzolised yellow-brown earths
Firm Brown Soils	Typic Dystrudepts	Haplic Cambisols	Yellow-brown earths
Orthic Brown Soils	Typic Dystrudepts	Haplic Cambisols	Yellow-brown earths

[a]The correlations given here are a guide only and for accurate classifications, the relevant soil classification documents should be consulted. The two major international soil classification systems are *Soil Taxonomy*, which was developed in the USA, and *World Reference Base*, which was developed primarily in Europe. The NZ genetic soil classification was used in NZ prior to 1992

Cambisols is the better correlative of most of the Brown Soils, until such time that the *World Reference Base* soil groups are specifically studied in New Zealand.

4.7 Use and Management of Brown Soils

4.7.1 Introduction

Because of the wide range within the Brown Soils, it is useful to first consider the management and use of each soil group.

Allophanic Brown Soils occur most extensively in the South Island mountains where the land is used for extensive grazing, or managed for catchment protection. In northern Southland hill country, Allophanic Brown Soils are formed in tuffaceous greywacke and their fertility and structural character support highly productive uses. Allophanic Brown Soils on the fringes of the tephra-derived Allophanic Soils and Pumice Soils in the North Island on rolling to hilly land also have high productive capability, aided in part by the allophane-dominated soil physical properties. However, the allophane will provide the soils with higher anion retention, therefore it is better to apply smaller amounts of P fertiliser more often to ensure its availability to plants.

Sandy Brown Soils range from soils that are weakly weathered to soils with well-developed weathered B-horizons. They are predominantly in coastal sand dunes and associated sand plains. Challenges include rapid permeability and limited profile available water capacity. Risk of soil-water deficits may necessitate irrigation. Groundwater may be accessible for deep rooting plants in sand plains. Wind erosion is a risk and maintaining vegetation cover has to be a priority. Available reserve nutrients are low. Dune lands are commonly under some form of land conservation protection. Pine forests are established in some areas, particularly in the Manawatu sand country on the west coast near Foxton and Himatangi.

Oxidic Brown Soils are transition soils between Brown and Oxidic Soils in which there are significant weatherable minerals remaining, and soils are expected to have higher inherent (chemical) fertility than Oxidic Soils. The Oxidic Brown Soils are frequently in hill or steepland sites where there is sufficient disturbance to partly counteract the full expression of Oxidic Soils. While more fertile than Oxidic Soils, the Oxidic Brown Soils are more strongly weathered than other Brown Soil groups. Thus Oxidic Brown Soils will likely have higher clay contents and lower fertility, and so need higher fertiliser inputs and more careful management (e.g. to avoid compaction from mob-stocking or ploughing at high moisture contents) than other Brown Soils. On the other hand, the oxidic character can provide a stronger soil structure.

Mafic Brown Soils have favourable topsoil structure and soil aggregates are stable. The dark volcanic (basalt or andesite) parent rocks or tephra impart good inherent fertility. Mafic Brown Soils, such as those on Mt Cargill near Dunedin, often form on hilly or steep slopes. However, where slopes are suitable, they have high capability for productive use.

Acid Brown Soils have very strongly acid subsoils and high risk of aluminium toxicity to many plants. The acidity and associated high rainfall limits their capability for productive use. Inputs of lime can lead to greatly increased productivity as the acidity is neutralised.

Acidic Brown Soils tend to occur at higher altitudes on steeper land which may limit their preferred use to sheep grazing or forestry.

Firm Brown Soils have a subsoil layer that is frequently slowly permeable and may be an impediment for some deep rooting plants. However, in the Southland lowlands, Firm Brown Soils are a productive mainstay for a range of pastoral or cropping uses. Subsurface drainage is often installed on Firm Brown Soils, in flatter areas, providing long-term improvement in soil productivity and improved resilience to pugging damage during wet periods.

Orthic Brown Soils are widely used for pastoral agriculture. They have few productive limitations, other than fertility requirements, and lack the firm subsoil layers of the Firm Brown Soils.

4.7.2 Soil Fertility

Pastoral farming on Brown Soils across the extensive hill country is one of New Zealand's major land uses. Pastoral farming was made possible with the advent of aerial topdressing, from around 1946, which provided a key to the productivity of hill country farms, both for initial development and continuing maintenance. The advent of aerial topdressing was a major boost to production in the hill country as nutrient deficiencies could be readily remedied, low pH addressed, and the nutrients removed in products could be efficiently replaced to ensure a sustainable farming system could be developed. The improved pasture cover also helps reduce erosion.

Many Brown Soils, and particularly the Acid Brown Soils, respond to addition of lime. Phosphorus fertilisers are regularly applied to build or maintain productivity and to support the clover-based pastures where clover provides a source of nitrogen. As with all fertiliser use, soil sampling and analysis should be undertaken to determine the soil nutrient status and needs of the site prior to any fertiliser addition.

Potential for nutrient, along with sediment, runoff from hill country is gaining wide attention and a number of mechanisms may be considered to maximise fertiliser use efficiency and minimise loss (and wastage of expensive fertiliser materials) to runoff. Where practical, stream riparian fencing with a good grass cover to filter sediment and nutrients from runoff can be effective. Use of improved GPS in aircraft can ensure that fertiliser is placed on productive areas, and stream gullies and unproductive areas are avoided. Ideally, fertilisers should be applied ahead of gentle rainfall to maximise fertiliser movement into the soil, but not ahead of high intensity storms where runoff is likely.

4.7.3 Hill Country Erosion

One of the major issues of the hill and steepland is hillslope erosion. Some of New Zealand's worst hillslope erosion occurs on the mudstone hill country in higher rainfall areas where Brown Soils often predominate, particularly on the east coast of the North Island from the Wairarapa through to East Cape, and in the Whanganui and Rangitikei River catchments. Periodic high intensity rainstorms have caused a long history of severe erosion events. A series of storms in the 1930s first brought erosion of the North Island east coast hill country to national attention and Cyclone Bola, in March 1988, caused catastrophic damage across the Gisborne-East Cape area. About 20% of the land surface was affected. Similar effects were seen in the southern North Island in February 2004 when a severe storm resulted in widespread erosion. The hill country is also prone to major earthquakes that may set off large-scale slumping and landslides. Workers from Landcare Research and AgResearch have shown that in the south Wairarapa pasture production on eroded sites recovered reasonably rapidly in the first 20–30 years after a landslide event but was unlikely to recover to more than about 80% of the production on non-eroded sites.

To some degree the erosion of the soft rock hill country is an ongoing geologic process (Fig. 4.7), but it has been shown that erosion, in major rainfall events, is greater under grass than forest vegetation. Some of the steeper, most erosion prone, areas are probably best left to revert to native forest vegetation. Forestry is an option across virtually all the Brown Soils. Where pastoral agriculture is the preferred land use, much can be done to prevent or minimise erosion effects. Retirement, or reforestation, of steep slopes and stream margins is a good option that also contributes to biodiversity. Use of simple debris dams within gullies to prevent ongoing stream down-cutting and undermining of foot slopes can help. Paired willow planting in association with small debris dams has provided effective medium-term gully control. Flood detention dams and establishment of wetland areas at the foot of slopes to intercept and filter runoff before it reaches a watercourse can greatly enhance the quality of runoff water. Slope stabilisation using scattered tree planting, often using poplars, has been effective in stabilising many slopes. Suitable tree species may also be trimmed to provide fodder for animals during summer droughts. Improved soil fertility and strong pasture growth also help prevent surface soil erosion; however, careful stock management is needed as the benefits may be counteracted

Fig. 4.7 Major landslide on Tertiary sedimentary rocks and weathered materials on State Highway 4 alongside the Mangawhero River in October 2019. The whole hillside is gradually moving down, aided by the stream constantly removing material from the toe of the slope. Prolonged heavy rainfall, and/or earthquakes, periodically reactivate such mass movements. Photo: Mark Brimblecombe

by stock trampling on hill-slopes in winter or by close grazing during periods of feed shortage.

The hill country landscape tends to present a mosaic of landform units. Thus a mosaic approach to land use and management, with pasture, productive forest, and protection forest, matched to land capability is likely to be the best way to maintain sustainable food and fibre production, economically sustainable rural communities, and environmental protection.

Further Reading

Almond PC, Tonkin PJ (1999) Pedogenesis by upbuilding in an extreme leaching and weathering, environment, and slow loess accretion, south Westland, New Zealand. Geoderma 92:1–36

Almond PC, Shanun FL, Rieser U et al (2007) An OSL, radiocarbon and tephra isochron-based chronology for Birdlings Flat loess at Ahuriri Quarry, Banks Peninsula, Canterbury, New Zealand. Quat Geochronol 2:4–8

Barringer JRF, Hewitt AE, Lynn IH, et al (2007) National mapping of landform elements for New Zealand in support of S-Map, a New Zealand soils database. Benmore Range & Motueka Catchment, Landform element mapping, Soil series mapping. https://lris.scinfo.org.nz/document/9367-landform-elements-barringer-et-al-2007/. Accessed 4 Mar 2020

Basher LR (2013) Erosion processes and their control in New Zealand. In: Dymond JR (ed) Ecosystem services in New Zealand—conditions and trends. Manaaki Whenua Press, Lincoln, pp p363–374

Brash DW, Beecroft FG (1987) Soil resources of Central Otago. Proceedings of New Zealand Grassland Association 48:23–30

Bruce JG (1984) Soils of the Gore-Waikaka District, South Island, New Zealand. NZ Soil Survey Report 74. New Zealand Soil Bureau, DSIR, Lower Hutt. 86 p

Bruce JG (1984) Soil resources of the Southland Region. NZ Soil Survey Report 76. New Zealand Soil Bureau, DSIR, Lower Hutt. 27 p

Campbell IB (ed) (1985) Proceedings of the soil dynamics and land use seminar, Blenheim, May 1985. New Zealand Society of Soil Science, and New Zealand Soil Conservators Association

Childs CW (1992) Ferrihydrite: a review of structure, properties and occurrence in relation to soils. Zeitschrift fuer Pflanzenaehrung Bodenkunde 154:441–8

Churchman GJ, Lowe DJ (2012) Alteration, formation, and occurrence of minerals in soils. In: Huang PM et al (eds) Handbook of soil sciences, 2nd edn. Vol. 1: Properties and processes. CRC Press, Boca Raton, pp 20.1–20.72

Crozier MJ (2005) Multiple-occurrence regional landslide events in New Zealand: hazard management issues. Landslides 2:247–256

Davies TRH (2017) Mountain process geomorphology: conceptual progress in the Southern Alps. In: Shulmeister J (ed) Landscape and Quaternary environmental change in New Zealand. Springer/Atlantis Press, pp 205–233

Dymond JR, Derose RC, Harmsworth GR (1995) Automated mapping of land components from digital elevation data. Earth Surf Proc Land 20(2):131–137

Gibbs HS, Raeside JD, Dixson JK, et al (1945) Soil erosion in the High Country of the South Island. DSIR Bulletin No. 92. 72 p

Gomez B, Rosser B (2018) Slip slidin' away: a post-glacial environmental history of the Waipaoa River Basin. Geomorphology 307:65–76. https://doi.org/10.1016/j.geomorph.2017.08.035

Grab SW, Dickinson KJM, Mark AF et al (2008) Ploughing boulders on the Rock and Pillar Range, south-central New Zealand: their geomorpology and alpine plant associations. J Royal Soc N Z 38:51–70

Heaphy MJ, Lowe DJ, Palmer DJ et al (2014) Assessing drivers of plantation forest productivity on eroded and non-eroded soils in hilly land, eastern North Island, New Zealand. N Z J Forest Sci 44:24–33

Hewitt AE (1982) Soils of Waipori farm settlement. NZ Soil Survey Report 65. DSIR. 45 p. http://doi.org/10.7931/DL1-SSR-65

Hewitt AE (2012) New Zealand Soil Classification, 3rd edn. Landcare Research Science Series No. 1. Manaaki Whenua Press, Lincoln, New Zealand, 136 p

Ives DW, Cutler EJB (1972) A toposequence of steepland soils in the drier high–country yellow–brown earth (dry–hygrous eldefulvic) region, Canterbury, New Zealand. N Z J Soil Sci 15:385–407

Jane GT (1986) Wind damage as an ecological process in mountain beech forests of Canterbury, New Zealand. N Z J Ecol 9:25–39

Lambert MG, Trustrum NA, Costall DA et al (1993) Revegetation of erosion scars in Wairarapa hill country. Proceedings of the New Zealand Grasslands Association 55:177–181

Leathwick J, Wilson G, Rutledge D et al (2003) Land Environments of New Zealand. David Bateman Ltd, Auckland

Leckie HD, Almond PC (2015) Evidence of prehistoric wind erosion of the Mackenzie Basin, South Island, New Zealand: an assessment based on ^{137}Cs and Kawakawa-Oruanui tephra. Soil Res 53:56–66

Lee JM, Bland KJ, Townsend DB, et al (2011) Geology of the Hawke's Bay area. Geological Map 8. Scale 1: 250,000. IGNS, Lower Hutt. 93 pp + map

Lowe DJ, Tonkin PJ, Palmer AS, et al (2015) Dusty horizons [Loess]. In: Graham IJ (ed) A continent on the move: New Zealand geoscience revealed, 2nd edn. Geoscience Society of New Zealand with GNS Science, Wellington, GSNZ Misc Publ, vol 141, pp 286–289

Lynn I, Basher L (1994) Principles underlying land systems in resource assessment of hill and mountain lands in New Zealand. In: Webb TH (ed) Soil-landscape modelling in New Zealand. Landcare Research Science Series 5:38–51

Lynn IH, Tonkin PJ (1985) Use of soil-landscapes models to enhance land resource assessment. In: Campbell IB (ed) Proceedings of the soil dynamics and land use seminar, Blenheim, May 1985. New Zealand Society of Soil Science, and New Zealand Soil Conservators Association, pp 1–18

Mackay A, Rhodes AP, Power I, Wedderburn ME (2011) Has the eco-efficiency of sheep and beef farms changed in the last 20 years? Proc N Zealan Grassland Association 73:119–124

McCaskill LW (1973) Hold this land. A history of soil conservation in New Zealand. AH & AW Reed, Wellington, 274p

McIntosh PD (1994) Guide to the soils of the Kaihiku-Hokonui land region. Landcare Research Science Series No 4. Manaaki Whenua Press. http://doi.org/10.7931/DL1-LRSS-4

McIntosh PD, Hunter GG (1994) Portability of a soil-landscape key in hilly and steep land. In: Webb TH (ed) Soil landscape modelling in NZ: proceedings of a workshop at Aokoutere, New Zealand, Feb. 1993, Manaaki Whenua Press

McIntosh PD, Lee WG, Banks T (1983) Soil development and vegetation trends along a rainfall gradient on the east Otago uplands. N Z J Soil Sci 26:379–401

McLeod M, Rijkse W (1994) Development of a soil–landscape model for a close jointed mudstone land system. In: Webb TH (ed) Soil landscape modelling in NZ: proceedings of a workshop at Aokoutere, New Zealand, Feb. 1993, Manaaki Whenua Press

Mew G, Hunt JL, Froggatt PC et al (1986) An occurrence of Kawakawa Tephra from the Grey River valley, South Island, New Zealand. N Z J Geol Geophys 29:315–322

Page MJ, Trustrum NA, Dymond JR (1994) Sediment budget to assess the geomorphic effect of a cyclonic storm, New Zealand. Geomorphology 9(3):169–188

Page MJ, Reid LM, Lynn IH (1999) Sediment production from Cyclone Bola landslides, Waipaoa catchment. J Hydrol N Z 38(2):289–308

Palmer RWP (1974) Soil variation on hillslopes at Whatawhata. MSc thesis, University of Waikato, Hamilton

Parfitt RL, Van der Gaast SJ, Childs CW (1992) A structural model for natural siliceous ferrihydrite. Clays Clay Miner 40(6):675–681

Parfitt RL (1990) Allophane in New Zealand – a review. Aust J Soil Sci 28:343–360

Parfitt RL, Childs CW (1988) Estimation of forms of Fe and Al: a review, and analysis of contrasting soils by dissolution and Moessbauer methods. Aust J Soil Res 26:289–96

Parfitt RL, Kimble JM (1989) Conditions for formation of allophane in soils. Soil Sci Soc Am J 53:971–977

Parfitt RL, Webb TW (1984) Allophane in some South Island yellow-brown shallow and stony soils and high country yellow-brown earths. N Z J Sci 27:37–40

Raeside JD, Vucetich CG, Cox JE, et al (1968) Soils of South Island. In: soils of New Zealand Part 1. N Z Soil Bureau Bull 26(1):67–88

Robertson SM, Mew G (1982) The presence of volcanic glass in soils on the West Coast, South Island, New Zealand. N Z J Geol Geophys 25:503–507

Roche MM (1994) Land and water: water and soil conservation and central government in New Zealand. Department of Internal Affairs, Wellington, 219 p

Rosser BJ, Ross CW (2011) Recovery of pasture production and soil properties on soil slip scars in erodible siltstone hill country, Wairarapa, New Zealand. N Z J Agr Res 54(1):23–44. https://doi.org/10.1080/00288233.2010.535489

Sattler K, Anderson B, Macintosh A et al (2016) Estimating permafrost distribution in the maritime Southern Alps, New Zealand, based on climatic conditions at rock glacier sites. Front Earth Sci 4:4. https://doi.org/10.3389/feart.2016.00004

Soil Survey Staff (2014) Keys to Soil Taxonomy twelfth edition. USDA Natural Resources Conservation Service. 360p. http://www.nrcs.usda.gov/wps/PA_NRCSConsumption/download?cid=stelprdb1252094&ext=pdf. Accessed on 4 March 2020

Tonkin PJ (1984) Studies of soil development and distribution in the eastern hill country, central South Island, New Zealand. PhD thesis, University of Canterbury

Tonkin PJ, Basher LR (1990) Soil-stratigraphic techniques in the study of soil and landform evolution across the Southern Alps, New Zealand. Geomorphology 3:547–575. https://doi.org/10.1016/0169-555X(90)90020-Q

Turner BL, Condron LM (2013) Pedogenesis, nutrient dynamics, and ecosystem development: the legacy of T.W. Walker and J.K. Syers. Plant Soil 367:1–10

Webb TH (ed) (1994) Soil–landscape modelling in New Zealand. Landcare Research Science Series, Vol 5. Lincoln, New Zealand, Manaaki Whenua Press

Webb TH, Campbell AS (1986) Effect of rainfall on pedogenesis in a climosequence of soils near Lake Pukaki, New Zealand. N Z J Geol Geophys 29:323–334

Webb TH, Fahey BD, Giddens KM et al (1999) Soil–landscape and soil–hydrological relationships in the Glendhu experimental catchments, east Otago uplands, New Zealand. Aust J Soil Res 37:761–785

Wilde RH (1972) An age sequence of parent materials and the soils formed from them in Central Otago, New Zealand. N Z J Sci 15:637–664

World Reference Base (2015) http://www.fao.org/3/i3794en/I3794en.pdf. Accessed 2 Mar 2020

Wyse SV, Wilmshurst JM, Burns BR et al (2018) New Zealand forest dynamics: a review of past and present vegetation responses to disturbance, and development of conceptual forest model. N Z J Ecol 42:87–106

Gley Soils

5

In places where water lays around
sodden soggy Gley Soils are found
with subsoil colours of grey or blue
Gley Soils are deficient in O-2.

Abstract

Gley Soils form where soils are saturated for long periods due to high or perched water tables, or slow drainage. Gley Soils occur throughout New Zealand but are more common in areas of high rainfall such as the West Coast of the South Island. Distinctive pale blue-grey colours in Gley Soils indicate a lack of oxygen and associated reducing conditions. When water saturates the soil, microbes use up the available oxygen leading to anaerobic (or reducing) conditions where the iron in rust-coloured iron oxides (containing Fe^{3+}) is reduced to a colourless, soluble form (containing Fe^{2+}). Manganese oxides are similarly reduced from black or brown compounds (containing Mn^{4+}) to a pale, soluble form (containing Mn^{2+}). The soluble, reduced cations may migrate to places where oxygen is available, such as along root channels or at profile texture changes. There they may be oxidised back to Fe^{3+} forming rust-coloured (or Mn^{4+} giving brown/black) mottles, concretions, or iron pans. Many Gley Soils occur on fertile, flat, flood plains, and terraces. Where drainage is installed they can be highly productive for pastoral and cropping land uses. Careful management is needed to ensure drains are maintained and to prevent pugging and compaction during wet periods.

5.1 Important Features of Gley Soils

5.1.1 Concept and Key Features of the Soil Order

Gley Soils form in areas where the soil is saturated as a result of periodic or permanent ground- or perched-water tables. Gley Soils occur in predominantly mineral parent materials as distinct from Organic Soils that form from the accumulation of plant material in wet conditions. In an undrained state, Gley Soils are often low in oxygen (anaerobic), which limits the oxygen available to plant roots, resulting in vegetation that is tolerant of wet conditions. Anaerobic (oxygen depleted) conditions are indicated by pale blue-grey soil colours in areas of the profile where the soil remains saturated and soil minerals are in chemically reduced forms. Where oxygen is periodically available, for example, along root channels or in areas where there is seasonal wetting and drying, areas of oxidised minerals will accumulate forming rust-coloured or orange mottles and/or small black to brown spots or lumps (concretions or nodules, Fig. 5.1).

Large areas of Gley Soils have been artificially drained to form productive agricultural land. When a Gley Soil has been drained, if the mainly grey colours remain, it is still classified as a Gley Soil.

A. E. Hewitt et al., *The Soils of Aotearoa New Zealand*, World Soils Book Series,
https://doi.org/10.1007/978-3-030-64763-6_5

Fig. 5.1 Gley Soil and associated landscape. Left: Gley Soils often form in swampy, low lying, areas where there is a high groundwater table such as this foot slope position. Such areas often have water-tolerant vegetation such as rushes (*Juncus* sp.). Right: Gley Soil, with mottled and grey (low-chroma) colours that indicate that saturation has occurred intermittently in the topsoil and for prolonged periods at depths below about 60 cm. The water table is evident at 95 cm depth

Gley Soils have 'reductimorphic features' which are evidence of oxygen depleted (chemically reduced) conditions that generally extend to at least 90 cm depth. Reductimorphic features are defined as including:

(1) *redox depletions*, which are pale (low chroma, ≤ 2), predominantly bluish-, greenish-, or whitish-grey, colours, in the soil matrix or on ped faces or pore walls, signifying long-term water saturation; and/or
(2) *redox concentrations* which are rust-coloured or orange mottles and/or small black to brown spots, concretions, or nodules, in the soil matrix together with dark Mn-rich or Fe-rich coatings on ped faces or pore walls signifying short-term water saturation (and oxygen depletion) interspersed with periods of oxidation.

5.1.2 Areas of Occurrence

Gley Soils are located throughout New Zealand comprising about 3% of the country (Fig. 5.2). Gley Soils form in the landscape wherever there is prolonged wetness that has led to oxygen depletion. Commonly Gley Soils form in low-lying areas with high groundwater tables, or where water accumulates in sloping gullies, seepages, or at the base of slopes (Sect. 5.3). Gley Soils can also occur in elevated situations (on hills) where upper horizons are slow draining because water is held up (perched) on an impeding pan or slowly permeable horizon. Gley Soils are more common in areas of high rainfall, such as south-Westland. Together with

Fig. 5.2 Map showing main areas of occurrence of Gley Soils. There are many small areas of Gley Soil throughout New Zealand that do not show on a map of this scale

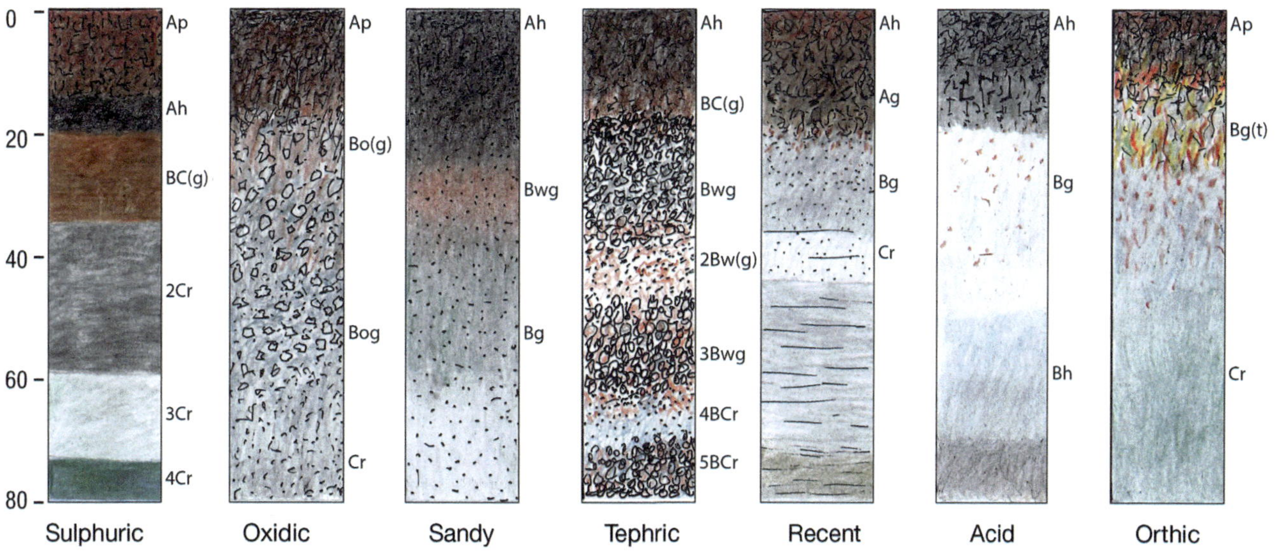

Fig. 5.3 Groups within the Gley Soil order. Vertical axis is depth in cm

Organic Soils, Gley Soils represent the original extent of New Zealand wetlands.

5.1.3 Variation Within the Gley Soil Order

The following soil groups (Fig. 5.3) are defined in the Gley Soil order in the *New Zealand Soil Classification*:

Sulphuric Gley Soils—marine estuarine soils that contain sulphur,

Oxidic Gley Soils—share some properties for Oxidic Soils,

Sandy Gley Soils—dominated by sand or loamy sand to depth,

Tephric Gley Soils—contain tephra accumulated from volcanic eruptions or volcanogenic alluvial redeposition,

Recent Gley Soils—on young land surfaces, mainly alluvial or estuarine,

Acid Gley Soils—strongly or extremely acid, and

Orthic Gley Soils—other (usually simple) Gley Soils that do not have any of the properties above.

In Gley Soils, all of the subgroups are intergrades to other soil groups or soil orders. The intergrading subgroups are Fluid, Saline, Peaty, Sandy, Argillic, Melanic, Ironstone, Placic, Humose, Ultic, Mottled, Oxidic, Calcareous, Granular, Nodular, Concretionary, Acidic, and Ultic. All of the subgroups depart from the central concept of the soil group in some properties. The central concept soil group is identified as the Typic subgroup. There are many defined subgroups in the Gley Soil order, reflecting the wide range of environments where gley soil features are superimposed on the characteristic soil processes and properties of another soil order.

5.1.4 Origin of the Soil Order Name

The term 'gley' is derived from the Ukrainian 'glei' for sticky or waterlogged blue clay. The name gley is commonly used internationally for wet soils with dominant grey matrix colours.

5.2 Soil Profile Genesis

Water saturation, for an extended period, is the key factor in formation of Gley Soils. Saturation is important as it impacts on both soil hydrological and biochemical processes. In a saturated soil, water can move rapidly through macropores carrying any dissolved or suspended materials with it. Water saturation may also make a soil more prone to surface compaction and pugging damage as well as increasing the possibility of hill slope mass movement by reducing soil cohesion and increasing the mass of material. Under saturated conditions, dissolved oxygen in the water becomes depleted because of the respiration of soil microbes. The lack of available oxygen (anaerobic conditions) restricts many living organisms and leads to chemical reduction of a range of compounds. The chemical reduction reactions include:

(1) converting nitrate (NO_3^-) to nitrite (NO_2^-) and, eventually, to nitrogen gas (N_2);
(2) reducing black or brown manganese oxide minerals (commonly with Mn^{4+} as insoluble MnO_2) to very soluble, pale Mn^{2+} minerals;

(3) reducing red-brown coloured ferric iron minerals (with Fe^{3+} as insoluble FeOOH) to soluble, colourless, ferrous (Fe^{2+}) iron minerals; and
(4) sulphate (SO_4^{2-}) reduction to hydrogen sulphide (H_2S).

Both Mn and Fe in their soluble (reduced) states are mobile (dissolve in water, Mn especially so) and so can easily move with the soil water to other parts of the profile or possibly to adjacent soils in the landscape. As Mn is oxidised, it may precipitate around pre-existing nuclei forming concentric patterns within concretions or non-concentric nodules. Such concretions or nodules may additionally contain Fe oxides.

Saturation is not mapped directly because of its high temporal and spatial variability. Instead the presence of pale, grey, colours (redox depletions or low-chroma mottles) and bright orange/brown mottles (redox concentrations or segregations), together known as reductimorphic or redoximorphic features, are used as indicators of prolonged soil saturation. The connection between soil colour and water saturation is a result of biochemical reduction. The presence of reducing conditions, which can sometimes be uncertain where the parent materials are themselves pale-coloured (such as white rhyolitic pumiceous alluvium), can be easily shown by using the Childs' test for ferrous iron in solution (developed by Cyril Childs). A dye solution α,α'-dipyridyl is sprayed on to the soil matrix or a ped and a positive response is indicated by the development of pink to red colours. If the presence of Mn-oxides (e.g. MnO_2) is uncertain, rapid effervescence following the addition of hydrogen peroxide (H_2O_2) will confirm their occurrence and abundance. Alternatively, black MnO_2 concretions give a brown streak on paper (whereas black charcoal fragments make black marks).

Where there is a lack of oxygen (anaerobic conditions) then the reduction process, the gain of electrons (the opposite process to oxidation, the loss of electrons), is triggered which causes oxygen to be removed from a number of soil minerals. Soil in an oxygenated state is normally pigmented in red, yellowish-brown, brown, or other ochreous colours. The pigments are derived from Fe, Mn, and Al species that occur as stable oxides or hydroxy-oxide minerals, including, for Fe^{3+} the polymorphs goethite, lepidocrocite, ferrihydrite, or haematite; and for MnO_2 the polymorphs birnessite or pyrolusite. Under reducing (anaerobic) conditions the Fe, Mn, and Al species become soluble and colourless. In areas where rust brown, insoluble ferric (Fe^{3+}) iron minerals are reduced to colourless, soluble ferrous (Fe^{2+}) iron minerals the soil colour changes from a rusty brown colour to grey or bluish grey. Similarly, black or brown, insoluble Mn^{4+} or Mn^{3+} minerals are reduced to soluble pale Mn^{2+} minerals. Once in the soluble form, the iron and manganese may migrate to zones of higher oxygen content where oxidation returns the minerals to the insoluble, coloured, form which thus precipitates out of solution.

Permanently high water tables result in soils that have continuously reduced forms and pale grey colours (moist chromas ≤ 2, or ≤ 3 with value ≥ 6). The precipitated manganese or iron minerals accumulate in pores and along cracks or in root channels where oxygen is available, as finely dispersed particles, as discontinuous coatings on ped surfaces, or they may form concentrations that are observed as red or brown mottles (redox segregations), concretions, or nodules. In some Gley Soils an iron pan may form if there is a horizontal layer in which oxygen is present. The reduced iron may migrate to and precipitate out in a distinct layer that typically marks a textural change (e.g. from silt loam to sandy gravel) in the profile. A pan may also form at a point where a seasonally fluctuating water table repeatedly rises and falls. If all the solubilised manganese and iron have migrated out of the soil then the grey colours may remain even if a soil is drained and oxidised.

Thus in Gley Soils, the redox depletions (pale colours) together with the redox concentrations (reddish brown Mn- and/or Fe-rich spots, concretions, or ped coatings) reflect the depth of the water table or the persistence of perching, and the length of time the soil is saturated (Fig. 5.4). Gley Soils generally have a dark-coloured topsoil that may be quite high in organic matter. To qualify as a Gley Soil, generally, the soil must have a soil horizon with at least 50% pale grey colours (i.e. a reductimorphic horizon which usually extends to at least 90 cm depth).

5.3 Soil-Landscape Relationships

Gley Soils form in any area of the landscape where water accumulates in the soil such that saturation occurs reasonably frequently, or for at least a few weeks per year. Thus the occurrence of Gley Soils is strongly influenced by topography and rainfall, and is, therefore, somewhat predictable. Gley Soils form in areas with high groundwater tables or in areas where slowly permeable layers result in perched water tables (Fig. 5.5). Flat, low-lying areas, and hollows that prevent surface runoff of excess water are often saturated for extended periods. Back swamp areas on river flood plains accumulate water during floods and if a levee (raised flood-depositional area on the river margin) prevents the flood water from returning to the river, then water may lie on the soil surface for several weeks after a flood event, saturating the soil and leading to development of a Gley Soil. Such 'back swamp' areas are invariably low lying and are likely to have high water tables near the soil surface at times in winter and spring.

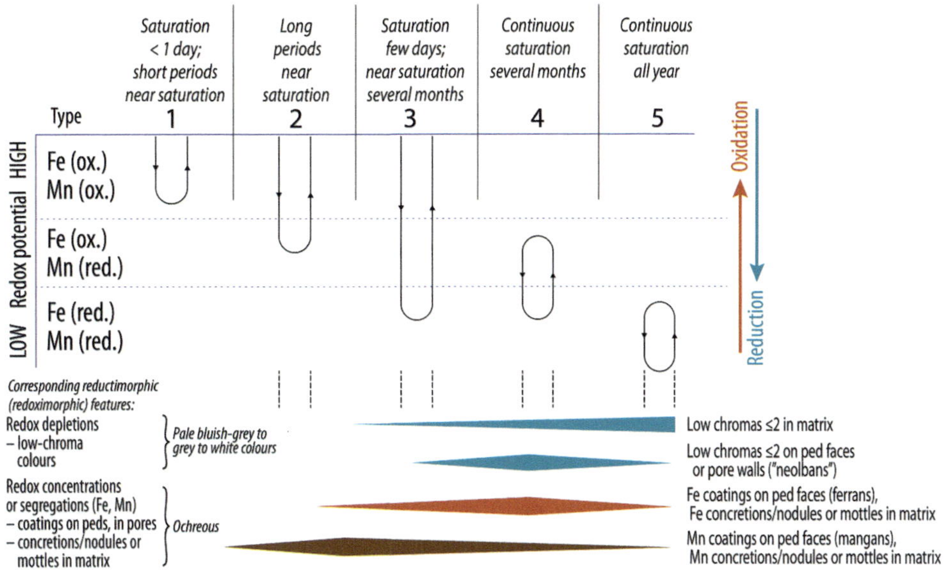

Fig. 5.4 The length of time a soil is saturated in relation to the oxidation/reduction state of the soil (upper part of diagram) and the soil morphological features (soil colour and occurrence of iron and manganese coatings and concentrations) that indicate gleying (lower part of the diagram). (Redrawn after Bouma, 1983)

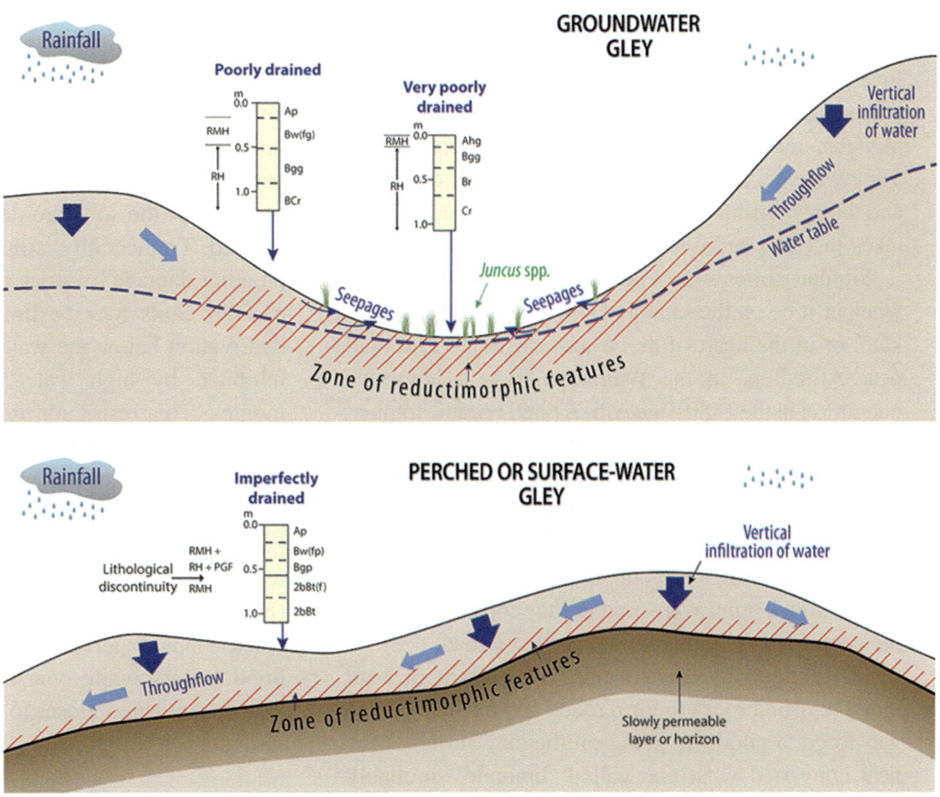

Fig. 5.5 Examples of landscape positions where Gley Soils occur. Top: groundwater-gley soils in a seepage zone at the foot of a hill. Bottom: perched-gley soils formed above a slowly permeable horizon. Typical soil horizonation and associated drainage classes are depicted in the examples of soil profiles. Diagnostic horizons for the *New Zealand Soil Classification* are also shown: RMH = redox mottled horizon, RH = reductimorphic horizon, PGF = perch-gley features

In hill country water often accumulates in the colluvial toe-slopes or foot slopes at the base of hill-slopes. Water may be derived from runoff on the slopes as well as seepage from groundwater to the land surface in toe-slopes. Thus an area of Gley Soil is often evident at or near the base of hill-slopes. Gley Soils also occur on lake, stream, and coastal margins where the height of the water table is controlled by the height of the adjacent water-body.

On some flat surfaces where runoff is limited and water accumulates within the soil above a slowly permeable layer (such as a pan or clay-rich horizon) a perched water table forms. Such soils may qualify as Gley Soils, but often they are recognised as a perch-gley group (or mottled subgroup) in another soil order because the extent or depth of gleying is insufficient to qualify for the Gley Soils order.

Gley Soils are more common in areas with higher rainfall, particularly on the West Coast of the South Island, as there is more opportunity for soils to be saturated for long periods of time. Because of the saturated conditions and limited oxygen, Gley Soils are often marked by the presence of water-tolerant vegetation such as *Juncus* species (rushes).

Traversing the lowlands in the nineteenth century

Travelling in early New Zealand was always challenging. In the nineteenth century, travellers struggled against rugged alpine mountain lands, deep steep-sided gorges, and swift rivers, where the prevalent cause of death was drowning (known at the time as 'the New Zealand death'). However, travel in the lowlands was also challenging due to extensive areas of saturated soil. Thomas and John Royds wrote, in 1861, of their nine-day journey from Dunedin to Invercargill. They had a one shaft cart, with two horses, carrying five tonnes of freight including ploughs, tools, a stove, and other farm equipment. They were severely tested by frequent river crossings and boggy bushland tracks. The same journey now takes just two and a half hours.

Similar stories were told throughout New Zealand, whether it be on the saturated soils of Northland, Waikato, or in the heart of the Wairarapa Bush. The road near Alfredton, in the Wairarapa, was a notorious quagmire. In the 1880s one Alfred Shell became trapped up to his thighs in mud and as night fell was sure he would die before morning. However, he was fortunate to be rescued some 20 h later. On another occasion one man was not so lucky and died before help arrived. Horses were known to be lost, saddle and all, into the mud. Bullock trains were used as they coped better with the deep, saturated, mud that served as a road.

In the north Wairarapa there was a lack of hard metal (crushed rock) to form roads and one tactic used was to get a good fire going on the clay surface—the heat converted what was called 'limonite' (probably goethite, a soft, common, iron oxide mineral) into haematite which provided a harder material for the road surface providing improvement for some time. However, after a while the haematite would weather and revert to goethite. Heating in a fire was also used by Māori to convert ferrihydrite (an iron oxide 'gel') to haematite to make red ochre (kokowai). To be effective the temperature had to reach at least 750 °C.

5.4 Key Soil Properties

5.4.1 Soil Composition

Gley Soils are most commonly formed in alluvial or colluvial parent materials, but parent material is not a defining characteristic of Gley Soils. Thus Gley Soils have wide ranging clay types that reflect the mineralogy of the parent materials. The most common mineralogy groups are mixed, illitic, smectitic, and kaolinitic. The one common mineralogical feature of Gley Soils is the s*egregation of iron and manganese oxides, as a result of the redox processes* (Sect. 5.2). Particles in reduced parts of the soil are not coated or stained by iron and manganese oxides. The oxides migrate along biochemical gradients to congregate in redox concentrations (mottles, concretions, and pans) at points where oxygen is available within the soil.

5.4.2 Physical Properties

The physical properties of Gley Soils are wide ranging as a result of the diverse parent materials and topographic situations in which Gley Soils form (Fig. 5.6). The common feature of Gley Soils is the presence of high groundwater or perched water tables and periodically saturated soil conditions. Thus the soils are described as poorly or very poorly drained. Trafficability is limited in most soils when wet and pugging damage by stock is a common risk. Rooting depths of Gley Soils may be limited for many plants by oxygen deprivation below the water table. Most crop plants will be inhibited by high water tables, at least through winter months. This restriction may be exacerbated by higher soil dry bulk density and associated low macroporosity. Soil dry bulk densities are likely to be higher in Gley Soils than in well drained soils in similar soil materials. Gley Soils, particularly those on river flood plains and terraces, tend to be relatively fine-textured with a moderate clay content (Fig. 5.6). Gley Soils often form in the back swamp areas where finer materials are carried by, and settle out from, flood waters while coarser gravel and sand materials are mainly deposited closer to river margins forming levees.

5.4.3 Chemical Properties

The topsoils in Gley Soils generally have moderate levels of organic matter and some are peaty as the wet conditions inhibit biodegradation of plant materials (Fig. 5.7, Table 5.1). The mean cation exchange capacity in subsoils of Gley Soils is high to very high (ranges from 26 to 56 $cmol^{(+)}\ kg^{-1}$) and phosphorus retention is low to medium.

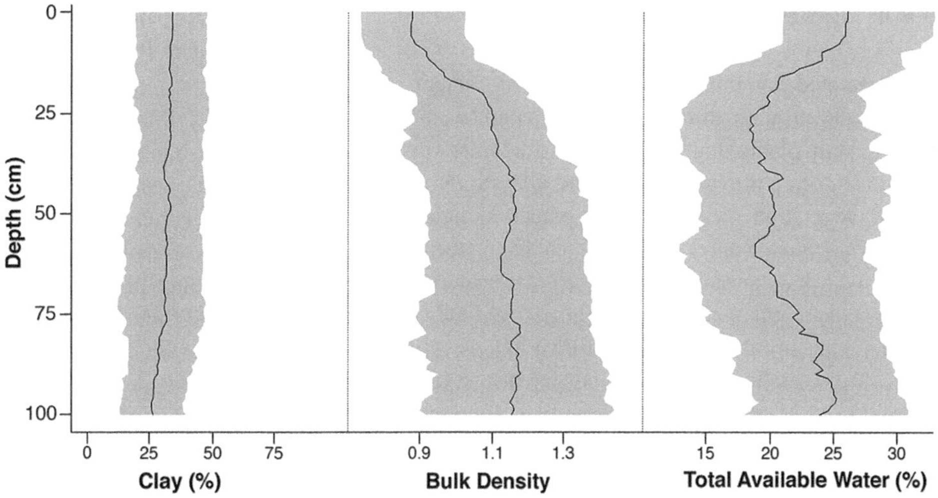

Fig. 5.6 Median and upper and lower quartiles of clay content, soil dry bulk density (g cm^{-3}), and soil available water capacity for Gley Soils in the New Zealand Soil Data Repository

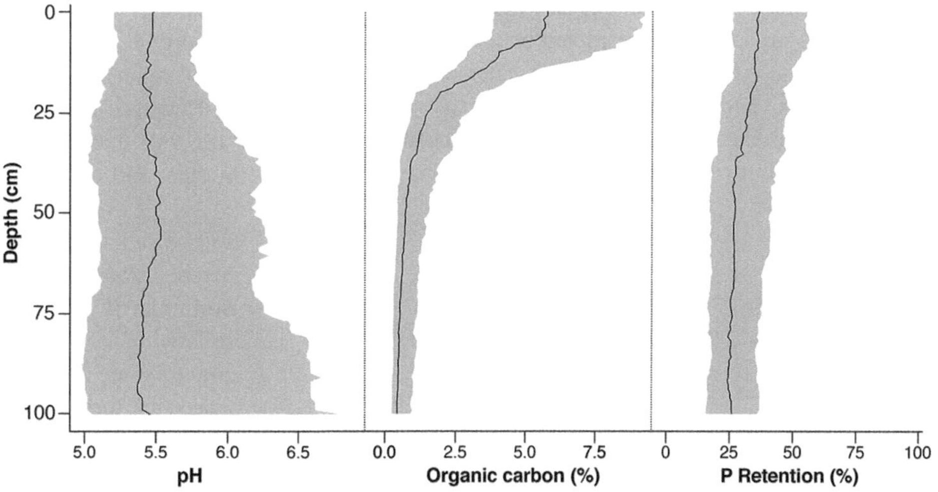

Fig. 5.7 Median and range of soil pH, carbon, and P retention for Gley Soils in the New Zealand Soil Data Repository

Table 5.1 Typical example of soil chemical properties of a Typic Orthic Gley Soil: Kairanga series SB08572

Horizon	Depth cm	Carbon %	Nitrogen %	CEC[a] $cmol^{(+)}\ kg^{-1}$	Sum bases $cmol^{(+)}\ kg^{-1}$	P retention %
Ap1	0–8	6.3	0.61	25.5	18.1	28
Ap2	8–25	4.3	0.45	25.4	14.9	32
Bg	25–48	0.8	0.1	17.6	14.1	24
Cg	48–69	0.6	0.07	17.3	14.4	26

[a]CEC = cation exchange capacity

Thus Gley Soils are generally fertile as they can retain and make nutrients potentially available to plants.

A key feature of Gley Soils is the presence of anaerobic conditions which means a number of compounds occur in their chemically reduced form. Of particular note is nitrogen. In anaerobic conditions, nitrate (NO_3^-) is reduced to nitrite (NO_2^-) and then to nitrogen gas (N_2) as anaerobic microbes utilise nitrates as an oxygen source. Thus, providing that there is sufficient carbon in the soil, nitrate is less likely to be leached from Gley Soils than other soils, as the nitrogen is instead converted to nitrogen gas and returned to the atmosphere (which is 78% N_2). On the down side, some of the nitrogen may also be returned to the atmosphere as nitrous oxide (N_2O), which is a greenhouse gas.

5.4.4 Biological Properties

The saturated conditions and low oxygen content of many Gley Soils provide difficult growing conditions for many plants. Thus plants that are adapted to tolerate waterlogged conditions often prevail in undrained Gley Soils. In areas of native vegetation, water-tolerant plants include harakeke (flax, *Phormium tenax*), ti kouka (cabbage tree, *Cordyline australis*), kahikatea (white pine, *Dacrycarpus dacrydioides*), and pukatea (*Laurelia novae-zealandiae*), which occur (or occurred before pastoralism) extensively on Gley Soils. Trees such as kahikatea and pukatea have buttressed roots (Fig. 5.8) which give the large trees support in the soft wet substrate. Many of the water-tolerant species will also grow in other situations but their water tolerance gives them a competitive advantage in wet areas so they tend to dominate vegetation on Gley Soils. In pasture, the presence of rushes (*Juncus* sp.), buttercup (*Ranunculus* sp.), and docks (*Rumex obtusifolius*) may indicate the presence of a Gley Soil.

Many soil organisms are restricted because of anaerobic conditions. Anaerobic bacteria are associated with low soil oxygen levels and conditions of chemical reduction. Anaerobic microbes use nitrate and iron, manganese, and sulphur minerals as their energy (oxygen) sources, rather than oxygen gas.

Fig. 5.8 The buttressed roots of a pukatea tree provide support for a large tree in a sodden substrate

5.5 Distinguishing Between Gley Soils and Related Soil Orders

The Gley Soil order was defined for poorly or very poorly drained soils in which distinct marks of other soil-forming processes were absent. Gley Soils therefore comprise soils that are either too young to display the distinct marks of other soil-forming processes, or in which the marks of those soil-forming processes have been retarded or destroyed by gleying. Where features associated with strong gleying occur together with the distinct marks of other soil-forming processes that define other soil orders, then the soils are recognised as either Perch-gley or Groundwater-gley groups of the dominant orders.

In the *New Zealand Soil Classification*, Gley Soils key out second (after Organic Soils). Gley Soils are generally defined as having a gley profile form with at least 50% pale colours (due to water saturation) that extend from within 30 cm of the soil surface to at least 90 cm depth (i.e. a reductimorphic horizon) , or a substantial iron pan and associated reducing conditions (see the NZ Soil Classification of Hewitt 2010 for details). However, there are exceptions for soils that have some gley soil features but fit best within other soil orders. The exceptions are for:

(1) soils with a fragipan (a dense subsoil layer that causes a perched water table in winter), which fall into the Pallic Soil order (Chap. 10);
(2) soils with a densipan (Ed-horizon) or podzolic-B horizon (organic matter-rich Bh and Fe-Al-rich Bs horizons) that form below an E-horizon which fall into the Podzol Soil order (Chap. 11); and
(3) soils that have 'fluid' material within the soil, such as in tidal mud-flats, where the soil would be classified as a Raw Soil (Chap. 13).

The following soil groups are not included in the Gley Soil order but have gley characteristics that are superimposed on other important soil properties:

Perch-gley Allophanic Soils,
Gley Allophanic Soils,
Perch-gley Granular Soils,
Perch-gley Melanic Soils,
Perch-gley Oxidic Soils,
Perch-gley Pallic Soils,

Perch-gley Podzols,
Groundwater-gley Podzols,
Perch-gley Pumice Soils,
Gley Raw Soils, and
Perch-gley Ultic Soils.

As water saturation tends to slow biodegradation of organic matter there is an intergrade between Gley Soils and Organic Soils. Gley Soils may contain high levels of organic matter and include thin layers of peat, but they are predominantly formed in mineral material. Organic Soils in wet conditions have at least 30 cm of peaty material within the soil profile.

Gley Soils are also recognised where there is an ironstone-pan within 30 cm of the soil surface and evidence for reducing conditions in some part of the year.

5.6 Correlation with Other Classification Systems

In the *World Reference Base*, gleyed soils are recognised at the highest level as Gleysols. However, in *Soil Taxonomy* gleyed soils are recognised mainly at the second level, namely, suborder, through the presence of an aquic soil moisture regime (Table 5.2). In Europe (*World Reference Base*), perch-gley soils are recognised as 'pseudogleys'. Like New Zealand, various other soil classifications that recognise gley soils at order level do not incorporate all gley soils into that one class. The Soil Classification for England and Wales and the Canadian system of soil classification both place strongly gleyed podzols with Podzols rather than with Gley Soils. Thus, the precedent is established that, for some soils, features other than gleying are so marked that they determine order assignment. The *New Zealand Soil Classification* applies this not only to the Podzols but also to nine other orders.

5.7 Use and Management of Gley Soils

5.7.1 Introduction

Wetlands underlain by Gley Soils were extensive throughout the New Zealand lowlands. However, once drained, with pasture established, Gley Soils are generally regarded as highly productive and they are commonly used for sheep and beef farming, though in many regions dairy farming now predominates on flat areas. While some crops, such as maize and peas, can be successfully grown on drained Gley Soils, such soils are not so widely used for other horticultural crops (such as kiwifruit) that may be sensitive to periods of high water tables or water logging.

Drainage is an important component of management of Gley Soils to ensure trafficability and productivity. Drainage may also be managed to control nitrogen leaching to waterways. Increasingly the role of wetlands in flood attenuation, removal of nutrients and sediments from water, carbon storage, and to protect biodiversity is becoming recognised. Thus wetland restoration is also becoming an important aspect of the management of Gley Soils.

5.7.2 Soil Drainage Classes

The presence or absence of grey colours and mottles are used in New Zealand (and most other countries) to define five soil drainage classes: very poorly, poorly, imperfectly, moderately well, and well drained (which were defined in the "Soil Description Handbook" by Milne and others in 1995). Drainage classes indicate the depth to the top of a water table, from the ground surface, and the duration of saturated conditions. The soil drainage classes are widely used in soil survey to indicate soil wetness. They are generalised classes that provide a guide for making land management decisions. For intensive land development, or to

Table 5.2 Correlation[a] between Gley Soils and equivalent classes of *Soil Taxonomy*, *World Reference Base*, and the earlier NZ genetic soil classification

New Zealand Soil Classification	*Soil Taxonomy*	*World Reference Base*	NZ genetic soil classification
Sulphuric Gley Soils	Sulfaquents	Thionic gleysols	Not recognised
Sandy Gley Soils	Psammaquents	Arenic gleysols	Gley soils
Acid Gley Soils	Endoaquepts	Dystric gleysols	Gley soils
Oxidic Gley Soils	Haplaquox	Ferralic gleysols	Gley soils
Recent Gley Soils	Fluvaquents	Gleysols	Recent gley soils
Orthic Gley Soils	Endoaquepts	Gleysols	Gley soils

[a]The correlations given here are a guide only and for accurate classifications the relevant soil classification documents should be consulted. The two major international soil classification systems are *Soil Taxonomy*, which was developed in the USA, and *World Reference Base*, which was developed primarily in Europe. The NZ genetic soil classification was used in NZ prior to 1992

design an effective drainage system, detailed soil profile sampling and site analysis should be undertaken to assess local variability and to support precision modelling and mapping.

The *New Zealand Soil Classification* defines a 'mottled profile form' which identifies imperfectly drained soils, and a 'gley profile form' which identifies poorly and very poorly drained soils. However, there are uncertainties when soil colour patterns are interpreted, as inference of wet conditions may be misleading where artificial drainage, or landscape change, has caused water tables to recede leaving the gleyed soil colours preserved. In cases where iron is still present in a soil, upon drainage, and entry of oxygen into the soil atmosphere, colours may quickly (within a few minutes or days, depending on the rate of oxygen diffusion) return to the ochreous brownish colours we most often see in soil.

5.7.3 Soil Drainage Design and Management

In the nineteenth century, as land was cleared of native vegetation and developed for agriculture, the need for drainage on the more easily farmed flat lands became immediately evident in many regions. River mouths were cleared to enhance flow to the sea, regional outlet drains or canals were established, often along roadsides, and then surface or tile drains were installed in paddocks. Because drainage cuts across neighbourhoods, districts, and regions, people had to work together to instal effective drainage systems. Drainage boards were quickly established to support community drainage efforts, and many have remained in action for a long period. The 1876 Public Works Act formally empowered the government to control waterways and drainage in that any natural watercourse of swamp could be declared a 'drain'. The Land Drainage Act of 1908 (last amended in 2012) brought together a range of earlier legislation pertaining to land drainage. It provided for drainage boards to be established to construct and maintain drains and water courses with funding from rates. The 1908 Land Drainage Act also set out rights and responsibilities of landowners. However, according to Geoff Park (2013), application of the 1876 Public Works Act and the 1915 Swamp Drainage Act by the government resulted in its compulsory acquisition of large areas of wetlands (to enable drainage operations to be completed), and the value or otherwise of the drainage has thus been (and remains) controversial for multiple reasons. Since implementation of the Resource Management Act in 1991 the work of drainage boards is managed by regional and district councils. For example, the Waikato Region has over 50 land drainage schemes, the largest of which is the 'Thames Valley Drainage Area' which includes about 5000 ratepayers.

The drained lands generally proved to be highly productive. G.R. Reid of the Thames Valley Drainage Board noted in 1978 that 'Probably there is no single factor which gives such an early or quick economic response as artificial drainage, nor any which, under attentive management, will have such a permanent effect'. Drainage removes water from the soil allowing oxygen to penetrate the soil, and drier soils warm more quickly in spring. Thus drained soils are better able to support microbial and plant growth as well as vehicle and animal traffic.

In designing a drainage system the overall objective is to keep soils from becoming saturated. There are various possible approaches depending on the site, the soil, and the causes of water logging. Where water tables are high, lowering the water table using surface and/or subsurface drains is likely to be the best option. Drain design has to take into account the need for slope on drains and the location and height of an outlet to the drainage system. In areas where runoff accumulates at the foot of slopes then enhancing or preventing water infiltration by diverting surface water flow away with a surface, or subsurface drain is an option. Land smoothing may be used to enhance runoff and remove hollows, thus preventing surface ponding in hollows. In some cases techniques such as deep ripping may break up a slowly permeable layer or iron pan within the soil, increasing the infiltration rate and thus preventing prolonged surface saturation. Care needs to be taken to prevent over-drainage. It may be useful to maintain a water table near the base of the plant root zone to provide water in drier summer months.

Fencing off open drains will help support riparian vegetation and prevent stock access and damage to the drain. A good grass cover provides the best filtering for sediments and nutrients on drain margins. Planting of trees and shrubs will provide shade but should be kept at least three metres from the drain and constrained to one side to ensure digger access for drains that will need to have ongoing maintenance. It is wise to consult with regional and district councils and drainage engineers for advice—in many cases a resource consent will be required for any new drainage development.

Due to the often slow permeability of heavy textured Gley Soils, even when drainage has been installed, the soils are frequently wet following periods of high rainfall, particularly in winter and spring. Thus a risk that must be considered is the potential for soil compaction and damage to topsoil structure by vehicle traffic, or animal treading, when the soils are wet. Severe animal trampling, causing pugging (Fig. 5.9), has been shown to result in a marked decrease in pasture production over the following season. During wet periods, options for farmers include using stand-off pads so that grazing is restricted to just a couple of hours and pugging is minimised, or keeping stock to paddocks on better drained soils.

Fig. 5.9 Severe pugging of a Gley Soil in the Waikato leads to soil compaction and a prolonged decrease in pasture production. Note: this was a scientific experiment and the cows were held at a high stocking rates for a short time, so were ethically managed. Photo: Karsten Zegwaard

Sulphuric Gley Soils occur in tidal marine estuarine environments and although now they are usually reserved as part of the protected coastal environment, some areas have been drained and developed for agriculture or other coastal infrastructure. Drainage of such areas is an expensive undertaking. Usually flood stop banks need to be built to prevent flooding during storms. Due to the proximity to sea level, in order to lower the water table, pumping of the drainage water is usually needed. An additional hazard following drainage of Sulphuric Gley Soils is the development of extremely low pH (pH of 2 to 4) due to the oxidation of sulphides (mainly derived from iron sulphides such as pyrite or from sulphidic sediments such as in shallow coastal marine settings) to sulphuric acid when the formerly saturated soils become exposed to air and by biochemical reactions. Sulphuric acid forms only slowly in an abiotic environment, but the process is greatly accelerated in the presence of Fe-oxidising bacteria. Such soils are commonly known as acid-sulphate soils generally or Sulfaquents in *Soil Taxonomy*.

An unusual yellow coloured hydrous sulphate mineral called jarosite—$KFe_3(OH)_6(SO_4)_2$—may also form in acid-sulphate environments (it has also been identified on Mars). Jarosite has been identified in soils formed on reclaimed estuarine sediments such as those adjacent to Hokianga Harbour (Fluid-saline Sulphuric Gley Soil). Another sulphate mineral, schwertmanite—$Fe_8O_8(OH)_8$-2x (SO_4)x—can also occur in acid-sulphate soils and in strongly acid seepages such as from mine tailings.

As well as being lethal to many plants and animals, and corrosive, the very strong acidity in Sulphuric Gley Soils can attack clays, releasing Al and Fe, which in turn may generate Al, Fe, and heavy metal toxicity, and nutrient deficiencies. Thus for agricultural production, large inputs of lime are needed to neutralise the acid. It is best to keep water tables as high as possible, while still maintaining trafficability, in order to minimise the amount of oxygen entering the subsoil, and thus the amount of acid produced. The low pH environment can also lead to corrosion of infrastructure, such as concrete or metal foundations, and water or sewage pipes, as well as toxicity problems. Thus choice of materials needs to be carefully considered for engineering developments on Sulphuric Gley Soils.

Sandy Gley Soils commonly occur in low-laying areas among coastal dune sands, such as those in the Manawatu sand country that lies within about 20 km of the west coast between Bulls and Levin. In the hollows between sand dunes water is trapped, with no stream outlets, and thus high water tables occur. Artificial drainage may be difficult because of lack of fall to a drainage outlet. Lenses of peat materials may occur and therefore, bearing capacity should be tested for weight bearing structures. Some areas are probably best left to water-tolerant native vegetation and to support biodiversity. However, in some areas the Sandy Gley Soils can be usefully managed in association with the adjacent dunes (where care is needed to prevent severe wind erosion and dune remobilisation). The well-drained dunes provide support for pasture and stock during wet periods, while grazing of the moister sandflats can be supported during summer dry periods.

It is wise to not take the continuing performance of drainage systems for granted. To remain effective, and thus sustainable, drainage systems must be maintained to ensure that outlets from subsurface drainage do not become blocked and open drains continue to flow. In most cases an annual inspection is appropriate, but keeping an eye out for water ponding, or other evidence of drainage failure, is a constant task for the diligent land manager. In parts of the groundwater-iron-rich Waikato, for example, ferrihydrite 'gel-like' deposits regularly clog drainage pipes, so that the cost of cleaning them out is such that open drains are usually preferred. Another option is to pass iron-rich drainage water through oxidising systems (including such arrangements running drainage water through beds of conifer bark) to remove the iron materials and therefore reduce the build-up of ferrihydrite gel in downstream drainage systems.

In areas where water is pumped over flood stop banks, pump maintenance is a regular requirement. In many regions local councils are responsible for management of community outlet drains and arrange for a 'drain digger' to clear vegetation and sediment from drains once every few years. Land owners often have requirements on their land-titles to allow access for drain maintenance activities, as maintenance of downstream drains is important to prevent flooding for upstream neighbours.

5.7.4 Soil Fertility

Following drainage of Gley Soils, pasture or crop production is supported from the natural soil fertility and good yields can be obtained. However, the initial flush of productivity will not last as nutrients are exported with the farm products and hence establishment of sustainable farm production systems requires fertiliser input to maintain soil nutrient levels.

Acid Gley Soils have a subsoil pH $\leq$ 4.8 and generally occur in areas of high rainfall. The soils may be strongly weathered and strongly leached and in such cases natural fertility is low. Inputs of lime and other fertilisers will improve productivity. Development of the soils for acid-sensitive deep rooting plants may be precluded. Nitrogen requirements are likely to be higher than in associated well drained soils.

5.7.5 The Value of Wetlands—Their Maintenance and Enhancement

Manaaki Whenua – Landcare Research has shown that about 85% of New Zealand's original wetland areas, which comprise Gley Soils, along with Organic Soils (Chap. 8), have been drained and developed for agricultural production. In the Bay of Plenty Region, only about 7.7% of the wetlands remain in an unmodified state. Land drainage has led to a range of ecological changes including loss of habitat for a number of rare or endangered species. There may also be wider effects, for example, the reduced incidence of fog in the Hamilton area of the central Waikato basin has been attributed to the drainage of wetlands. Although many drained wetlands are agriculturally productive, the loss of such large tracts of wetlands (which more broadly include fens and bogs as well as swamps) is now lamented by many because wetlands have been recognised as having a range of important values including supporting a wide range of biodiversity. Wetlands can reduce flood peaks as they absorb and hold water, releasing it slowly to keep streams flowing in dry periods. Wetlands trap suspended sediments and nutrients, and may convert up to 90% of nitrogen in runoff to nitrogen gas, returning it to the atmosphere, and thus preventing eutrophication of downstream water bodies. Wetlands may also be effectively designed and managed to treat wastewaters.

In many areas there is value in maintaining and restoring wetlands. The first step in protecting wetlands is to ensure that adjacent drainage is not going to lower the water table of the wetland. This may mean a weir or high-level culvert is needed to control the water level. To protect wetland vegetation and prevent damage to ground surfaces and wetland margins, it is preferable to prevent stock (especially cattle) access. However, in some instances light grazing by sheep can be effective for weed control. Replanting areas with suitable native wetland plants can increase the rate of recovery. A number of nurseries provide a range of suitable plant material and excellent advice on what would be suitable for a particular site. Until wetland plants have achieved canopy closure, ongoing weed control is likely to be needed to prevent invasive weeds including privet and blackberry, among others, from taking over.

Further Reading

Adcock I (1973) A goodley heritage. Eketahuna and districts 100 years 1873–1973. Eketahuna Borough and County Councils. 354 p

Ausseil A-G, Gerbeaux P, Chadderton WL et al (2008) Wetland ecosystems of national importance for biodiversity: criteria, methods and candidate list of nationally important inland wetlands. Landcare Research Contract Report LC0708/158 for the Department of Conservation, 162 p

Ausseil A-GE, Chadderton WL, Stephens RTT et al (2011) Applying systematic conservation planning principles to palustrine and inland saline wetlands of New Zealand. Freshw Biol 56(1):142–161

Belliss S, Shepherd J, Newsome P et al (2017) An analysis of wetland loss between 2001/02 and 2015/16. Landcare Research Contract Report LC2798 for the Ministry for the Environment, 37p. Birkeland PW (1999) Soils and gneomorphology, 3rd ed. Oxford Univ. Press, New York, 429 p

Bouma J (1983) Hydrology and genesis of soils with aquic moisture regimes. In: Wilding LP, Smeck NE, Hall GF (eds) Pedogenesis and Soil Taxonomy: I. Concepts and interactions. Dev Soil Sci 11:253–281

Childs CW (1981) Field tests for ferrous iron and ferric-organic complexes (on exchange sites or in water-soluble forms) in soils. Aust J Soil Res 19:175–180

Childs CW (1992) Ferrihydrite: a review of structure, properties and occurrence in relation to soils. Zeitschrift für Pflanzenernährung und Bodenkunde 155:441–448

Childs CW, Palmer RWP, Ross CW (1990) Thick iron oxide pans in soils of Taranaki, New Zealand. Aust J Soil Res 28:245–257

Churchman GJ, Lowe DJ (2012) Alteration, formation, and occurrence of minerals in soils. In: Huang PM, Li Y, Sumner ME (eds) Handbook of soil sciences, 2nd edn. Vol. 1: Properties and Processes. CRC Press, Boca Raton, pp 20.1–20.72

Clarkson BR, Ausseil AE, Gerbeaux P (2013) Wetland ecosystem services. In: Dymond K (ed) Ecosystem services in New Zealand—conditions and trends. Manaaki Whenua Press, Lincoln, pp 192–202

Cowie JD (1978) Soils and agriculture of Kairanga County. NZ DSIR Soil Bureau Bulletin 33. 92 p

Dent DL (1978) Saline gley soils and acid sulphate soils. In: Soil groups of New Zealand Part 3 gley soils. New Zealand Society of Soil Science. 127 p

Fitzpatrick RW, Taylor RM, Schwertmann U et al (1985) Occurrence and properties of lepidocrocite in some soils of New Zealand, South Africa and Australia. Aust J Soil Res 23:543–567

Gibbs M (2007) Best practices environmental guidelines—land drainage. Environment Waikato. 33p. ISSN:1772-4005

Hewitt AE (1993) Methods and rationale of the New Zealand Soil Classification. Landcare Research Science Series No. 2. Manaaki Whenua Press, Lincoln, New Zealand. 71 p

Hewitt AE (2003) New Zealand Soil Classification—purposes and principles, p 179–186. In: Eswaran H, Rice T, Ahrens R et al (eds) Soil classification, a global desk reference. CRC Press, Washington, DC, 263 p
Hewitt AE (2010) New Zealand Soil Classification, 3rd edn. Landcare Research Science Series No. 1, Manaaki Whenua Press, Lincoln, New Zealand
Hudson HR (ed) (2005) H20-DSS: Hillslopes to ocean: a decision support system for sustainable drainage management. New Zealand Water Environment Research Foundation, Wellington
Milne D, Clayden B, Singleton P et al (1995) Soil description handbook, revised edition. Manaaki Whenua Press Ltd, Lincoln, 157 p
Metson AJ, Janica GE, Cox TE et al (1977) The problem of acid-sulphate soils with examples from North Auckland, New Zealand. N Z J Sci 20:371–94
Mew G, Webb TH, Ross CW et al (1975) Soils of Inangahua Depression, South Island. New Zealand, NZ Soil Survey Report, p 17
Moore TR (1974) Gley morphology and soil water regimes in some soils in South-Central England. Geoderma 11:297–304
Nguyen L, Singleton P (2005) The roles of wetland areas, seepage zones and vegetated drainage ditches in dairy farms to protect water quality in our waterways. National Institute of Water and Atmospheric Research Ltd; Environment Waikato, Hamilton
Park G (2013) "Swamps which might doubtless easily be drained": swamp drainage and its impact on the indigenous. In: Pawson E, Brooking T (eds) Making a New Land: Environmental Histories of New Zealand. Otago UP, Dunedin, pp 174–189
Rijkse WC (ed) (1978) Soil groups of New Zealand, Part 3, Gley Soils, New Zealand Society of Soil Science. Wellington, New Zealand, 127 p
Roberts RC, McConchie J (2017) Preliminary assessment of the acid sulphate soils hazard in the Auckland region. In: Alexander GJ, Chin CY (eds) Proc. 20th NZGS Geotechnical Symposium. Napier, 8 p
Schaetzl RJ, Thompson ML (2015) Soils: genesis and geomorphology, 2nd edn. Cambridge University Press, 778 p
Schwertmann U, Rechter H, Taylor RM et al (1995) A lecture and demonstration for students on iron oxide formation. In: Churchman GJ, Fitzpatrick RW, Eggleton RA (eds) Clays: controlling the environment. Proceedings 10th International Clay Conference, Adelaide, Australia (1993). CSIRO Publishing, Melbourne, pp 11–14
Schoeneberger PJ, Wysocki DA, Benham EC et al (2012) Field book for describing and sampling soils, Version 3.0. Natural Resources Conservation Service, National Soil Survey Center, Lincoln, NE
Schofield JC (1972) Ground water of the Hamilton lowland. New Zealand Geological Survey Bulletin 89. 71p
Vaughn D, Wheatley RE, Ord BG (1984) Removal of ferrous ions iron from field drainage water by conifer bark. J Soil Sci 35:149–154
Vepraskas MJ, Bouma J (1976) The physical significance of soil mottling in a Wisconsin toposequence. Geoderma 15:103–118
Vespraskas MJ, Sprecher SW (1997) Overview of aquic conditions and hydric soils. In: Aquic conditions and hydric soils: the problem soils, vol 50. Soil Science Society of America Special Publication, pp 1–22
Vespraskas MJ, Craft CB (eds) (2015) Wetland soils: genesis, hydrology, landscapes and classification, 2nd edn. CRC Press, Boca Raton, FL, 508 p
Webb TH, Fahey BD, Giddens KM et al (1999) Soil-landscape, and soil-hydrological relationships in the Glendhu experimental catchments, east Otago uplands, New Zealand. Aust J Soil Res 37:761–785
Wheatley RE (1988) Ochre deposits and associated bacteria in some field drains in Scotland. J Soil Sci 39:253–264

6 Granular Soils

Near Pukekohe on Granular Soil
the market gardens feed us all
'tis not the onions that make one cry
but the expanding city drawing nigh.

Abstract

Granular Soils form on old (≫c. 45,000 years), strongly weathered, tephra mainly on rolling hills and terraces in the northern and western Waikato and South Auckland areas, and on mafic volcanic rocks in Northland. Granular Soils develop on the older tephras where young tephra layers are thin or absent. The soils are dark reddish brown, moderately acidic, sticky, and clay rich. The clays dry to form strongly developed aggregates that give the soils their distinctive granular structure, and their name. Granular Soils comprise just 1% of New Zealand soils and are important for vegetable production. 'Pukekohe onions' are a well-known product of these soils along with potatoes, salad greens, and brassicas. Granular Soils around Pukekohe provide about one-third of New Zealand's fresh vegetable production. Some Granular Soils have been used continuously for horticulture for over 70 years. Granular Soils are extraordinarily resilient to frequent ploughing, although erosion needs to be managed. They can have reduced workability after heavy rain and are naturally low in nutrients. The Granular Soils of Pukekohe and Bombay benefit from a frost-free climate, allowing vegetable production in late winter. There is concern that the productive Granular Soils in the Pukekohe/Bombay area are being lost to urban development.

6.1 Important Features of Granular Soils

6.1.1 Concept and Key Features of the Soil Order

Granular Soils are clayey, strongly structured, soils formed mainly from highly weathered rhyolitic and andesitic tephras, chiefly the Hamilton Ash beds, which are considerably older than c. 45,000 years. Some Granular Soils are formed on strongly weathered basaltic to andesitic lava flows where the Hamilton Ash beds do not occur, either where tephra was not deposited or has been eroded off. The soils have high clay contents and when wetted, and rubbed between the fingers, the soils become sticky and plastic and may be easily remoulded with little cracking. Granular Soils are typically strongly structured (as the name implies) because the iron oxides and aluminium hydroxides are just sufficient to modify the stickiness and improve the soil fabric, so that dry or moist materials may be easily parted into small aggregates.

The subsoils have dark, or bright, reddish brown to brown colours (Fig. 6.1). At some sites a number of altered tephra beds of the Hamilton Ash sequence, buried clay-rich soil horizons (paleosols), are discernible (Sect. 6.2). Clay also accumulates in the subsoil by moving down through the profile in suspension (a process known as clay illuviation or

A. E. Hewitt et al., *The Soils of Aotearoa New Zealand*, World Soils Book Series,
https://doi.org/10.1007/978-3-030-64763-6_6

Fig. 6.1 Granular Soil. Left: seed bed prepared on Granular Soil near Bombay. The brown-coloured surface soil indicates that the topsoil is somewhat low in organic matter, probably as a result of a combination of loss from erosion, and biodegradation of organic matter due to regular ploughing. Right: Granular Soil profile, 1.4m high, showing typical strong soil aggregation (pedality) in upper horizons and different colours relating to paleosols on different (weathered) parent materials at depth

illimerisation) forming clay rich (argillic, cutanic, or cutanoxidic) horizons. The clay-rich horizons restrict plant rooting depth and have slow soil permeability which, combined with compaction at the base of the plough layer, can lead to water saturation in the topsoil after heavy rain. Where the soils have been continuously cropped they, almost everywhere, lack a strongly developed dark topsoil. Although nutrient levels are generally low, with appropriate fertiliser inputs, the soils remain highly productive.

6.1.2 Areas of Occurrence

Granular Soils cover 1% of New Zealand and occur only in the northern North Island (Fig. 6.2). Granular Soils are most common in the lowlands of the northern Waikato and on rolling land in the western Waikato and South Auckland regions. They also occur on rolling land on the foothills of the Hapuakohe Range in the Matamata-Piako District, on foothills of the Coromandel and Kaimai Ranges near Paeroa, and on the elevated land in western Waikato around Raglan and north towards Naike. Granular Soils also occur in Auckland and Northland, associated with basalt and andesite lava flows.

6.1.3 Variation Within the Granular Soil Order

Four soil groups are defined within the Granular Soils in the *New Zealand Soil Classification* (Fig. 6.3):

Perch-gley Granular Soils—periodic wetness caused by a perched water table,
Melanic Granular Soils—less acid and higher natural fertility than other soil groups, transitional to Melanic Soils,
Oxidic Granular Soils—transitional to Oxidic Soils, and
Orthic Granular Soils—other Granular Soils.

6.1.4 Origin of the Soil Order Name

The name 'Granular' is a reference to the well-formed, fine polyhedral or spheroidal soil structure that is frequently present in the soils. The term 'granular' was used as part of the name for these soils (brown granular loams and clays) in the 'General Survey of the Soils of North Island' by N. H. Taylor and I. J. Pohlen in 1954 and then adopted in their New Zealand genetic soil classification in 1962. Earlier, the soil survey of Waipa County, published in 1939, defined the Hamilton series and included the Hamilton clay loam (an important Granular Soil) and a related phase within a group named 'brown granular clays'.

6.2 Soil Profile Genesis

Marked typically by rolling or gently undulating slopes, relatively stable land surfaces, and a warm, humid climate, the parent materials of Granular Soils have been weathered to clays. Most Granular Soils are formed in the strongly

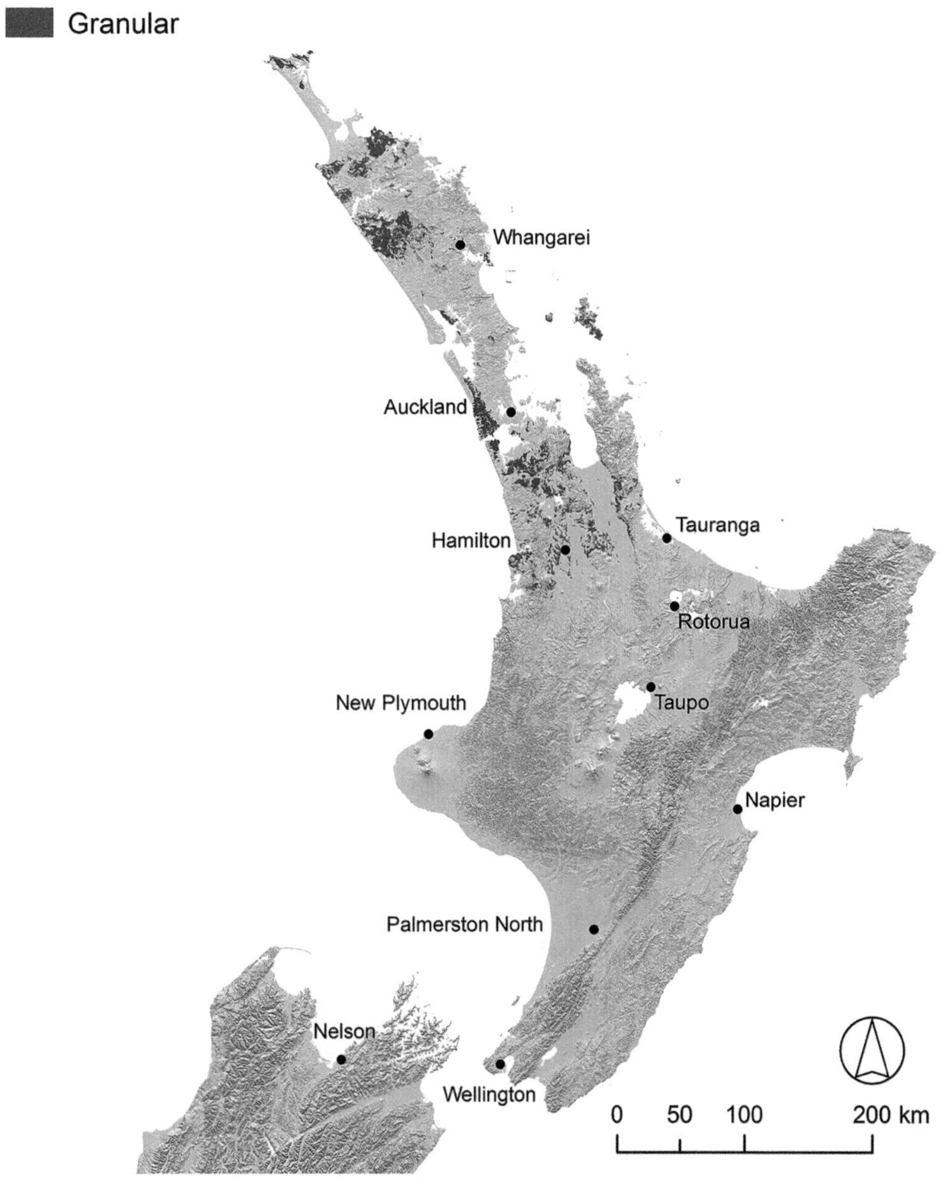

Fig. 6.2 Map of distribution of Granular Soils

weathered tephras of the Hamilton Ash Formation (derived predominantly from eruptions in central North Island), which have no, or less than a few decimetres of, later tephra deposits.

The Hamilton Ash Formation comprises a series of now-altered tephras deposited in the northern North Island from about 340,000 to c. 75,000 years ago. The upper beds in the sequence are inferred to be between c. 125,000 and 75,000 years old, but have not been dated directly. Based on tephrochronology, however, they are considerably older than c. 45,000 years, the age of Rotoehu Ash that overlies them unconformably. Apart from the sandy basal unit (bed H1 or Rangitawa Tephra, c. 340,000 years old), the entire Hamilton Ash sequence, up to c. 6 m thick, is dominated by clay (c. 60–90%) and beds often contain relict clay coatings that represent (paleo) argillic horizons, and also often coatings of MnO_2 (mangans).

In the subsoil, clay coatings (clay skins or cutans) can be seen on some aggregate surfaces, indicating that downward movement of clay through the soil has occurred. Such movement is also evident from increasing ratios of fine-clay to total-clay with increasing depth. The soils also become more weathered with depth suggesting that 'developmental upbuilding pedogenesis' (Chap. 2) dominated in forming the soils. In developmental upbuilding, the incremental accumulation of thin tephra materials has allowed weathering and soil formation to commence on each tephra at the soil surface before it is gradually buried by new material.

In some places in the Pukekohe area the Hamilton Ash-derived soil may be covered by a thin mantle of

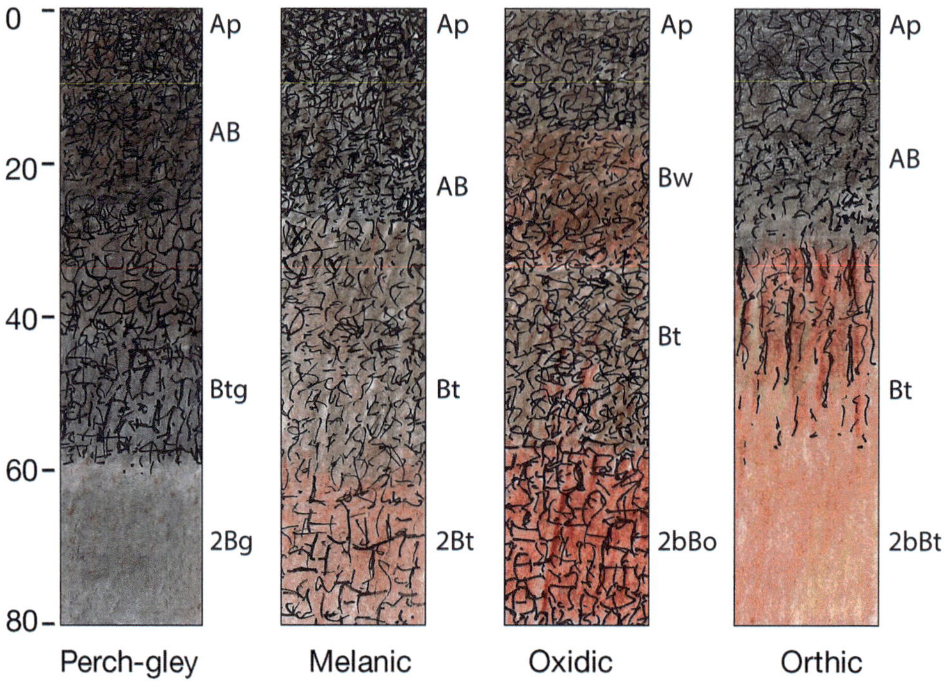

Fig. 6.3 Groups within the Granular Soil order. Vertical axis is depth (cm)

indistinct, younger tephra deposits ≤ c. 45,000 years old. Where the thickness of younger tephra is ≤ 25 cm the soils remain as Granular Soils because the bulk of the profile is dominated by the clayey and well-structured Hamilton Ash beds.

Underlying the Hamilton Ash Formation are older, strongly weathered tephra beds termed the Kauroa Ash Formation that range in age from about 2.3 to ≥ c. 0.8 Ma (million years). The oldest Kauroa Ash beds (which were first described by Bill Ward in 1967) were possibly derived from volcanic centres of the Coromandel Volcanic Zone but the remainder (erupted since c. 2 Ma) were likely derived from centres in the Taupo Volcanic Zone. The strongly altered Kauroa Ash beds are extremely clay rich and are marked by strongly developed paleosols that are among the oldest known in New Zealand, forming remnants of a remarkably ancient paleo-landscape dating to about 1 million years in the northern Waikato and South Auckland regions. Thus the Kauroa Ash beds form an extended, composite 'giant' welded paleosol that represents up to about 1.5 million years of developmental upbuilding pedogenesis. In a few places, Granular Soils encompass both Hamilton and Kauroa ash beds in the profile within the upper 2 m of the land surface. In some places older tephra beds, and/or underlying weathered geological materials, have been exposed by the erosion of younger cover tephra beds, a process that can lead to 'exhumed paleosols' being exposed at the modern soil surface (Fig. 6.4).

In situations where the Hamilton Ash beds were not deposited, such as in Northland, the parent materials of Granular Soils may include basaltic and andesitic rocks (aged from about 2–20 million years). In the Pukekohe-Bombay area magnesium- and iron-rich basaltic lavas (c. 1.6–0.5 million years old) underlie the soil-forming Hamilton Ash beds and occasionally, on steep slopes, are exhumed (exposed at the surface due to erosion), forming contemporary Granular Soils.

The soils on lava flows will likely have more oxidic properties than those formed entirely on altered Hamilton Ash beds so may form Oxidic Soils (Chap. 9), or Oxidic Granular Soils, alongside the more common Orthic Granular Soils. At some sites the soil profiles include both tephra and rock materials with reddish brown to brown, 1–3.5 m thick, Hamilton Ash beds, over buried soils. The buried soils, developed on the underlying weathered basaltic deposits, are usually a distinct dusky reddish brown colour. Additions of aeolian or reworked materials may be locally important.

Although similar in most respects, the Granular Soils of the Pukekohe-Bombay area have more allophane, gibbsite, and crystalline iron oxides (haematite and goethite) than those in the Waikato. As a consequence, the soils in the Pukekohe area tend to be less sticky and have better tilth, than the equivalent mainly halloysitic and heavy Granular Soils in the Waikato region. Why the differences in clay mineralogy occur is uncertain but the variation, which impacts the soil properties, may be a result of different climatic and vegetational histories associated with the development of the soils (through upbuilding pedogenesis) over at least 125,000 years.

The Pukekohe soils have probably always been under mainly forest cover (Sect. 6.4.4), whereas the soils in the

Fig. 6.4 Surface erosion has led to development of a series of soils (exhumed paleosols) formed on sequences of strongly weathered tephra beds and colluvium at Waiterimu in the north Waikato. **A** Granular Soil on Hamilton Ash Formation. **B** The pale-coloured H1 bed at the base of the Hamilton Ash Formation (Rangitawa Tephra, 0.34 Ma). **C** Extremely weathered dark reddish brown Granular Soil on Kauroa Ash Formation (>0.8 Ma), one of the oldest soils in New Zealand. **D** Ultic Soil on weathered greywacke colluvium (?). Photo: H. S. Gibbs

Waikato have also experienced a substantial period of drier conditions with shrub- and grassland dominated cover during cold periods including the last glacial maximum. Thus the soils in the Pukekohe area have effectively had continuous, wetter, and warmer, conditions which favour the formation of allophane and gibbsite through enhanced silicon leaching. In contrast, the drier conditions in the Waikato area, indicated by grassland/shrubland vegetation during the coldest periods, may have favoured halloysite formation because of reduced silica leaching.

6.3 Soil-Landscape Relationships

In the Hamilton Basin, Granular Soils are formed on the mid-Quaternary-aged rolling hills, such as the low hill in the suburb of Hillcrest where the University of Waikato is situated in Hamilton, and northward and north-eastward of Hamilton including in the vicinity of Morrinsville. The hills in the Hamilton Basin are old, Mid-Quaternary landforms that represent a paleo-topography that was formed before c. 340,000 years ago.

To the west of the Waikato Basin, and in Northland, Granular Soils form mainly on basaltic and andesitic lava flows that form rolling land to hill country, such as those of the extensive basalts of the Tutumoe Plateau north of Dargaville where Granular Soils merge with Oxidic Soils. The old basaltic landscapes of South Auckland around Pukekohe and Bombay (developed on volcanic landforms dating from c. 1.6 to 0.5 Ma) are for the most part draped with between c. 1 and 3.5 m of altered Hamilton Ash beds. Thus, most of the highly productive soils in the area are *not* formed on the basaltic deposits, as is commonly assumed, because the pre-Hamilton basaltic soils (now paleosols) are buried up to 3.5 m below the land surface by Hamilton Ash beds.

In steeper landscape positions, such as on the upper slopes of lava flows around Bombay and Pukekohe East where the land steepens towards ridge crests, the modern surface soils are developed in the reddish brown-coloured exhumed paleosols on the basaltic materials, the mantling brown-coloured Hamilton Ash beds having been removed by erosion. Thus the landscape in the Pukehohe-Bombay area comprises a mosaic dominated by brownish soils (Granular Soils on Hamilton Ash) and much smaller areas, near steep crests, of markedly reddish (paleo)soils (Oxidic Soils on exhumed paleosols on basaltic lavas and associated deposits).

The occurrence of Granular Soils in the northern Waikato is spatially patchy and dependent on the upper weathered Hamilton Ash beds being at the land surface, with the cover-bed of young ($\leq$ c. 45,000 years) tephras $\leq$ c. 25 cm (Naike soil). Where the soils have a thicker cover-bed of young tephras (c. 60 cm or so) that are dominated by halloysite (not allophane), the soils are classed as a special two-storeyed Ultic Soil (Chap. 16), classified as a Buried-granular Yellow Ultic Soil (Kainui soil). The classification recognises the subsurface 'granular' character associated with the weathered Hamilton Ash in the lower profile. In places, both the 'Buried-granular' Ultic Soils are visible directly alongside Granular Soils over very short

Fig. 6.5 High soil variability over short distances, the result of tree overturn, exposed on a road cutting just north of Hamilton. The areas where dark reddish brown soil material extends to within ~25 cm of the land surface are Typic Oxidic Granular Soils. Buried-granular Yellow Ultic Soils occur where there is more than about 60 cm of paler, cover-bed tephras over the reddish brown material (e.g. immediately left of the person). The soils are underlain by older paleosols (yellowish brown, cream, or brown beds) in the Hamilton Ash sequence

distances (metres to tens of metres) in the same outcrop or road cutting (Fig. 6.5).

On steeper slopes including on the higher hills and steep lands to the west of Hamilton, and steeper terrain of the Kaimai and Coromandel Ranges, the Hamilton Ash beds have generally been eroded, leaving the underlying basement rocks as the soil parent material, usually forming Ultic Soils or strongly developed Brown Soils. On the lower lying flatter areas much of the central Waikato region has younger, river deposited, materials, some with a thin cover-bed of post-20,000-year-old tephras, and peat accumulations. Hence the Granular Soils are confined to the low rolling hills. South of Hamilton, the mantle of younger tephras ($\leq$ c. 45,000 years), being closer to central North Island and Taranaki source volcanoes, is sufficiently thick ($\geq$ c. 80 cm) to bury the weathered Hamilton Ash beds so that the ensuing soils are Allophanic Soils (Chap. 2) rather than Granular Soils.

6.4 Key Soil Properties

6.4.1 Soil Composition

Strongly weathered tephras, >45,000 years old, dominate the parent materials of most of the Granular Soils (Sects. 6.2 and 6.3). The weathering has severely altered the parent materials. The soils are dominated by kaolin-subgroup clays, kaolinite and halloysite, together with subordinate iron oxides (likely to be goethite and haematite on the basis of colour) and oxyhydroxides and gibbsite with occasional allophane, vermiculite, and secondary silica polymorphs.

6.4.2 Physical Properties

The defining properties of Granular Soils are as follows:

(1) the polyhedral (also known previously as nut) structure, which is usually well developed in the topsoil and upper subsoil, with blocky structures that break readily to finer polyhedral peds in the lower subsoil (Fig. 6.1), and

(2) the presence of clay skins, and usually an associated increase of clay, in the subsoil (forming cutanic, argillic, or cutanoxidic horizons).

The polyhedral soil structure is unusually stable in the upper soil horizons. The Patumahoe soil, for example, has a well-developed microstructure with 90% of the material that is finer than 20 μm formed into aggregates including many small, irregular iron oxide nodules (10–50 μm). Thus the topsoil can be formed into fine seed beds and provides a great medium for plant root growth and water movement and storage.

However, in subsoils the soil dry bulk density may be high with associated firm soil strength and limited porosity, with macroporosity values of less than 10% and as low as 4%. The restricted porosity, in deeper subsoil horizons may limit access to plant available water in dry periods (Fig. 6.6, Table 6.1).

The polyhedral soil aggregates in Granular Soils are resistant to breakdown from ploughing. Instead, the Granular Soils are effectively self-mulching, i.e. after cultivation the surface tends to re-aggregate into a strong granular structure which is quite persistent, making this a particularly resilient soil which can sustain continuous cropping.

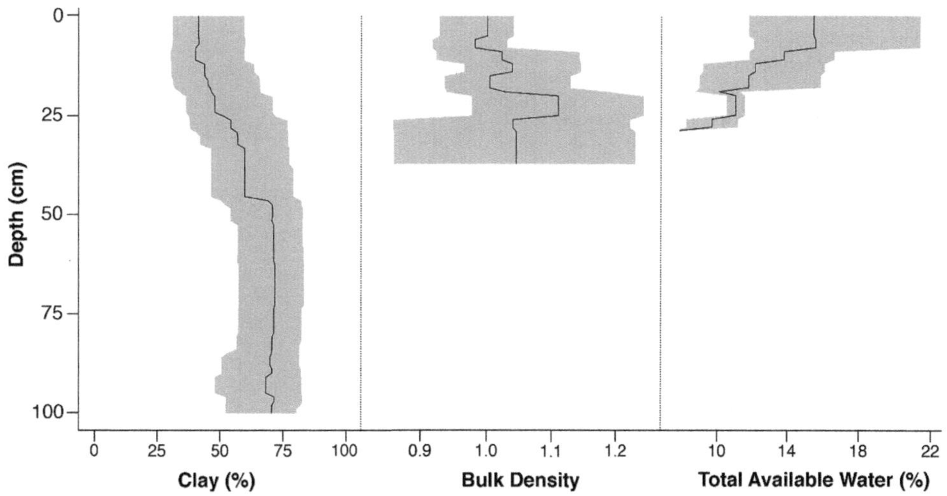

Fig. 6.6 Median, and upper and lower quartiles, of clay content, soil dry bulk density (t m^{-3}), and total available water-holding capacity for Granular Soils in the New Zealand Soil Data Repository

Table 6.1 Soil physical properties for a Typic Oxidic Granular Soil (Naike, SB9579)

Horizon	Depth	Sand (%)	Silt (%)	Clay (%)	Dry bulk density (t m^{-3})
Ap	0–12	8	39	52	0.89
Bw	12–20	5	26	68	0.95
Bt1	20–48	1	16	83	1.14
Bt2	48–73	1	14	84	1.11
Bt(f)1	73–97	1	17	82	0.94
Bt(f)2	97–131	1	16	82	0.93
Bw(f)	131–153	3	23	73	1.01
Cu	153–165	7	27	64	1.01
Cu(f)	165–181	8	27	64	1.02

Despite the high clay contents (up to 90% in subsoils) the soil remains friable with low plasticity and fine soil structure. However, when wet, topsoils have limited workability and trafficability, especially in the Waikato area. The soil dry bulk density is often highest at about 20–30 cm depth, indicating compaction at the base of the plough layer in market garden soils, and just below the depth of pugging in soils supporting cattle grazing during wet periods (Fig. 6.6).

Water infiltration rates in topsoils can vary from slow to rapid depending on management. However, the subsoils of Granular Soils are moderate or slowly permeable so perched water tables and surface soil saturation may occur in heavy rain. The soil features that restrict drainage may also affect root extension. Root extension in subsoils may be limited by high penetration resistance and aluminium toxicity as well as wetness and associated poor aeration.

6.4.3 Chemical Properties

In Granular Soils the soil pH varies greatly (Fig. 6.7 and Table 6.2). In undeveloped soils the pH is generally moderately acid, but under intensive horticulture the pH is generally managed (using lime addition) to ensure suitable conditions for crop growth. The strong weathering in the Granular Soils means they have naturally low reserves of nutrients such as phosphorus, potassium, and magnesium, particularly in the oxidic group. Phosphorus retention is medium in topsoils and medium to high in subsoils, due to iron (haematite) and aluminium oxides (gibbsite) derived by weathering especially where the soils are developed mainly on mafic parent materials. The high anion retention in subsoils means that sulphate also tends to be strongly adsorbed in B-horizons.

The cation exchange capacity is medium to high in topsoils and medium in the subsoil (Table 6.2). Base saturation in topsoils may be high or very high as a result of fertiliser use, whereas in subsoils base saturation ranges from medium to low. Clay activity ranges from moderate to low with the cation exchange capacity spanning the 16 $cmol^{(+)}kg^{-1}$(clay) limit of low activity clay.

Low nutrient availability is a manifestation of the weathered nature of the soil with weatherable minerals having been broken down, and dissolved constituents

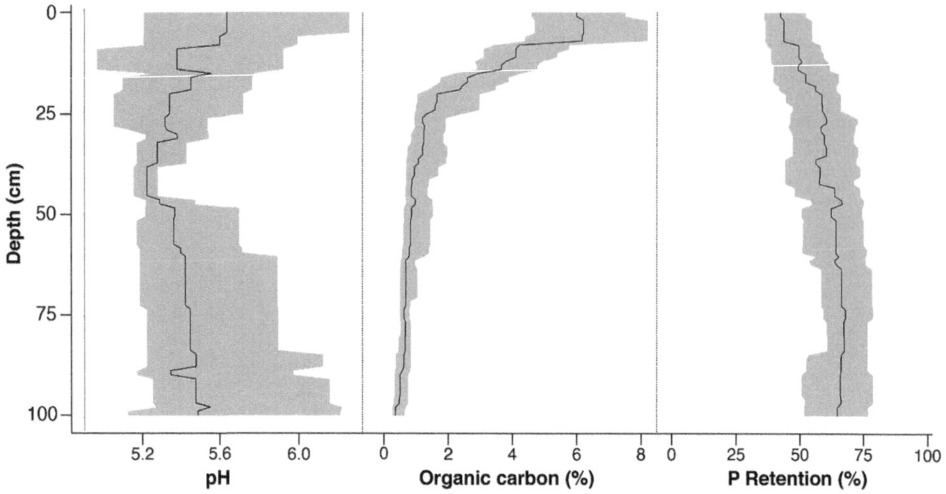

Fig. 6.7 Median and upper and lower quartiles of soil pH, organic carbon, and P retention for Granular Soils in the New Zealand Soil Data Repository

leached from the soil, over a long time (hundreds of thousands of years). When soil formation began the parent material would have been rich in minerals and volcanic glass shards that were readily weathered and easily broken down by hydrolysis with release of nutrients for storage and uptake by plants. The advanced weathering of Granular Soils does not significantly limit crop production provided normal fertiliser management is applied.

Granular Soils—an endangered species?

Granular Soils are of particular importance for horticulture in the area around Pukekohe. A favourable climate, resilient soil, and proximity to half of New Zealand's population make the area important for food growing. It is the only area that produces quantities of fresh green vegetables in the late winter season. However, greater Auckland city is rapidly encroaching. Much of the flatter or gently sloping land that comprises the roads, industrial development, and housing, in South Auckland is on Granular Soils. During major earthworks, that are undertaken during road construction and modern urban subdivision, topsoils are removed and the land surface is recontoured. The changes are such that many areas that were Granular Soils could now be classified as Anthropic Soils in much of the newer urban environment.

Gone are the days of the 'quarter acre section' and a large productive home garden. Under current intense urban development much of the land is covered by buildings and sealed surfaces with minimal soil left exposed to support any form of plant production. In 2017 it was reported that 16% of the Granular Soils around Pukekohe had been taken over for housing in just the *previous 10 years* with a further 5000 houses in the pipeline and a further 9000 planned in the next ten years.

While food prices remain relatively low the market for horticultural land cannot compete with that for housing, thus without some strong intervention we are in danger of losing some of our most productive soils to intense urban development as well as 'rural fragmentation' for lifestyle blocks and rural industry such as extensive glasshouses.

At the time of writing the New Zealand government was working on a proposal for a National Policy Statement (NPS) to help protect our most highly productive land. In 2019 the New Zealand Society of Soil Science stated, in their submission to the proposed NPS, that '*highly productive land and soil are a strategic national non-renewable resource that is finite in supply and needs to be sustained for this generation, and for generations to come. Sustaining food supply, enabling versatility in crops that can be grown, and minimizing impact on wider environmental values are core to NZ community wellbeing*'. It will likely be some years before any NPS will be implemented and take effect.

Table 6.2 Soil chemical properties of a Typic Oxidic Granular Soil (Naike series, SB9579)

Horizon	Depth	pH (in H_2O)	Carbon (%)	Nitrogen (%)	CEC[a] (cmol$^{(+)}$ kg^{-1})	Sum bases (cmol$^{(+)}$ kg^{-1})	P retention (%)
Ap	0–12	5.6	6.6	0.48	25.4	13.8	52
Bw	12–20	5.6	2.5	0.16	17.8	9.34	63
Bt1	20–48	5.3	1	0.06	16.1	5.88	70
Bt2	48–73	5.2	0.7	0.03	16.0	4.03	75
Bt(f)1	73–97	5.3	0.6	0.03	15.5	3.55	75
Bt(f)2	97–131	5.5	0.4	0.01	14.0	3.64	70
Bw(f)	131–153	5.4	0.3	0.01	12.7	3.08	64
Cu	153–165	5.3	0.2	0.01	11.4	2.43	58
Cu(f)	165–181	5.3	0.2	0.01	10.1	2.26	51

[a]CEC = cation exchange capacity

6.4.4 Biological Properties

In pre-human times (i.e. before c. 750 years ago) in the Holocene (post-glacial) period, Granular Soils would have supported a diverse lowland rainforest with dominant trees including tawa (*Beilschmiedia tawa*), totara (*Podocarpus totara* and *Podocarpus laetus*), rimu (*Dacrydium cupressinaum*), miro (*Prumnopitys ferruginea*), and karaka (*Corynocarpus laevigatus*). However, over the more than 100,000 years in which the Granular soils have been forming, the vegetation cover in northern North Island differed regionally. Pollen studies show that forest (with some shrubland) cover remained more or less intact north of Auckland during glacial periods, the composition changing somewhat between cold and warm intervals. In the Waikato region to the south, however, grassland and shrubland, with only small patches of forest, endured during the cold periods, especially the last glacial maximum (c. 32,000–18,000 years ago).

Burrowing creatures, including kiwi and the large native earthworms, would have been active in turning over leaf litter and soil material. The value of the timber, lowland topography, proximity to areas of greatest human population growth, and relative accessibility, mean that most areas of Granular Soil are now in pasture, horticulture, or urban or peri-urban development.

6.5 Distinguishing Between Granular Soils and Related Soil Orders

Granular Soils are one of three orders of strongly weathered clay-enriched soils (Granular, Oxidic, and Ultic Soils) that occur mainly in northern New Zealand. Granular Soils have increased illuvial clay and clay coatings in the subsoil. Most horizons of Granular Soils lack the weak strength, friable failure, and low plasticity of the Oxidic Soils, although such properties can occur in lower subsoils of Granular Soils (forming the oxidic group of the order). Granular Soils are also distinguished from Oxidic Soils by the absence of low activity clays (indicated by a cation exchange capacity of less than 16 cmol$^{(+)}$ kg^{-1}(clay)). Low activity clays are characteristic of Oxidic Soils but do not serve as a clear differentiation from Granular Soils. Instead, the presence of an oxidic horizon (which contains sesquioxides as well as kaolin-subgroup minerals and hence has weak ped strength and friable failure) within the upper 60 cm of the soil profile defines the Oxidic Soil.

Granular Soils generally have dark reddish brown or dark brown colours, unlike the paler yellow colours of Ultic Soils. An important 'hybrid', the acidic Kainui soil in and north of Hamilton, has features pertaining to both Ultic and Granular Soils and is separated into a unique subgroup, a Buried-granular Yellow Ultic Soil. The upper profile comprises halloysite-rich young tephra deposits ($\leq$c. 45,000 years old), about 60 cm thick, overlying a buried 'granular' soil formed on weathered Hamilton Ash beds.

Where the mantle of young tephras exceeds about 80 cm in thickness, the upper profile is typically allophanic, not halloysitic. Where the allophanic soil material is at least 35 cm thick, within the upper 60 cm, the soils are classed as Allophanic Soils rather than Granular Soils. Such soils generally occur to the south (and southwest and southeast) of Hamilton nearer recently active volcanic centres.

6.6 Correlation with Other Soil Classification Systems

The Granular Soils generally fall within the Ultic Soils of *Soil Taxonomy* and Acrisols in the *World Reference Base* (Table 6.3).

Although New Zealand soils are rarely sampled to depths of two metres, the correlation with Ultisols in *Soil Taxonomy* can usually be confirmed. Internationally, soils with argillic horizons frequently have matching clay depletion horizons, commonly albic horizons, in the upper subsoil. In New Zealand soils the depletion horizons are uncommon, or not

Table 6.3 Correlation[a] between Granular Soils and *Soil Taxonomy*, *World Reference Base*, and the New Zealand genetic soil classification

New Zealand Soil Classification	*Soil Taxonomy*	*World Reference Base*	NZ genetic soil classification
Orthic Granular Soils	Kandiudults, Hapludults	Haplic Nitisols	Brown granular loams and clays
Oxidic Granular Soils	Kandiudults, Palehumults, Kandihumults	Ferralitic Nitisols	Strongly leached brown granular clays
Melanic Granular Soils	Hapludults	Mollic Nitisols	Brown granular loams and clays
Perch-gley Granular Soils	Epiaquults	Epistagnic Nitisols	Brown granular loams and clays

[a]The correlations given here are a guide only and for accurate classifications the relevant soil classification documents should be consulted. The two major international soil classification systems are *Soil Taxonomy*, which was developed in the USA, and *World Reference Base*, which was developed primarily in Europe. The NZ genetic soil classification was used in NZ prior to 1992

as pale as their counterparts in other countries (EB- or pale Bw-horizons rather than E-horizons), and this can cause difficulties in arriving at clear correlations with international soil classes. One reason for the difference is that in the New Zealand environment there has been continuing accumulation of material (mainly tephra and loess) on much of the land surface over a considerable time. Another is that, broadly, the high average rainfall and mainly non-deciduous vegetation (conservative nutrient cycling) tend to favour a podzolization rather than argilluviation pathway.

Granular Soils, with moderate clay activity and low base saturation, should correlate with *World Reference Base* Acrisols but they will fail to meet the strong clay contrast between an upper albic horizon and underlying clayey argic horizon. Granular Soils more closely correlate with Nitisols, though they do not develop the shiny ped faces common in Nitisols. This might be explained by the weathered tephric component in many New Zealand soils. Tephra may be responsible for the flocculated surface appearance of ped faces.

6.7 Use and Management of Granular Soils

6.7.1 Uses for Granular Soils

Granular Soils are important for New Zealand vegetable production. Extensive areas under horticulture at Pukekohe and Bombay demonstrate the productivity of these soils when they are well managed despite high clay contents, stickiness when wet, and low natural fertility. The soils are, however, subject to erosion under long-term cultivation. The area around the Pukekohe Hill was cleared of forest between about 1876 and 1900. The dense timber from puriri trees (*Vitex lucens*), which were common in the area, was in demand for railway sleepers. Specialised growing of onions and potatoes began in the 1890s. Many Granular Soils, around Pukekohe, have been used for horticulture continuously or semi-continuously for >40 years, some for >70 years, and some for >110 years, often with more than one crop each year.

Vegetables grown around Pukekohe help to feed Auckland and other regions of New Zealand and are also exported. The 'Pukekohe long-keeper onions' are a particularly well known product of Granular Soils with an annual production around Pukekohe of about 120,000 tonnes, 90% of which are exported, particularly to the Northern Hemisphere off season. Also widely grown are potatoes, salad greens, and brassicas. Production yields are among the highest in New Zealand: onions 55 t/ha; potatoes 60 t/ha; cabbage 30 t/ha. The vegetables grown in the Pukekohe area provide about one-third of New Zealand's fresh vegetable production. A report in 2018 by Deloitte showed that vegetable production generated about $330 million per year on the 4,359 ha growing vegetables around Pukekohe on predominantly Granular Soils.

The Granular Soils of Pukekohe benefit from a unique climate, especially the frost-free status and thermic soil temperature regime, allowing vegetable production in late winter when the area is the main supplier of green vegetables to New Zealand. In the Waikato, most Granular Soils are in pastoral, predominantly dairy, farming with some associated maize cropping.

6.7.2 Urban and Peri-urban Development

Granular Soils have no major limitations for urban development. They provide adequate substrate for road and building construction. However, there is concern that the productive Granular Soils of South Auckland are being lost to urban development and the fragmentation of the soil resource to lifestyle blocks (see grey box). Growers are also concerned about increasing 'reverse sensibility' as urban neighbours object to necessary horticultural activities that generate noise or odour. Rising land values mean that growers have to keep increasing intensification to maintain economic operations. This need can lead to some sustainable practices such as crop rotation being utilised less, and higher fertiliser and pesticide inputs, to maintain production of only the most lucrative crops. With growing urban populations there is also increased pressure on water resources with a need to balance urban water requirements against irrigation to support food production.

6.7.3 Soil Water Management

Topsoils are usually free draining with moderate permeability. Subsoils can show some evidence of redox reactions with veins of low-chroma colours, marking localised reduced iron common in Granular Soils in the Waikato region. MnO_2-rich coatings on ped faces are also common including in soils in the Pukekohe area. The Perch-gley group of the Granular Soils can, however, become waterlogged and poorly aerated during wet periods of intense rainfall. When wet, the soil may be susceptible to livestock treading and cultivation damage but the risk is usually lower than in Ultic Soils. There is a risk of soil water deficit in summer, thus, for reliable crop production, irrigation is desirable.

6.7.4 Soil Erosion

Soil erosion is not generally an issue on Granular Soils under pasture because the landforms are commonly gently undulating to rolling rather than hilly or steep. However, when they are ploughed, given the generally rolling topography, Granular Soils are prone to surface erosion under heavy rainfall. Les Basher and Craig Ross have carried out extensive research into the occurrence, and effects, of soil erosion in the Pukekohe vegetable growing areas. In 2002 they showed, using ^{137}Cs as a marker (since the accumulation of ^{137}Cs from atmospheric nuclear testing, starting in the 1950s), that soil had been removed from the upper slopes of many paddocks and redeposited at the lower slopes (Fig. 6.8). They found that in the paddocks studied there was an overall loss of soil of 7–30 t $ha^{-1}yr^{-1}$, due mainly to water erosion of ploughed soil. However, within individual fields the losses could be up to 92 t $ha^{-1}yr^{-1}$ from upland areas with deposition of as much as 100 t $ha^{-1}yr^{-1}$ near the foot of slopes. The redeposition of the soil reflects the strong aggregate stability with larger particles redeposited when water flow slowed (Fig. 6.6). Of the soil lost from the fields only about 1% was estimated to be 'lost' from the catchment (to the Manukau Harbour). Much of the rest was redeposited in areas marginal to the paddocks, or in roadside ditches. At some sites farmers have moved soil back from the lower slopes up to the higher areas of the paddocks.

There was a notable storm in 1999 (considered to be a 2% annual probability rainfall event) which caused severe runoff. Soil and onions were carried as a slurry and memorably deposited in the streets of Pukekohe. Although the erosion from ploughed fields was severe in places, areas under pasture or cover crops were not badly impacted by the 1999 storm.

Much of the erosion takes the form of surface rill or sheet erosion. A compacted layer forms at the base of the plough layer which restricts water infiltration, and the increasing clay content with depth in the subsoils would exacerbate this restriction. The soil in the plough layer is often cultivated to a fine tilth with a high infiltration rate. When a high intensity rainfall event occurs, the surface soil can become saturated, and then the water, and soil, flow along the top of the plough layer (Fig. 6.9). Basher and Ross noted that large areas of

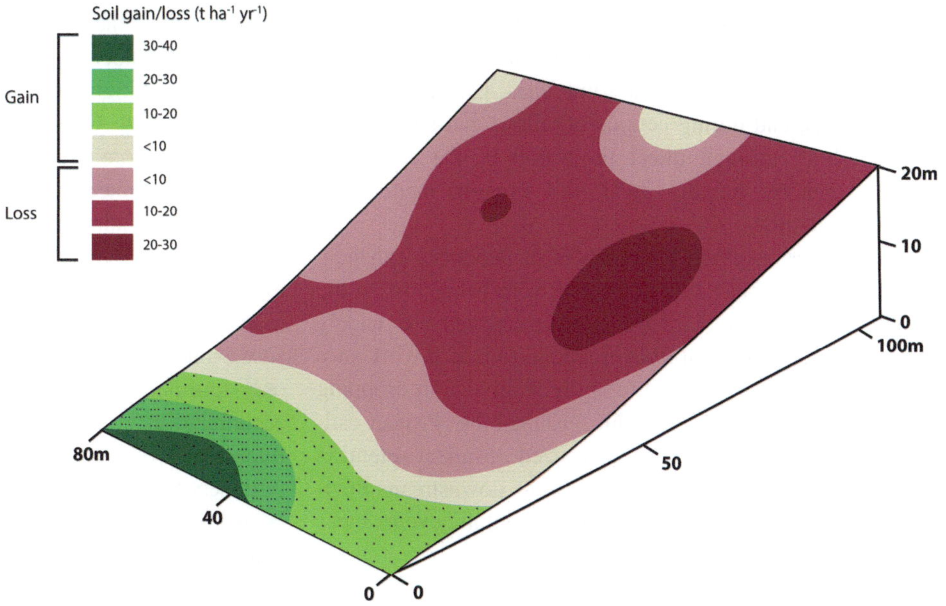

Fig. 6.8 Soil redistribution and loss from a field in Pukekohe (adapted from work by Craig Ross and Les Basher)

Fig. 6.9 Granular Soils under horticulture near Pukekohe. Left: rill erosion following a rainstorm. Right: soil profile showing compaction at base of plough layer. Water saturates the soil above the compacted layer and then it moves downslope. Photos: Craig Ross

these soils have effectively lost the full depth of their topsoil. The extent of this loss is visually obvious quite commonly because the cultivated fields can look brown or reddish brown because upper subsoils (Bt horizons) rather than darker A horizons (topsoils) are exposed (Fig. 6.1). However, in effect a new topsoil is established, albeit with a reduced soil carbon content and without the darkish colours. The use of fertiliser, and the largely favourable friability and porosity of the upper subsoils, appears to mask any discernible impact on productivity.

Compacted wheel tracks, especially when they run up and down the slope, rather than along contours, are also sources of runoff and erosion during intense rainstorms. Ripping with a tined implement (or cultivation) can break up the compacted soil and increase infiltration and has been shown to reduce erosion by 95% at some sites. Cover crops, on wheel tracks, and during periods of fallow, can help reduce erosion, both through direct protection of the soil from erosion and by helping build soil organic matter and aggregate stability.

Grass filter strips on lower edges of paddocks can help trap eroded soil, and thus prevent it from being lost to the wider environment. Sediment retention ponds are widely used to prevent sediment from escaping from earthwork sites and there could be potential to apply them also to cropping sites. One study by the Franklin Sustainability Project estimated that when an appropriately sized sediment retention pond was used, the annual soil loss to downstream environments following high intensity rainfall events was reduced by two thirds.

6.7.5 Sustainable Soil and Environmental Management

Maintaining soil structure, preventing compaction and erosion, and maintaining soil fertility are all important aspects of sustainable management of the Granular Soils under intensive horticulture. Although Granular Soils are exceptionally resilient as already noted, frequent ploughing leads to reduced soil organic matter (Fig. 6.10) and degraded soil structure. Incorporating compost, green manure crops, and other organic materials, or including periods of pasture within the crop rotation, could help ameliorate the loss of soil carbon.

When wet, the dominant sticky clays can also reduce workability. Under intensive market gardening there has been a tendency for high fertiliser use leading to some soils having excessively high Olsen P values. Thus soil tests should be undertaken before fertiliser addition. Cadmium accumulation is also high at many sites as a result of past high phosphorus fertiliser use. Cadmium occurs naturally in some of the rock deposits from which phosphate fertilisers are made. Cadmium is strongly adsorbed in the soil, hence amounts can increase gradually as more phosphate fertiliser is applied. At sites with high cadmium there is now a requirement to use only fertilisers that are low in cadmium to prevent further accumulation in the soil. Most recently, accumulations of fluorine, derived from long-term applications of phosphate fertilisers to farmlands (including on Granular Soils) in the Waikato region have essentially doubled the average total F concentrations in surface soils (from ~220 to 440 mg/kg).

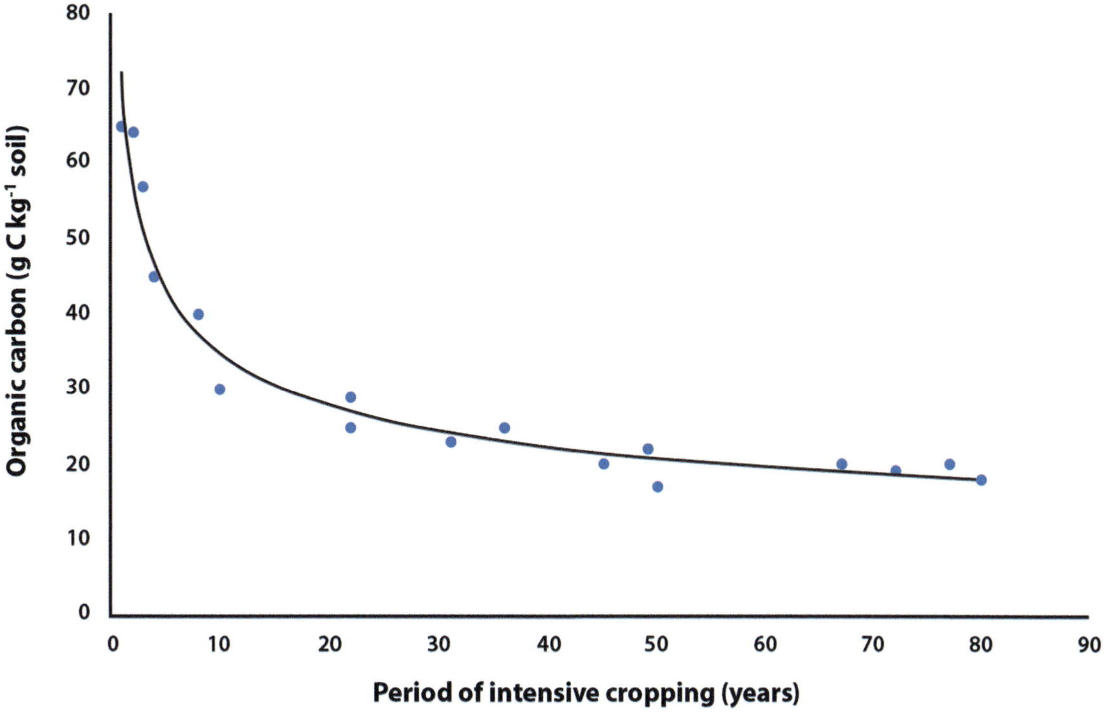

Fig. 6.10 Decline in organic carbon with time in Allophanic Oxidic Granular Soils (Patumahoe soils) under intensive vegetable cropping, Pukekohe (adapted from Haynes and Tregurtha, 1999)

Nick Kim and Matthew Taylor have suggested that possible impacts could include increased toxicity to grazing animals (e.g. chronic fluorosis), altered soil chemistry and functioning (e.g. reduced organic matter turnover, accelerated weathering), and increased deleterious effects on the wider environment (e.g. contamination of groundwater).

There is increasing concern about off-site environmental effects, not only related to sediment in waterways and harbours, but also issues of nutrients (nitrogen and phosphorus) and pesticides leaching to groundwater or reaching waterways through runoff. Various strategies to reduce N leaching, especially during winter, are useful. One is to apply N fertiliser about a month after planting potatoes, not at the time of planting, because potatoes only start taking up N after 30 days. Similarly, growing cover crops during autumn fallow periods adds organic matter to the soil and the cover crops will also take up potentially leachable N from the soil.

Further Reading

Bakker L, Lowe DJ, Jongmans AG (1996) A micromorphological study of pedogenic processes in an evolutionary soil sequence formed on Late Quaternary rhyolitic tephra deposits, North Island, New Zealand. Q Int 34–36:249–261

Basher LR, Ross CW (2002) Soil erosion rates under intensive vegetable production on clay loam, strongly structured soils at Pukekohe, New Zealand. Australian J Soil Res 40(6):947–961

Basher LR, Ross CW (2010) Progress in understanding erosion rates and management at Pukekohe. In: Lowe DJ, Neall VE, Hedley M et al (eds) Guidebook for pre-conference North Island, New Zealand 'Volcanoes to Ocean' field tour (27–30 July, 2010). 19th World Soils Congress, International Union of Soil Sciences, Brisbane. Soil and Earth Sciences Occasional Publication No 3, Massey University, Palmerston North, pp 1.4–1.11

Basher LR, Moores J, McLean G (2016) Scientific basis for erosion and sediment control practices in New Zealand. Report prepared for Tasman District Council. Landcare Research Ltd, 121p. http://www.tasman.govt.nz/tasman/projects/environmental-projects/erosion-and-sediment-control-science-review/

Briggs RM, Okada T, Itaya T et al (1994) K-Ar ages, paleomagnetism, and geochemistry of the South Auckland volcanic field, North Island, New Zealand. N Z J Geol Geophys 37:143–153

Bruce JG (1978) Soils of Part Raglan County, South Auckland, New Zealand. New Zealand Soil Bureau Bulletin 41. DSIR Wellington. Maps + 102p

Bruce JG (1979) Soils of Hamilton City, North Island, New Zealand. New Zealand Soil Survey Report 31. 65 pp + 1 sheet 1: 20,000

Curran-Cournane F, Fraser S, Hicks D et al (2013) Changes in soil quality and land use in grazed pasture within rural Auckland. N Z J Agricu Res 56:102–116

Curran-Cournane F, Vaughan M, Memon A et al (2014) Trade-offs between high class land and development: recent and future pressures on Auckland's valuable soil resources. Land Use Policy 39:146–154

Curran-Cournane F, Golubiewski N, Buckthought L (2018) The odds appear stacked against versatile land: can we change them? N Z J Agricu Res 61:315–326. https://doi.org/10.1080/00288233.2018.1430590

Deloitte (2018) New Zealand's food story: The Pukekohe hub. Report prepared for Horticulture New Zealand, August 2018. http://www.knowledgeauckland.org.nz/assets/publications/New-Zealands-food-story-Pukekohe-hub-Hort-NZ-Deloitte-Aug-2018.pdf

Department of Scientific and Industrial Research, and Department of Agriculture (1954) General survey of the soils of the North Island of New Zealand. Soil Bureau Bulletin 5. DSIR. Wellington. Maps + 285p

Edbrooke SW (compiler) (2001) Geology of the Auckland area. Institute of Geological and Nuclear Sciences 1:250,000 Geological Map 3. 1 sheet + 74p IGNS, Lower Hutt

Environment Waikato (2008) The condition of rural water and soil in the Waikato region—risks and opportunities. Waikato Regional Council (Environment Waikato), Hamilton, 60 pp. https://www.waikatoregion.govt.nz/assets/PageFiles/10480/Soil%20and%20water%20issues.pdf

Flynn JJ (2005) Growing up on the Hill. J.J. Flynn, Pukekohe, 160p

Francis GS, Trimmer LA, Tregurtha CS (2003) Winter nitrate leaching losses from three land uses in the Pukekohe area of New Zealand. N Z J Agricu Res 46:215–224

Gibbs HS, Cowie JD, Pullar WA (1968) Soils of North Island. In: Soils of New Zealand Part 1. N Z Soil Bur Bull 26(1):48–67

Grange LI, Taylor NH, Sutherland CF et al (1939) Soils. in Soils and Agriculture of Part of Waipa County. DSIR Bulletin No 76, pp 30–63

Haynes RJ, Tregurtha R (1999) Effects of increasing periods under intensive vegetable production on biological, chemical and physical indices of soil quality. Biol Fertil Soils 28:259–266

Hayward BW (2017) Out of the Ocean, into the Fire. History in the rocks, fossils and landforms of Auckland, Northland and Coromandel. Geosci Soc N Z Misc Publ 146, 336 p

Holland P, Rahman A (1999) Review of trends in agricultural pesticide use in New Zealand. MAF Policy Technical Paper 99/11. http://www.maf.govt.nz/mafnet/rural-nz/sustainable-resource-use/resourcemanagement/pesticide-use-trends/pestrends.htm

Kim ND, Taylor MD (2017) A tale of two metals. In: Massey C (ed) The New Zealand land and food annual: no free lunch: can New Zealand feed the world sustainably? Massey University Press, Palmerston North, pp 101–118

Kim ND, Taylor MD, Drewry JJ (2016) Anthropogenic fluorine accumulation in the Waikato and Bay of Plenty regions of New Zealand: comparison of field data with projections. Environ Earth Sci 75:147. https://doi.org/10.1007/s12665-015-4897-2

Lowe DJ (2008) Kainui silt loam and Naike clay, Gordonton Rd [Hamilton Basin]. In: Lowe DJ (compiler), Guidebook for Pre-conference North Island Field Trip A1 “Ashes and Issues”, 28–30 November, 2008. Australian and New Zealand 4th joint soils conference, Massey University, Palmerston North. New Zealand Society of Soil Science, Christchurch, pp 55–72

Lowe DJ (2010) Pukekohe silt loam, Pukekohe Hill. In: Lowe DJ, Neall VE, Hedley M et al Guidebook for pre-conference North Island, New Zealand ‘Volcanoes to Ocean’ field tour (27–30 July, 2010). 19th world soils congress, international union of soil sciences, Brisbane. Soil and Earth Sciences Occasional Publication No 3, Massey University, Palmerston North, pp 1.12–1.23

Lowe DJ (2019) Using soil stratigraphy and tephrochronology to understand the origin, age, and classification of a unique Late Quaternary tephra-derived Ultisol in Aotearoa New Zealand. Quaternary 2(1):9. https://doi.org/10.3390/quat2010009

McCraw JD (2011) The wandering river: landforms and geological history of the Hamilton Basin. Geosci Soc N Z Guideb 16:85p

McLeod M (1984) Soils of the Waikato lowlands. NZ Soil Bureau District Office Report HN 11. 32p + 2 sheets 1: 31,680

McLeod M (1992) Soils of the Hauraki Plains County, North Island, New Zealand. New Zealand Department of Scientific and Industrial Research Land Resources Scientific Report 31; 162 pp + 1 map 1: 50,000

Molloy L, Christie Q (1998) Soils in the New Zealand Landscape, the living mantle. Second edition. N Z Soc Soil Sci 253 p. ISBN 0-908783-37-x

Newnham RM, Lowe DJ, Green JD (1989) Palynology, vegetation, and climate of the Waikato lowlands, North Island, New Zealand, since c. 18 000 years ago. J R Soc N Z 19:127–150

Newnham RM, McGlone MS, Moar NT et al (2013) The vegetation cover of New Zealand at the Last Glacial Maximum. Q Sci Rev 74:202–214

Orbell GE (1974) Soils of Government horticultural research areas, Pukekohe, New Zealand. NZ Department of Scientific and Industrial Research Soil Survey Report 20. 15 pp + map 1: 2000

Orbell GE (1977) Soils of part Franklin County, South Auckland, New Zealand. NZ Department of Scientific and Industrial Research Soil Survey Report 33

Parfitt RL, Pollok J, Furkert RJ (1981) Guide book for tour 1, North Island. International ‘Soils With Variable Charge’ conference, Palmerston North, Feb 1981

Richardson BF (2021) The price we pay for land: the political economy of Pukekohe’s development. J N Z Pac Stud 9(1) (in press)

Salter RT (1979) A pedological study of the Kauroa Ash Formation at Woodstock. Unpublished MSc thesis lodged in the Library, University of Waikato, Hamilton

Shepherd TG (1984) A pedological study of the Hamilton Ash Group at Welches Road, Mangawara, north Waikato. Unpublished MSc thesis lodged in the Library, University of Waikato, Hamilton

Singleton PL (1991) Soils of Ruakura. DSIR Land Resources Scientific Report 5. 127 p + map 1:5000

Taylor NH, Pohlen IJ (1962) Soil survey method: a New Zealand handbook for the field study of soils. N Z Soil Bur Bull 25:242 p

Ward WT (1967) Volcanic ash beds of the lower Waikato Basin, North Island, New Zealand. N Z J Geol Geophys 10:1109–1135

Wilson AD (1980) Soils of Piako County, North Island, New Zealand. NZ Soil Survey Report 39. 171 p + maps

7 Melanic Soils

The Melanic Soil looks just fine
with a pH near to alkaline
formed on limestone or on marble
the black topsoil is a marvel

Abstract

Melanic Soils make up about 1% of New Zealand's soils and are scattered throughout the country where soils are developed on calcium- or magnesium-rich rocks. Melanic Soils form on limestone, marble, calcareous sandstone, and mafic to ultramafic rocks. Melanic Soils occur on limestones that outcrop throughout New Zealand, but particularly along the east coast in Hawke's Bay, Wairarapa, Marlborough, and Otago. They also form on ultramafic rocks in northwest Nelson. The most extensive areas of Melanic Soils are in coastal North Otago on limestone and mafic volcanic rocks (basalts and tuffaceous greywackes). Topsoils are black (giving the soil order its name), deep, and rich in soil organic matter, and smectite clays, that bind the soil together forming strong stable aggregates. The upper subsoil is often relatively thin grading into the underlying parent rocks. Melanic Soils are rich in calcium and/or magnesium which contribute to high natural fertility and near neutral to slightly alkaline soil pHs. The underlying rock forms a physical barrier, limiting root penetration, but water may drain through the cracks. The clayey materials of Melanic Soils are used for cricket pitches because of their strong shrink-swell character. Truffles can form a niche crop on some alkaline Melanic Soils.

7.1 Important Features of Melanic Soils

7.1.1 Concept and Key Features of the Soil Order

Melanic Soils are formed from parent materials rich in calcium carbonate (lime), mainly limestone, marble, and calcareous sandstone, or dark-coloured basic (magnesium- and iron-rich) volcanogenic rocks (basalt, tuffaceous greywacke) or ultramafic rocks (dunite, serpentinite, peridotite).

Melanic Soils are well structured and naturally fertile with particularly dark to black, deep, A horizons (topsoils), thin B or AB horizons, and parent rock (R or CR horizons) within about 60 cm of the soil surface (Fig. 7.1). The soil clay is dominated by smectites, notably montmorillonite, which cause the soil to shrink and crack markedly on drying, and to swell on wetting. The clay mineralogy, together with the high soil organic matter content, promotes a well-developed resilient soil structure.

The high natural fertility is indicated by high saturation of the cation exchange capacity, especially with calcium and magnesium ions. Due to the lime-rich parent materials, soil acidity is generally mild and some Melanic Soils are even slightly alkaline, helping enable humus to spread through the topsoil.

A. E. Hewitt et al., *The Soils of Aotearoa New Zealand*, World Soils Book Series,
https://doi.org/10.1007/978-3-030-64763-6_7

Fig. 7.1 Melanic Soil landscape. Left: vegetables growing on a Vertic Melanic Soil near Oamaru. Weathered mafic rock outcrops are visible on the hill in the distance. Although the soil is dry, the dark colour of the topsoil is still evident. Right: a Rendzic Melanic Soil formed on Oamaru limestone has a black A horizon, with a relatively thin (about 20 cm) Bw horizon over the weathering rock material (photo courtesy of NZ Society of Soil Science)

7.1.2 Areas of Occurrence

Melanic Soils occupy small areas throughout New Zealand (Fig. 7.2) and comprise about 1% of New Zealand soils. Melanic Soils are widely scattered from Northland to Southland where they occur mainly on the dispersed outcrops of Tertiary age limestones or calcareous sandstones and marbles, such as the limestones of the east coast from Hawke's Bay through to Otago, and on marble at Takaka Mountain and Mount Owen in northwest Nelson. They also occur on some basaltic outcrops, particularly in Canterbury and Otago, and on ultramafic rocks. The main outcrop of ultramafic rocks is in the Dun Mountain ophiolite belt in northwest Nelson.

7.1.3 Variation Within the Melanic Soil Order

Five soil groups are recognised within the Melanic Soil Order (Fig. 7.3):

Vertic Melanic Soils—clayey with high capacity for shrink-swell,
Perch-gley Melanic Soils—periodic wetness caused by a perched water table,
Rendzic Melanic Soils—limestone or calcareous rock at shallow depth,
Mafic Melanic Soils—on dark, base-rich igneous or sedimentary rocks, and
Orthic Melanic Soils—other Melanic Soils.

Intergrading subgroups recognise transitions from the order to other related soil orders or soil groups. They are Mottled (to Gley Soils), Vertic (to Vertic Melanic Soils), Peaty (to Organic Soils), and Magnesic (to Mafic Melanic Soils). Other subgroups depart in some specific way from the soil group (Calcareous, Weathered, and Pedal). The subgroup closest to the central concept of the group is identified as Typic.

7.1.4 Origin of Soil Order Name

The term 'Melanic' refers to the characteristic black or very dark topsoil colours (from the Greek *melan*, dark-coloured, black). The term 'melanic' was originally used in a soil name by the South African soil classification to describe soils similar to New Zealand's Melanic Soils.

7.2 Soil Profile Genesis

Melanic Soils form as a result of the weathering processes that occur on limestone, marble, calcareous sandstones, and mafic rocks that are dominated by calcium or magnesium carbonates and sulphates. During weathering, such rocks are readily dissolved with much of the rock mineral material being converted to CO_2, leaving calcium and magnesium ions dissolved in water. Thus, there is little remaining mineral material and hence strongly developed, thick, B horizons do not usually form. The dark-coloured and deep A horizon forms where

Fig. 7.2 Distribution of Melanic Soils in New Zealand

organic matter is strongly accumulated along with some less soluble weathering products of the rocks. Due to the alkaline nature of the parent materials, the resulting soils tend to have higher pHs than most other New Zealand soils.

The clays that form are dominantly smectites (notably montmorillonite) that may shrink and swell by up to 50% during wetting and drying. Thus the soils form strongly aggregated structures, typically very stable, with large cracks forming when the soils dry (Fig. 7.4). In contrast, when wet, Melanic Soils are very sticky and they become massive as the soil materials swell to fill any voids. In Vertic Melanic Soils, distinctive slickensides (smoothly polished surfaces where clayey soil masses have sheared through swelling pressure) are formed in subsoils because of the dominant shrink-swell clays (Fig. 7.4). The strong aggregation in topsoils is also supported by the predominance of dissolved calcium and magnesium ions in the soil solution. The cation exchange sites are thus also dominated by calcium and

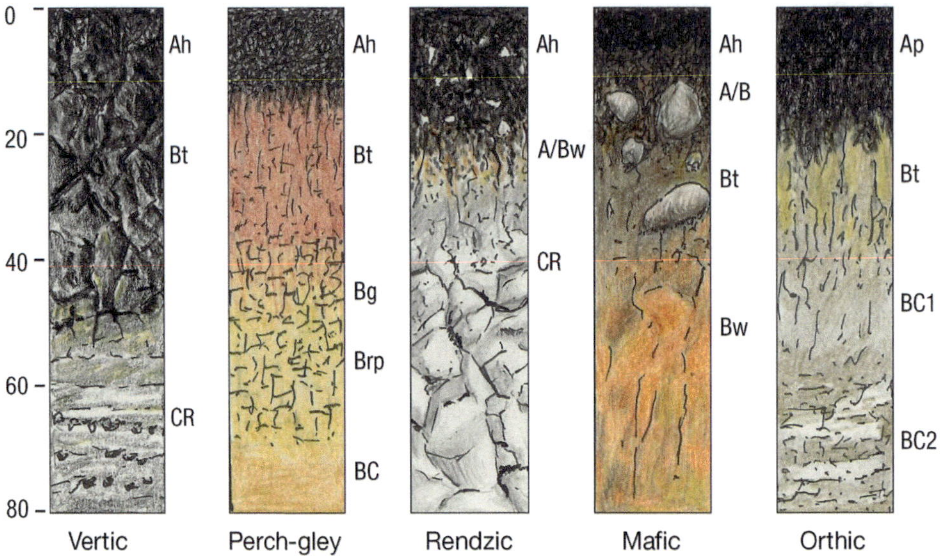

Fig. 7.3 Soil groups within the Melanic Soil order. Vertical scale is depth (cm)

Fig. 7.4 Structural features of Melanic Soil. Left: a strongly aggregated Vertic Melanic Soil (Waipara series) with the cracks between aggregates clearly visible. Photo: Phil Tonkin. Right: shiny slickenside clay surfaces in the subsoil of an Orthic Melanic Soil

magnesium (rather than the monovalent cations sodium, potassium, and hydrogen), giving a relatively thinner cation exchange layer on the surface of negatively charged particles. The thinner exchange layer allows the clay particles and organic materials to approach each other closely, forming strongly bonded, hence stable, aggregates.

The polyhedral peds are self-mulching, meaning they re-form seasonally on wetting and drying, recovering from the impacts of cropping, for example. The strong attraction to the cation exchange layer allows large quantities of organic matter to be adsorbed and stabilised which contributes to the high organic matter content and dark colour of the surface horizon which is often much thicker than in other soils, typically ~40 cm or so in depth.

7.3 Soil-Landscape Relationships

7.3.1 Introduction

The geographic distribution of the Melanic Soils is controlled by underlying rock types and the distribution of surficial materials. Where calcium or magnesium-rich rocks are exposed at the ground surface, Melanic Soils form. Two areas of New Zealand show a concentration of Melanic Soils: the Waipara region of North Canterbury and the hinterland of Oamaru in North Otago and its northern extension into South Canterbury. Melanic Soils also occur on the volcanic terrain around Banks Peninsula and Dunedin, on tuffaceous greywacke in Southland, and on other widely dispersed small outcrops of limestone/calcareous mudstone or ultramafic rocks throughout the rest of New Zealand.

However, not all limestone, calcareous, or basaltic outcrops in a landscape support Melanic Soils. Other materials may cover the limestone or basalt and prevent Melanic Soil formation. The cover beds include loess, tephra, and colluvial debris derived from various materials occurring upslope. For example, in the Waikato region, there are widespread limestone deposits, such as near the world-famous Waitomo Caves, but the soils are predominantly Allophanic Soils formed mainly in the overlying tephra deposits. Areas of limestone and mafic gabbro rocks also occur in the Fiordland region, and the Mt. Arthur region of Nelson, where we might expect to find Melanic Soils. However, the formation of Melanic Soil has been prevented at many sites by erosion or by extreme acid leaching under acid-litter-forming native forest under high rainfall.

7.3.2 Melanic Soils Associated with Limestone Terrain: Waipara and Oamaru

In the Waipara region of North Canterbury, the major geological features comprise limestone ridges (supporting Rendzic Melanic Soils), associated with calcareous sandstone on hills and rolling land with Orthic Melanic Soils. Clayey alluvial silts, clays, or gravels occur at downland margins on foot slopes, and on fans (with Orthic Melanic Soils and Vertic Melanic soils). It has been suggested that the Argillic Orthic Melanic Soils (Tokarahi soils) described in both North Canterbury and North Otago were developed on loess that was originally calcareous.

In the Oamaru hinterland, substantial areas of limestone, marl, and calcareous mudstones occur but Melanic Soils have been suppressed by quartzofeldspathic loess cover. Where loess is absent or thin, Orthic Melanic Soils prevail. Vertic Melanic Soils in Waipara occur on clayey basin-infill alluvium that incorporates calcareous gravels. A comparable geological setting results in a similar suite of soils in Oamaru, but there, the Vertic Melanic Soils occur on weathered basaltic, and calcareous-basaltic, tuffs.

7.3.3 Melanic Soils in Hilly Volcanic Terrain: Otago and Banks Peninsula

The soils of Banks Peninsula and Otago Peninsula, and their pattern in the landscape, have similarities. The hills are dominantly formed from a basaltic suite of mafic volcanic rocks (basalt and andesite in Banks Peninsula, and basalt, phonolite, and trachyte in Otago Peninsula). The volcanos were active 11–6 million years ago. The soil pattern on these volcanic hills is controlled by

- deposition of a loess blanket that has smothered Melanic Soil genesis on lower ridges and flanks of the hills, leading to the formation of Pallic Soils,
- exposure of weathered volcanic mafic rocks with the formation of Melanic Soils,
- high leaching and depression of pH on higher altitude sites, with consequent increase of acidity and loss of Si in soil solution, and thus prevention of smectite clay formation, leading to the development of Brown Soils instead.

The Melanic Soils are thus confined to the intermediate altitude zone. The zone has irregular boundaries because it has been controlled by deposition of loess subject to topographic effects on wind transport and deposition, as well as by rainfall and associated leaching.

7.3.4 The Waiareka Loess Gap

The highly productive market gardens and cropping land on the easy-rolling landscape immediately to the south of Oamaru (Fig. 7.1) owe their productivity to the Waiareka series of Vertic Melanic Soils. The Waiareka clay soils are formed in weathered mafic volcanic rocks. There are also minor areas of the Oamaru series (Rendzic Melanic Soils) on limestone ridges. Surrounding the Melanic Soils are soils formed in a blanket of loess (mainly Fragic and Laminar Pallic Soils). The loess blanket is near continuous in the surrounding landscape apart from the area of Waiareka soils where there appears to be a 'hole' in the loess blanket. The hole has been informally called the Waiareka loess gap, but what is the origin of this apparent gap? One favoured hypothesis was that loess may have settled on the Waiareka soil material but its slow permeability caused overland flow that would have eroded the loess.

During one Waiareka soil sampling effort, a large digger was used to prepare a deep sampling pit. At a depth greater than the usual sampling depth, the investigators were surprised to find buried loess at the base of the weathered mafic-rock soil material. In the midst of the so-called loess gap there was indeed loess—though deeply buried beneath the Waiareka material. This prompted the alternative hypothesis that the weathering mafic rock exposed on slopes was shattered by freeze and thaw processes and transported by solifluction downslope, covering the loess after its deposition. Despite these findings the origin of the loess gap remains largely unresolved and awaits a keen examination of field evidence and paleoenvironmental reconstruction.

7.3.5 Southland Tuffaceous Greywacke

A series of hilly ridges divide the southern plains of Southland from the northern Southland plains of the upper Mataura River valley. The hills are generally known as the Kaihiku-Hokonui region. The rocks are termed tuffaceous greywacke. Unlike the greywacke rocks (Torlesse terrain) and Otago schists further north, the primary materials were lain down as sea-floor sediments while andesitic volcanoes were active in the vicinity. Consequently, the sediments incorporated andesitic eruptives. The soils, described by Peter McIntosh, have higher iron, manganese, titanium, potassium, aluminium, and magnesium than most other New Zealand soils formed on greywackes and schists.

Two soil series in the Melanic Soil order have been defined in this landscape. The Kaihiku soils (Argillic Orthic Melanic Soils) and Stonycreek soils (Mottled Orthic Melanic Soils) are both mapped on dry, north-facing hill slopes in stony colluvial slope debris.

7.3.6 Nelson and Otago-Southland Dunite/Serpentinite

An unusual Melanic Soil is found on ultramafic and mafic rocks in a few places in New Zealand where such rocks, associated with the Dun Mountain ophialite belt, occur at the land surface. This 'belt' comprises a narrow zone that extends through New Zealand where the upper mantle has become exposed at the surface because of tectonism. The ultramafic/mafic rocks form Mafic Melanic Soils on Dun Mountain and Red Hills in the Nelson-Marlborough area, West Dome in Southland, and in Otago and Northland. The serpentinite rocks are also found in the Piopio area in King Country but there the pervasive tephra mantle limits the occurrence of genuine Mafic Melanic Soils.

The ultramafic parent rocks of Mafic Melanic Soils comprise mainly dunite, serpentinite, peridotite, and associated basalts and dolerites, and they are enriched in minerals such as olivine and pyroxene. They have low concentrations of most major plant nutrients and high concentrations of metals such as nickel, chromium, manganese, cobalt, and magnesium, and a high pH. On Dun Mountain (for example, Fig. 7.5), the metal enrichment has meant that the native *Nothofagus* forest has been unable to colonise the soils. The vegetation instead is dominated by stunted shrubs and grasses, which can be toxic to sheep due to their high content of metals. In a study on soils on an ultramafic scree in northwest Otago, led by Bill Lee, it was found that plant diversity and vegetation stature increase with declining heavy metal availability and increasing soil nutrient status. The boundary between the ultramafic and non-ultramafic soils and associated vegetation is often sharp, a notable exception to the usual gradational boundary paradigm associated with many other soils in the landscape.

7.4 Key Soil Properties

7.4.1 Soil Composition

Most Melanic Soils have smectite clays or clay minerals with inter-stratifications involving smectite. Melanic Soils usually have smectitic, illitic, or kandic mineralogy groups. 'Smectite' is the name of a group of clay minerals which includes montmorillonite and bentonite, which are usually associated with soils with strong shrink/swell character, and beidellite which is also found in some New Zealand Podzols. Melanic Soils often also have the clay minerals illite, kaolinite, and vermiculite interstratified with smectite.

7.4.2 Physical Properties

Melanic Soils have a wide range of clay contents (Fig. 7.6, Table 7.1). Some soils, such as the Waiareka (Table 7.1), have high (50–70%) clay contents and they are outside the 75th percentile illustrated in Fig. 7.5. However, all Melanic Soils have sufficient clay to impact the soil properties and the soil materials are sticky and plastic, and are sensitive to water content. Melanic Soils generally have low readily available, but high (20–25%) total available, water contents reflecting the moisture-holding effect of the high clay content. Except for shallow soils on rock, or soils affected by high water tables, potential rooting depths are relatively large giving good overall profile available water.

Fig. 7.5 Landscape near Dun Mountain (peak in background) showing the sharp vegetation boundary that relates to differences in the underlying soils and parent materials. The dark green *Nothofagus* (beech) forest, mid-distance right, is on non-ultramafic rocks and has been unable to colonise the Magnesic Mafic Melanic Soils (Dun soils) on ultramafic rocks which instead support shrubs and grasses

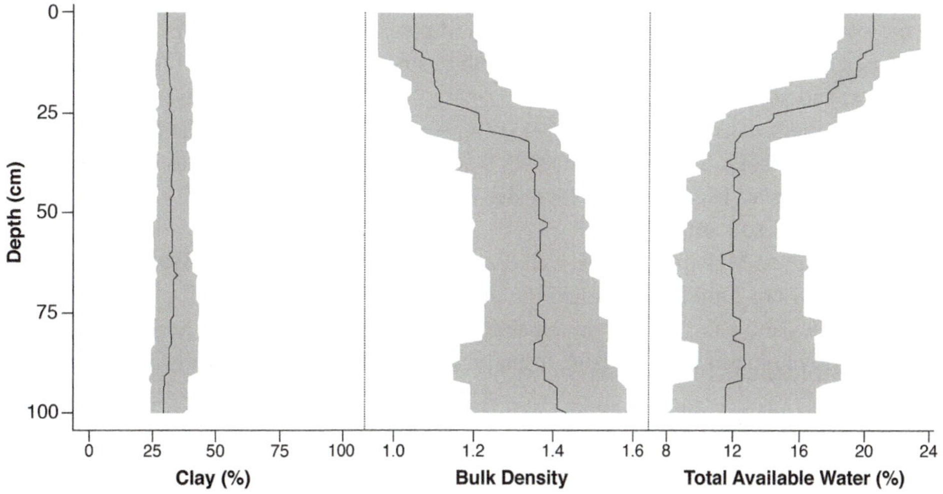

Fig. 7.6 Median and upper and lower quartiles of clay (%), soil dry bulk density (t m^{-3}), and total available water capacity (%) for Melanic Soils in the New Zealand Soil Data Repository

Table 7.1 Physical properties of a Typic Vertic Melanic Soil (Waiareka soil SB09893 and SWAMP)

Horizon	Depth	Sand (%)	Silt (%)	Clay (%)	Dry bulk density (t m^{-3})	RAW[a] (%)	TAW[a] (%)
Ap1	0–21	3	31	65	1.01	3	23
B/A	21–37	3	27	69	0.96	5	22
Bw1	37–64	3	26	70	0.98	6	25
Bw2	64–92	8	42	59	1.1	5	22
BCk	92–113	14	35	50	1.2	5	20
Ck	133–120	10	37	52			

[a]RAW = readily available water, TAW = total available water

An important feature of Melanic Soils is their well-formed soil structure resulting from the presence of swelling clays. During winter or spring, the clays become wet and swell. In summer, the soils dry and shrink, and the soil mass cracks with fissures extending from the surface downward. The cracks extend throughout the soil profile, opening up the planar pores between soil aggregates. If heavy rainfall occurs during summer, when the soils are dry and cracked, much water may be lost to the soil as by-pass flow down the cracks.

As the soils rewet in winter, the soil mass swells and the cracks close. Pressure comes on the peds as the soil heaves upward and sideways. The compression forms the characteristic strong polyhedral, blocky, and prismatic structure. The structure has good stability, which is aided by high soil organic matter, and the stabilising effect of calcium or magnesium bivalent cations in the exchange complex. Structural stability usually also means stability of the soil pores, which is important for aeration, drainage, water storage, and root penetration.

An indicator of shrink/swell is the presence of slickensides. As noted earlier, these are polished surfaces or striations on ped faces caused by lateral ped-on-ped smearing action (Fig. 7.4). The orientation of slickensides shows the relative direction of the internal soil movement in response to the swelling pressure. Significant shrink/swell potential is measured by a coefficient of linear expandability, strong soil structure, and slickensides.

Soil dry bulk density is about 0.75–1.0 t m^{-3} in topsoils and lower in subsoils. The values are lower than expected for the amount of clay because of the porosity associated with the well-formed soil structure.

Although most Melanic Soils are well drained, Perch-gley Melanic Soils are found where gleying (shown by reddish brown mottles and pale, low chroma, matrix colours) occurs above a slowly permeable horizon, usually in the deeper subsoil.

7.4.3 Chemical Properties

Melanic Soils are derived from rocks with significant components of either calcareous or mafic minerals that, by weathering, supply high levels of calcium or magnesium cations (or both) to the developing soil. The calcium and magnesium cations are bivalent cations meaning they have two positive charges per atom.

The cation exchange capacity (CEC) of Melanic Soils is very high (Table 7.2) due to the organic matter content and the smectite and vermiculite clay minerals that provide negatively charged surfaces for exchange. Because of the exchangeable Ca^{2+} and Mg^{2+} being readily weathered from the parent materials, the base saturation is also high. The pH ranges from weakly acid to weakly alkaline. KCl-extractable Al, which indicates soil acidity and the risk of aluminium toxicity, is very low as a result of the medium to high pH derived from the parent materials.

Soil carbon levels are medium in topsoils (Fig. 7.7 and Table 7.2). The nitrogen content of the soil organic matter is relatively high. Thus the resulting C/N ratios are very low which indicates that the soil organic matter is highly decomposed.

Nutrient levels are usually naturally high in calcium, magnesium, and potassium in the well-stocked exchange complex (as indicated by high base saturation). Total phosphorus levels are generally moderate to high. Phosphate retention is low to medium, and thus phosphorus is readily available to plants. However, sulphur may be deficient.

7.4.4 Biological Properties

Melanic Soils are probably the most naturally fertile and biologically active soils in New Zealand, and their structure has the potential to provide good habitats for soil fauna.

Table 7.2 Chemical properties of a Typic Vertic Melanic Soil (Waiareka soil, SB09893)

Horizon	Depth	pH (in H_2O)	Carbon (%)	Nitrogen (%)	CEC[a] ($cmol^{(+)}kg^{-1}$)	Sum bases ($cmol^{(+)}kg^{-1}$)	P retention (%)
Ap1	0–21	6.5	4	0.51	72.2	74.7	36
B/A	21–37	6.8	1.6	0.21	79.1	83.5	42
Bw1	37–64	7.1	0.5	0.08	76	87.4	39
Bw2	64–92	7.8		0.04	66.5		28
BCk	92–113	8.1	0.1	0.03	60.3		26
Ck	133–120	8.2		0.03	58.9		25

[a]CEC = cation exchange capacity

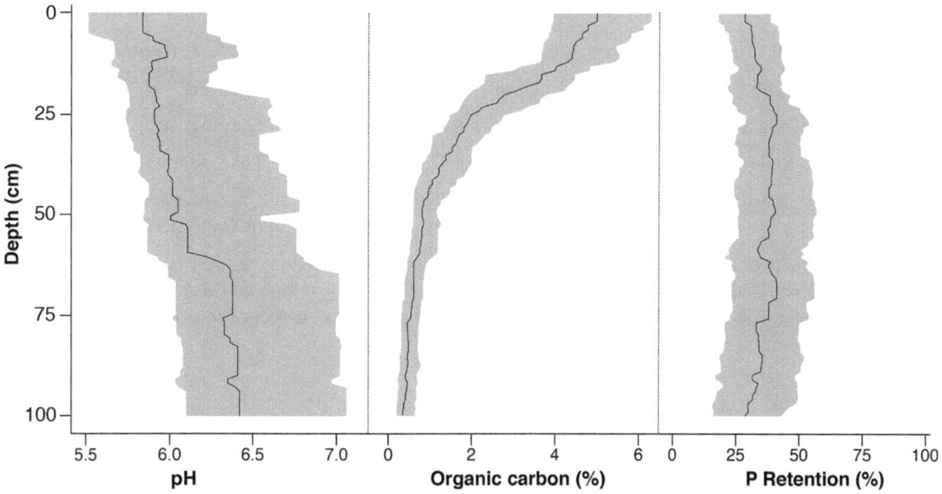

Fig. 7.7 Median and upper and lower quartiles of soil pH, organic carbon (%), and P retention (%) for Melanic Soils in the New Zealand Soil Data Repository

However, in cropping land, soil cracks that open in dry weather may also provide a habitat for crop pests.

Soil influences on grapes and truffles

Melanic Soils have been particularly sought after for the establishment of truffles as it was thought that the 'rendzina' or limestone-rich soil was a key to successful truffle establishment, requiring high pH (≥ 7) as well as warm dry summers and no competing fungi. Truffles are the fruiting bodies of a type of fungus which grows on or around the roots of hazelnut and oak trees. Hence truffle growing enterprises or truffière are established on the Melanic Soils in North Canterbury. However, as the management of the soil mycorrhiza that lead to truffle formation has become better understood, truffles are now grown on a wider range of soils. The first successful truffière were established in New Zealand by mycologist, Dr Ian Hall. The first Périgord black truffles were harvested in July 1993 in the Gisborne area.

Both truffles and grapes are notable because their success in the market sits astride the application of science and the subjectivity of the human pallet. Finding truffles at harvest time is also dependant on the olfactory capabilities of pigs or dogs. Our appreciation of the subtle environmental influences of soil and climate on both wine and truffles is growing. The Waipara district of North Canterbury enjoys a wide variety of soils (including Melanic Soils) that provide many potential outcomes of soil and *terroir*.

Soil and environmental characteristics are highly interdependent so there are few simple relationships we can use to form clear linkages between soils and wine or food qualities. However, it is understood that wine qualities respond to soil moisture and temperature. Heat storage in soil is related to wetness, stoniness, porosity, and drainage. The drier a soil is the faster it will warm up, whereas a wet, poorly drained, soil will be slower to warm up. Soil water storage is related to porosity, horizon thicknesses, and stoniness. For example, a high stone content reduces the volume of soil available to hold water.

7.5 Distinguishing Between Melanic Soils and Related Soil Orders

Melanic Soils stand out from all other New Zealand soil orders because of their dark topsoil colour and strongly developed soil structure. The other diagnostic features relate to the influence of either calcareous or mafic/ultramafic rocks in the parent material. Calcareous influence is indicated by a visible (fizzing) reaction to 10% HCl which reacts with carbonates to give off CO_2. The presence of the mafic rock influence is indicated by a pH of 5.9 or more in the lower B horizon, strong soil structure, and stickiness. Although Oxidic and Granular Soils are both strongly weathered with well-developed soil aggregates, they lack the smectite clays, dark and deep topsoils, and relatively high pH of Melanic Soils.

7.6 Correlation with Other Classification Systems

The Melanic Soils are generally similar to soils known globally as 'Rendzinas'—a Russian-adopted Polish term for soils formed in carbonate-rich parent materials. Melanic Soils mainly correlate with Vertisols and Mollisols in *Soil Taxonomy* (Table 7.3). Globally, there is renewed interest in

Table 7.3 Correlation[a] between Melanic Soil groups and *Soil Taxonomy*, *World Reference Base* and the New Zealand genetic soil classification

New Zealand Soil Classification	*Soil Taxonomy*	*World Reference Base*	NZ genetic soil classification
Vertic Melanic Soils	Usterts	Vertisols	Brown granular loams and clays
Perch-gley Melanic Soils	Aquerts or Aquolls	Gleysols	Gley soils
Mafic Melanic Soils	Ustolls or Udolls	Phaeozems	Brown granular loams and clays
Orthic Melanic Soils	Ustolls or Udolls	Chernozems or Calcisols	Rendzina and related soils

[a]The correlations given here are a guide only and for accurate classifications the relevant soil classification documents should be consulted. The two major international soil classification systems are *Soil Taxonomy*, which was developed in the USA, and *World Reference Base*, which was developed primarily in Europe. The NZ genetic soil classification was used in NZ prior to 1992

so-called 'Black soils' (which include some New Zealand Melanic Soils) as a consequence of their high carbon contents and high soil fertility. Black soils are important as they are globally extensive and productive soil orders. In contrast, in New Zealand the Melanic Soils, which incorporate important features of both Vertisols and Mollisols, are relatively uncommon. The Vertisols are extensive and diverse in Australia where they are informally known as black cracking clay soils, and formally designated as Vertosols which Ray Isbell defined as 'clay soils with shrink-swell properties that exhibit strong cracking when dry and at depth have slickensides and/or lenticular structural aggregates'. Vertisols are dominated by smectite group clays (particularly the mineral montmorillonite). The Mollisols are extensive in natural grasslands regions such as the Great Plains of the USA and the Russian Steppe, where they are known as Chernozems (a term also used in the *World Reference Base* classification), and characteristically have thick, black, A horizons and are well structured, highly fertile soils.

In New Zealand, the Melanic Soils were formerly mainly classified as Rendzinas or Brown granular loams and clays (Table 7.3).

7.7 Use and Management of Melanic Soils

Melanic Soils are generally highly versatile. They are used for intensive cropping in the area south of Oamaru, particularly for potatoes, cabbages, and other vegetables. If you live in the South Island you can often tell that store-bought potatoes have been grown in a Melanic Soil due to the sticky black soil material that adheres to them. Melanic Soils have been sought after for truffle growing as they have some qualities particularly suitable for boutique, high value, truffle production (see grey box).

In many areas, the Melanic Soils occupy small areas bounded by Brown or Pallic Soils and are used, along with the adjacent soils, predominantly for pastoral agriculture to which they are highly suited. Some Melanic Soils have limitations that need to be taken into account, including wetness (Perch-gley group), shallow rooting depth and low readily available soil water, and potential cultivation difficulties due to shallow stony soils.

Soil erosion is not normally considered a significant risk in Melanic Soils. However, because Melanic Soils are particularly suited to cropping, they are often exposed to potential wind and water erosion following ploughing. Exposure of the tilth to desiccating winds may cause fine structure units to dry and shrink with a high risk of loss by wind erosion. In the Vertic Melanic Soil (Waiareka soil) near Oamaru, topsoil has been observed to be blown by north-westerly winds and piled up along fence lines. Like all soils, if left exposed on slopes, rill or sheet erosion may occur in rainstorms because of overland water flow.

The clay-rich Melanic Soils have particularly high shrink-swell behaviour which maintains their soil structure despite intensive use, a property referred to as 'self-mulching' (i.e. the structure re-forms seasonally on wetting and drying so that the soil pedality recovers from cropping impact). The soils are likely to have relatively high resistance to structural damage under cropping unless organic matter is reduced significantly. The structural stability of Vertic Melanic soils (Waiareka soils, Oamaru) was studied by Graham Sparling, Louis Schipper, and others in a sequence of vegetable cropping fields near Oamaru. Fields under continuous cropping were arranged in a land use chronosequence of sites that had been cropped for up to 40 years. It was concluded that where soil structure had been impacted by intensive use, two mechanisms had acted to protect the soils. Firstly, Melanic Soils show resistance to loss of soil organic matter through organic-to-organic, and organic-to-mineral, bonds that form a tough stabilising network. Secondly, Melanic Soil aggregates also have strong mechanical strength which provides resistance to physical rupture of soil structural units.

An expectation prior to the study was that Vertic Melanic Soils would be highly resilient to the impacts of long-term cropping. However, continuous cropping was observed to lead to development of a 'plough pan' (a compacted layer at the bottom of the plough depth) that limits the available rooting depth of the soil, and thus the water and nutrients that are available to the plant.

Soil organic matter is protected by its inclusion within soil aggregates. Even though working on the soil to a fine tilth provides an excellent seedbed, the breaking up of aggregates exposes small pockets of protected soil organic matter to the air. Thus organic decomposition occurs rapidly with the release of carbon dioxide to the atmosphere and loss of carbon from the soil. However, given time under pasture or a cover crop, the soils are likely to recover their good structure. The time to recover a strong soil structure needs to be further researched.

Vertic Melanic Soils around Oamaru are locally known as 'tarry soils' because of their black colour and sticky consistence. When wet, the soils can be a challenge to plough as clods will adhere strongly to the plough sheer and tractor tyres, thus more care than usual needs to be taken to ensure tillage is undertaken at suitable soil moisture contents. Perch-gley Melanic Soils are regularly wet and so it is particularly important to protect them from sustained high impacts of stock trampling, wheel traffic, or cultivation when the soils are wet.

The strong shrink-swell capacity of the Melanic Soils needs to be considered when undertaking foundation works. The soils will shrink and crack when dry, then swell when wet. The capacity for shrink and swell with wetting and drying can sometimes be used to advantage—the Waikari is known to cricketers because of its use for constructing cricket wickets: the highly compacted hard playing surface or pitch provides a fast bounce for cricket balls when dry, and the pitch maintains a smooth surface if it has to absorb moisture.

Rendzic Melanic Soils are usually shallow and on slopes, so are not suited to arable cropping. They may be used for pasture, trees, or perennial crops that do not require mechanised operations. Truffle cultivation is a possibility. The shallow soil depth means they dry out readily in summer and can lack available soil water. The Mafic Melanic Soils are developed in dark, base-rich igneous rocks or derivative sediments. Mafic Melanic Soils generally have higher levels of iron oxides than in other Melanic soil groups which may impart higher soil structural stability than other Melanic soil groups.

Further Reading

Campbell IB (1977) Soils of Waikouaiti County, Otago. In: Soil Bureau Bulletin 37. DSIR Soil Bureau, 59p

Churchman GJ, Velde B (2019) Soil clays—linking geology, biology, agriculture, and the environment. CRC Press, Boca Raton, 250p

Coombs DS, Landis CA, Norris RJ et al (1976) The Dun Mountain Ophiolite Belt, New Zealand, its tectonic setting, constitution, and origin, with special reference to the southern portion. Am J Sci 276:562–603

Isbell RF and National Committee on Soil and Terrain (2016) The Australian soil classification, 2nd edn. CSIRO Publishing, Clayton South, Vic, 141p

Kear BS, Gibbs HS, Miller RB (1967) Soils of the Downs and Plains, Canterbury and North Otago, New Zealand. In: New Zealand Soil Bureau Bulletin 14. Government Printer, Wellington, 92p

Lee WG (1992) New Zealand ultramafics. In: Roberts BA, Proctor J (eds) The ecology of areas with serpentinized rocks. A world view. Kluwer, The Netherlands, pp 375–418

Lee WG, Hewitt AE (1982) Soil changes associated with development of vegetation on an ultramafic scree, northwest Otago, New Zealand. J R Soc N Z 12:229–242

McIntosh PD, Lee WG (1986) Soil-vegetation relationships on the Dun Mountain Ophiolite Belt at West Dome, Southland, New Zealand. J R Soc N Z 16:363–379

Renowden G (2005) The truffle book. Limestone Hills Publishing, Amberley

Robinson BH, Brooks RR, Kirkman JH et al (1996) Plant-available elements in soils and their influence on the vegetation over ultramafic ("serpentine") rocks in New Zealand. J R Soc N Z 26:457–468

Scott JM (2020) An updated catalogue of New Zealand's mantle peridotite and serpentinite. NZJ Geol Geophys 63:428–449

Soil Survey Staff (2014) Keys to Soil Taxonomy twelfth edition. USDA Natural Resources Conservation Service, 360p. http://www.nrcs.usda.gov/wps/PA_NRCSConsumption/download?cid=stelprdb1252094&ext=pdf. Accessed 10 March 2020

Sparling GP, Schipper LA, Hewitt AE et al (2000) Resistance to cropping pressure of two New Zealand soils with contrasting mineralogy. Aust J Soil Res 38(1):85–100

Tomlinson PR, Leslie DM (1978) Soils of Dunedin City and environs, New Zealand. N.Z. Soil Survey Report 37. DSIR, Wellington, New Zealand, 70 pp. ISSN:0110–2079

Tonkin PJ, Webb T, Almond P et al (2015) Geology, landforms and soils of the Waipara and Waikari regions of North Canterbury with an emphasis on lands used for viticulture. Lincoln University and Landcare Research, 220p

Trangmar BB, Cutler EJB (1983) Soils and erosion of the Sumner region of the Port Hills, Canterbury, New Zealand. Part 1 Environment and soils, Part 2 Erosion and its implications for urban use. NZ Soil Survey report 70. NZ Soil Bureau, DSIR, Lower Hutt

Webb TH (2008) Soils. In: Winterbourn M, Knox G, Burrows C et al (eds) The natural history of Canterbury. Canterbury University Press, Christchurch, New Zealand, pp 89–118

Wardle P (1991) Vegetation of New Zealand. Cambridge University Press, Cambridge, p 672p

Wilson JB, Lee WG, Mark AF (1990) Species diversity in relation to ultramafic substrate and to altitude in southwestern New Zealand. Vegetation 86:15–20

World Reference Base (2015 update). http://www.fao.org/3/i3794en/I3794en.pdf. Accessed 11 March 2020

8 Organic Soils

Down in the swamp where we get wet feet
Organic Soils are formed in peat.
Plants may die but they're not decomposed
unless fertiliser and drains are imposed.

Abstract

Organic Soils form from partly decomposed remains of plants in peat-forming wetlands, or in deep accumulations of forest litter. Organic Soils comprise about 1% of New Zealand with the largest areas in the Waikato, on the high rainfall South Island West Coast, and in some moist mountainous areas where soils are often saturated. Organic Soils are acidic (pH $\leq$ ~4), have a low dry bulk density, high C:N ratios, high shrinkage potential, and exceptionally high soil water holding capacity. Organic Soils have been developed mainly for pasture, maize, potato, and blueberry production, and are also mined to extract peat. Organic Soils have naturally low nutrient contents so need inputs of fertiliser and lime, along with drainage, for productive farming. Such inputs hasten the biodegradation of Organic Soils. To slow the rate of degradation, water tables need to be kept high to prevent oxygen from entering the soil. Although about 70% of Organic Soils have been heavily impacted by drainage and development for farming, or by peat mining, some natural areas remain in reserves. The Whangamarino wetland and the rainfed Kopouatai and Moanatuatua bogs are recognised as wetlands of international importance with unique indigenous vegetation, including the tall cane rush (*Sporodanthus ferrugineus*).

8.1 Important Features of Organic Soils

8.1.1 Concept and Key Features of the Soil Order

Organic Soils (Fig. 8.1) are formed where organic matter is dominant in the soil profile in both wetland and forest environments. Organic Soils occur where the partly decomposed remains of wetland plants (peat) is $\geq$30 cm in thickness (within 60 cm of the soil surface), having accumulated under wet, anaerobic (oxygen-depleted) conditions. In forests, Organic Soils form in acidic and/or nitrogen-poor, partly or well decomposed, forest litter that is $\geq$40 cm deep. Organic Soil profiles comprise organic soil materials (defined as having $\geq$18% organic carbon) that may range from undecomposed plant remains through to humic material where the original plant material is no longer recognisable. All parts of the plant may be biodegraded and incorporated into Organic Soils including roots, trunks, branches, leaves, flowers, and seeds. Organic soil material, used to define Organic Soils, specifically excludes fresh litter (L horizons) and living plant material.

The predominant plants at a site influence the structure and nature of the soil. For instance, some peats are formed exclusively from sphagnum moss while others may form from a range of wetland plants including restiads (jointed-rush

A. E. Hewitt et al., *The Soils of Aotearoa New Zealand*, World Soils Book Series,
https://doi.org/10.1007/978-3-030-64763-6_8

plants of the Southern Hemisphere Restionaceae family, which are common in New Zealand), sedges, shrubs, and/or trees. Although some mineral material (such as flood deposited alluvium or tephra-fall deposits) may be mixed through the organic material, or present as thin interbedded mineral layers, the soil as a whole is dominated by organic matter. Thus the soil properties are markedly different from those of mineral soils. Soils on peat are characterised by O horizons and those on forest litter by L, F, and H horizons.

8.1.2 Areas of Occurrence

Organic Soils occur in wetlands, or under forests that produce thick acid litter, most commonly in areas of high rainfall and/or with a high frequency of rain-days. Organic Soils cover 1% of New Zealand, with an estimated total area of 260 000 ha, of which 190000 ha (72%) are under agricultural production (Fig. 8.2). The most extensive areas are in the Waikato lowlands, Westland, Southland, and the Otago uplands. However, Organic Soils occur in small pockets throughout New Zealand where water is trapped at the land surface. Together with Gley Soils, the peat-derived Organic Soils represent the original extent of New Zealand wetlands.

8.1.3 Variation Within the Organic Soil Order

Four soil groups (Fig. 8.3) are defined within the Organic Soils in the *New Zealand Soil Classification*. The groups distinguish between litter-derived (forest) and peat-derived (wetland) soils and further distinguish the degree of decomposition of the soil organic matter as follows:

Litter Organic Soils—partly or fully decomposed litter $\geq$40 cm thick that has accumulated under forest,
Fibric Organic Soils—in peat with plant fibres that are only weakly decomposed,
Mesic Organic Soils—in peat that is moderately decomposed, and
Humic Organic Soils—in peat that is strongly decomposed where plant material is no longer recognisable.

The degree of decomposition of organic soil material is determined by the proportion of a sample that can be recognised as plant fibres. A judgement of the amount of fibres can be made by first rubbing the fibres in the hand, then manipulating the sample in the fist, and then squeezing it firmly. The decomposed material will be forced between the fingers as a black, or dark-brown, slurry leaving the fibrous plant material in the hand. For humic soil materials, less than 15% of the material is left as fibres in the hand. For fibric soil materials, 75% or more of the soil remains in the fist as fibres. Peats with between 15 and 75% fibres are identified as mesic. Although the test is easy it can be somewhat messy.

Soil subgroups include Acidic (with pH predominantly less than 4.5 in top 60 cm of soil) and mellow (pH >4.5). Sphagnic subgroups are dominated by sphagnum moss. Buried-podzol, Buried-gley, and Orthic subgroups are identified within the Litter Organic Soils, formed in forests, to recognise the underlying mineral soils.

Fig. 8.1 Organic Soil and landscape. Left: Mellow Mesic Organic Soil profile from a dairy farm in the Hauraki Plains on the margins of the Kopouatai bog. Pumice of the Taupo Tephra is visible in the profile marking the ground surface in AD 232±10 years. Photo: Anne Wecking. Right: landscape of an Organic Soil used for dairy farming (and occasional maize cropping) in the Waikato (Rukuhia bog). Prior to farm development, both soils are likely to have been Fibric Acidic Organic Soils

Fig. 8.2 Distribution of Organic Soils. Many areas of Organic Soils are too small to feature on a map of this scale

8.1.4 Origin of Soil Order Name

'Organic' refers to the dominance of soil organic matter over mineral soil materials. 'Organic Soil' is commonly used for these soils internationally except for *Soil Taxonomy* and the W*orld Reference Base* that both use the term 'Histosols' from the Greek word *histos* meaning plant tissue.

8.2 Soil Profile Genesis

Organic soils form where conditions inhibit degradation of dead plant material and thus the organic matter gradually accumulates. In New Zealand, most Organic Soils are formed in saturated wetland conditions where lack of oxygen is the

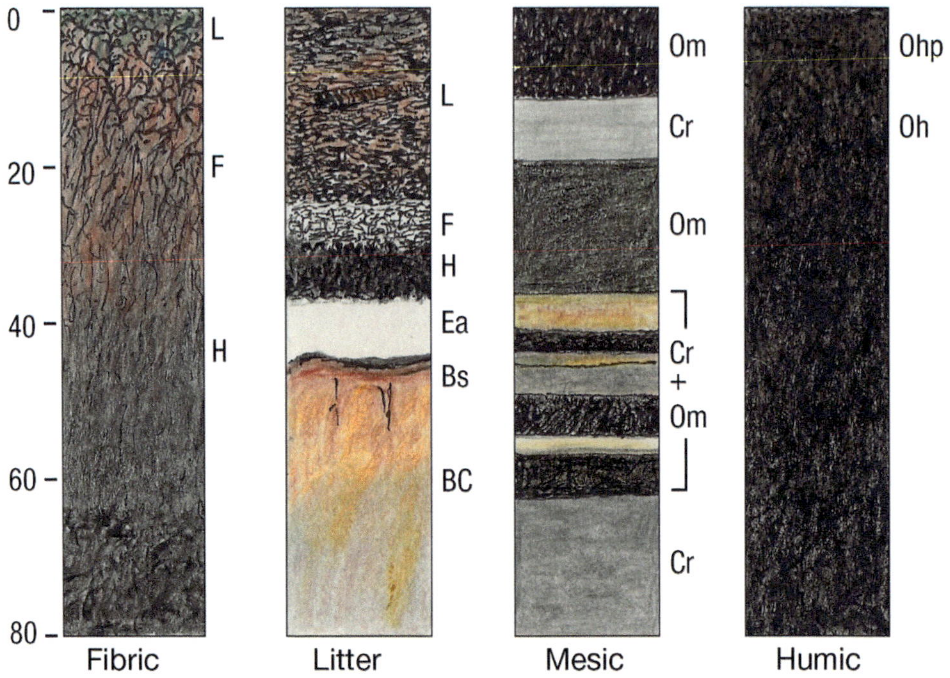

Fig. 8.3 Groups within the Organic Soil order. Vertical axis is depth (cm)

main limit to biodegradation of plant materials—air is replaced by stagnant or slowly moving water. The process is sometimes called paludisation (from the Latin *palus* meaning swamp or marsh). In forest litter environments, such as under kauri or pine trees, it is most often the availability of nutrients, especially nitrogen, along with low pH, and sometimes also lack of moisture, which limit biodegradation. Thus Organic Soils form where limits to degradation of plant material lead to organic matter accumulation. Plant materials that may accumulate to form Organic Soils vary widely from soft-tissue sphagnum and rush to woody material.

Many Organic Soils form as a natural progression from a lake or wetland where water lies permanently on the ground surface. Plants may grow in or near the shallow water margins and gradually fall into the water. When they start to decompose, the microbes deplete the available oxygen, and thus the degradation process slows or stops and the plant materials gradually accumulate. Plants tolerant of the wet, nutrient-poor environment become established on the surface and, in turn, eventually become incorporated into the peat. Over a long period, shallow lakes may be gradually infilled as the peat material accumulates around the lake margins developing zones of 'quaking bog', ultimately leading to extinction of the lake. Thus there may be a progression from a lake to a mire then ultimately, if climatic conditions are suitable, to a raised bog.

However, infilling and extinction do not always occur and some lakes that have been in existence for ~20,000 years or more can remain open despite being surrounded by peat, as is evident in some of the Waikato peat lakes such as Lake Maratoto near Hamilton (Sect. 8.3.2). Counterintuitively, John Green and David Lowe showed that Maratoto maintained its area and increased in depth despite massive peat growth in the catchment. To explain this unusual status, they suggested that the lake receives mineral- and nutrient-rich water from adjacent land that has enhanced microbial breakdown of encroaching peat; wind-induced wave action (produced by prevailing southwest and northerly winds) causes erosion of peat margins especially at the north and south ends of the lake; and wind-induced currents stir and aerate the lake waters for most of the year and so the lake waters are usually saturated with oxygen which will enhance organic breakdown.

It is often possible to date materials within an Organic Soil profile using radiocarbon dating, and in North Island identification of tephras of known ages within the peat (tephrochronology), to gain an estimate of the rate and timing of peat accumulation. Such methods may also help to determine the rate of subsidence following drainage. Soil organic matter decomposition is important as it is part of the process of cycling materials from dead organisms and releasing the nutrients to support new life. This process is true of all soils, but especially Organic Soils because the outputs of partial decomposition produce the very substance of the soil material itself. Decomposition is the breakdown of dead plant material and biochemical transformation of complex organic molecules into simpler inorganic molecules. The process releases methane, carbon dioxide, and water to the environment as well as making plant nutrients available for reuse.

Generally, there are three parts to the decomposition process:

assimilation or uptake of organic material by microbial decomposers, whereby plant or animal material becomes part of the microbial biomass;
mineralisation, which is the conversion of organic molecules to inorganic (or mineral) molecules, which are then available for plant uptake, or may be released to the wider environment, predominantly as carbon dioxide and water, and
immobilisation, which is the conversion of mineral elements to organic matter (in Organic Soils particularly to stable forms of humus that do not break down readily), thus making the nutrients unavailable for plant uptake.

The main conditions for decomposition of plant material are favourable temperature, pH, moisture, and available oxygen and nutrients. Therefore, if an Organic Soil is drained and has fertiliser and lime added, to increase the ability of the soil to support plant growth, microbial growth is also enhanced and biodegradation of the organic material will greatly increase. The decomposition process will ultimately lead to breakdown of the organic matter, with the main products being water and carbon dioxide which escape from the soil. Biodegradation of the organic matter thus leads to losses of organic material to the atmosphere and water table, as well as further surface subsidence because of shrinkage and consolidation of the remaining organic material.

8.3 Soil-Landscape Relationships

8.3.1 Introduction

Organic Soils occur in a wide range of wet environments. Nine wetland types are widely recognised (Table 8.1). The wetland types have different water and nutrient sources and varying plant assemblages, with a range of resulting soils formed on them. Organic Soils are most commonly formed on raised and blanket bogs, fens, and swamps.

Domed or raised bogs, dominated by restiad species, occur predominantly in the Waikato and Northland regions. It takes an abundant water supply (and sufficiently frequent rain-days) to produce an accumulation of peat. Most Organic Soils form in places where the soil water table is close to the land surface, as occurs with fens. However, a dome-shaped raised bog is special because the only water source is from rainfall (referred to as ombrogenous). Many such sites have now been drained and developed, primarily for pastoral agriculture.

Blanket bogs occur in cool, wet climates on flattish to rolling uplands (at relatively high altitude), such as the Lammerlaw Range of east Otago. Blanket bogs are formed from restiad plants (including wire rush) and mosses

Table 8.1 Wetland types in New Zealand and the inferred relationship to the *NZ Soil Classification*

Wetland type	Description	Likely Organic Soils and associated soils
Bog (raised or blanket)	Peatland that receives its water solely from precipitation	Fibric Organic Soils, Acid Mesic Organic Soils
Fen	Peatland that receives water from the surrounding land. Water table is near the land surface. Low to moderate pH	Mellow subgroups of Mesic or Humic Organic Soils Peaty Orthic Gley Soils
Swamp	Mineral or peaty wetland in topographically low sites, slightly acid to near neutral pH	Fluvial Gley Soils. Mellow Mesic Organic Soils
Marsh	Mineral wetland, moderate pH, fed by groundwater, surface fluctuates from wet to inundated, often at margins of water bodies	Fluvial Gley Soils
Seepage	Wet patches on hillsides (often footslopes) releasing modest volumes of groundwater by slow leakage to the surface	Orthic Gley Soil, Mellow Mesic Organic Soils
Shallow water	Standing water most of the time, less than a few metres deep, rivers, lake margins, streams	Gley Raw Soils
Ephemeral wetland	Closed depressions intermittently wet in winter and spring	Gley Raw Soils, Recent Gley Soils
Pakihi	Extremely low fertility formed under high (metres/year) rainfall	Acid Gley Soils Perch-Gley Podzols, Acid Fibric Organic Soils
Saltmarsh	Tidal influence, including saline mudflats	Gley Raw Soils, Fluvial Raw Soils, Saline, Sandy, or Orthic Gley Soils

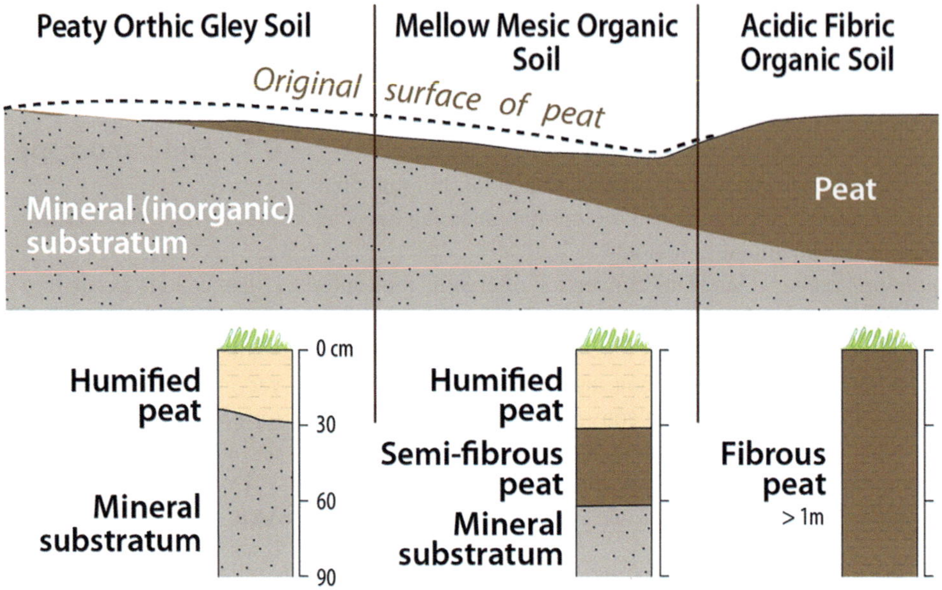

Fig. 8.4 Soil-landscape relationship of Organic Soils in the Waikato Basin

(including Sphagnum) under high rainfall and frequent rain-days (> ~ 160 per year). Fens tend to form in valleys or at the footslopes of hills where runoff and a high water table keep them moist while swamps form in topographic hollows where water ponds. Swamps commonly occur in 'back swamps' on river flood plains where drainage to the river is cut off by a levee along the river bank.

Litter Organic Soils have been little studied and are relatively uncommon. A depth of partly or well-decomposed litter ≥ 40 cm (i.e. F or H horizons) is needed to qualify a soil as a Litter Organic Soil. They occur adjacent to mature kauri trees (*Agathis Australis*) which accumulate exceptionally thick litter layers near the tree. The litter helps kauri to repel competitors, and to conserve nutrients and water and thus recycle them through the root system back to the tree. Disturbance of the litter may compromise a kauri tree's longevity. Litter Organic Soils may also develop under other native forest species, and under *Pinus radiata* plantations where conditions allow the litter to accumulate (although the 40-cm thickness threshold may not always be met). However, the litter layer under pines will rapidly biodegrade following tree harvest, and therefore such soils, if they had been considered Organic Soil, would soon revert to the underlying mineral soil order following tree removal.

8.3.2 Organic Soils in the Waikato Lowlands

About half of New Zealand's Organic Soils are in the Waikato region (which includes the lowlands of the Hauraki Plains and the Waikato Basin). The Waikato lowlands lie in geological grabens that have gradually infilled with river deposits (alluvium) and pyroclastic material (fragmental volcanic deposits including ignimbrite and tephra-fall deposits). Widespread low-lying and poorly drained areas on alluvial surfaces, such as the lower lying parts of the Hinuera Formation, would have supported a patchwork of gleyed soils, swamps, and possibly small shallow lakes. Over time, and with changing climatic conditions that led to rising water tables, many of the wetlands expanded. Peat accumulation rates increased, particularly in the early Holocene, and the wetlands coalesced to form a massive peat bog with a range of associated Organic Soils. On thick peat (>1 m), Acid Fibric Organic Soils are formed (Rukuhia soils). Towards the bog's margins, Mellow Mesic Organic Soils about 1 m thick (Kaipaki soils) give way to Peaty Orthic Gley Soils (Te Rapa soils) as the peat thins to <30 cm (Fig. 8.4).

Many former wetlands have been drained and developed for agriculture, such as the Rukuhia bog south of Hamilton, but some extensive natural wetland areas remain. The most important wetlands include:

- the Torohape and Kopouatai raised bogs (peat domes) in the Hauraki Plains in an area between the Piako and Waihou Rivers (Fig. 8.5);
- the Whangamarino wetland along with a wider mosaic of shallow lakes and mineralised swamps associated with the lower reaches of the Waikato River;
- the Moanatuatua raised bog east of Ohaupo, at one time similar in size to Kopouatai. However, most of the area has been developed for agriculture with now only a remnant of about one square kilometre remaining in natural vegetation which is protected as a scientific reserve (see Fig. 8.12);
- numerous peat lakes situated on the margins of the low hills within the central Waikato (Hamilton) Basin.

Fig. 8.5 Peat dome vegetation and view of the landscape. Left: tall cane rush (*Sporodanthus ferrugineus*) (left, c. 2 m high) and short wire rush (*Empodisma robustum*) (right) on Moanatuatua bog reserve. Manuka shrubs are also evident in background. Right: view of expansive area of the Kopouatai peat dome (photo: Jordan Goodrich)

The Whangamarino wetland and the Kopouatai bog are recognised as wetlands of international importance. The Whangamarino wetland is a 7000 ha mosaic of swamps, fens, and bogs. Baumea (rush), manuka, and wire rush are the most common plant species in the peat bogs, whereas the swamp areas support a wider range of vegetation with introduced willow over much of the area. Many species of mosses, including 13 species first described in the Whangamarino wetland, have been recorded.

The Kopouatai bog is the largest (10,201 ha, 102 km^2) unaltered restiad bog in New Zealand and is recognised internationally (as Ramsar Site 444). The Kopouatai bog predominantly comprises organic matter but does include thin mineral layers from the deposition of tephra-fall deposits and a thin marine-derived layer of mud in northern parts reflecting a period in the mid-Holocene of slightly higher sea level. The total depth of organic material (including minor thin tephra layers) is about 14 m and the peat has been accumulating for about 15,000 years. The Kopouatai bog is the last secure stronghold for the tall (up to 2.8 m) cane rush *Sporadanthus ferrugineus* (Fig. 8.5). Only three Sporadanthus sites remain: Kopouatai and remnants at Moanatuatua and Torehape. There are other modified bog remnants from which the Sporadanthus has now been lost, for example, Otakairangi near Whangarei, and Tangonge, near Kaitaia. However, Sporodanthus is being more commonly used for the urban rain garden and enhancement plantings, particularly in Hamilton city, which may help its survival. Two species, the wire rushes *Empodisma robustum* (north of 38 °S) and *Empodisma minus* (south of 38 ° S), are the main peat-formers in many New Zealand Organic Soils (Fig. 8.5). A different species, *Sporadanthus traversii*, is endemic to Chatham Islands.

Because the only water source for the restiad plants is rainfall, the nutrient content of the soils is very low and the soils are strongly acidic. However, the bog species are highly efficient at extracting their nutrient needs from the meagre soil stores. The dominant jointed wire rushes (*Empodisma robustum*) form a tangled mat of dead litter, roots and stems with tiny leaves. The jointed wire rush, aided by the taller *Sporadanthus ferrugineus*, form a dense layer at the soil surface that shades the underlying surface of the bog water table. This reduces evapotranspiration from the bog surface to levels normally expected of a semiarid site and such areas have been described as 'a wet desert'! Therefore, while it is difficult for plants to grow on the surface, the water content is maintained within the bog, thus preserving the peat.

Bev and Bruce Clarkson, and Peter de Lange, have researched the taxonomy of the restiad bogs, their stratigraphy, and depth. Dave Campbell, Louis Schipper, and their collaborators have worked extensively on the hydrology, carbon balance, and gas exchange between the bog and the atmosphere. Work on tephra deposits and past environments and climates has been undertaken by researchers including David Lowe, Rewi Newnham, de Lange, and Maria Gehrels.

On the margins of Rukuhia bog, profiles obtained by ground-penetrating radar have enabled the peat-substratum to be readily mapped, along with the sub-peat materials including tephric alluvium and older volcanogenic deposits. Some radar reflections on the sub-peat surface were interpreted as tree stumps or logs and confirmed by coring.

Prior to about 25,000 years ago, the central Waikato lowlands, including the area around what is now Hamilton (Fig. 8.6), comprised a hilly landscape. The ancestral Waikato River was absent from the lowlands at this time—it was

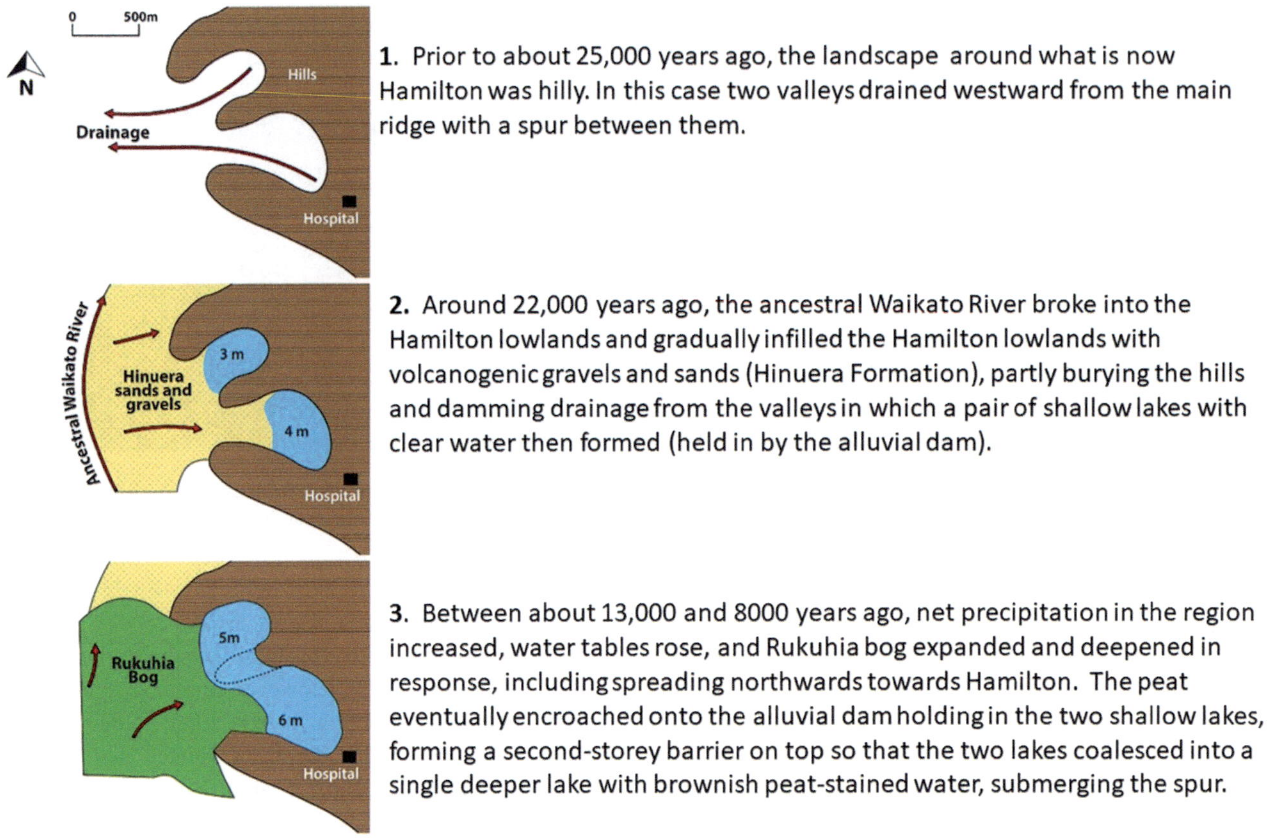

Fig. 8.6 Example of the multi-step process of formation of a peat lake in the Waikato lowlands: the development of Lake Rotoroa (Hamilton Lake) which is overlooked by Waikato Hospital. Most of the lakes were formed in a similar way, but the extent and impact of peat growth varies from lake to lake. New work suggests the river entered the Hamilton lowlands c. 23,500 years ago

flowing via the Matamata-Hauraki lowlands into the Pacific Ocean. However, following the Oruanui super-eruption of Taupo volcano c. 25,400 years ago, which generated huge volumes of loose pyroclastic materials that were then reworked by the choked Waikato River, sediment built up at Piarere. Eventually, several thousand years after the eruption, the river was forced to switch course, c. 23,500 years ago, at Piarere and flow into the Hamilton lowlands (and out to the Tasman Sea). Thus much of the alluvium carried into the Hamilton lowlands was derived from the reworked Oruanui eruptives, referred to as 'volcanogenic'. At the coldest point of the last glacial maximum, the ancestral Waikato River (at that time a high energy, braided river system) gradually infilled the Hamilton lowlands with the volcanogenic gravels and sands (Hinuera Formation), partly burying the hills and damming drainage from the valleys, and forming an extensive plain.

The thick sediments acted as small dams, generating small basins into which drainage from the hills then formed lakes. A series of around 30 lakes (many now surrounded by peat) formed on the margins of the low hills within the Hamilton lowlands. Most of the lakes formed about 20,000 years ago. Hence the lakes are most accurately described as riverine in origin.

Peat then began forming in scattered low-lying swampy hollows on the surface of Hinuera sediments, and around the margins of many of the lakes. From about 13,000 years ago net rainfall increased markedly and the regional water tables rose (accelerating from c. 8000 years ago) so that the scattered peats rapidly expanded, coalescing to form large bogs, including Rukuhia and Moanatuatua. Peat growth around lakes, such as Lake Rotoroa (Hamilton Lake) (Fig. 8.6) and Lake Maratoto (Fig. 8.7), formed a second dam on top of the alluvium and so these, and other lakes, became deeper and peaty in character. Such lakes thus can be classified as both riverine and peat-formed in origin.

Lake Maratoto (Fig. 8.7) is especially important because its sediments contain a world reference locality for the Pleistocene-Holocene boundary (dated at 11,700 years BP). Most of the lake and surrounding peat is in a QE II National Trust covenant which provides protection, in perpetuity, even though it is on privately owned land.

Fig. 8.7 Lake Maratoto, a Waikato peat lake, viewed looking westward. The land near the lake at the top of the photo is thick peat (8–11 m deep) of the Rukuhia bog. The peat dam is holding in the water of the upper part of the lake. Alluvium (beneath the peat) is damming the lower lake column

8.3.3 The Waihola-Waipori Wetland system—a Glimpse of the Taieri Plain in Pre-Human Times

The Waihola-Waipori wetland system comprises Lakes Waihola and Waipori and associated wetlands at the southern end of the Taieri Plain in Otago. It features the open waters of the two main lakes, many small lakes and extensive wetland areas. Active sedimentary inputs come from the Taieri and Waipori rivers forming deltas at the entrance to the wetlands (Fig. 8.8). The area is tidal with fluctuating water levels controlled by the tidal influence of a connection to the ocean through the Waihola gorge. The area comprises a complex system of rivers, lakes, tides, shifting sediment, lake-margin flora, organic infill of swamps, fluvial erosion, deltaic sedimentation, and an array of habitats where wetland plants thrive and accumulate organic matter. The variation is driven by river volume and velocity which control the sediment passing through the system. The relatively quiet backwaters sustain a rich biodiversity and an outdoor 'laboratory' of wetland processes.

Organic Soils are formed in back swamp areas where organic matter has accumulated with occasional inputs of alluvial material during floods. The soils are predominantly Mellow Fibric Organic Soils formed in materials derived mainly from sedges (*Carex secta*) and flax (*Phormium tenax*). In areas that have been subject to previous attempts at drainage and farming, the soils are more strongly decomposed and Mellow Humic Organic Soils occur. In some of the previously farmed (in the 1940s) areas, the peat was biodegraded and paddock-shaped shallow lakes remain (Fig. 8.8). Most areas of peat overlie unconsolidated lake-bed sediments at 1–2 m depth. Where less than 30 cm of organic material has accumulated, the soils are Raw Soils if the sediments remain fluid, Gley Soils if the material is consolidated and reducing conditions predominate, or Recent Soils on the wetland margins and higher parts of the levees.

Further up the Taieri Plain, where the Silver Stream adjoins the Taieri River, is a shallow basin with an area of Organic Soils that has been developed for farming. At the central topographic low point of this basin is an extensive area of deep Acidic Mesic Organic Soils. At its southern margin, loamy flood deposits have spilled over from the Taieri River (Fluvial Recent Gley Soils) partly burying the Organic Soils. At its northern margin, the organic layer thins to the point that it can no longer qualify as an Organic Soil and instead is a Recent Soil with a peaty topsoil. In this area and further north on the Taieri plain (Chap. 14), the alluvial deposits indicate deposition by deltaic fluvial sedimentation.

It is tempting to view the current Waihola-Waipori wetlands, and variation in plant and hydro-sedimentary systems, observed in the southern end of the Taieri, as glimpses of the former deltaic lacustrine environment of the mid and northern Taieri Plain long before human occupation.

8.3.4 Southern Sphagnum Wetlands of the Lammerlaw Range, East Otago

The Lammerlaw Range and other similar ranges of the east Otago tussock grassland plateau have extensive sphagnum-dominated wetlands comprising Acidic Mesic Organic Soils draped, like a blanket, over the broad ridges and plateaux of the easy rolling uplands (above 500 m altitude). The landforms are recognised as bogs because for

Fig. 8.8 Part of the Waihola-Waipori wetland. The somewhat rectilinear-shaped areas near the top of the wetland are areas where previous drainage and farming have led to loss of the peat

most of the areas the water is supplied solely from rainfall. In marked contrast to the Central Otago dry lands further west, the upland landscapes are well watered from frequent cloud, rain, and snow.

Throughout the Otago upland landscape, there are extensive Organic Soils associated with hollows at the heads of first-order streams in the head waters of the Waipori River, Deep Stream, and other major streams. The sites qualify as fens because they comprise peat, receive water from the surrounding land, and have water tables near the land surface. The boundary between the bog soils and the fen soils, in terms of the nature of the wetland water source is fuzzy. However, a boundary can be defined using downslope curvature, where the bog has convex curvature and the fen has concave curvature.

Sphagnum is the prominent wetland species where the plateau rises above about 600 m altitude. However, sedges and tussock are also common along with a range of other plants. Tony Davoren and co-workers reported a number of sites that generally had 1–2 m of peat over bedrock. The peat tended to be formed predominantly from sphagnum moss, often with an increasing portion of sedge material at depth. Tussocks *(Chinochloa rubra* and others) were more dominant near the edges and on better-drained sites.

8.3.5 Pakihi Environments on the West Coast of the South Island

The term 'pakihi' is used to describe generally flat, naturally open, spaces in Westland beech forest. The Māori word 'pakihi' means 'open or barren land'. The pakihi soils form in conditions of extreme high (several metres per year) rainfall in the Westland region of the South Island and support an impoverished vegetation that is tolerant of wet, low nutrient conditions. Pakihi soils often have a hard pan, restricting tree growth, and are extremely wet. Much of the pakihi land comprises Perch-Gley Podzol Soils (Chap. 11). However, where saturated conditions predominate, Organic Soils develop mainly from sphagnum moss, and *Gleichina/Empodisma minus* with manuka, rushes *(Baumea sp), Schoenus, Orobolus*, red tussock *(Chionochloa rubra)*, and *Dracophyllum* sp. also often present. Many pakihi contain areas of Organic Soil dominated by the restiad rush *Empodisma minus*.

8.4 Key Soil Properties

8.4.1 Introduction

The properties of Organic Soils, being dominated by organic matter, are decidedly different from those of any other Soil Order. Consequently, such soils are always separated from the mineral or 'inorganic' soils in soil classification systems. Organic Soils have been mapped and sampled but data are limited, except in areas of research interest such as the Waikato and Hauraki lowlands. A national survey of organic soil resources was undertaken for the National Soil and Water Conservation Authority in 1978 by Tony Davoren together with John McCraw and Keith Thompson. More recently, organic resources have been investigated as part of New Zealand's national carbon accounting project by the Ministry for the Environment.

8.4.2 Soil Composition

Organic Soils are, by definition, composed of organic, predominantly plant, material in various stages of decay. It is the plant species, hydrological setting (including rainfall and rain-days), and landscape position that strongly influence the composition of Organic Soils. Any plant material can potentially become an Organic Soil parent material if it is prevented from biodegradation, usually as a consequence of saturated (anaerobic) conditions. The effect of plant species is important as the nature of the peat fabric is conditioned by the structure and composition of the plant material. For example, sphagnum bogs of the east Otago uplands are highly porous with very low bearing strength, whereas peat from sedges may have greater density and moderate load-bearing strength. Woody materials may be slower to decompose because of the presence of lignin which has a particularly high C/N ratio and so nitrogen, as well as oxygen, availability limits microbial biodegradation. Thus relatively 'intact' wood material, including large tree stumps and logs, may be found within some Organic Soils.

Layers of mineral (sand, silt, and clay) material may occur within an Organic Soil, especially in areas where river flood deposits, or tephra-fallout, occur. The composition of the mineral matter depends on the source of the material and mode of deposition. The mineralogy of the sand, silt, and clay within an Organic Soil is likely to be similar to that of associated Recent and Gley Soils in the same environment.

8.4.3 Physical Properties

The depth of Organic Soils is highly variable, both spatially and temporally. If conditions remain favourable for extended periods (thousands of years), the organic material can gradually accumulate. At some sites, the depth of organic material may run to tens of metres (for example, Kopouatai bog is up to 14 m deep, the base being below sea level). However, Organic Soils are vulnerable to loss by oxidation and biodegradation which converts the organic matter predominantly to carbon dioxide and water. Fire or drainage may lead to rapid loss, or consolidation, of peat material, and thus a rapid decrease in the soil depth.

The rate of carbon loss on farmed soils is faster in the initial years following development but continues at a lesser rate while the peat resource remains. Dave Campbell and co-workers reported a net loss of about 300 g C m^{-2} yr^{-1} on Waikato dairy farms on Organic Soils with a water table at about 50 cm depth. Louis Schipper, Malcolm McLeod, Jack Pronger, and others, have estimated a long-term loss of 370 ± 130 g C m^{-2} yr^{-1} on farmed peat. They reported surface subsidence rates of about 3–3.5 cm yr^{-1}, over the 40 years following initial development and about 2 cm yr^{-1} between 2000 and 2014. About two-thirds of the subsidence was attributed to consolidation with the balance due to carbon loss.

Most of the Organic Soils in the NZ Soil Data Repository contain some clay material (Fig. 8.9), most commonly derived from tephra or alluvium. Soil dry bulk densities are very low, usually in the range of 0.03–0.4 t m^{-3}. With the low dry bulk density comes an exceptionally high porosity and water holding capacity (Fig. 8.9, Table 8.2), with the soil often containing a much greater weight of water than soil dry matter, thus often giving gravimetric moisture contents much greater than 100%. The volumetric water content of Organic Soil can be as high as 90% and walking on the surface of a peat bog can be unnerving as the 'ground' may quiver like jelly.

The high water holding capacity of Organic Soils is due to the 'sponge' effect of both sphagnum moss and the root system of the restiads. The water holding capacity in moss is provided by the close-packed leaves, whereas in the restiads it is the dense root mat that traps and contains water. Although total available-water capacity is high, in *drained* Organic Soils the plant-available-water capacity may only be moderate. This is because the coarse pores in the Organic Soils, which supply readily available water, may be lost as the soil compacts following drainage. Organic Soils shrink markedly upon drying, and may also lose organic matter through oxidation. Consequently, following drainage, the soil water holding capacity, and indeed the soil classification group, may change.

Organic Soils formed in peats are usually described as very poorly drained. However, Litter Organic Soils, formed on forest floors, may range from well drained to very poorly drained. Organic Soils warm and cool slowly due to the high heat capacity of the large volume of water held within the soil.

8.4.4 Chemical Properties

The chemical properties of Organic Soils reflect the predominance of organic material within the soil profile. Strong acidity (strongly and extremely acid) is common in New Zealand Organic Soils with pH often at or below 4.5 (Fig. 8.10, Table 8.3). In the Kopouatai bog, Peter de Lange (with Peter Hodder) reported pHs as low as 3 in the upper 3 m of the bog, increasing to 4 from 3 to 8 m depth, and to pH 5 from 8 to 10 m. Mellow subgroups have pHs of more than 4.5 and are usually associated with calcareous sediments in the catchment or calcic mineral material in the soil profile.

Organic Soils can contain 35–100% organic matter, and thus comprise about 30–60% organic carbon (Fig. 8.10, Table 8.3). The organic material provides a large surface

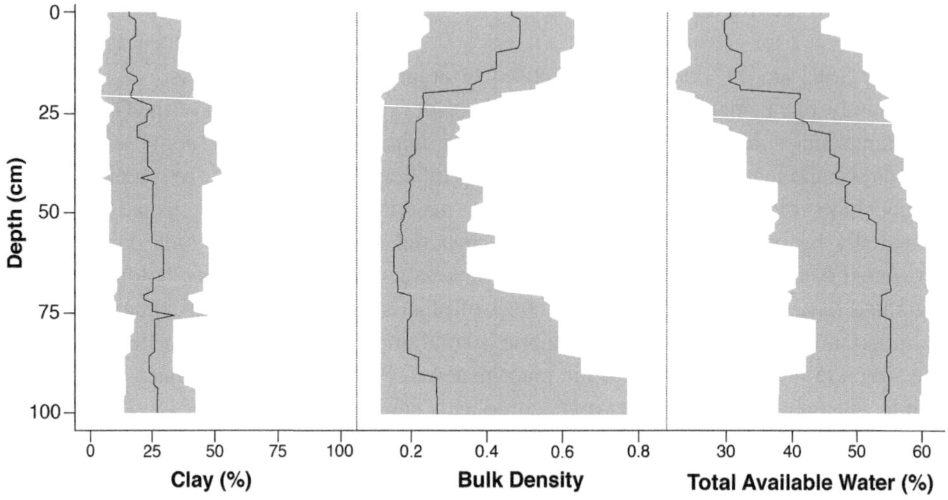

Fig. 8.9 Median and upper and lower quartile of clay content, soil dry bulk density (t m^{-3}), and soil total available water holding capacity for Organic Soils in the New Zealand Soil Data Repository

Table 8.2 Example of physical properties of an Acid Fibric Organic Soil under pasture (Rukuhia series, SB09948, SWAMP)

Horizon	Depth (cm)	Dry bulk density (t m^{-3})	RAW[a] (%)	TAW[a] (%)	Total porosity (%)
Of1	0–7	0.25	19.0	51.7	84
Of2	7–15	0.23	16.7	52.0	85
Of3	15–37	0.10	25.5	63.2	93
Om	37–49	0.15	26.1	61.8	93
2C	49–52	0.16	25.2	58.9	91
3Of	52–80	0.08	27.2	64.3	94

[a]RAW = Readily available water, TAW = total available water, percentage by volume

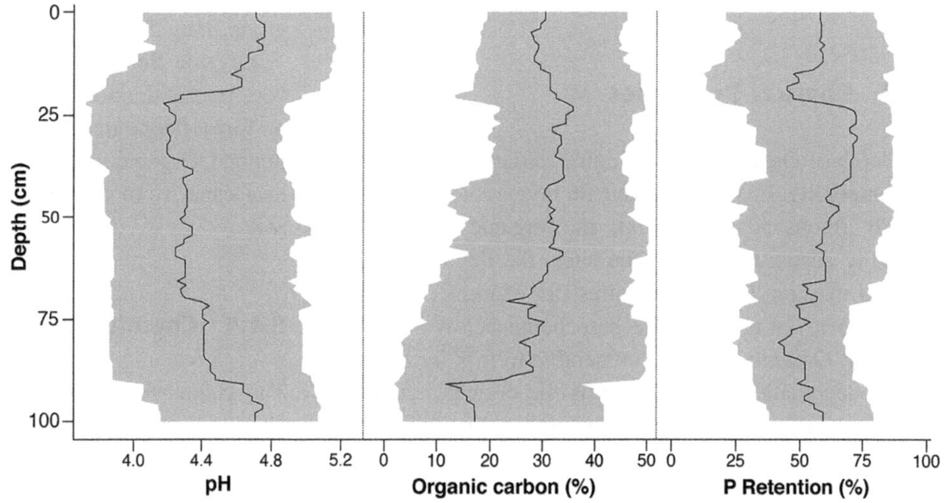

Fig. 8.10 Median and upper and lower quartiles of soil pH, organic carbon, and phosphorus retention for Organic Soils in the NZ Soil Data Repository

area which gives the Organic Soils high cation exchange capacities (CEC) mostly ranging from 40 to 170 $cmol^{(+)}$ kg^{-1}. The high net negative charge varies markedly with pH. Soil nitrogen levels are usually low and hence C/N ratios range from 18 to as high as 70 in unfertilized and uncultivated Organic Soils.

Nutrient deficiencies are common (including nitrogen, phosphorus, potassium, and sulphur). The trace elements copper, selenium, and molybdenum are also frequently deficient for crops and pasture.

8.4.5 Biological Properties

Restiad bogs were historically present from near Kaitaia to the low-lying land south of Hamilton and between

Table 8.3 Example of chemical properties of an Acid Fibric Organic Soil under pasture (Rukuhia series, SB09948, SWAMP)

Horizon	Depth	pH (in H_2O)	Carbon (%)	Nitrogen (%)	CEC[a] ($cmol^{(+)}$ kg^{-1})	Sum bases ($cmol^{(+)}$ $kg^{-})^1$	P retention (%)
Of1	0–7	4.4	50	2.41		89.2	
Of2	7–15	4.1	52	1.66		99.2	
Of3	15–37	3.7	55	1.27		45.4	
Om	37–49	3.9	51	1.21		29.5	
2C	49–52	4.0	28	0.61	56.6	5.57	20
3Of1	52–80	3.9	57	1.32		30.4	
3Of2	80–110	3.9	53	1.31		37.7	

[a]CEC = cation exchange capacity

Cambridge and Te Awamutu. Restiad bogs also occur in the Chatham Islands and Australia. There are three restiad bog species: *Empodisma robustum* the jointed wire rush (northern North Island); *Empodisma minus* (much of North, South, and Stewart islands); and *Sporadanthus ferrugineus* (northern North Island). Further south, sphagnum moss is often the dominant peat-forming plant. Fens, swamps, and seepages originally supported water-tolerant native plants including flax (harakeke, *Phormium tenax*), cabbage tree (ti kouka, *Cordyline australis*), kahikatea (*Dacrycarpus dacridioides*), pukeatea (*Laurelia novae-zelandiae*), raupo (*Typha orientalis*), and manuka (*Leptospermum scoparium*). The low pH and saturated conditions that limit oxygen availability, as well as the lack of nutrients, such as nitrogen and phosphorus, limit biological activity within Organic Soils.

Peat bogs and lake sediments as paleoenvironmental records

Bogs and lake sediments provide a unique archive for understanding environmental change over thousands of years because the anaerobic conditions enable the remains of plants and animals to be preserved. Analyses of materials such as pollen, plant macrofossils, and plant cuticles, along with peat humification, carbon content, and dust input, provide a basis for reconstructing past environments. Vegetation, temperature, wind patterns, and rainfall over thousands of years can all be determined. For example, a detailed study on the Kopouatai bog gave a 14,000-year-long record of environmental change using both pollen analysis (which gives regional vegetation cover) and plant macrofossils (which provide a local signal) (Fig. 8.11).

As well as Kopouatai, the sediments in many lakes have been cored and analysed including Lake Pupuke and other Auckland crater lakes, peat lakes in the Waikato, lakes Tutira and Poukawa in Hawke's Bay, Adelaide Tarn near Nelson, and Lake Ohau in Otago/Canterbury. In addition to pollen and plant macrofossils, lake sediments may also contain the remains of midges, microcrustacea, algae, diatoms, and organic compounds. A climate-event stratigraphy for New Zealand from 30,000 to 8000 years ago has been developed using pollen and tephra studies that linked records from three key sites: Pukaki Lagoon (Auckland), Galway Tarn (Westland), and Kaipo bog (Te Urewera National Park) in the INTIMATE (INTegration of Ice-core, MArine, and TErrestrial records) project undertaken by the entire New Zealand paleoclimate community.

8.5 Distinguishing Between Organic Soils and Related Soil Orders

It is the amount of carbon (or the presence of organic soil material) that distinguishes Organic Soils from all other soil orders. Organic Soils formed on peat have more than 30 cm of organic soil material within a depth of 60 cm. In forests, the qualifying thickness of (decomposed) litter is ≥ 40 cm. The organic material in both cases must contain at least 18% carbon (or approximately 30% soil organic matter) to qualify as an Organic Soil. The *New Zealand Soil Classification*, where possible, uses soil morphological properties including colour, deformable failure, weight loss on oven-drying, and unrubbed fibre content to enable the organic material to be identified when carbon analyses are not easily available.

Other soil orders (e.g. Gley Soil, and Recent Soils), may have peaty topsoils identified by a 'Peaty' subgroup. The peaty topsoil is similar to organic soil material but too thin to meet the thickness and depth requirements for the Organic Soils order.

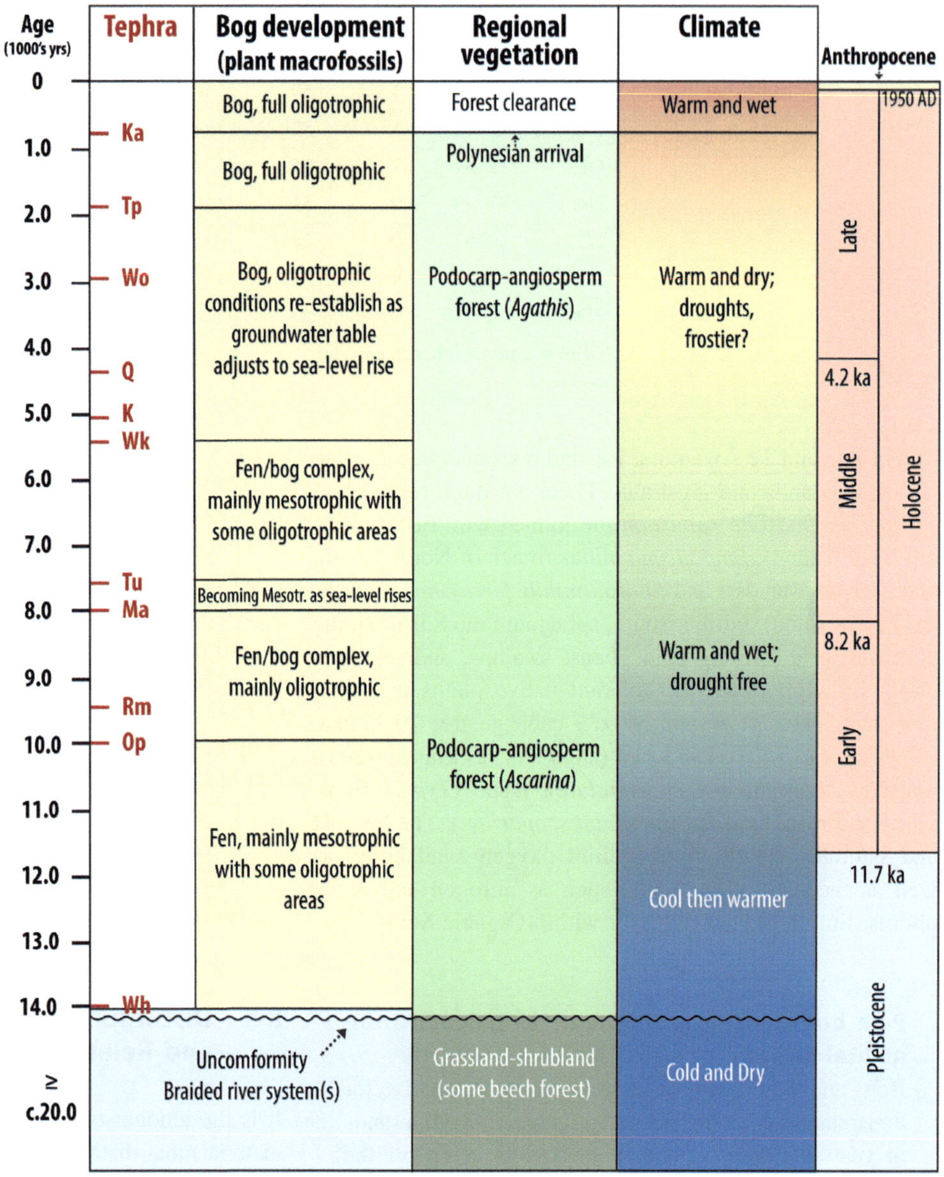

Fig. 8.11 Record of past environments in northern New Zealand determined using pollen and plant macrofossils, radiocarbon dates and tephra ages in Kopouatai bog. Red dashes and letters represent tephra deposits of known age. Note that AD1950 is used informally as the start of the Anthropocene in this diagram; ka = × 1000 years ago. The tephra layers and their ages are (source volcano in brackets): Wh, Waiohau (Tarawera) c. 14 ka; Op, Opepe (Taupo) c. 10 ka; Rm, Rotoma (Haroharo) c. 9.4 ka; Ma, Mamaku (Haroharo) c. 8.0 ka; Tu, Tuhua (Mayor Island) c. 7.6 ka; Wk, Whakatane (Haroharo) c. 5.5 ka; K, Unit K (Taupo) c. 5.1 ka; Q, Unit Q (Taupo) c. 4.3 ka; Wo, Whakaipo (Taupo) c. 2.8 ka; Tp, Taupo (Taupo), c. 1.8 ka (AD 232 ± 10); Ka, Kaharoa (Tarawera) c. 0.7 ka (AD 1314 ± 12) (ages after Lowe et al. 2013). Diagram redrawn after Newnham et al. (1995)

8.6 Correlation with Other Classification Systems

Organic Soils in New Zealand are essentially similar to soils formed on peat, or in deep and partly decomposed forest litter, in other parts of the world. Hence, there is a clear correlation between the *New Zealand Soil Classification* and the Histosols in both *Soil Taxonomy*, and the *World Reference Base* (Table 8.4).

8.7 Use and Management of Organic Soils

8.7.1 Introduction

About 30% of the Organic Soils in New Zealand are in reserves where they provide habitat for a unique set of flora and fauna and are an important component of New Zealand's biodiversity. The reserves also provide invaluable sites for studying past environments and climates (see grey box)

Fig. 8.12 Productive land uses and a remaining reserve area on the Moanatuatua bog near Hamilton. The linear features are drains installed across the landscape. Photo: courtesy of Reece Hill, Waikato Regional Council

Table 8.4 Correlation between Organic Soil groups and *Soil Taxonomy*, the *World Reference Base*, and the NZ genetic soil classification[a]

NZ Soil Classification	*Soil Taxonomy*	*World Reference Base*	NZ genetic Soil classification
Litter Organic Soils	Folists	Lignic Histosols	Not recognised
Fibric Organic Soils	Fibrists	Fibric Histosols	Organic soils
Mesic Organic Soils	Hemists	Hemic histosols	Organic soils
Humic Organic Soils	Saprists	Sapric histosols	Organic soils

[a]The correlations given here are a guide only and for accurate classifications, the relevant soil classification documents should be consulted. The two major international soil classification systems are *Soil Taxonomy*, which was developed in the USA, and *World Reference Base*, which was developed primarily in Europe. The NZ genetic soil classification was used in NZ prior to 1992

as well as useful sites for research into past volcanism (because bogs in North Island preserve distal tephra-fall deposits, hence a record of volcanic events).

About 70% of New Zealand Organic Soils have been developed for agriculture, supporting pastoral farming (predominantly dairy farming), maize, and potato cropping, and permanent horticultural crops, particularly blueberries (which are well adapted to the low pH and high moisture environment of Organic Soils).

Some areas of Organic Soils have been mined for peat resources, often for use in potting mix and mushroom-growing compost. Sphagnum moss is also harvested and used for potting mixes. Moss may be dried and used for insulation in buildings and has also been used in wound dressing.

Organic Soils have a role in the hydrological functioning of catchments. Because they act as a giant sponge, Organic Soils can absorb water during periods of high rainfall, slowly releasing it over time. Thus they help minimise flood peak flows and maintain water flow in rivers during dry periods.

Organic Soils provide a reservoir of soil carbon. Dave Campbell, Louis Schipper, Joss Ratcliffe and colleagues and students have been particularly involved in research on the rates of losses and gains in soil carbon from the Organic Soils in the Waikato region. They have found that the short-term C budget for pasture on peat (used for dairying) to be approximately in balance, i.e., net gains of C from photosynthesis are essentially matched by losses as milk solids and CH_4 and CO_2 emissions.

8.7.2 Management of Organic Soils for Conservation, Biodiversity Protection, Carbon Sequestration, and Catchment Hydrological Protection

Maintaining Organic Soils in their natural state has a number of important benefits, including biodiversity protection, carbon sequestration, absorbing flood waters, and maintaining stream flows during drier periods. In order to maintain Organic Soils, to prevent loss of carbon dioxide to the atmosphere, and to keep the water holding capacity of wetlands intact, the peat material needs to be protected from fire, and also from drainage.

The biodiversity value of wetland ecosystems in New Zealand includes 27 species of protected birds, rare mud fish, long-finned eels, insects, and a number of threatened plants. However, simply fencing such areas off is not sufficient to maintain all their values. Invasive plants are a threat in many wetland areas containing Organic Soils, willows and some invasive grasses (such as 'reed sweet grass', *Glyceria maxima*) have overtaken large areas of wetlands in both the North and South islands. While efforts to control willows, using herbicides, have proven reasonably successful (though somewhat unsightly for some years as the dead trees take a long time to break down), no practical way to deal with some weeds has been identified. If weeds are invading it is best to remove a small infestation early, rather than wait until it becomes a major problem.

Activities on adjacent land that cause lowering of water tables are a problem in maintaining areas of Organic Soils. There are many examples around New Zealand where drainage of adjacent farmland may threaten the viability of a wetland. For example, in the Waikato region drainage on farms near the peat lakes (Sect. 8.3.2) has led to lowering of water tables, biodegradation, and consolidation of the peat, and thus lowering the lake level as well. Several lakes near Ohaupo, and two lakes north of Hamilton, have been drained and completely lost since the 1930s as a result of drainage of neighbouring land. Thus, drainage of the peatlands surrounding peat lakes needs to be carefully managed to prevent drainage and further loss of the unique Waikato peat lakes. It is important to develop understanding and co-operation between neighbours for wetlands to be protected.

Fire is an ever-present threat to Organic Soils, destroying vegetation, combusting the organic soil material, and releasing CO_2 into the atmosphere. Once started, fires in peat may be very difficult to extinguish. Peat fires have been known to smoulder for years, with potential to break out as a spreading fire if conditions become favourable (hot, dry windy conditions when water tables are seasonally low). While fires occasionally occur through natural causes, such as lightning strikes (evident as charcoal fragments in the peat records back through time), fires in peatland have become more common due to both deliberate and accidental human activities.

Peat fires were a regular occurrence in the farmed peatlands of the Waikato in the early-mid-twentieth century (https://teara.govt.nz/mi/video/27060/peat-fire). In days past, burning to clear vegetation and sparks from steam engines were common causes of fires, along with careless disposal of cigarettes. In 2011, a fire in the Rukuhia Peat near Hamilton was started by sparks from welding repairs on a railway line. The natural vegetation of peatlands is somewhat adapted to fire. Research by Bruce and Bev Clarkson, among others, has shown that over decades the vegetation will slowly recover to pre-fire conditions. However, repeated fire will undoubtedly gradually degrade the peat resource and fire has a detrimental impact on wildlife, which for threatened plants or animals could be devastating. On the other hand, some early and mid-successional species, including some orchids, benefit from an occasional fire to provide them with habitat. Overall, people need to be particularly vigilant to ensure that no fires are started in or near areas of Organic Soils.

Increased inflow of sediments and nutrients from surrounding farmland can adversely impact the conservation values of wetlands by changing the soil conditions to favour invasive plants. While over time many wetlands gradually naturally infill and change, such processes may be greatly enhanced by inputs of runoff from surrounding farmland. In the area of the Whangamarino wetland in the Waikato there have been long-running differences over management of the wetland between farmers who want water tables lowered, conservationists who want to maintain natural conditions, and regional council flood management programmes. The flood management works divert (potentially nutrient and sediment rich) flood water into the wetland area to protect property downstream by lowering the flood peaks in the Waikato River. The issues are complex and proving difficult to resolve.

Carbon dioxide is lost to the atmosphere when plants or organic materials are biodegraded or burned; conversely, carbon is sequestered (or locked up) away from the atmosphere when plant material is preserved in a peatland. However, within wetlands some biodegradation continues in anaerobic conditions and results in the release of methane (CH_4), which is also a greenhouse gas. Work by Dave Campbell and others has shown that the undisturbed Kopouatai restiad peat bog has relatively high overall rates of carbon sequestration (from about 135 to 250 g C

m^{-2} yr^{-1}) compared to peatlands in many other parts of the world. They also reported that Kopouatai lost about 14–22 g m^{-2} yr^{-1} of carbon in methane to the atmosphere. The year-round relatively warm, moist climate ensures a greater rate of plant growth than occurs in much of the northern hemisphere.

Ratcliffe and others demonstrated that the fallout of rhyolitic and andesitic tephra has stimulated C accumulation rates in the Kopouatai bog by increasing the availability of phosphorus and thus supporting periods of high carbon uptake. Such phosphorus is potentially derived partly from fallout of PO_4^{3-} generated as aerosols during the tephra-producing eruptions, partly from leachates generated by aerosol-derived acids attacking particles of volcanic glass and apatite crystals within the eruption plume, and partly from the rapid dissolution of tephra-derived glass shards and apatite that fell on the bog surface.

8.7.3 Management of Organic Soils for Pastoral Farming and Cropping

Large areas (about 70%) of Organic Soils are utilised in productive agriculture (Fig. 8.12). Dairy farming is undertaken in the Waikato, and also in other regions including the Manawatu and Southland. Maize and potato cropping are also common in Organic Soils. Blueberries are especially well adapted to peat substrates and New Zealand has a niche market supplying blueberries to the Northern Hemisphere at Christmas time.

For agricultural or horticultural use, Organic Soils require some drainage to enable practical management. However, drainage will inevitably lead to a loss of the soil resource. The Waikato Regional Council recommends control of the water table at a level of 50 cm depth or less in Organic Soils (https://www.waikatoregion.govt.nz/environment/natural-resources/land-and-soil/managing-land-and-soil/managing-peat/). Deep drains should be avoided as they enhance the rate of peat subsidence and biodegradation. Also, if peat is allowed to dry, then it is difficult to get it to rewet and much rainfall may flow rapidly through cracks in the soil to drains, and thus be lost to plant production. In summer it may be beneficial to put weirs or blocks in drains to keep water tables high and soils moist. Drains will need to be cleared out less frequently if weeds are controlled (by summer spraying) and stock are fenced out of drains.

Due to the initial low soil pH, and low nutrient status, large inputs of lime and fertiliser were applied to establish productive farming on Organic Soils. A substantial dose of lime, to raise pH levels from $\sim$4 to $\sim$5, will be effective for about four or five years. Both drainage and inputs of fertiliser enhance microbial activity and thus biodegradation of the soil resource. Therefore, soil tests need to be undertaken to determine fertiliser and lime requirements for a given site and crop in order to avoid the use of more fertiliser than necessary.

In the long term, agriculture or horticultural development will lead to degradation of the peat, and lowering of the ground surface, until the farming/horticultural operation intercepts the underlying mineral material. This has already occurred on many farms in parts of the Hauraki Plains and Hamilton Basin, where the peat has gone with the only evidence remaining in a 'peaty topsoil'. Thus farming is now undertaken in soils formed from the last remnants of the peat (giving peaty, organic matter-rich, topsoils) and the underlying inorganic sediments. In some regions that are close to sea level, such as on the Hauraki Plains, to maintain drainage as the surface subsides, water has to be pumped up, over river stop-banks into the Waihou and Piako Rivers.

The high carbon content, and anaerobic conditions in Organic Soils favour denitrification, and so there is a relatively low risk of nitrogen escaping to groundwater, provided that rapid flow through macropores is avoided. Thus farmed Organic Soils can be suitable for effluent irrigation during periods when the soil is not saturated to the surface.

Cultivation of Organic Soils greatly increases the rate of shrinkage and biodegradation compared to that under established pasture. Thus the less cultivation that is undertaken the longer the peat resource will last. It is preferable to use no-till (direct drill) methods for pasture renewal or crop establishment. If ploughing is deemed necessary, it is best to avoid chopping the peat material finely, which accelerates biodegradation. Maintaining a dense pasture sward will help protect the peat. Overgrazing, pugging, or other activities that leave patches of bare soil exposed, will enhance the degradation of the organic matter resulting in an uneven ground surface.

8.7.4 Management of Infrastructure on Organic Soils

Organic Soils have low bearing strength and are often saturated with water, thus they are generally not suitable for building infrastructure such as roads and housing without special engineering design. Wheeled vehicles such as tractors need twin sets of rear wheels and soft tyres to avoid sinking. The low pH means that concrete and steel may both be prone to corrosion. If weighty structures are built, the soil may be compressed unevenly. Roads and railway lines, even when constructed with large amounts of imported base materials, are prone to uneven subsidence, which results in

Fig. 8.13 Power poles with extra support to try to keep them vertical in Organic Soils on Collins Road near Hamilton

hazards to vehicles and trains. Buildings on peat material are usually built on piles (that reach through to the underlying mineral substrate) or on a concrete base designed to 'float' on the surface of the peat. Pre-loading with large piles of rock materials to compress sites is now commonly undertaken for months, if not years, before road construction is undertaken on sites with peat that is too deep to economically remove. One example of the challenges of maintaining infrastructure on Organic Soils is the 'drunken' power poles on some of the road sides on Organic Soils near Hamilton (Fig. 8.13). Extra support is provided to try to keep the poles upright in the soft substrate.

Further Reading

Aislabie J, Deslippe J (2013) Soil microbes and their contribution to soil services. Section 1.12. In: Dymond JR (ed) Ecosystem services in New Zealand: condition and trends. Manaaki Whenua Press, Lincoln, New Zealand

Balks MR, Beecroft FG (1987) Contributions to Waihola/Waipori draft management plan. A report to the Department of Conservation. NZ Soil Bureau Contract Report 87/18. NZ Soil Bureau, DSIR, Dunedin

Barrell DJA, Almond PC, Vandergoes MJ et al (2013) A composite pollen-based stratotype for inter-regional evaluation of climatic events in New Zealand over the past 30,000 years (NZ-INTIMATE project). Quatern Sci Rev 74:4–20

Campbell DI, Jackson R (2004) Hydrology of wetlands. In: Harding JS et al (eds) Freshwaters of New Zealand. NZ Hydrological Society and New Zealand Limnological Society, Christchurch, pp 20.1–20.14

Campbell DI, Smith J, Goodrich JP et al (2014) Year-round growing conditions explain large CO_2 sink strength in a New Zealand raised peat bog. Agric For Meteorol 192(3):59–68

Campbell DI, Wall AM, Nieveen JP (2015) Variations in CO_2 exchange for dairy farms with year-round rotational grazing on drained peatlands. Agric Ecosyst Environ 202:68–78

Clarkson BR, Stanway EA (1994) Vegetation recovery following fire in the Whangamarino wetland. Science and Research Series no. 73. Department of Conservation Wellington. 24p. ISSN 0113–3713. ISBN 0–478–01613–1

Clarkson BR, Clarkson BD (2013) Restiad bogs in New Zealand. In: Rydin H, Jeglum JK (eds) The biology of Peatlands, 2nd edn. Oxford University Press, Oxford, The Biology of Habitat Series, pp 241–248

Clarkson B, Merrett M, Downs T (2002) Botany of the Waikato. Waikato Botanical Society, Hamilton, 136 p

Cranwell LM (1939) Native vegetation. In: Soils and Agriculture of Part of Waipa County. N.Z. DSIR Bull 76:23–30

Davoren A, McCraw JD, Thompson K (1978) A survey of New Zealand's peat resources. Water and Soil Technical Publication No. 14. 157p. + maps

de Lange PJ (1989) Late Quaternary development of the Kopouatai peat bog, Hauraki lowlands, and some palaeoenvironmental inferences. Unpublished M.Sc. thesis, lodged in the Library, University of Waikato, Hamilton

de Lange PJ, Heenan P, Clarkson BD et al (1999) Taxonomy, ecology, and conservation of Sporadanthus (Restionaceae) in New Zealand. NZ J Bot 37:413–431

de Lange PJ, Lowe DJ (1990) History of vertical displacement of Kerepehi Fault at Kopouatai bog, Hauraki lowlands, New Zealand, since c. 10 700 years ago. NZ J Geol Geophys 33:277–283

Environment Waikato (1999) For peat's sake: good management practices for Waikato peat farmers. Environment Waikato, Hamilton

Gehrels MJ, Lowe DJ, Hazell ZJ et al (2006) A continuous 5300-yr Holocene cryptotephrostratigraphic record from northern New Zealand and implications for tephrochronology and volcanic-hazard assessment. The Holocene 16:173–187

Goodrich JP, Campbell DI, Roulet NT et al (2015) Overriding control of methane flux temporal variability by water table dynamics in a Southern Hemisphere, raised bog. J Geophys Res—Biogeosciences 120. https://doi.org/10.1002/2014JG002844

Goodrich JP, Campbell DI, Schipper LA (2017) Southern Hemisphere bog persists as a strong carbon sink during droughts. Biogeosciences 14:4563–4576. https://doi.org/10.5194/bg-14-4563-2017

Grange LI, Taylor NH, Sutherland CF et al (1939) Soils and agriculture of part of Waipa County. N.Z. DSIR Bull 76. 85 p

Green JD, Lowe DJ (1985) Stratigraphy and development of c. 17,000 year old Lake Maratoto, North Island, New Zealand, with some inferences about postglacial climatic change. NZ J Geol Geophys 28:675–699

Green JD, Lowe DJ (1994) Origins and development. In: Clayton JS, de Winton MD (eds) Lake Rotoroa: change in an Urban Lake. NIWA Research Ecosystems Publication 9:13–23

Haenfling C, Newnham R, Rees A et al (2017) Holocene history of a raised bog, northern New Zealand, based on plant cuticles. The Holocene 27(2):309–314. https://doi.org/10.1177/0959683616658524

Harris WF (1968) Peat classification by pedological methods, applied to peats of western Wellington. NZ DSIR Bull 189

Hodder APW, de Lange PJ, Lowe DJ (1991) Dissolution and depletion of ferromagnesian minerals from Holocene tephras in an acid bog, New Zealand, and implications for tephra correlation. J Quat Sci 6:195–208

Hogg AG, Lowe DJ, Palmer JG et al (2012) Revised calendar date for the Taupo eruption derived by ^{14}C wiggle-matching using a New Zealand kauri ^{14}C calibration data set. The Holocene 22:439–449

Hogg AG, Wilson CJN, Lowe DJ et al (2019) Wiggle-match radiocarbon dating of the Taupo eruption. Nat Commun 10:4669. https://doi.org/10.1038/s41467-019-12532-8

Jara IA, Newnham RM, Alloway BV et al (2017) Pollen-based temperature and precipitation records of the past 14,600 years in northern New Zealand (37°S) and their linkages with the Southern Hemisphere atmospheric circulation. The Holocene. https://doi.org/10.1177/0959683617708444

Johnson P, Gerbeaux P (2004) Wetland types in New Zealand. Department of Conservation, Te Papa Atawhai PO Box 10–420 Wellington. https://www.doc.govt.nz/documents/science-and-technical/WetlandsBWa.pdf

Lambie SM, Ratcliffe JL (2020) Multi-substrate induced respiration (functional capacity) in agriculturally degraded and intact restiad bogs: implications for carbon and nitrogen cycling. Mires and Peat 26: article 21

Lowe DJ (1985) Application of impulse radar to continuous profiling of tephra-bearing lake sediments and peats: an initial evaluation. NZJ Geol Geophys 28:667–674

Lowe DJ (2014) How the lake [Rotoroa] was formed: influence of the ancestral Waikato River. Taylor J Hamilton Lake—City Playground. Published by Jeff Taylor, Hamilton, pp 6–13

Lowe DJ, Green JD (1992) Lakes. In: Soons JM, Selby MJ (eds) Landforms of New Zealand, 2nd edn. Auckland, Longman Paul, pp 107–143

Lowe DJ, Shane PAR, Alloway BV et al (2008) Fingerprints and age models for widespread New Zealand tephra marker beds erupted since 30,000 years ago: a framework for NZ-INTIMATE. Quatern Sci Rev 27:95–126

Lowe DJ, Blaauw M, Hogg AG et al (2013) Ages of 24 widespread tephras erupted since 30,000 years ago in New Zealand, with re-evaluation of the timing and palaeoclimatic implications of the late-glacial cool episode recorded at Kaipo bog. Quatern Sci Rev 74:170–194

Manville V, Wilson CJN (2004) The 26.5 ka Oruanui eruption, New Zealand: a review of the roles of volcanism and climate in the post-eruptive sedimentary response. NZ J Geol Geophys 47:525–547

McCraw JD (2011) The wandering river—landforms and geological history of the Hamilton Basin. Geoscience Society of New Zealand Guidebook No. 16. 88 p

McGlone MS (2009) Postglacial history of New Zealand wetlands and implications for their conservation. NZ J Ecol 33:1–23

McLay CDA, Allbrook RF, Thompson K (1992) Effect of development and cultivation on physical properties of peat soils in New Zealand. Geoderma 54:23–37

Moar NT (1952) A botanical survey of a peat area between Raumati South and Paraparaumu. NZ J Sci Technol A33(5):78–89

Moar NT (1953) Vegetation and peat of a small mire at Plimmerton, North Island, New Zealand. NZ J Sci Technol A34(6):479–486

Moar NT (1954) Peat profiles, Wharaeroa Block, Paekakariki, N.Z. NZ J Sci Technol A36(3):221–231

Newnham RM, Lowe DJ, Green JD (1989) Palynology, vegetation and climate of the Waikato lowlands, North Island, New Zealand, since c. 18,000 years ago. J Roy Soc NZ 19:27–150

Newnham RM, de Lange PJ, Lowe DJ (1995) Holocene vegetation, climate and history of a raised bog complex, northern New Zealand,

based on palynology, plant macrofossils and tephrochronology. The Holocene 5:267–282
Newnham RM, Hazell ZJ, Charman DJ et al (2019) Peat humification records from Restionaceae bogs in northern New Zealand as potential indicators of Holocene precipitation, seasonality, and ENSO. Quatern Sci Rev 218:78–394
Nieveen JP, Campbell DI, Schipper LA et al (2005) Carbon exchange of grazed pasture on a drained peat soil. Glob Change Biol 11 (4):607–618
Pronger J, Schipper LA, Hill RB et al (2014) Subsidence rates of drained agricultural peatlands in New Zealand and the relationship with time since drainage. J Environ Qual 43(4):1442–1449
Ratcliffe JL, Campbell DI, Clarkson BR et al (2019) Water table fluctuations control CO_2 exchange in wet and dry bogs through different mechanisms. Sci Total Environ 655:1037–1046
Ratcliffe JL, Campbell DI, Schipper LA et al (2020) Recovery of the CO_2 sink in a remnant peatland following water table lowering. Sci Total Environ 718:134613
Ratcliffe JL, Lowe DJ, Schipper LA et al (2020) Rapid carbon accumulation in a peatland following Late Holocene tephra deposition, New Zealand. Quat Sci Rev 246. https://doi.org/10.1016/j.quascirev.2020.106505
Rapson GL, Sykes MT, Lee WG et al (2006) Subalpine gully-head ribbon fens of the Lammerlaw and Lammermoor Ranges, Otago, New Zealand. NZ J Bot 44:351–375
Schipper LA, McLeod M (2002) Subsidence rates and carbon loss in peat soils following conversion to pasture in the Waikato Region, New Zealand. Soil Use Manag 18:91–93
Selby MJ, Lowe DJ (1992) The middle Waikato Basin and hills. In: Soons JM, Selby MJ (eds) Landforms of New Zealand, 2nd edn. Longman Paul, Auckland, pp 233–255
Soil Survey Staff 2014. Keys to Soil Taxonomy, 12th edn. USDA Natural Resources Conservation Service, p 360. https://www.nrcs.usda.gov/wps/PA_NRCSConsumption/download?cid=stelprdb1252094&ext=pdf
Sorrell B, Reeves P, Clarkson BR (2004) Wetland management and restoration. In: Harding JS et al. (eds) Freshwaters of New Zealand. NZ Hydrol Soc NZ Limnol Soc, Christchurch, pp 40.1–40.13
Vandergoes MJ, Hogg AG, Lowe DJ et al (2013) A revised age for the Kawakawa/Oruanui tephra, a key marker for the Last Glacial Maximum in New Zealand. Quatern Sci Rev 74:195–201
Wagstaff SJ, Clarkson BR (2012) Systematics and ecology of the Australasian genus *Empodisma* (Restionaceae) and description of a new species from peatlands in northern New Zealand. Phytokeys 13:39–79
Walker M, Johnsen S, Rasmussen SO et al (2009) Formal definition and dating of the GSSP (Global Stratotype section and point) for the base of the Holocene using the Greenland NGRIP ice core, and selected auxiliary records. J Quat Sci 24:3–17
Williams PW (2017) New Zealand landscape—behind the scene. Elsevier, Amsterdam, 470 p
Wilson CJN (2001) The 26.5 ka Oruanui eruption, New Zealand: an introduction and overview. J Volcanol Geoth Res 112:133–174
World Reference Base (2015 update). https://www.fao.org/3/a-i3794e.pdf

Oxidic Soils

9

In the far north where it's wet and warm
weathered old Oxidic Soils form.
In oxide clays just the tough survive
add nutrients to ensure plants thrive.

Abstract

Oxidic Soils occupy less than 1% of New Zealand, occurring in small areas throughout Northland, but concentrated near Kerikeri. New Zealand's Oxidic Soils represent a southern limit of the globally extensive Oxisols which occur mainly in the tropics and subtropics. Oxidic Soils form in New Zealand where conditions favour the most intense weathering. Thus Oxidic Soils occur mainly on readily weatherable, iron-rich, basaltic to andesitic volcanic or mafic/ultramafic igneous rocks and scoriaceous deposits on sites that have been stable, potentially for millions of years, in the warm, humid environment of Northland. The parent materials of Oxidic Soils have been almost totally transformed to clay minerals, oxides, and oxyhydroxides, including gibbsite, goethite, haematite, and ferrihydrite. The soils have a friable, fine polyhedral, structure although some may include horizons comprising over 50% iron- or aluminium-rich nodules giving a gravelly texture. The soils are highly leached and low in nutrients, but with good fertiliser management are used to successfully grow citrus and other subtropical crops. The first grape vines in New Zealand were probably planted on Oxidic Soils by an associate of Samuel Marsden at Kerikeri in 1817, and the first British-style farm was established on Oxidic Soils inland from Kerikeri in 1830.

9.1 Important Features of Oxidic Soils

9.1.1 Concept and Key Features of the Soil Order

Oxidic Soils (Fig. 9.1) form in warm, humid, geologically stable environments in the far north of New Zealand on iron- and magnesium-rich (mafic) volcanic rocks, dolerite (a mafic igneous rock), ultramafic rocks (including peridotite), and basaltic tephras (scoria and associated deposits). Such conditions favour weathering and thus large proportions of the primary minerals have weathered mainly to oxides, hydroxides, and oxyhydroxides (referred to as 'oxides' or 'oxide clays' hereafter for brevity).

Oxidic Soils are clay-rich soils that have high concentrations of iron and/or aluminium oxides which form oxidic horizons (Bo). Such oxidic horizons must be at least 30 cm thick and have an upper boundary between 20 and 60 cm deep to define an Oxidic Soil in the *New Zealand Soil Classification*. Soil aggregates in Bo horizons slake (fall apart) readily, when immersed in water, to form stable microaggregates which show no, or only slight, dispersion. In contrast, aggregates in similar horizons in the related Granular or Ultic soils both slake and disperse more readily.

The Oxidic Soils, alongside some of the clay-rich Ultic Soils (Chap. 16), are the most weathered soils in New

A. E. Hewitt et al., *The Soils of Aotearoa New Zealand*, World Soils Book Series,
https://doi.org/10.1007/978-3-030-64763-6_9

Fig. 9.1 Typic Nodular Oxidic Soil (Okaihau soil) from Northland and the landscape with which this soil is associated (photos courtesy of NZ Society of Soil Science)

Zealand and are similar to Oxisols of subtropical and tropical environments throughout the world. Just like old iron having been left outside too long—they have effectively 'rusted'. Oxidic Soils are about as weathered as a soil can get, the Al- and Fe-rich oxides being thermodynamically very stable in the prevailing conditions.

Oxidic Soils contain appreciable amounts of iron and aluminium oxides. Despite high clay contents the soils are friable with low plasticity. The soils typically have a well-formed, stable, fine polyhedral structure that crushes, with friable failure, to microaggregates with diameters of 2 mm or less. The fine structure is stabilised by the iron and aluminium oxides, thus providing good physical properties for plant growth. Some soils are characterised by the formation of iron- and/or aluminium-rich nodules that are segregations of hard, rounded, subrounded, or irregularly shaped materials, which are typically ~0.5–5 cm in size and which (unlike concretions) lack internal concentric banding. In some soils, the nodules are sufficiently abundant to be referred to as 'ironstone' or, where aluminium is dominant, to be considered as 'bauxite' (an aluminium-rich ore). Nodular horizons are ~45 cm thick on average (range 20–75 cm), and usually occur below ~40 cm from the surface, and have ~15–75% nodules by volume.

Natural fertility is low and so Oxidic Soils require added fertiliser, and good fertility management, to be productive. Because they occur in the warm and humid climates of Northland and Auckland, and have good physical properties, Oxidic Soils are valued for horticulture especially kiwifruit, citrus, tamarillo, and, increasingly, avocado.

9.1.2 Areas of Occurrence

Oxidic Soils are only known in the Auckland and Northland regions (Fig. 9.2). They cover <1% of New Zealand. The Oxidic Soils form in small pockets throughout Northland, with the largest areas concentrated in the Bay of Islands near Kerikeri.

9.1.3 Variation Within the Oxidic Soil Order

The following soil groups are defined within the *New Zealand Soil Classification* (Fig. 9.3):

Perch-gley Oxidic Soils—periodic wetness caused by a perched water table,
Nodular Oxidic Soils—prominent segregated accumulations mainly of iron oxides and/or gibbsite, and
Orthic Oxidic Soil—other Oxidic Soils.

The subgroups include Nodular (containing a layer at least 10 cm thick with at least 15% iron- or aluminium-rich nodules), Mottled (with a rust-coloured (redox-mottled) horizon, or a pale-coloured greyish horizon above a slowly

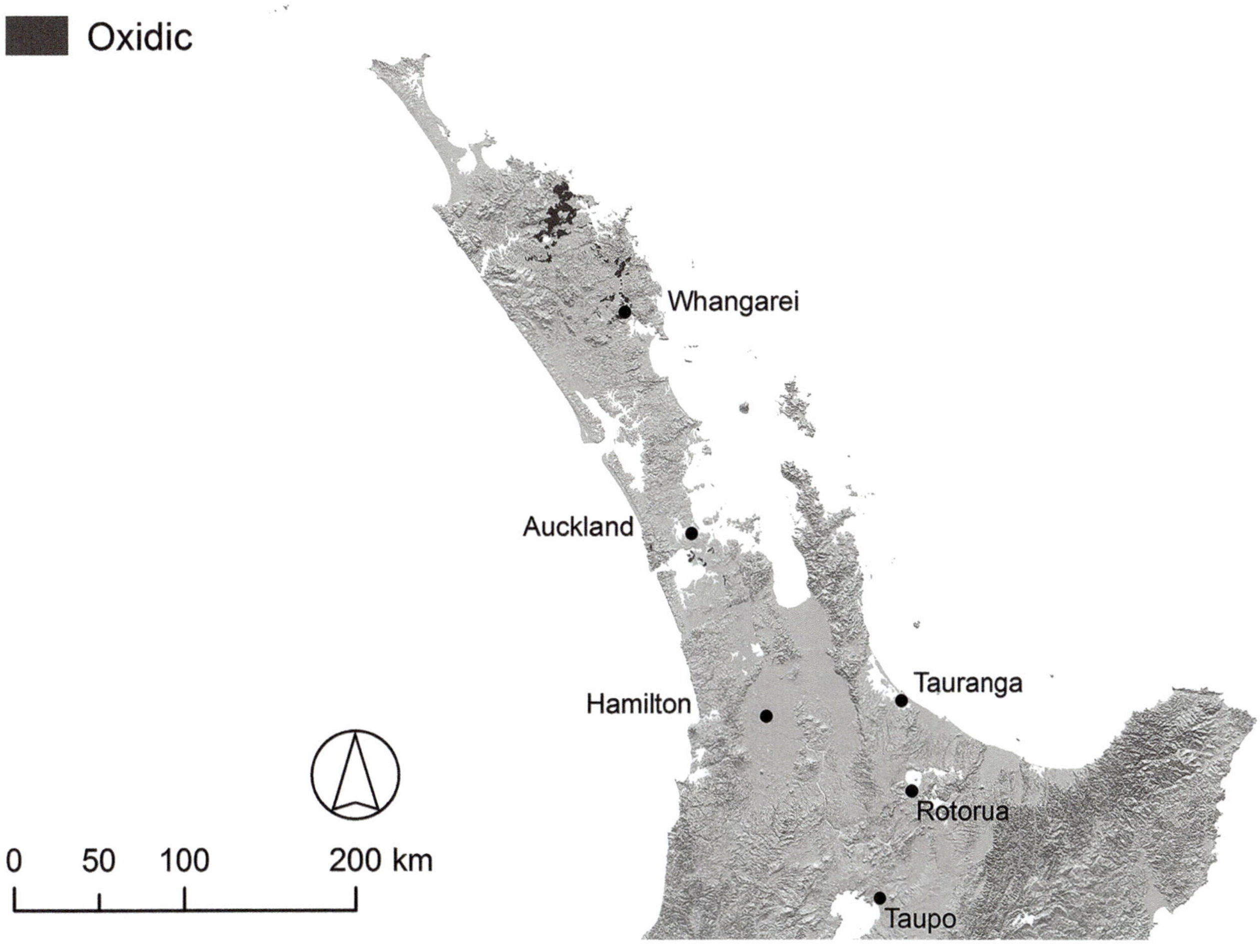

Fig. 9.2 Distribution of Oxidic Soils Photo: Allan Hewitt

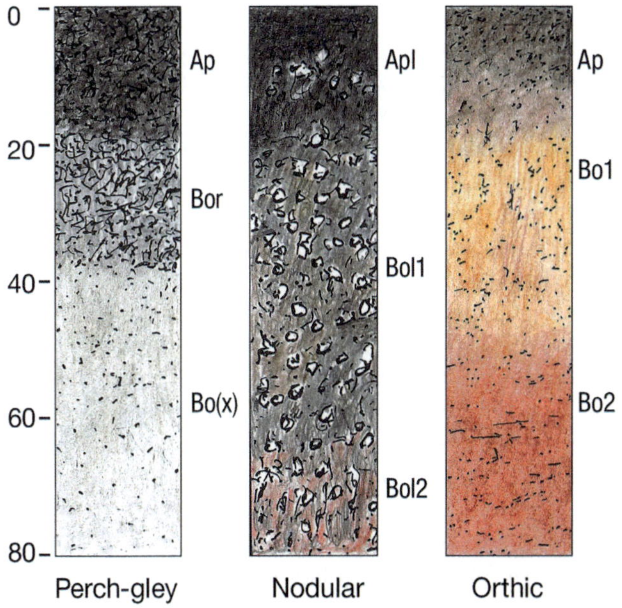

Fig. 9.3 Groups within the Oxidic Soil order. Vertical axis is depth (cm) Photo: Allan Hewitt

permeable layer, or where there is intermittent groundwater), Brown (where Oxidic Soils share some properties with Brown Soils), and Typic as the central concept of the Orthic Oxidic Soils group.

9.1.4 Origin of the Soil Order Name

'Oxidic' is derived from the *Soil Taxonomy* name Oxisols (originally Greek 'oxus' and French 'oxide'), and reflects the high proportion of iron and aluminium oxides present.

9.2 Soil Profile Genesis

In the relatively geologically stable Northland region the mafic lavas and associated deposits have been exposed to weathering for long periods (often millions of years) in landscapes that have remained stable under conditions of strong weathering and leaching, and potentially with marked movements in water table in lower landscape positions. Thus the weathering has advanced to the stage of forming oxides (especially goethite, haematite, and gibbsite) as well as subordinate kaolin-subgroup clays (halloysite and kaolinite).

The processes that form the soil may be viewed as a game of 'winners and losers' and are centred around so-called ferralitic weathering or ferralisation. The soil composition depends on the availability, and relative mobility, of the elements that make up the soil parent material. A range of factors influence an element's ability to either remain and build up as a residual concentration (arguably a "winner"), or be broken down and leached from the soil and so become depleted (or "lost"). The fate of elements can thus be best understood by focusing on the factors that promote mobilisation with consequent loss from the soil, or factors that cause elements to be left behind and thus become concentrated in the soil as newly formed secondary minerals.

In the iron-rich soil-forming environment, the main elements of interest are iron, aluminium, silicon, and the non-acid cations (also known as base cations, or bases), namely, Ca^{2+}, Mg^{2+}, K^+, and Na^+. Where the parent material does not contain calcium carbonate (natural lime), leaching has the potential to increase soil acidity, especially where soils are free draining and rainfall is high. At more acid (lower) pH levels, silicon, along with the non-acid cations, become dissolved in the soil solution and then are leached from the soil profile, which, in turn, leads to a residual concentration of mainly iron and aluminium then, if conditions are favourable, form less soluble oxides (sometimes together with lesser amounts of kaolinite and halloysite) that remain in the soil.

In some soils, the processes have led to the formation of nodules that may comprise 50% of the soil volume in a distinct coarse-textured, often gravelly, horizon. Analyses show a nodule typically to contain ~50% Fe oxides (haematite, goethite, ferrihydrite), ~35% gibbsite, and ~15% kaolinite and/or halloysite. Significant organic matter is found in nodules (~20% loss-on-ignition) with major elements dominated by Fe (34%), Al (13%), Si (3%), and Ti (~1%). Iron- and manganese-rich concretions may also form in Oxidic Soils where redox conditions prevail.

Thus, in Oxidic Soils, the major factors influencing soil formation are an iron-rich parent material and strong weathering and acidity with the subsequent concentration of iron and aluminium. In contrast to Oxidic Soils, Gley Soils (Chap. 5) are controlled by acidity and biochemical reduction which leads to the mobilisation, and potential loss by leaching, of iron under reducing conditions. Podzol Soils (Chap. 11) are controlled by marked acidity with iron, silicon, and organic materials moved down and precipitated in the subsoil, leaving a distinct, bleached E horizon near the soil surface.

Despite being formed mainly on basaltic lavas, which do not contain quartz, Oxidic Soils may contain small amounts of quartz, mainly in the upper profile. Bob Stewart showed, using oxygen isotope analyses, that the quartz grains in an Oxidic Soil were probably derived mainly from fallout of distal rhyolitic tephra, such as Rotoehu Ash (c. 45,000 years old) and Kaharoa Tephra (c. 700 years old), erupted from central North Island volcanic centres. Sparse deposits of these and other thin rhyolitic tephras have been recorded in the sediments of Lake Omapere near Kaikohe. Small amounts of the clay mineral vermiculite are thought to derive from local loess or aeolian dust from Australia.

One opinion is that the well-formed structure of Oxidic Soils is strongly influenced by bioturbation. This seems likely in areas of high termite populations such as Australia, but could such levels of bioturbation be relevant to New Zealand Oxidic Soils? We do have native termites in Northland but have they ever been sufficiently active to have contributed to the friable fine structure that is characteristic of our Oxidic Soils?

9.3 Soil-Landscape Relationships

The Oxidic Soils in New Zealand form at the end of the range of tropical and subtropical Oxisols and so occur only in conditions that maximise weathering to form mainly oxidic clays. Thus Oxidic Soils form on flat to rolling, stable, land surfaces on iron, magnesium, and aluminium-rich (readily weatherable) parent materials, with (by New Zealand standards) a warm and humid climate. Oxidic Soils are associated with the oldest mafic and ultramafic igneous rocks (including dolerite, peridotite, and basaltic to andesitic volcanic lavas and associated scoriaceous deposits) in the Northland and Auckland landscapes. Oxidic Soils are formed on thick saprolite (weathered rock) and the soil minerals mainly reflect the mineralogy of the underlying rocks.

Leslie I. Grange, a pioneer of New Zealand soil science, was the first person we have found to record a scientific study related to the Oxidic Soils—he published a description of the 'ironstone soils' (Typic Nodular Oxidic Soils) of 'North Auckland' in 1934. Norman Taylor had worked closely with Grange and in about 1937 Taylor started trying to understand the complex soil pattern of Northland, including the soils now known as Oxidic Soils. Thus the first soil survey of Northland by Taylor (along with Charles Sutherland and Charles Wright) was undertaken between 1937 and 1951, resulting in a series of provisional 1:63 360 scale soil maps. From 1979 to 1985, Ted Cox updated the soil maps of Northland which were published as part of the New Zealand Land Inventory of Northland at a 1:100 000 scale. Alistair Wilson and Warwick McDonald sampled some suites of soils in the 1980s with results available in the National Soils Data Repository. Limited data are also available in international field trip guidebooks. However, research on the soil-landscape relationships of Oxidic Soils and Granular Soils in Northland is still in its infancy with relatively little published on the subject in the last 30 years.

The Oxidic Soils are most concentrated around, and north and west of, Kerikeri in the Kaikohe-Bay of Islands Volcanic Field, and near Whangarei. The basalt lavas and scoria materials (Kerekeri Volcanic Group basalts) are Late Miocene to Quaternary in age (ranging from c. 9 million to c. 300,000 years old). Oxidic Soils are also likely to occur on Early Miocene (c. 20 million years old) basalt lava flows (Waipoua basalt) in the Waipoua area south of Hokianga Harbour, and on basaltic volcanic and mafic to ultramafic igneous deposits in the Karikari Penisula, North Cape, and Surville Cliffs areas. The soils occur on flattish, stable plateau and interfluve surfaces. The variable topography and geology of Northland mean that the Oxidic Soils are likely to occur in small pockets in soil complexes with Granular, Gley, and Ultic soils as well as Oxidic Brown Soils and Allophanic Soils. Granular Soils (Chap. 6) or Oxidic Brown Soils (Chap. 4) tend to prevail on younger basalts and where weathering has not become so advanced. Local incidences of Allophanic Soils (Chap. 2) occur on well-drained Late Quaternary scoria cones. Ultic Soils (Chap. 16) form on the more silica-rich parent materials including saprolite on Tertiary sedimentary rocks and dacites and rhyolites.

Work associated with the Smap project in the Pukekohe-Bombay region has helped to elucidate the topographic relationship between uncommon (probable) Oxidic Soils, formed on basalt lavas re-exposed (exhumed) at the land surface, by erosion, in upper slope positions in the South Auckland Volcanic Field, and other adjacent soils. Oxidic Granular Soils and Oxidic Brown Soils are formed mainly on lower slopes and flat terraces where clay-rich, strongly altered, rhyolitic tephra beds (Hamilton Ash) overlie paleosols formed on the basalts. The composite soils will be Oxidic Soils where the buried oxidic horizon on the underlying weathered basalt is within 60 cm of the modern land surface. The accumulated understanding of the 'Red and Brown loams' (some of which are Oxidic Soils) originally demarcated by Taylor was published in 'New Zealand Soil News' in December 1957, with the same material largely repeated (often word for word) in the few publications since then.

Stan Buol and Hari Eswaran described three landscapes where Oxisols occur in Africa and South America. Firstly, many Oxisols are associated with stable slopes on igneous (especially basaltic volcanic) rocks and associated deposits that are weathered sufficiently to produce low activity clays, similar to the New Zealand Oxidic Soils. Secondly, Oxisols are formed on younger landscapes, but only on the most stable land surfaces in mafic deposits that are readily weatherable. New Zealand Oxisols and Granular Soils may have some genetic elements in common with the second kind of Oxisols. Allophanic soil materials are sometimes found in upper horizons and may be weathered from the local parent

Fig. 9.4 Concentrations of iron oxide nodules in an Oxidic Soil at North Cape. Photo: Peter de Lange

materials, or are possibly derived from weathering of small amounts of tephra blown from central North Island volcanoes and thus fit Buol and Eswaran's second landscape type. The third type occurs on late Tertiary high plateaux of mafic igneous deposits with deep weathering profiles. They are distinguished by thick resistant caps of plinthite or ferricrete (a hard, cemented, iron-rich layer) that may have formed through landscape inversion. Such landscapes are not a feature in New Zealand, except at Cape Reinga, North Cape, and Surville Cliffs where ultramafic igneous materials are present (Fig. 9.4). Peter de Lange reports that the (pedogenic) nodules can grow to boulder size and that the Oxidic Soils, of the Surville Cliffs portion of North Cape Peninsula, host a number of endemic plant species.

International connections

Although the Oxidic Soils in New Zealand cover only a small area, they are a southerly limit of an extensive and important group of global soils known as *Oxisols* (or *Ferralsols*). Globally, Oxisols cover about 750 million ha (8% of the ice-free land area) on mostly flat to gentle slopes in the humid tropics. They occur onthe stable continental shields of South America and Africa where land surfaces are among the oldest on Earth, with exposure to weathering and leaching for prolonged periods of time. Oxisols are also extensive in humid tropical areas of easily weatherable basic volcanic rocks, such as South East Asia and on basalts in northern Australia.

Large areas of Oxisols provide only meagre food production. While their physical properties are to be appreciated, the soil chemistry presents challenges to nutrient management. Under low input, subsistence, agriculture of the soil-plant system will recycle nutrients by returns to the soil via litter fall and root extension. However, the low nutrient storage capacity of the soil means careful soil management is important to maintain soil organic matter and prevent erosion.

More intensive production requires correction of multiple deficiencies including nitrogen, potassium, calcium, magnesium, sulphur, zinc, boron, and copper, along with considerable management finesse. Research is refining nutrient management. Since about 2010, Brazil has successfully developed vast areas of Oxisols into highly productive, sustainably managed, crop-land with carefully managed inputs of lime and fertiliser. As a result, Brazil has moved from being an importer of grain to one of the larger global exporters. Brazil is providing an example which undoubtedly others will follow with potentially world-changing effect.

9.4 Key Soil Properties

9.4.1 Soil Composition

The Oxidic Soil parent materials are mainly derived from dark coloured mafic or ultramafic rocks and associated deposits, especially basalts, andesites, dolerites, and peridotites of Tertiary to Quaternary age. Oxidic Soils may also form in sediments derived from the mafic rocks. The soils are strongly weathered with 50–90% of parent material altered to clay. The clay fraction comprises a mixture of minerals including substantial amounts of iron oxides (especially goethite together with haematite and ferrihydrite), aluminium oxide (gibbsite), and kaolin-subgroup clays (halloysite and/or kaolinite). Vermiculite may also be present, mainly in upper profiles and is likely to be derived from aeolian dust.

9.4.2 Physical Properties

Topsoils are usually friable with low plasticity and well developed fine spheroidal or polyhedral structure despite high clay contents. The upper profile is generally free draining with medium soil dry bulk densities and moderate or rapid infiltration (Fig. 9.5 and Table 9.1).

Good rooting volume is provided by the friable structure in the topsoil and upper B horizons. However, in lower subsoils, the higher soil dry bulk densities, firm soil strength, high penetration resistance, and limited porosity, may limit

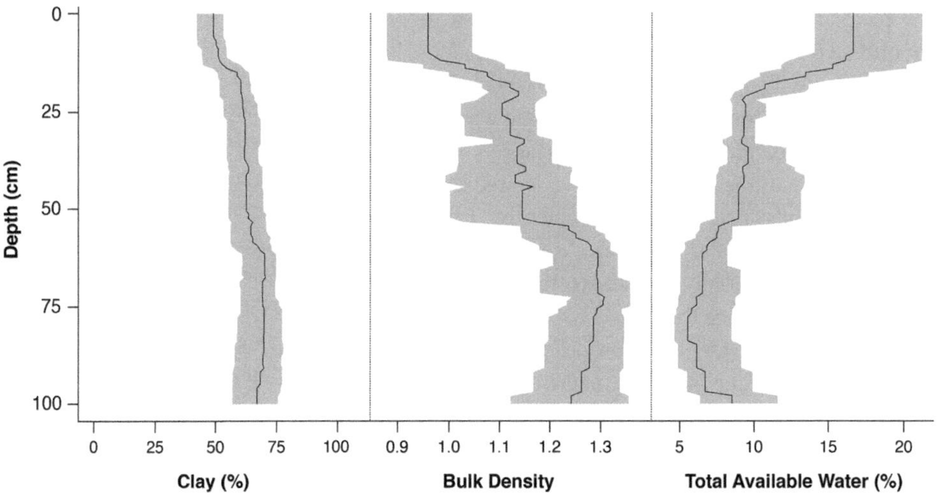

Fig. 9.5 Median and upper and lower quartiles of clay content, soil dry bulk density (t m^{-3}), and total available water capacity for Oxidic Soils in the New Zealand Soil Data Repository

Table 9.1 Physical properties of a Typic Nodular Oxidic Soil (Okaihau series, SB10057)

Horizon	Horizon depth (cm)	Sand (2.0–0.06 mm) %, w/w	Silt (0.06–0.002 mm) %, w/w	Clay (<0.002 mm) %, w/w
Ap	0–18	11	29	59
Bo1	18–41	26	12	61
Bi	41–59	32	14	53
Bo2	59–73	13	12	74
Bw1	73–97	2	11	85
Bw2	97–120	3	19	77

downward root extension particularly in well-drained soils. Such a limitation is indicated by macro-porosity values in some subsoils of less than 10% and as low as 4%.

Water logging and poor aeration may occur during wet periods in the Gley Oxidic group, and temporary perching (lasting 1–2 days in well-drained soils, and 4 days or more in less-well drained soils) may occur after intense rain. Permeability decreases with depth to low, or very low, with the consequence that perching may occur on lower horizons although high clay contents and limited root penetration may reduce plant access to stored water in dry periods.

9.4.3 Chemical Properties

The chemical properties of Oxidic Soils are dominated by their highly weathered and strongly leached status. The clays, dominated by Fe- and Al-oxides and kaolin-subgroup clays, may be referred to as 'low activity clays' because they have relatively few charged surface sites that can hold and exchange cations. Thus the soils are also weakly buffered. Due to the strong leaching, unless fertiliser has been added, the cation exchange sites are dominated by H^+ and Al^{3+} cations. As a result, the subsoil pH is strongly, and in places extremely, acid (Fig. 9.6 and Table 9.2). The pH of the topsoil may be higher due to lime and fertiliser addition. Where pH values are less than about 4.8 the soils are likely to be affected by high levels of KCl-extractable aluminium. About ten percent of Oxidic Soils in the New Zealand Soil Database have high to very high subsoil KCl-extractable Al (>2 $cmol^{(+)}\ kg^{-1}$) which indicates risk of aluminium toxicity in plants.

Phosphate retention is high, ranging from 60 to 90%. The high P retention is a consequence of the high content of iron and aluminium oxides, together with small amounts of allophane in some cases. High sulphate adsorption also occurs in B horizons.

Low, to very low, cation exchange capacities, at the natural pH of the soil, are an outstanding feature of the Oxidic Soils due to the abundance of low activity clays. Some subsoil horizons may have CEC values that are so low that the net soil charge is positive (i.e. the soils have variable charge properties where the cation exchange capacity varies with pH). Even with the low CEC, the subsoil percent base saturation is generally low or very low indicating that a high proportion of the CEC is occupied by Al^{+3} and H^+, with the non-acid cations (e.g. Ca^{2+}, Mg^{2+}, Na^+, K^+) being lost through leaching.

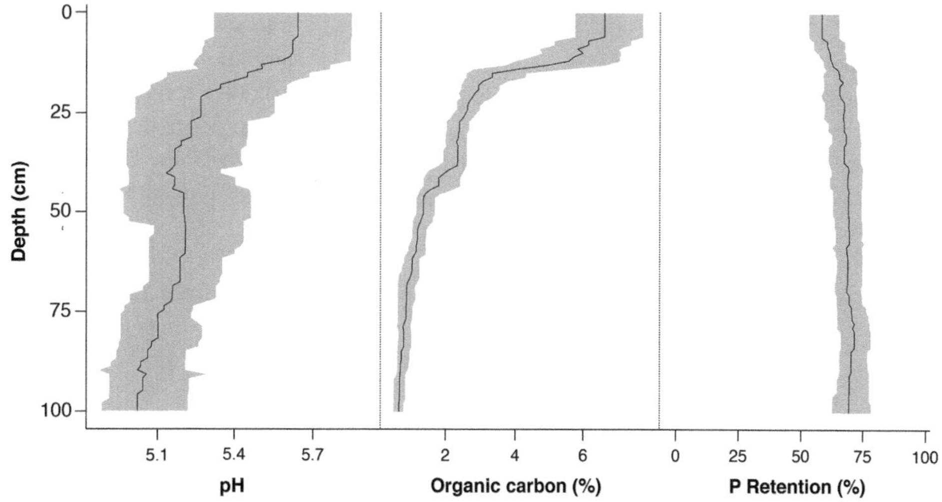

Fig. 9.6 Median and upper and lower quartiles of soil pH, organic carbon, and P retention for Oxidic Soils in the NZ Soil Data Repository

Table 9.2 Typical example of chemical properties of a Typic Nodular Oxidic Soil (Okaihau series, SB10057)

Horizon	Depth	pH (in H_2O)	Carbon (%)	Nitrogen (%)	CEC[a] (cmol$^{(+)}$ kg^{-1})	Sum bases (cmol$^{(+)}$ kg^{-1})	P retention (%)
Ap	0–18	5.7	8.3	0.57	27	24.7	54
Bo1	18–41	5.1	2.5	0.12	7.4	1.2	64
Bi	41–59	5.5	1.2	0.07	3.8	0.8	59
Bo2	59–73	5.4	1.3	0.05	3.7	1.3	67
Bw1	73–97	5.0	0.9	0.02	3.6	0.9	79
Bw2	97–120	5.0	0.7		5.2	0.8	82

[a]CEC = cation exchange capacity

In topsoils, where soil carbon contents tend to be moderate, the organic matter provides some exchange sites and the CEC is often in the medium range. As for many soils, base saturation levels of topsoils vary widely, being higher where fertiliser has been applied.

9.4.4 Biological Properties

Much of the area of Oxidic Soils was likely to have originally been under kauri/podocarp forest since the end of the last glaciation. Unlike the rest of New Zealand, forest cover was maintained in Northland during the preceding glacial period and comprised tall podocarp-broadleaf and beech forest. Fire and precipitation change (especially dry phases) have played key roles in vegetation development in the region during the last glacial-interglacial cycle. However, human-fire-induced, relatively impoverished, scrub vegetation dominated by manuka (*Leptospernum scoparium*), kumarahou (*Pomaderris kumaraho*), and wiwi (*Juncus krussii*), plants that are tolerant of low pH and low nutrients, predominated following Polynesian arrival. Most of the area of Oxidic Soils was reported as still being in 'scrub' in the 1960s and early 1970s.

In its natural soil state the soil biological habitat, characterised by low nutrient availability, and low to very low pH, the potential for aluminium toxicity in plants, and often with droughty conditions in summer, is a challenging environment for plants, and also for many soil organisms. Such conditions favour organisms that are adapted to the environment with reduced competition from less tolerant species. Use of fertiliser and lime, along with irrigation in dry periods, will greatly improve conditions for many plants and soil organisms in the Oxidic Soils.

9.5 Distinguishing Between Oxidic Soils and Related Soil Orders

Oxidic Soils differ from Granular Soils by having an oxidic horizon (>30 cm in thickness with its upper boundary between 20 and 60 cm from the soil surface) and very low cation exchange capacities which limit the soil's capacity to store nutrients. The soil aggregates in an oxidic horizon slake rapidly in water to become stable microaggregates but these essentially do not disperse, in contrast to the behaviour of aggregates in Granular and Ultic Soils. Granular Soils are also more porous and less likely to restrict root extension. Field identification of Oxidic Soil is based on the stable behaviour of the soil material from moist to dry states. Primary ped strength and soil strength are either weak or very weak. Most Oxidic Soils occur on the oldest, most stable basalt or mafic-ultramafic rock surfaces (or associated fragmental or reworked deposits) that usually lack surface stone outcrops because of their deep weathering. Granular Soils occur largely on the thick, weathered tephra deposits of the Hamilton Ash beds or younger volcanic surfaces, including basalt lavas, or on sloping areas where erosion has reinvigorated soil processes. On lavas, boulders often outcrop at the soil surface, and many have been collected up and assembled into stone walls. Oxidic Granular Soils provide a useful intergrade where oxidic horizons, although insufficiently thick or outside the control section, nevertheless confer oxidic properties to the Granular Soils. A similar intergrade, Oxidic Brown Soils, is common in the Pukekohe-Bombay area where limited Oxidic Soils can occur alongside Granular Soils.

Oxidic Soils differ from Ultic Soils by darker colours arising from the presence of the iron oxides, which have

Table 9.3 Correlation[a] between Oxidic Soils and equivalent classes of *Soil Taxonomy*, *World Reference Base*, and the earlier NZ genetic soil classification

New Zealand Soil Classification	*Soil Taxonomy*	*World Reference Base*	NZ genetic soil classification
Perch-gley Oxidic Soils	Aquic or Aeric subgroups of Aquox, or Kandiudox	Epigleyic Ferralsols	Brown loam, or friable clay
Nodular Oxidic Soils	Kandiudox, or Acrudox	Plinthic Acric Ferralsols	Brown loam, or friable clay
Orthic Oxidic Soils	Kandiudox, or Acrudox	Acric Ferralsols	Brown loam, or friable clay

[a]The correlations given here are a guide only and for accurate classifications the relevant soil classification documents should be consulted. The two major international soil classification systems are *Soil Taxonomy,* which was developed in the USA and *World Reference Base* which was developed primarily in Europe. The NZ genetic soil classification was used in NZ prior to 1992

strong pigmenting power, derived from the weathering of dark-coloured basaltic volcanic parent materials. Oxidic Soils are also generally more porous and more permeable than Ultic Soils, which do not have oxidic horizons, and Oxidic soils lack the pale uppermost subhorizons of Ultic Soils. Oxidic Soils differ from Allophanic Soils by their limited content of allophanic minerals as evidenced by lack of greasy texture, and absence of a positive reaction to the sodium fluoride test for allophane. The soil material in Oxidic Soils does not meet the requirements of 'allophanic soil material' .

9.6 Correlation with Other Classification Systems

The New Zealand Oxidic Soils are local representatives of the internationally extensive Oxisols (as defined by *Soil Taxonomy*) or Ferralsols (as defined by the *World Reference Base*) (Table 9.3). Oxisols dominate the Earth's tropical belt, and their occurrence in New Zealand is close to the subtropical limit of the Oxisols. The CEC in relation to clay content is the main classification criterion for recognition of the Oxisols. The New Zealand data demonstrate that our Oxidic Soils correlate well with the globally important Oxisols.

9.7 Use and Management of Oxidic Soils

The first grape vines in New Zealand were probably planted on Oxidic Soils by an associate of Samuel Marsden at Kerikeri in 1817. Not far inland and southwest of Kerikeri, along the first road to be constructed in New Zealand, is the Te Waimate Mission (at Waimate North). The surviving mission house, built by George Clark (and visited by Charles Darwin 23–24 December, 1835), marks the site of the first British-style farm in New Zealand that was instigated in 1830 on the Oxidic Soils in the area (Fig. 9.7).

The Oxidic Soils of the Kerikeri region, when managed appropriately, are moderately suitable for horticulture and the warm, generally moist climate favours a range of crops that are difficult to grow elsewhere in New Zealand. The rolling slopes are a limitation for cultivation but citrus and other tree crops including apples, avocados, tamarillos, and feijoas, as well as specialist crops such as bananas, are successfully grown on the Orthic Oxidic Soils. As well as providing small, but tasty, bananas, and requiring little irrigation, banana plants potentially provide a source of stock feed and a means to 'mop up' excess nutrients (N, P and K) in dairy riparian zones.

The Nodular Oxidic Soils may need deep ripping to break up the accumulation of iron and gibbsitic nodules to facilitate root and moisture penetration into the subsoil. Ripping is especially important prior to planting trees for forestry or tree crops. Perch-gley Oxidic Soils are limited by poor drainage in the subsoil.

The Oxidic Soils are generally productive when well-managed. Challenges include low soil water holding capacity and common summer soil water deficits, as well as low reserves of potassium, magnesium, calcium, and phosphorus along with trace elements, particularly molybdenum. Because the clays have low cation exchange capacity at the natural pH of the soil, and phosphate retention is moderate or high, regular additions of lime and fertiliser are needed to ensure reliable crop production. Addition of organic matter will improve both nutrient and water holding capacity.

Topsoil infiltration is normally good and will prevent runoff if the porous structure is not compacted. However when wet, after intense rain, the soil may be susceptible to livestock treading damage and compaction. Seasonal wetness occurs in the poorly and imperfectly drained Perch-gley Oxidic Soils, causing limited oxygen supply and prevalence of biochemical reduction as indicated by presence of reduced iron (Fe^{2+} ions) and manganese (Mn^{2+}) in the soil solution and formation of pale grey (low chroma) soil colours, and redox concentrations of iron (as Fe^{3+}) and manganese (as Mn^{4+}) in rust-, orange-, or black-coloured soil mottles or

Fig. 9.7 The Te Waimate Mission House near Kerikeri marks the site of the first British-style farm in New Zealand, established on Oxidic Soils in 1830 Photo: David Lowe

concretions. Intense, and occasionally prolonged, periods of rain can occur in Northland at any time of the year and will saturate the soil. Thus, all soils will benefit from artificial drainage, particularly the Perch-gley Oxidic Soils.

Bypass flow occurs when drainage water, along with nutrients and contaminants, escapes directly to the base of the soil down pores and fissures. Bypass flow is a risk in Oxidic Soils where there are slowly permeable horizons in the lower subsoil, particularly in Perch-gley Oxidic Soils.

Although Oxidic Soils have favourable porosity and structural qualities, sheet erosion is a risk where vegetation cover is lacking and if the protective topsoil is removed and less permeable subsoils are exposed then heavy rain will rapidly run off causing rill and then gully erosion.

Soil water deficits occur in summer because of the warm climate and high solar radiation (hence evapotranspiration is high), so that the region is susceptible to drought when rainfall is only a little less than usual. The soil water deficits are exacerbated by low readily available water and limited nutrient storage capacity, especially in the Nodular Oxidic Soils where up to 50% of the soil volume may be ironstone or gibbsitic gravels. Thus soils used for horticulture will require irrigation in dry periods in most years.

Further Reading

Beinroth FH, Eswaran H, Uehara G et al (2012) Oxisols. In: Huang PM, Li Y, Sumner ME, (eds) Handbook of soil sciences, 2nd edn. vol 1. Properties and Processes. CRC Press, Boca Raton, FL, USA, pp 33.177–33.190

Blanchon DJ, de Lange PJ, Galloway DJ (2015) New records of Ramalina (Ramalinaceae, Ascomycota) for mainland New Zealand. NZ J Bot 53:192–201

Buol SW, Eswaran H (2000) Oxisols. Adv Agron 68:155–195

Cox JE, Bowen FE, Davidson JM (compilers) (1973) Guidebook for excursion 4—Northland. IX INQUA Congress, Christchurch, New Zealand, 142p

Cox JE, McLeod M, Griffiths E (1981) Soil survey and horticultural potential of Kapiro Block, Kerikeri, Northland. Soil Bureau District Office Report AK1. DSIR Soil Bureau, 11p de Lange PJ (2003) *Pittosporum serpentinum* (de Lange) de Lange (Pittosporaceae), a new species combination for an ultramafic endemic from North Cape, New Zealand. NZ J Bot 41:725–726

de Lange PJ, Heenan PB (1997) *Carex ophiolithica* (Cyperaceae): a new ultramafic endemic from the Surville Cliffs, North Cape, New Zealand. NZ J Bot 35:429–436

Edbrooke SW, Brook FJ (compilers) (2009) Geology of the Whangarei area. IGNS 1:250,000 Geological Map 2. GNS Science, Lower Hutt, 68 pp + map

Elliott MB, Striewski B, Flenley JR, Sutton DG (1997) A palynological and sedimentological record of environmental change and human impact from Wharau Road Swamp, Northland, New Zealand. J R Soc N Z 27:401–418

Gibbs HS, Cowie JD, Pullar WA (1968) Soils of North Island. NZ Soil Bur Bull 26:48–67

Grange LI (1934) The ironstone soils of North Auckland. N Z J Sci Technol 16(1):9–18

Harmsworth GR (1996) Land use capability of the Northland Region. In: Landcare Research science series no 9. Maanaki Whenua Press. Lincoln, Canterbury, New Zealand, 269p

Hayward BW (2017) Out of the ocean, into the fire—history of rocks, fossils and landforms of Auckland, Northland and Coromandel. Geoscience Society of New Zealand Miscellaneous Publication 146, 336 p

Isaac MJ (compiler) (1996) Geology of the Kaitaia area. IGNS 1:250,000 Geological Map 1. GNS Science, Lower Hutt, 44 pp + map

Kear D, Waterhouse BC, Swindale LD (1961) Bauxite deposits in Northland. NZ Department of Scientific and Industrial Research Information Series 32, 58p

Newnham RM (1999) Environmental change in Northland, New Zealand, during the last glacial and Holocene. Quatern Int 57 (58):61–70

Newnham RM, Ogden J, Mildenhall D (1993) A vegetation history of the Far North of New Zealand during the Late Otira (Last) Glaciation. Quatern Res 39:361–372

Newnham RM, Lowe DJ, Green JD et al (2004) A discontinuous ca. 80 ka record of Late Quaternary environmental change from Lake Omapere, Northland, New Zealand. Palaeogeogr Palaeoclimatol Palaeoecol 207:165–198

Newnham RM, Alloway BV, Holt KA et al (2017) Last Glacial pollen–climate reconstructions from Northland, New Zealand. J Quat Sci 32:685–703

Orbell GE, Parfitt RL, Furkert RJ (1980) Guide Book for Tour 6—Specialist North Auckland. "Soils with Variable Charge" Conference, Palmerston North, February 1981, 105 p

Ross C, Wilson A (1994) Pedogenic nodules in an Oxidic Soil, Northland. Abstracts, New Zealand Society of Soil Science Conference, Lincoln University

Staffs of Soil Bureau, DSIR, and Extension Division, Department of Agriculture (1954) General survey of the soils of the North Island of New Zealand. Soil Bureau Bulletin 5. New Zealand Department of Scientific and Industrial Research, 286p

Stewart RB, Neall VE, Syers JK (1984) Occurrence and source of quartz in six basaltic soils from Northland, New Zealand. Aust J Soil Res 22:365–377

Smith IEM, Okada T, Itaya T et al (1993) Age relationships and tectonic implications of late Cenozoic basaltic volcanism in Northland, New Zealand. NZ J Geol Geophys 36:385–393

Swindale LD (1961) Appendix 3. Bauxite in the Red-Brown loam soils of New Zealand. NZ Department of Scientific and Industrial Research Information Series 32, pp 52–58

Taylor NH, Dixon JK, Seelye FT (1950) Soils of North Auckland Peninsula, New Zealand. Soil Bureau Publication no 30. Department of Scientific and Industrial Research, Wellington, New Zealand

Taylor NH, Sutherland CF (1942) Soils of North Auckland. Soil Bur Bull 3:25–38

Town and Country Planning Branch, Ministry of Works (1964) National resources survey Part III. Northland Region. Government Printer, Wellington, 202 p

Wilson AD, McDonald WS (1987) Soils of Northland 2. Awarua Soils. New Zealand Soil Bureau District Office Report KK2. DSIR, New Zealand. 23p

Pallic Soils

10

In regions with a summer drought
the Pallic Soils are about
with dense subsoils formed in loess
the winter wet may make one curse.

Abstract

Pallic Soils are prominent on the rolling downlands of the lower North Island and eastern South Island. Pallic Soils, the third-most extensive soil order, covering 12% of New Zealand, occur in regions where there is normally a dry summer with a mean annual rainfall of about 500–1000 mm in the South Island and $< \sim 800$ mm in the North Island. Pallic Soils are often formed in loess, a uniform silt-dominated material that is blown off river floodplains and deposited onto the surrounding landscape. A dense (soil dry bulk density $\geq \sim 1.5$ t m^{-3}) fragipan often forms in the subsoil which can obstruct root penetration and drainage. Thus Pallic Soils often suffer winter wetness, as water perches above the fragipan, and are also prone to summer drought. Subsoils are pale-coloured ("pallic") because of their low contents of iron oxides and, in some cases, gleying. Pallic Soils are most suitable for pastoral grazing, maize, and other summer crops. Tree crops may be limited by winter saturation and rooting depth. Subsoil drainage can be effective in preventing winter or spring saturation. Some Pallic Soils form in alluvium or slope debris and are friable, without a strong fragipan, hence are deep rooting and versatile.

10.1 Important Features of Pallic Soils

10.1.1 Concept and Key Features of the Soil Order

Pallic Soils are generally fine-textured soils with dense, pale-coloured subsoils (Fig. 10.1). Pallic Soils are formed mainly in loess, a uniform, silt-dominated material that has been blown off river floodplains and deposited onto the surrounding landscape, at a relatively slow rate, during glacial periods. A key feature that contributes to development of Pallic Soils is a summer drought. Thus, Pallic Soils

A. E. Hewitt et al., *The Soils of Aotearoa New Zealand*, World Soils Book Series,
https://doi.org/10.1007/978-3-030-64763-6_10

are dry in summer and many have an annual water deficit of about 90–200 mm. However, the soils may be wet in winter or spring when a surplus of soil water is common. Pallic Soils have pale-coloured subsoils, primarily because of low contents of iron oxides. In Pallic Soils that have a winter/spring perched water table, the consequent gleying also contributes to the pale colours.

As a result of the loess-dominated parent material, Pallic Soils typically have silt loam textures, weak structure, low porosities, and dense subsoils. In some situations, a distinct, high bulk density pan, called a fragipan, forms in the subsoil. The fragipan may be over a metre thick and generally has a distinct extremely coarse prismatic structure with grey, gleyed, material in the cracks between the prisms, sometimes with reddish, oxidised, material on the aggregate surfaces where oxygen moves down open cracks during summer dry periods. Under pasture, topsoils are dark coloured and distinct. The B horizons often contain evidence of clay movement (translocation) down the profile with clay coatings on ped surfaces in lower horizons. In soils with a fragipan, a perched water table often forms above the fragipan in winter. The resulting anaerobic conditions lead to gleying in the upper B horizons with mottles and/or a net-like (gammate or reticulate) pattern of pale grey veins evident. The degree of gammation and gleying increases with increasing rainfall.

10.1.2 Areas of Occurrence

Pallic Soils cover 12% of New Zealand (Fig. 10.2) and are the third-most commonly occurring soil order in New Zealand (after Brown and Podzol Soils). In the North Island they stretch from East Cape, southward along the east coast to the Wairarapa, and from the Manawatu to the Kapiti Coast. In the South Island, Pallic Soils occur through coastal Nelson, Marlborough, Canterbury, Otago, and Southland. Pallic Soils form predominantly on lower altitude rolling hills and terraces. They are prominent in plains, downlands, and inland basins and are also extensive in dry hill country.

10.1.3 Variation Within the Pallic Soil Order

Six soil groups are recognised within the Pallic Soil order (Fig. 10.3):

Perch-gley Pallic Soils—prolonged periodic wetness caused by a perched water table,
Duric Pallic Soils—silica-cemented pan (duripan) in the subsoil,
Fragic Pallic Soils—a dense, brittle fragipan in the subsoil,
Laminar Pallic Soils—clay accumulation (via translocation) as thin subsoil bands (lamellae),

Fig. 10.1 Pallic Soil downland landscape and soil profile. Photos reproduced with permission of NZ Society of Soil Science

Fig. 10.2 Map of distribution of Pallic Soils, most coincident with areas of substantial loess deposits

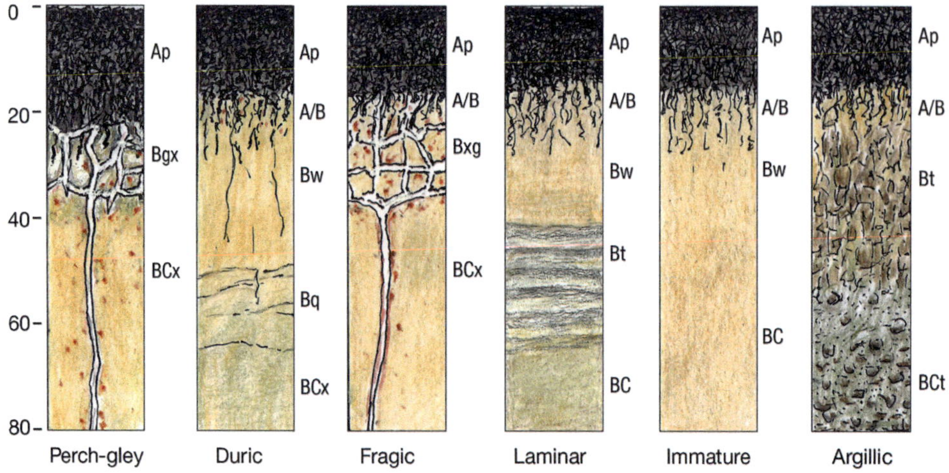

Fig. 10.3 Representative profiles of the six soil groups of the Pallic Soils. Vertical axis is depth (cm)

Argillic Pallic Soils—clay accumulation (via translocation) as thin coatings on peds or in pores,
Immature Pallic Soils—no fragipan, duripan, or argillic horizon, and weakly expressed Pallic Soil features.

Further distinctions are recognised as subgroups. The Immature group is the least developed soil with a simple morphology. Other groups are named according to horizons that are genetically distinctive, or have significant impacts on ecosystems or land-use management.

Clay accumulation features prominently in two groups, the Laminar Pallic Soils and the Argillic Pallic Soils. The Laminar Pallic Soils contain thin wavy laminations (many of them only a few millimetres in width) of clay. The laminations do not appear to have significant management implications, but they do indicate a degree of site stability, conditions that favour clay translocation (via illimerisation), and enhanced water holding capacity. The Laminar Pallic Soils tend to occur where the parent loess is fine or very fine sand, rather than silt, thus they occur close to the source of the loess where coarser material was deposited. In the Argillic Pallic Soils, clay accumulates in the form of thin (<0.5 mm) clay coatings on the surfaces of structural peds and on the sides of pores. Sodic Argillic Pallic Soils, found in the Awatere Valley area in Marlborough, are included in the sodic subgroup because of their sodic features (exchangeable sodium content >6%) in or beneath the argillic horizon. Fragic and Duric Pallic Soils may occur as complex soil associations. For instance in Hawke's Bay, Fragic Pallic Soils (defined by a dispersive fragipan) and Duric Pallic soils (defined by a non-dispersive duripan) may occur interspersed on the same landform within the same paddock.

Most Pallic Soils have subsoil impediments to root extension and water permeability. The Perch-gley Pallic Soils have evidence of prolonged saturation, and in other groups intermittent water saturation is indicated by 'Mottled' subgroups.

Soil subgroups, defined by the *New Zealand Soil Classification* within the soil groups, recognise soils that are related to other soil orders or soil groups (Duric, Argillic, Fragic, Mottled, Ultic), or soils that depart in some specific way from the soil group or soil order (Calcareous, Weathered, Pedal, Cemented, Sodic, or Aged). All of the subgroups depart in some properties from the central concept of the soil group which is identified as the Typic subgroup.

10.1.4 Origin of the Soil Order Name

The name 'Pallic' refers to the pale, straw-coloured subsoil (generally 2.5Y or 5Y hues) that is a distinctive feature of the Pallic Soil order. The word 'Pallic' was used for the order in the *New Zealand Soil Classification* in 1992. The term was previously applied in Taylor's Technical Soil Classification. Initially, Hewitt toyed with the term "Pallid" in developing NZSC, but rejected it because the possible 'sickly' connotation was not appropriate for these generally productive and extensive soils. In the New Zealand genetic Soil Classification (in 1962), Pallic Soils were termed yellow-grey earths.

10.2 Soil Profile Genesis

Most Pallic Soils are formed in loess material deposited from about 30,000 to 11,700 years ago during the coldest part of the last glaciation. However, small amounts of loess continue to be deposited today, particularly adjacent to the braided rivers in the South Island. The loess typically comprises quartzo-feldspathic and micaceous, uniform-sized, wind-deposited, silt or fine sand material blown from aggrading river floodplains and fans onto adjacent landscapes.

Most Pallic Soils form in regions where there is a summer drought. While the loess was accumulating, the land surface slowly built up (mean rate ~10–20 mm per century) and the loess was transformed into soil horizons that are weakly expressed because the material soon became buried too deeply for topdown pedogenic processes to be effective. Therefore, each part of the profile has at one time been at the land surface. Thus, the Pallic Soils have formed over long periods of time through the alternation of developmental upbuilding pedogenesis (generating weakly expressed soil horizonation during cold climate episodes when loess accumulated) and topdown pedogenesis (generating strongly developed soil horizon features when climate warmed and loess deposition effectively ceased) (Fig. 10.4). When loess is not accumulating (during warmer climate periods known as interglacials or interstadials), topdown processes modify the underlying material in a downward-moving front, forming more distinctive subsoil features including prismatic fragipans or argillic horizons, the edges of which are usually gleyed (Fig. 10.4).

Developmental upbuilding results in a deposit that has a soil 'vermiform' (earthworm formed) fabric throughout, inherited from when the loess was an A horizon at the land surface. The vermiform features even occur in fragipans, the interiors of which have a soil fabric comprising traces of faunal activity such as back-filled burrows as well as root traces (Fig. 10.5). The vermiform fabric is one of the features used to distinguish loess from other silty parent materials such as weathered siltstones.

Under soil-forming conditions that favour formation of Pallic Soils, the silt-dominated loess tends to develop into a dense subsoil. With sufficient summer drought, and moist winter or spring conditions, a fragipan forms. A fragipan is a distinct dense horizon with an extremely coarse columnar structure, defined by shrinkage cracks that are infilled with fine material (Fig. 10.6). The first description of a fragipan was made in New Zealand in about 1890 at the Dashing Rocks section near Timaru by John Hardcastle. The material in a fragipan, although very, to extremely, firm when dry, disperses (slakes) when saturated with water. The mechanisms of formation of the fragipan have been widely debated. Jim Pollok undertook a detailed comparison between the processes operating in Germany and those in New Zealand conditions. He attributed the lack of fragipans in German loess deposits to their calcareous nature, in contrast to most of the loess deposits in New Zealand. Instead of fragipans, argillic horizons were formed in Germany once the loess had been decalcified by leaching (see grey text box, Fig. 10.11).

Roger Langohr proposed that fragipan formation in New Zealand can be attributed to pressure exerted by ice forming within loess soils during episodes of freezing and thawing during Quaternary cold climate events. An alternative (more likely) explanation is that desiccation exerted by shrinkage within loess soils during summer drought, combined with dispersion during soil wetting cycles, may work to produce the dense fragipan. There is a clear association between fragipan and sites that have strong summer soil water deficits, but which are wet in winter or spring. Other prerequisites of the latter theory include (1) uniform particle size and (2) parent material in which there is a lack of potential bonding materials, such as iron oxides, calcium carbonate, or soil organic matter, which might otherwise prevent

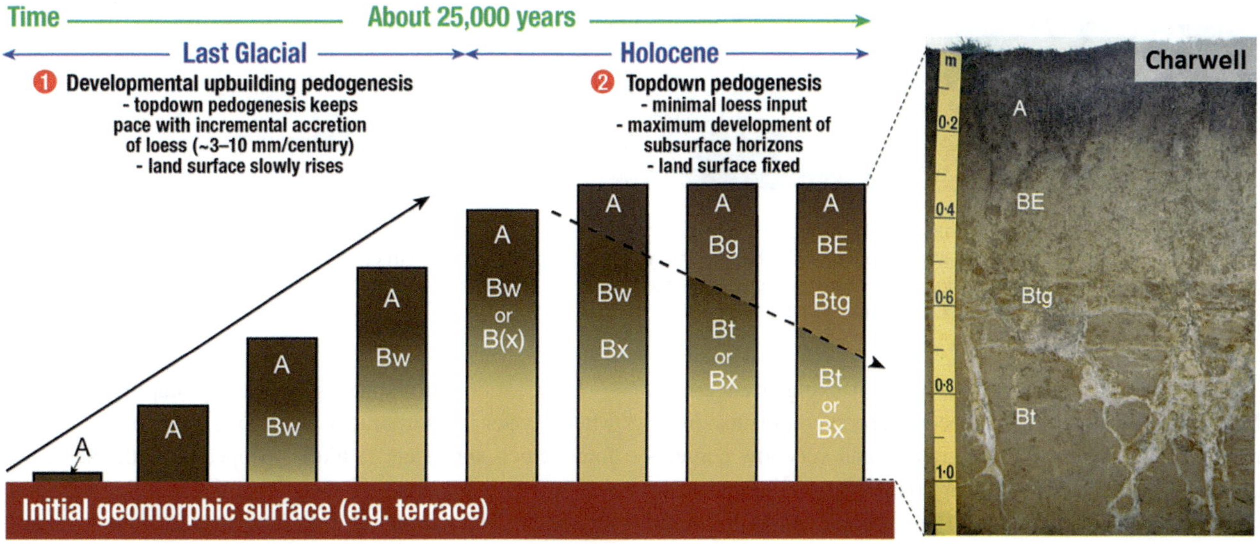

Fig. 10.4 Model of Mottled Argillic Pallic Soil development in loess since c. 25,000 years ago. Gleyed margins of prismatic structures in the lower subsoil are well expressed below c. 60 cm depth. Each part of the profile has been an A horizon at some point in the soil's history. Figure modified after Lowe and Tonkin (2010). Photo: P.J. Tonkin

Fig. 10.5 Soil fabric in loess-derived soils, South Island, showing that pedogenesis has occurred simultaneously with slow loess accumulation. A: vermiform (back-filled worm burrows) fabric in Bw horizon of Barrhill soil formed on loess near Rakaia River (scale marks 10 cm apart). B: worm holes in (now buried) Bt horizon in loess at Cust, north of Waimakariri River (scale in cm). C: calcium carbonate root pseudomorphs exposed in loess-soil fabric in Ahuriri section on Port Hills, Bank's Peninsula (lens cap 5 cm in diameter). Photos: P. J. Tonkin

Fig. 10.6 Loess exposure at Dashing Rocks, Timaru. Left: profile showing several periods of loess accumulation with the extremely coarse prismatic structure of an erosion resistant fragipan evident as the shelf extending out from the main face (ruler = 1.1 m) (photo: G. Vallender). Right: looking down on the shelf surface shows the tops of the fragipan prismatic structure with pale material in shrinkage cracks between the prisms (pocket knife for scale)

adjustment of the silt fabric to a denser packing arrangement. Most New Zealand loess meets these prerequisites in its narrow silt- to fine-sand-size grading and very low contents of soil organic matter, secondary oxide weathering products, clay, and calcium carbonates. Thus New Zealand loess has the ideal soil fabric and composition to enable densification in response to wetting and drying cycles.

Desiccation shrinkage is driven by capillary pressure where water is withdrawn by strong capillarity 'suction' force. Such shrinkage is likely to operate in the same way that wet soil materials or mud form the familiar hexagonal shrinkage cracks on drying, and hence initiate similar hexagonal large prismatic cracking in loess fragipans (Fig. 10.6). In addition, a suction of several bars can be exerted by plant roots during water deficits that average ~150 mm over 4–5 months. Roger Parfitt and Derek Milne proposed that the suction is transmitted through the water in pores in the uniform silt-sized material, causing shrinkage which is the equivalent of putting the soil under a compressive stress, leading to an increase in bulk density. The soils in drier areas will experience larger suctions and thus attain higher bulk densities. The dense fragipans, being well below the more porous, somewhat wetter, and biologically active surface horizons, will persist and possibly increase in density over time.

Dispersion occurs where clay, silt, or sand particles are separated from soil aggregates and behave as individual particles or as much smaller aggregates. In many Pallic Soils the particles are readily dispersed when the soil is saturated (a process known as slaking) and then easily move with the water. The relevance to fragipans is that dispersed particles may clog soil pores, and so increase the over-all density of the soil. Dispersion is counteracted in soils in which the fabric is stabilised by any soil component that 'glues' the particles together. Such components include soil organic matter, secondary oxide weathering products, and lime (calcium carbonate). The 'glues' are lacking in Pallic Soils and so the soil fabric is vulnerable to dispersion.

Clay movement down the soil profile is evident in many Pallic Soils, with clay skins often evident on aggregate surfaces. Such clay skins also contribute to blockage of soil pores and slowing of water movement through the soil profile. The clay skins accumulate to form argillic horizons (designated Bt, e.g. Fig. 10.4) including as thin subsoil clay bands (lamellae).

Once a dense subsoil layer, or a fragipan, forms then root and water penetration are limited. A perched water table forms in wet months, resulting in anaerobic conditions and reduction of iron minerals. Thus the pale grey colours of the soil matrix form with rusty brown iron- and manganese-rich mottles in areas associated with root channels or cracks where oxygen is available. Under pasture, organic-matter-rich topsoils form with strongly developed granular soil structures.

10.3 Soil-Landscape Relationships

10.3.1 Overview

Pallic Soils form mainly on the flattish to rolling landscapes (downlands) of the lower North Island and eastern South Island, in areas where loess mantles the landscape. Loess is extensive in New Zealand. That the deposition of loess occurs primarily during glacial periods has long been recognised, with Timaru-based John Hardcastle (1847–1927), the first in the world to make this climatic connection in about 1890. Consequently, deep loess sections may show a series of sediment-soil sheets with intervening (more-strongly developed) paleosols (Fig. 10.6). With the emplacement of each additional loess sheet on the landscape, flattish summits grew higher while the valley floors remained the same or lowered in response to base-level lowering (Fig. 10.7). The outcome is the smooth, undulating downland landscapes—for example, in the South Canterbury, Manawatu, and Whanganui areas (Fig. 10.1). Given the uniform parent materials on the near-flat terrace surfaces, the Pallic Soils usually form a relatively consistent soil across a landscape. However, on rolling surfaces a soil catena often prevails (Sect. 10.3.3).

On the youngest river terraces and flood plains, Pallic Soils do not form, due to insufficient time and the generally coarser alluvial parent material. At higher altitudes, and in parts of Southland, the cooler, wetter, environment results in formation of Brown Soils (Chap. 4). In the semiarid regions of Central Otago, where soils remain dry for longer periods, and are seldom saturated, semiarid, rather than Pallic Soils are formed. Loess may be observed in road cuttings and eroded faces (Fig. 10.6) as a prominent soil parent material which stands out because its density protects it from erosion of the cutting face, and it has a distinctive characteristic pale colour.

In the fine sandy material, nearer the loess source, Pallic Soils are not so dense, pores remain relatively open, and roots and water movement are not significantly restricted. New Zealand has some distinctive Pallic Soil soil-landscape patterns. Two examples are discussed here, the Rangitikei terrace sequence in the southern North Island, and the Claremont catena of the Canterbury region.

10.3.2 Rangitikei Loess Chronosequence

Although loess had first been recognised at Banks Peninsula by Julius (von) Haast in 1878, there was still much discussion in the 1950s about the source of the silt material that mantled river terraces in the lower North Island—was it fine homogenous alluvium, or loess? Des Cowie finally closed that argument in 1964 when he showed that a pale layer within the silt material was a tephra (the Aokautere Ash, now known more broadly as the Kawakawa Tephra). The Kawakawa Tephra was deposited by the Oruanui super-eruption of Taupo volcano about 25,400 years ago (sometimes referred to as the Kawakawa/Oruanui tephra). The presence of the tephra-fall deposit provided conclusive evidence that the encompassing silt material was in fact loess, also carried by the wind. The tephra could be accurately dated giving a time-line within the loess deposit.

Cowie observed that the depth of loess above Kawakawa Tephra increased from about 0.5 m at a distance from the

Fig. 10.7 Evolution of a characteristic loess landscape over three glacial (cold and warm) cycles (numbered at left from phase 1 to 5). After Lowe et al. (2015)

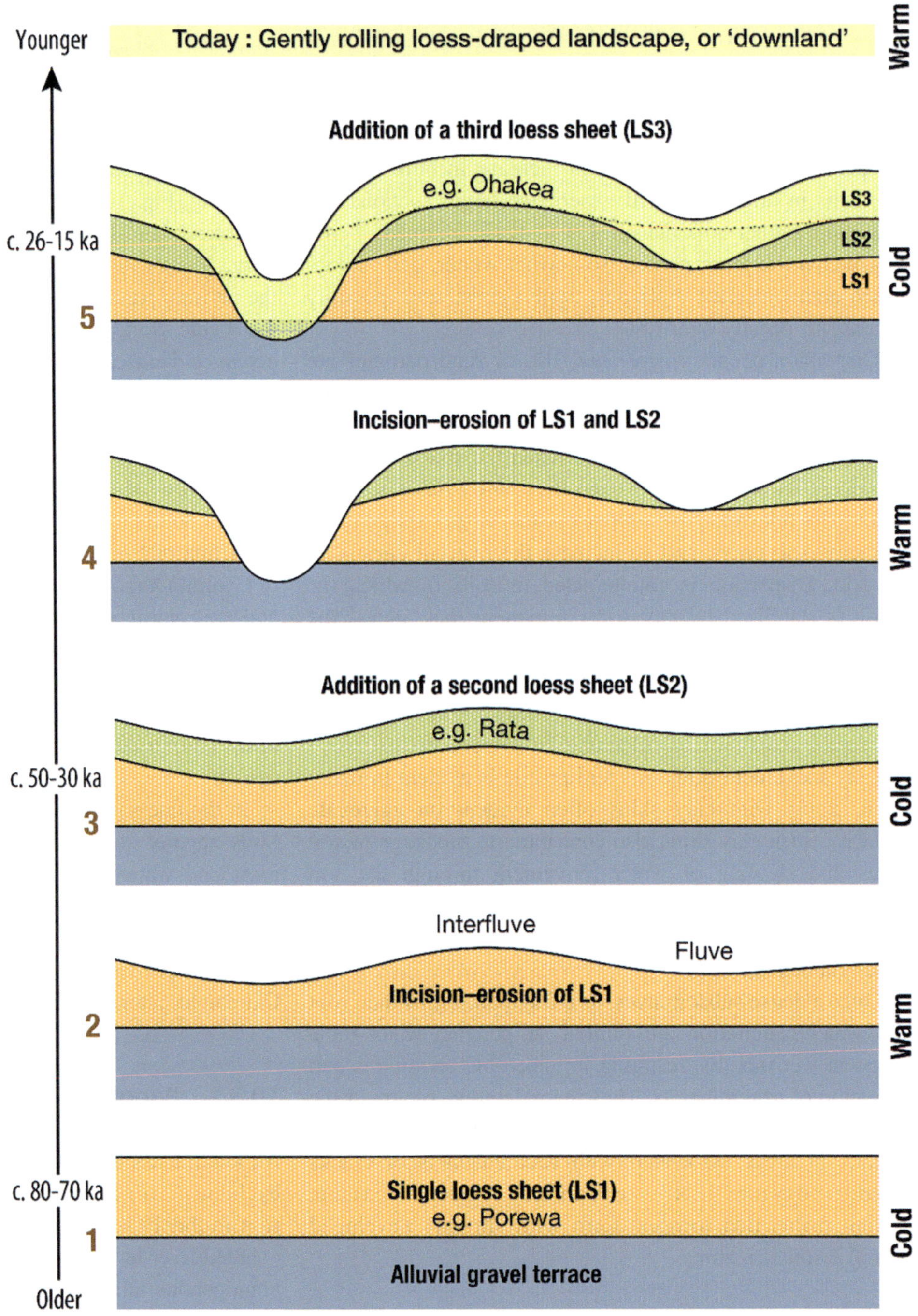

Manawatu and Rangitikei rivers to about 3 m on the south-eastern banks of the rivers (the down-wind side from prevailing nor'westerly winds). The particle size was also greater in the deeper loess deposits nearer the rivers. Cowie recognised a connection between the timing of loess deposition, the formation of river terraces, and the cold climate intervals of the last glaciation (as had been suggested by Hardcastle, 1889, in Timaru).

The Rangitikei River region is unique as it is being uplifted relatively rapidly and was also subject to changes in river morphology in response to the stadials (cooler periods) and interstadials (warmer periods) within the last glaciation dating from about 75,000 years ago. During cooler periods there was less vegetation, and high rates of erosion, in the upland river headwaters. Thus the rivers carried high bed-loads and became wide, energetic, braided rivers, depositing gravel, sand and silt sediments in the river valley and building (aggrading) a wide, flat, flood plain. The rivers have been described as 'dust engines'. The fine silt sediments were blown off the sparsely vegetated wide riverbeds and deposited onto the surrounding terraces and adjacent landforms, being driven by the belt of strong westerly winds that blew during the glacials. When the climate warmed, the vegetation cover improved, erosion declined, and the rivers, with reduced bedload, then started cutting down through, and removing, the previously deposited sediment. Thus a terrace was left behind and a new flood plain formed at a lower level. The gradual tectonic uplift of the land also allowed the rivers to cut to lower levels (Fig. 10.8).

The soil horizons continued to develop during the deposition of loess but the effects of pedogenesis were lessened as the land surface was slowly building up and the loess was soon buried too deeply for surface processes to be effective. During periods of warmer climate the down-cutting rivers carried far less silt and the rate of loess deposition slowed and practically ceased allowing the surrounding, now-stable, landscape to revegetate. Therefore, topdown soil-forming processes were able to exert a strong effect on the subsoils (Fig. 10.7, phase 2). When the next cold period arrived, the rivers again began to aggrade and build up. However, as the land had risen they did not fill up to cover their previous terrace, and a new layer of loess was deposited over the landscape, burying the soils formed in the previous warmer interval (Fig. 10.7, phase 3). Once again when a warmer interstadial arrived, the rivers cut down to new lower levels (Fig. 10.7, phase 4).

In the last glaciation, between about 75,000 and about 11,700 years ago, there were many subtle warmer and cooler cycles, especially from c. 30,000 years ago. A record of the major cooler and warmer intervals is left in the terrace and loess sequences preserved in the Rangitikei Valley (Fig. 10.8) and also in the Hawke's Bay, Wairarapa, and Canterbury regions of New Zealand.

The highest and thus the oldest terraces have the most sheets of loess on them (Fig. 10.7). Derek Milne's PhD study first elucidated much of the Rangitikei loess story. Three major terraces are recognised as having formed during the last glaciation. The lowest terrace in the Manawatu/Rangitikei region, known as the Ohakea terrace (actually a set of three terraces grouped together), formed between about 26,000 to $\leq$15,000 years ago (Kawakawa Tephra occurs in the loess derived from the Ohakea level that was deposited on the higher terraces). The next major terrace (the Rata terrace) most likely formed about 30,000–50,000 years ago and has one layer of loess (blown off the Ohakea terrace), at the base of which is Rotoehu Ash erupted c. 45,000 years ago. The Porewa terrace, which is higher again, formed in the colder period about 70,000–80,000 years ago and has two layers of loess on it (blown off the Rata and Ohakea terraces). Milne described a total of 14 terraces, some younger degradational terraces and also older higher terraces that date back to previous glaciations. The highest terrace, recognised at Stormy Point Lookout, is dated at c. 450,000 years old and the loess mantle includes the Rangitawa Tephra (dated at c. 340,000 years old). The older terraces have steep gradients as a result of tilting associated with the uplift.

The Ohakea terrace has minimal loess and Recent and Gley Soils are formed in alluvium. Pallic Soils form on the higher loess-covered terraces where there is a summer moisture deficit. To the north and at higher altitudes in the Rangitikei area, the soils merge into Brown Soils where the climate is wetter, and into Allophanic Soils (on well drained sites) where higher inputs of tephra from the central North Island volcanoes have occurred.

10.3.3 Claremont Catena

Where a sequence of soils down a slope, from summit to valley floor, are related to one another in a predictable pattern, the sequence is termed a "catena" (from a Greek work meaning 'chain'). Dave Leslie, Trevor Webb, and Stephen Burgham investigated catenas at Otokia in Otago, and

Fig. 10.8 Terrace sequences in the Rangitikei Valley. The oldest terrace is at the highest point (c. 450 m asl), with the terraces becoming younger as they descend to the tree-lined modern river (c. 130 m asl) in the centre of the photo

Claremont in Canterbury. We focus on the Claremont site and outline the soil variation and interconnections along the chain of soils.

On the summit (interfluve), the soil is wet in winter and Perch-gley Pallic Soils reign. The wetness is due to slow permeability in the loess at the summit, on near level slopes, which leads to prolonged periods of saturation. The dominant soils across the main rolling slopes of the landscape are Mottled-fragic Pallic Soils. Perch-gley Pallic Soils occur again further downslope, below the shoulder, where the slope changes from the convex shoulder to a straight or concave backslope.

Mottled Orthic Recent Soils are recognised on foot-slopes where sediment has accumulated at the base of the slope, derived from erosion upslope. The soil at the foot slope classifies as a Recent Soil because redeposition of material that has moved downslope has contributed to the topsoil which is over-thickened to more than 45 cm.

The main pedological processes that provide the genetic links between the catenary soils are as follows.

- *Water flows through the soils* both vertically downward across horizon boundaries, and laterally within horizons, guided by the shape of the land surface and the soil porosity. Water influences the movement of solutes and sediment, and is an important vector of soil formation in catenas. The catena displays two contrasting forms of gleying (Chap. 5). On the summits are Perch-gley Pallic Soils where the water table is perched on slowly permeable subsoil including a fragipan. Permeability measurements have shown values as low as a few millimetres over a 24-h period. In the shallow valleys are Recent Gley Soils with over-thickened topsoils due to accumulation of sediment eroded from the ridges. The Gley Soils are formed in groundwater fed by water draining from the rolling loess-covered landscape.
- *Sediment is redistributed* down the catena by erosion, especially on shoulder sites where topsoils are thin, and by sedimentation on lower slopes, especially on the foot slopes where the topsoil is thick. In similar uncultivated landscapes in the area of the Claremont study, topsoil thicknesses are comparable across all slopes, suggesting that the topsoil variation may be a result of enhanced erosion and deposition where and when soil is exposed by ploughing.
- *Leaching* usually occurs to varying degrees along the length of a catena, driven by surface water flow and water penetration through the soil on the slope. The expectation is that base cations would be higher in the foot slopes but, interestingly at Claremont, the indicators of leaching (base saturation, pH, exchangeable cations, and KCl-extractable Al) do not suggest differential leaching through the catena. It seems that the base cations have been leached beyond the foot slopes.

10.4 Key Soil Properties

10.4.1 Soil Composition

Pallic Soils are mainly formed in loess or other fine-grained sedimentary materials derived from the products of abrasion and breakdown of schist, greywacke, and mudstone rocks by energetic river action. The parent materials are dominated by quartz and feldspars with some mica, particularly from schist in Otago and Southland. Smaller amounts of fines originated by glacial grinding or alpine freeze-thaw processes, along with small inputs of dust blown from Australia. In central North Island, considerable loess derives from the sediment eroded from volcanoes and tephra deposits, which give rise to subsurface tephric loess deposits; however, the soils formed are generally Allophanic, rather than Pallic.

Weathering of the parent material, concomitant with the deposition of loess in glacial periods as described earlier, is low (or moderate at most in higher rainfall areas) because of the cool wet winters, dry summers, and limited time at the land surface during loess accumulation. The clay mineralogy reflects the limited weathering with clays dominated by illite, little altered from the original mica of the parent material.

Pallic Soils have low concentrations of secondary oxides. Extractable iron and aluminium values are normally low. Secondary iron oxides usually occur in redox segregations, and are not disseminated throughout the bulk of the soil (as in Brown Soils). Secondary calcium carbonate is notably low (zero) in most loess deposits in New Zealand although it does occur in some deposits including in parts of North Canterbury (Waipawa and Waikari areas) in the Waitaki Valley, near Timaru, north Otago, and on the flanks of Banks Peninsula.

10.4.2 Physical Properties

The distinctive soil physical properties of Pallic Soils are immediately evident in the National Soil Database summaries (Fig. 10.9 and Table 10.1). The very uniform particle size throughout the profile is illustrated by the clay content which is consistently about 20%, with most of the remaining particles in the silt or fine sand category. High total available water-holding capacity in the upper profile (Fig. 10.9) indicates high porosity and high available water storage. However, the marked drop in total water-holding capacity and accompanying rise in soil dry bulk density with depth are distinctive. The high lower profile soil dry bulk density (~ 1.5 to ~ 1.7 g cm^{-3}) points to the limited pore space at depth.

Pallic Soils are commonly summer droughty and winter-spring wet, aggravated by slow subsoil permeability and restricted root depth. Root exploration is limited (although less so in the Immature, Laminar, and Argillic Pallic groups), especially in subsoils with low porosity. The subsoil macro-porosity is frequently less than 10% and strongly restricts root aeration, and access to plant available water in dry periods. The same horizons that limit root penetration also limit water movement and hence perched water tables often form in winter and spring. Infiltration rates and hydraulic conductivities in the fragipans are very low ($\ll$1 mm hr^{-1}) to nil.

There is a high potential for slaking and dispersion in B horizons in most Pallic Soils. On the other hand, silica cementation is common in Duric Pallic Soils. Fragipans are generally regarded as uncemented but are relatively stable as a result of their high density and tight particle packing. Alan Palmer reports that some Pallic Soils (Tokomaru) can have a cemented iron pan c. 5–10 cm thick (ortstein-pan) at around 40–50 cm depth in addition to a dense fragipan below c. 70–80 cm.

10.4.3 Chemical Properties

The moderate pH, low carbon content, and low phosphorus retention (Fig. 10.10 and Table 10.2) all attest to the generally low leaching and weak weathering of Pallic Soils. Subsoil pH values are slightly to moderately acid. The pH of topsoil is, as with base saturation, strongly affected by fertiliser use. The cation exchange capacities of Pallic Soils are generally medium in topsoils and low in subsoils. Base saturation values in subsoils are generally medium with most values more than 50%, except in perch-gleyed soils, where values may be lower in horizons overlying fragipans.

Total P values are all very high, indicating the low weathering of the soil and parent material. A high proportion of the inorganic phosphorus is readily available (non-occluded) and a relatively high proportion of total phosphorus is in an organic form. P retention values are low (usually <30% in topsoils and subsoils) because of weak soil weathering and low contents of secondary iron oxides—as indicated by pale soil colours. The low P retention (an indicator of anion exchange) limits sulphur adsorption. An important sulphur input is by atmospheric deposition with higher amounts nearer the coast.

KCl-extractable-Al values are predominantly very low, except for high values in some of the Perch-gley Pallic and Duric Pallic Soils where some values are >2 cmol$^{(+)}$ kg^{-1}.

Although New Zealand soils are generally not impacted by sodium accumulation, significant sodium saturation ratios occur in many soils that have fragipans or duripans, particularly in coastal land, and in an argillic group in Marlborough. Calcium carbonate is not normally present in New Zealand's loess parent materials apart from a few exceptions (see grey text box and Sect. 10.4.1).

10.4.4 Biological Properties

The major potential impact on biological habitat in Pallic Soils is the strong annual cycle from summer water deficit to winter wetness. The perched water table in the upper profile during winter often leads to anaerobic conditions which impacts both soil microbial and fauna populations and on plant roots. Some plants that are sensitive to wet roots, for example, walnut trees, do not survive well on Pallic Soils that have fragipans. Pallic Soils are frequently strongly worm-mixed, with prominent *t*opsoil worm casts. A distinct worm-mixed horizon often occurs in the transition from the A to B horizons. Topsoil earthworm activity is greatly

Fig. 10.9 Median and upper and lower quartiles of clay content (percentage in <2 mm fraction), soil dry bulk density (t m^{-3}), and soil total available water holding capacity for Pallic Soils in the New Zealand National Soil Data Repository

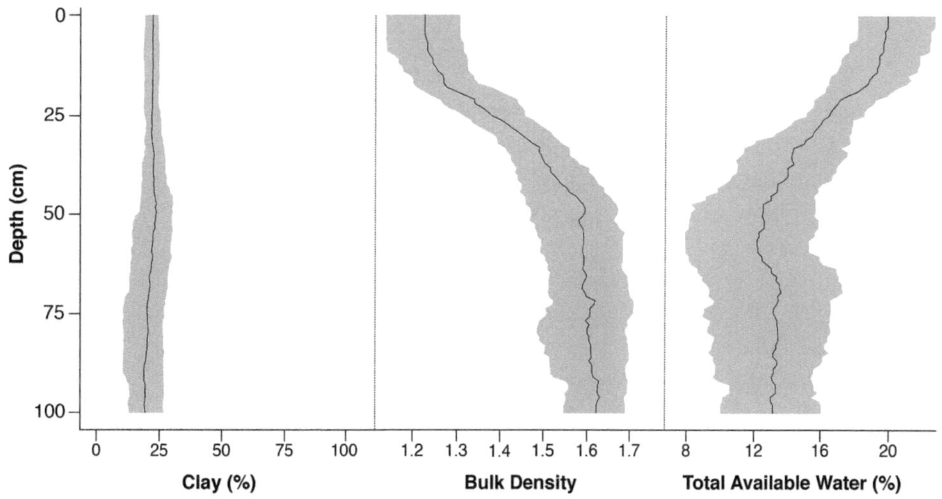

Table 10.1 Sand, silt and clay contents for a Mottled Duric Pallic Soil (Poporangi, SB09761) from the Hawke's Bay region

Horizon	Depth (cm)	Sand (%)	Silt (%)	Clay (%)
A	0–15	29	48	21
AB	15–25	30	44	25
Bwg1	25–38	32	47	20
Bwg2	38–49	24	55	19
2Btx	49–52	27	47	24
3Bmq1	52–65	32	53	13
3Bmq2	65–83	38	48	13
3Bmq3	83–102	38	47	13
4C1	102–112	80	11	7
4C2	112–137	88	7	3
5C	180–185	92	5	1

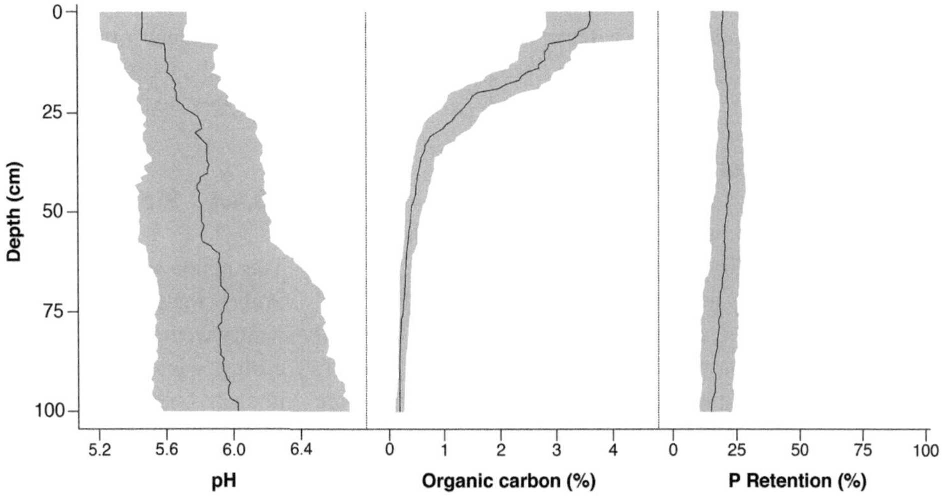

Fig. 10.10 Mean and upper and lower quartiles for pH, organic carbon, and phosphorus retention for Pallic Soils in the New Zealand National Soil Data Repository

Table 10.2 Soil chemical properties for a Mottled Duric Pallic Soil (Poporangi SB09761) from the Hawke's Bay region

Horizon	Depth (cm)	pH (in H_2O)	Carbon (%)	Nitrogen (%)	CEC[a] ($cmol^{(+)}\ kg^{-1}$)	Sum bases ($cmol^{(+)}\ kg^{-1}$)	P retention (%)
A	0–15	5.1	2.9	0.28	15.1	5.8	24
AB	15–25	5.6	1.3	0.14	10.7	6.3	22
Bwg1	25–38	6.2	0.6	0.08	8.7	6.6	17
Bwg2	38–49	6.5	0.3	0.05	9.0	7.7	14
2Btx	49–52	6.5	0.3	0.04	12.4	11.5	10
3Bmq1	52–65	6.7	0.2	0.03	10.1	10.7	5
3Bmq2	65–83	6.8	0.2	0.02	12.1	13.1	5
3Bmq3	83–102	6.9	0.2	0.02	12.0	13.5	6
4C1	102–112	6.6	0.2	–	–	–	–
4C2	112–137	6.4	0.2	–	–	–	–
5C	180–185	6.6	0.2	–	8.0	7.7	–

[a]CEC = cation exchange capacity

reduced during summer when the worms burrow deep to escape summer drought. Saturated soil conditions in winter may also be detrimental to earthworms.

Why were Pallic Soils regarded as "extremely peculiar"?

The Russian pedologist Glazovskaya made an extensive study of world soils and visited New Zealand. After examining the Pallic Soils she reported in her subsequent (1984) book "Soils of the World" that the yellowish grey (Pallic) soils of New Zealand were 'extremely peculiar'. This statement made after a survey of world soils strongly suggests that the New Zealand Pallic Soils are unique. What was it that Glazovskaya found so unusual? We cannot recover her reasoning, but we can search for factors that might point to uniqueness.

Probably the most unusual factor is the lack of calcium carbonate (lime) in most of New Zealand's Pallic Soils, in contrast to the lime-rich loess of equivalent soils in Europe and USA. Where natural lime is present, the soil starts its formation with lime being leached out of the upper subsoil. This decalcification stimulates the mobilisation of fine clay in suspension, forming an argillic horizon in the subsoil. Without the lime, New Zealand soils develop along a different genetic pathway, resulting in different soil properties (Fig. 10.11).

10.5 Distinguishing Between Pallic Soils and Related Soil Orders

A typical Pallic Soil profile has a distinctive morphology. However, in the landscape, Pallic Soils commonly merge laterally into Semiarid Soils (where long-term average rainfall is less than about 500 mm/year) and into Brown Soils (where rainfall is greater than about 1000 mm/year in the South Island and greater than about 800 mm/year in the North Island). Towards the central North Island, where tephra deposits occur, the Pallic Soils merge into Allophanic Soils. Gley Soils form in association with Pallic Soils where water tables are high and Recent Soils occur on younger surfaces adjacent to Pallic Soils.

Using the *New Zealand Soil Classification* key for identification first requires the elimination of ten other candidate orders before Pallic Soils appear. Many of the alternatives are easily discounted, by nature of the material—for example, Pumice Soils or Allophanic Soils are readily distinguished because of their distinctive parent materials and soil properties.

While Pallic Soils commonly form on loess this is not a determining feature as Semiarid, Brown, Podzol, Gley, Melanic, and Recent Soils may also form on loess parent materials. Long-term rainfall data showing a summer drought may be a useful clue, but are not always available. Thus positive identification of Pallic Soils must be based on observable features of the soil profile. The key identifying features, once competing orders have been eliminated, are

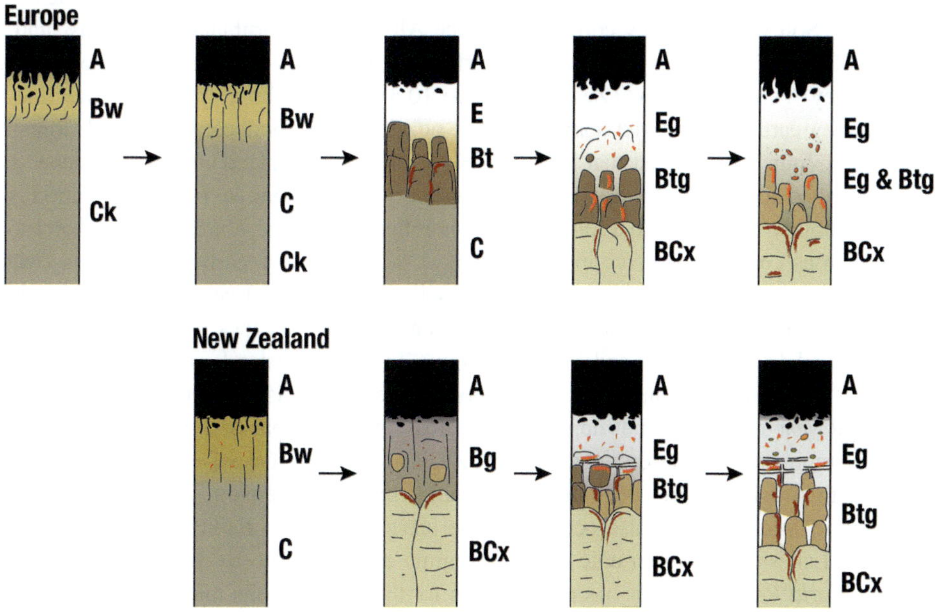

Fig. 10.11 Soil formation pathways for calcareous loess in Europe and non-calcareous loess in New Zealand

Table 10.3 Correlation between Pallic Soil Groups and *Soil taxonomy*, *World Reference Base* and the New Zealand genetic soil classification (These correlations are a general guide only. For definitive correlations, you should refer to the keys in the relevant classification manuals. The two major international soil classification systems are Soil Taxonomy, which was developed in the USA, and World Reference Base, which was developed primarily in Europe. The NZ genetic soil classification was used in NZ prior to 1992)

NZ *Soil Classification*	*Soil Taxonomy*	*World Reference Base*	NZ genetic soil classification
Perch-gley Pallic Soils	Epaquic Haplustepts or Epaquic Argustalf or Aqualfs, Aquepts	Epigleyic Cambisols, or Planosols	Yellow-grey earth
Duric Pallic Soils	Durustepts or Durustalfs or Duraqualfs	[Duric] Cambisols, or [Duric] Luvisols	Yellow-grey earth
Fragic Pallic Soils	[Fragic] Haplustepts, or [Fragic] Haplustalfs or Aeric/Typic Fragiaqualfs, Fragiochrepts	Cambisols Fragic, or Luvisols Fragic	Yellow-grey earth
Laminar Pallic Soils	Lamellic Haplustalfs	Lamellic Luvisols	Yellow-grey earth
Argillic Pallic Soils	Haplustalfs, Hapludalfs, Natrustalfs	Luvisols	Yellow-grey earth
Immature Pallic Soils	Haplustepts	Haplic Cambisols	Yellow-grey earth

1. evidence of soil development in the form of a weathered-B horizon;
2. evidence of a pale colour based on a soil colour chart, which can be used in conjunction with P retention and a test for reactive-Al, to separate Pallic from Brown Soils; and
3. features that typify Pallic Soil morphology— potentially including a fragipan, duripan, argillic horizon, clay laminations, or humus and/or clay coatings on aggregate surfaces (cutanic features).

10.6 Correlation with Other Classification Systems

The Pallic Soil groups mainly correlate with the Alfisols or Inceptisols of *Soil Taxonomy* and were formerly classified as yellow-grey earths in New Zealand (Table 10.3). The yellow-grey earths had proved somewhat of a puzzle taxonomically. John Bruce noted that around 43% of the order were ascribed to intergrades and associated soils, and this wide variability away from the modal profile is reflected in the numerous correlatives given in Table 10.3.

The relationship between fragipans and the soil moisture regime present a soil correlation problem between the *New Zealand Soil Classification* and Soil Taxonomy. In North America, fragipans occur in areas of udic moisture regimes (moist all year round) but not apparently in areas of ustic moisture regimes (summer drought). In New Zealand, Pallic Soils with fragipans are predominantly found in ustic moisture regimes but they also occur in some areas of the Southern Lakes district in an udic moisture regime on till.

The *World Reference Base* has no provision for the duric horizon. It is provisionally applied here (Table 10.3) as a prefix in square brackets. Similarly for the Perch-gley Pallic Soils, there is no expression of the perch-gleying horizon. Epigleyic is provisionally applied here as a prefix in square brackets. Epigleyic is not necessary for Planosols as perched-gleying is incorporated in the group concept.

10.7 Use and Management of Pallic Soils

10.7.1 Uses for Pallic Soils

On Pallic Soils, the native vegetation from around 11,000 years ago (the Holocene), as shown by pollen, sub-fossil wood, and charcoal studies, was podocarp-dominated forest in drier South Island sites, together with beech at wetter inland sites (the forest had replaced shrubland-grasslands, and sparse forest patches in refugia, which prevailed before the Holocene). High country sites above the natural treeline supported alpine tussock. Mixed podocarp-broadleaf forest dominated the southern North Island in the Holocene. Because of the seasonal drought, native forest vegetation was relatively easily burned and large areas were subject to repeated and large-scale burning by early Polynesians (Māori) within decades of arrival c. 750 years ago. By the time of European arrival, the landscapes were dominated by manuka scrublands in the North Island, and tussock grasslands or 'grey scrub' in the South Island. However, because most Pallic Soils are suited to pastoral agriculture and were easily converted from the tussock or scrub vegetation, few areas of native vegetation remain.

Pallic Soils are dominantly used for pastoral grazing, mostly for sheep and beef farming but also dairy farming. Soils under grazing are also used for fodder crops, often as part of a pasture renewal programme. Significant areas are utilised by cereal cropping in South Canterbury. Dense subsoils and winter perched water tables make many of the Fragic, Duric, and Perch-gley Pallic Soils less suited to such uses as orchard crops and intensive vegetable production. The Tokomaru soil in Manawatu and the Poporangi soil in Hawke's Bay are both widely used for pastoral agriculture and seasonal fodder crops such as maize.

Fragic Pallic Soils are extensive in the New Zealand sub-humid lowlands. Apart from a few reserves under remnants of the original native forest, the soils are predominantly in pastoral use, and have been for over a century. Peter McIntosh and collaborators compared Fragic Pallic Soils under native forest reserves with those at adjacent sites that had been under pasture for 80–100 years. The results showed that the native forest soils were different from the pasture soils. One important difference was that the soils under forest had more subsoil carbon than similar soils under pasture. It could be argued that Fragic Pallic Soils under forest are a truly endangered soil, and that we should protect remnants of these with reserve status.

The Immature Pallic Soils have high versatility, because of their deep rooting potential, which enables exploitation of the high available water capacity. Intermittent wetness may occur where there is a permeability restriction, or where groundwater is periodically high in topographically low positions in the landscape. Immature Pallic Soils are highly productive where they occur on flat to gently undulating land in deep loamy alluvium, recent loess, and on colluvium in hill country. The Templeton soils of the Canterbury Plains are versatile with productive Immature Pallic Soils that support a wide variety of cropping and intensive pastoral uses.

The Argillic Pallic Soils are also versatile with deep rooting potential, typified by the Pahau soils of Canterbury. The Laminar Pallic Soils are also productive versatile soils, again due to deep rooting potential and satisfactory permeability. They are exemplified by the Ngapara soil of North Otago, and the Otama soils of northern Southland.

10.7.2 Soil Water Management

All Pallic Soils are subject to summer drought in most years, and unless there is irrigation, animal farming will require summer feed strategies that involve reduced stock numbers or grazing off-farm. The soils are suited to crops that require summer senescence for ripening, for example, significant areas of cereals have been cropped on Pallic Soils in the loess downs near Timaru. Depending on water availability and crop needs, summer irrigation may be appropriate for all Pallic Soils. Irrigation is likely to be most efficiently used on the Immature Pallic, Argillic Pallic, and Laminar Pallic soil groups where there is enough available soil water holding capacity in the rooting zone for efficient water storage and access by roots.

The Perch-gley soil groups have a water table that forms above a slowly permeable layer. Most often the slowly permeable layer is a fragipan or other another poorly structured, dense, horizon such as an iron pan or a well-developed argillic horizon. The perched-gleying causes topsoil wetting in winter and commonly into spring, and heavy grazing can develop pugging and promote runoff of sediment including faecal matter into water bodies. Mottled subgroups normally have a shorter, intermittent, period of wetness. The periodic high water table in mottled subgroups may result either from perching on a slowly permeable layer, or from high groundwater on relatively low-lying sites.

Tile and mole drains (formed by dragging a 'mole' through the soil to leave a small open cavity) are widely used to improve winter soil conditions on perch-gley and mottled Pallic Soils. The mole drains have been known to remain effective for decades at some sites where maintenance of outfall drains is undertaken. However, it is generally recommended that mole drains be regularly replaced.

The hard pans in the Fragic and Duric Pallic Soil groups severely restrict potential root depth. Deep ripping has been trialled in soils with hard pans and benefits have been documented. However, in Pallic Soils from loess, the benefits are likely to be limited because the original high soil density may recover following ripping.

Topsoil structures may break down under prolonged impact by heavy stock trampling, machinery use, or continuous tillage. Structural decline is attributed to the low contents of stabilising iron, aluminium oxides, and organic matter. Lack of strong structure renders the soils susceptible to erosion, especially when exposed to water runoff.

10.7.3 Soil Erosion

Both the silt and fine sand loess materials are dispersive. Some coastal Pallic Soils also have a high sodium level which also contributes to clay dispersion when soils are saturated. Thus Pallic Soils on slopes are susceptible to tunnel gully erosion. Tunnel gulley erosion is initiated in slopes by entry of water into soil that is, to some degree, dispersive. Once the downward pathway is established through the soil (for example, a crack between structural prisms in a fragipan), the path steadily widens as saturated flow disperses and removes soil material and the flow is able to accelerate. Initially the only sign of the process will be a 'swallow hole' in the soil surface as a downslope tunnel forms. A sequence of events follows involving the discharge of soil-coloured water at lower points on the slope, surface soil collapse above the tunnel, and formation of a gully, upslope migration of the gully head, and formation of a major disrupting erosion feature (Fig. 10.12).

The Wither Hills near Blenheim is an area that has been subject to repeated tunnel gully erosion events. Following severe erosion in the early 1980s, the land was recontoured to fill in the gullies and pasture was established with a policy of only grazing cattle to ensure a reasonable grass cover was maintained to prevent erosion. Sheep graze much closer to the land surface and leave soil more exposed. However, 30 years later it was evident that new gullies were again forming.

In the lower North Island, Immature or Argillic Pallic Soils often form over a mudstone substrate. The mudstones are particularly prone to landslide erosion—controlled more by the nature of the rock than the soils. Severe erosion events occasionally occur as a result of major storms. Forest cover greatly reduces the erosion damage on the mudstone hill country.

When Pallic Soils are ploughed they are particularly prone to wind erosion, especially as most Pallic Soils occur in regions that are often subject to strong dry north-westerly winds. On sloping sites, water erosion is another hazard of ploughed soils. Maintenance of vegetation cover, establishment of wind breaks, and use of no-till crop establishment are important methods to prevent surface soil erosion.

10.7.4 Contaminant Behaviour

The Pallic Soils have low phosphate adsorbency indicated by low P retention, and can be a source of P losses to groundwater. Pallic Soils generally have a risk of dispersion due either to the lack of stabilising soil organic matter or iron/aluminium oxides (or both), or the presence of sufficient sodium in the cation exchange complex to promote slaking when the soil is saturated. This potential dispersion may be exacerbated by inputs

Fig. 10.12 A collapsed tunnel forming tunnel gully erosion in Pallic Soils, in loess on exposed sunny slopes on the Wither Hills near Blenheim. Photo courtesy of New Zealand Society of Soil Science

of sodium from sodium-rich applications of effluent. The dispersed material is readily carried by water and thus can become a source of sediment in waterways. Bypass flow is a risk for most Pallic Soils because it is promoted by their restricted permeability and tendency for continuous cracks to form as a result of soil shrinkage during dry periods.

Further Reading

Almond PJ, Shanun FL, Rieser U, Shulmeister J (2007) An OSL, radiocarbon and tephra isochron-based chronology for Birdlings Flat loess at Ahuriri Quarry, Banks Peninsula, Canterbury, New Zealand. Quat Geochronol 2:4–8

Barrell DJA, Almond PC, Vandergoes MJ et al (2013) A composite pollen-based stratotype for inter-regional evaluation of climatic events in New Zealand over the past 30,000 years (NZ-INTIMATE project). Quatern Sci Rev 74:4–20

Bruce JG (1983) Effect of climate on morphological features of soils from loess in the southern part of South Island, New Zealand. Aust J Soil Res 21:359–371

Bruce JG (ed) (1984) Soil groups of New Zealand. Part 7. Yellow-grey Earths. NZ Society of Soil Science, Lower Hutt, 123p

Bruce JG (1996) Morphological characteristics and interpretation of some polygenetic soils in loess in southern South Island, New Zealand. Quatern Int 34–36:205–211

Bruce JG, Ives DW, Leamy ML (1973) Maps and sections showing the distribution and stratigraphy of South Island loess and associated cover deposits, New Zealand. New Zealand Soil Survey Report 7

Cowie JD (1964a) Aokautere Ash in the Manawatu district, New Zealand. NZJ Geol Geophys 7:66–77

Cowie JD (1964b) Loess in the Manawatu district, New Zealand. NZ J Geol Geophys 7(2):389–396. https://doi.org/10.1080/00288306.1964.10420185

Cowie JD, Milne JDG (1973) Maps and sections showing the distribution and stratigraphy of North Island loess and associated cover deposits, New Zealand. New Zealand Soil Survey Report 6

Eden DN, Furkert RJ (eds) (1988) Loess—its distribution, geology and soils. A.A. Balkema, Rotterdam 245 p

Eden DN, Hammond AP (2003) Dust accumulation in the New Zealand region since the Last Glacial Maximum. Quatern Sci Rev 22:2037–2052

Eden DN, Froggatt PC, McIntosh P (1992) The distribution and composition of volcanic glass in late Quaternary loess deposits of southern South Island, New Zealand, and some possible correlations. NZ J Geol Geophys 35:69–79

Eger A, Almond PC, Condron LM (2020) Upbuilding pedogenesis under active loess deposition in a super-humid, temperate climate—quantification of deposition rates, soil chemistry and pedogenic thresholds. Geoderma 189–190:491–501

Fagg R, Smalley I (2019) Loess in New Zealand: observations by Haast, Hutton, Hardcastle, Wild and Speight, 1878–1948. Quatern Int 502(A):173–178

Glazovskaya MA (1984) Soils of the world, vol 11. Soil Geography Russian Translation Series 10. A.A. Balkema, Rotterdam

Hardcastle J (1889) The origin of the loess deposits on the Timaru plateau. Trans Proc N Z Inst 22:406–414

Hardcastle J (1890) On the Timaru loess as a climate register. Trans Proc N Z Inst 23:324–332

Hesse PP (1994) The record of continental dust from Australia in Tasman Sea sediments. Quatern Sci Rev 13:257–72

Hewitt AE (1997) Are New Zealand soils distinctive? A subterranean view of NZ ecosystems. Norman Taylor Memorial Lecture November 1996. NZ Soil News 45:7–16

Hewitt AE (2010) New Zealand Soil Classification, 3rd edn. Landcare Research Science Series No1, Manaaki Whenua Press, Lincoln, New Zealand

Hughes MW, Almond PC, Roering JJ et al (2010) Late Quaternary loess landscape evolution on an active tectonic margin, Charwell Basin, South Island, New Zealand. Geomorphology 122:294–308

Leslie DM (1980) Characterisation and comparative classification of soils of a catena from loess, East Taieri, Otago, New Zealand. New Zealand Journal of Science 23:19–26

Litchfield LJ, Rieser U (2005) Optically stimulated luminescence age constraints for fluvial aggradation terraces and loess in the eastern North Island, New Zealand. N Z J Geol Geophys 48:581–589

Lowe DJ, Neall VE, Hedley M et al (2010). Guidebook for pre-conference North Island New Zealand, "Volcanoes to Ocean" field tour. 26th–30th July 2010. 19th World Congress of Soil Science. ISSN-1175-4966

Lowe DJ, Tonkin PJ, Palmer AS et al. (2015) Dusty Horizons. In: Graham I (ed) A continent on the move: New Zealand geosyncline revealed (2nd ed). Geoscience Soc NZ Misc Publ 141, p 286–289

McCraw JD (1975) Quaternary airfall deposits of New Zealand. Bull R Soc N Z 13:35–44

McGlone MS (1988) New Zealand. In: Huntley B, Webb T Ill (eds) Vegetation history. Kluwer Academic Publishers, Dordrecht, pp 557–599

McWethy DB, Whitlock C, Wilmshurst JM et al (2009) Rapid deforestation of South Island, New Zealand, by early Polynesian fires. The Holocene 19(6):883–897

Milne JDG (1973) Map and sections of river terraces in the Rangitikei Basin, North Island, New Zealand. New Zealand Soil Survey Report 4. 4 sheets. DSIR, Wellington

Milne JDG (1973) Upper Quaternary Geology of the Rangitikei Drainage Basin, North Island, New Zealand. PhD thesis. Victoria University, Wellington

Milne JDG, Smalley IJ (1979) Loess deposits in the southern part of the North Island of New Zealand. Acta Geologica Academiae Scientarium Hungaricae 22:197–204

Newnham RM, McGlone MS, Moar NT et al (2013) The vegetation cover of New Zealand at the Last Glacial Maximum. Quatern Sci Rev 74:202–214

Palmer AS, Pillans BJ (1996) Record of climatic fluctuations from ca. 500 ka loess deposits and paleosols near Wanganui, New Zealand. Quatern Int 34–36:155–162

Perry GLW, Wilmshurst JM, McGlone MS et al (2012) Explaining fire-driven landscape transformation during the Initial Burning Period of New Zealand's prehistory. Global Change Biology 18:1609–1621

Pollok JM (1975) A comparative study of certain New Zealand and German soils formed from loess. University of Bonn, 312p Published D. Agr. thesis

Poulsen D (2013) The hydrogeological significance of loess in Canterbury. Canterbury Regional Council Report No R13/60

Raeside JD (1964) Loess deposits of the South Island, New Zealand, and soils formed on them. NZ J Geol Geophys 7:811–838

Runge ECA, Walker TW, Howarth D (1974) A study of late Pleistocene loess deposits, south Canterbury, New Zealand. I. Forms and amounts of phosphorus compared with other techniques for identifying paleosols. Quatern Res 4:76–84

Schmidt J, Almond P, Basher L et al (2005) Modelling loess landscapes for the South Island, New Zealand, based on expert knowledge. NZ J Geol Geophys 48:117–133

Sharp-Heward S, Almond PC, Robinson B (2014) Soil disturbance and salinisation on a vineyard affected by landscape recontouring in Marlborough, New Zealand. Catena 122:170–179

Smalley IJ (1982) The New Zealand loess in a world setting. In: Wasson RJ (ed) Quaternary Dust Mantles of China. Australian National University, New Zealand and Australia, pp 97–109

Smalley IJ, Davin JE (1980) The first hundred years—a historical bibliography of New Zealand loess 1878–1978. New Zealand Soil Bureau Bibliographic Report 28:166p

Smalley I, Fagg R (2015) John Hardcastle looks at the Timaru loess: climatic signals are observed, and fragipans. Quatern Int 372:51–57

Soil Survey Staff (2014) Keys to Soil Taxonomy. Twelfth edition. USDA Natural Resources Conservation Service, 362 p http://www.nrcs.usda.gov/wps/PA_NRCSConsumption/download?cid=stelprdb1252094&ext=pdf

Taylor NH, Pohlen IJ (1979) Soil Survey Method. Soil Bureau Bulletin 25. Soil Bureau, DSIR, Lower Hutt, 242p

Taylor NH, Pohlen IJ (1968) The classification of New Zealand soils. In Soils of New Zealand. NZ DSIR Research Bulletin 26(1):142p

Tonkin PJ, Runge ECA, Ives D (1974) A study of late Pleistocene loess deposits, south Canterbury, New Zealand. II. Paleosols and their stratigraphic implications. Quatern Res 4:217–31

Tonkin P, Webb R, Almond P, et al (2015) Geology, landforms and soils of the Waipara and Waikari regions of North Canterbury with an emphasis on lands used for viticulture. Lincoln University and Landcare Research, 220p

Webb TH, Burgham SJ (1994) Catenary relationships of downland soils derived from loess, South Canterbury, New Zealand. Aust. J. Soil Res. 32:1–11

Webb TH, Burgham SJ (1997) Soil–landscape relationships of downland soils formed from loess, eastern South Island, New Zealand. Aust J Soil Res 35:827–842

World Reference Base (2015 update). http://www.fao.org/3/a-i3794e.pdf

Podzol Soils

11

The Podzol Soil's one to feast your eyes on
with a leached and bleached E horizon
It 's large scale chromatography
caused by acids from a kauri tree

Abstract

Podzol Soils are formed in association with podocarp and beech forests in areas with good drainage and rainfall $\geq \sim 1200$ mm yr^{-1}, which provides an environment of acid leaching. Podzol Soils cover 13% of New Zealand and are most widespread on the South Island West Coast, the high country of the North Island, and in Northland. Distinctive Perch-gley Podzol Soils occur under high rainfall on the South Island West Coast. The Podzol's most distinctive feature is a pale grey, or white, E horizon beneath the topsoil. The E horizon is pale because iron, aluminium, and silicon-bearing materials, and organic compounds, have been moved downwards and redeposited beneath the E horizon as dark-coloured podzolic-B horizons. The podzolic-B horizons usually comprise an organic-rich (Bh) horizon over a reddish iron-mineral-rich (Bs) horizon. Podzol Soils are defined by the podzolic-B horizon or a humus- or ortstein-pan. Podzols are extremely acidic with pH often as low as 4. The acidity, and E-Bh-Bs horizonation, are caused by acidified rainwater seeping downwards, mobilising and leaching or translocating organic materials, iron, aluminium, and silicon compounds, and other nutrients down the profile. Enleached materials subsequently accumulate in lower horizons where increased pH, and microbial activity, support adsorption, and precipitation.

11.1 Important Features of Podzol Soils

11.1.1 Concept and Key Features of the Soil Order

Podzol Soils form under a regime of intense acid leaching, associated with acid-forming vegetation and high rainfall and typically good drainage. Under natural forest vegetation, the uppermost part of a Podzol Soil normally comprises one or more organic horizons: a thick litter (L) horizon accumulated from material discarded by trees including leaves, twigs, bark, and needles, usually together with partly decomposed (F), and well-decomposed (H), litter horizons on top of the mineral soil surface. The total thickness of organic horizons under kauri trees (*Agathis australis*) can attain ~ 1.5 m (a range of ~ 0.3–1 m is common).

Within the underlying mineral soil, the acid leaching leads to the formation of a distinct, pale-coloured E horizon below the litter-derived organic horizons and the topsoil (Fig. 11.1). Below the E horizon are dark, organic-matter-rich, Bh horizons and reddish to yellowish iron-aluminium-silicon complex-rich Bs horizons where much of the leached material is redeposited. Thin iron, humus, or densipans may also form, some of the fine humus potentially being translocated in suspension. In some cases, the E horizon may have been obliterated by tree overturn or activities such as ploughing. Therefore, Podzol Soils are defined by the presence of organic-matter-rich (Bh) and iron-aluminium-silicon complexes (Bs) horizons in the lower subsoils, sometimes cemented as pans. The Bs horizons are typically reddish in

A. E. Hewitt et al., *The Soils of Aotearoa New Zealand*, World Soils Book Series,
https://doi.org/10.1007/978-3-030-64763-6_11

Fig. 11.1 Podzol Soil and associated landscape. Left: kauri forest in Northland where podzols form under the acid litter of the large, long-lived kauri (Photo courtesy of NZ society of Soil Science). Right: a Densipan Podzol Soil (Te Kopuru series) formed in aeolian sand in Northland with characteristic Ed, Bh, and Bsm horizons (photo: Andrew Hosken).

hue (i.e. usually 7.5YR and 5YR, and occasionally 2.5YR) and Bh horizons are black (Fig. 1.1).

11.1.2 Areas of Occurrence

Podzol Soils cover 13% of New Zealand, the second-highest coverage after Brown Soils and are common in the high country and mountains of both the North and South islands (Fig. 11.2). However, unlike the Semiarid, Pallic, and Brown Soils, the distribution limits for Podzol Soils are not easily defined. Podzol Soils occur in Fiordland, much of the West Coast of the South Island, the Otago and Southland uplands, the Nelson region, and South Island high country, as well as in places in Northland, and in high altitude areas ($\geq \sim 500$ m) in the North Island, including the tephra-draped Volcanic Plateau. Podzol Soils occur in areas of high rainfall and are usually associated with forest trees with an acid litter, or areas that have previously been forested. The soils occur mainly in parent materials comprising silicic unconsolidated (sand or tephra) deposits or rocks that normally allow moderate or better drainage. Podzol Soils are moist throughout the year with annual rainfall of more than about 1250 mm and often over 2000 mm yr^{-1}.

11.1.3 Variation Within the Podzol Soil Order

The soil groups recognised within the Podzol Soil order (Fig. 11.3) are;

Densipan Podzol Soil—High density, pale-coloured, massive pan (Ed horizon) occurs in the upper subsoil beneath the topsoil,
Perch-gley Podzol Soil—Prolonged wetness in the upper horizons caused by a water table perched on a slowly permeable layer (usually a pan). The upper profile may be formed in a different material from that of the lower profile which may also contribute to the development of a perched water table,
Groundwater-gley Podzol Soil—Prolonged wetness in lower horizons caused by the groundwater table,
Pan Podzol Soil—Firm or indurated layer that restricts root penetration and permeability,
Orthic Podzol Soil—Moderately well, or well drained with no cemented or high-density pan,

Because of the complexity of Podzol Soils many subgroups are defined in the *New Zealand Soil Classification*. Podzol Soil subgroups recognise transitions from the Podzol Soil order to other related soil orders or soil groups (termed intergrading subgroups). Podzol Soil intergrades include Mottled and Groundwater gley that are intergrades to Gley Soils.

Podzol Soil subgroups also recognise the types of pans (Humus-pan, Humus-ortstein, Humus-placic, Placic, Fragic, and Firm), and the nature of materials (Fluid, Silt-mantled, Peaty-silt-mantled, and Humose). As for all soil orders, the central concept soil (with no, or minor, aberrant properties) is identified as the Typic subgroup. Because Podzol Soils typically contain the same secondary minerals that occur in

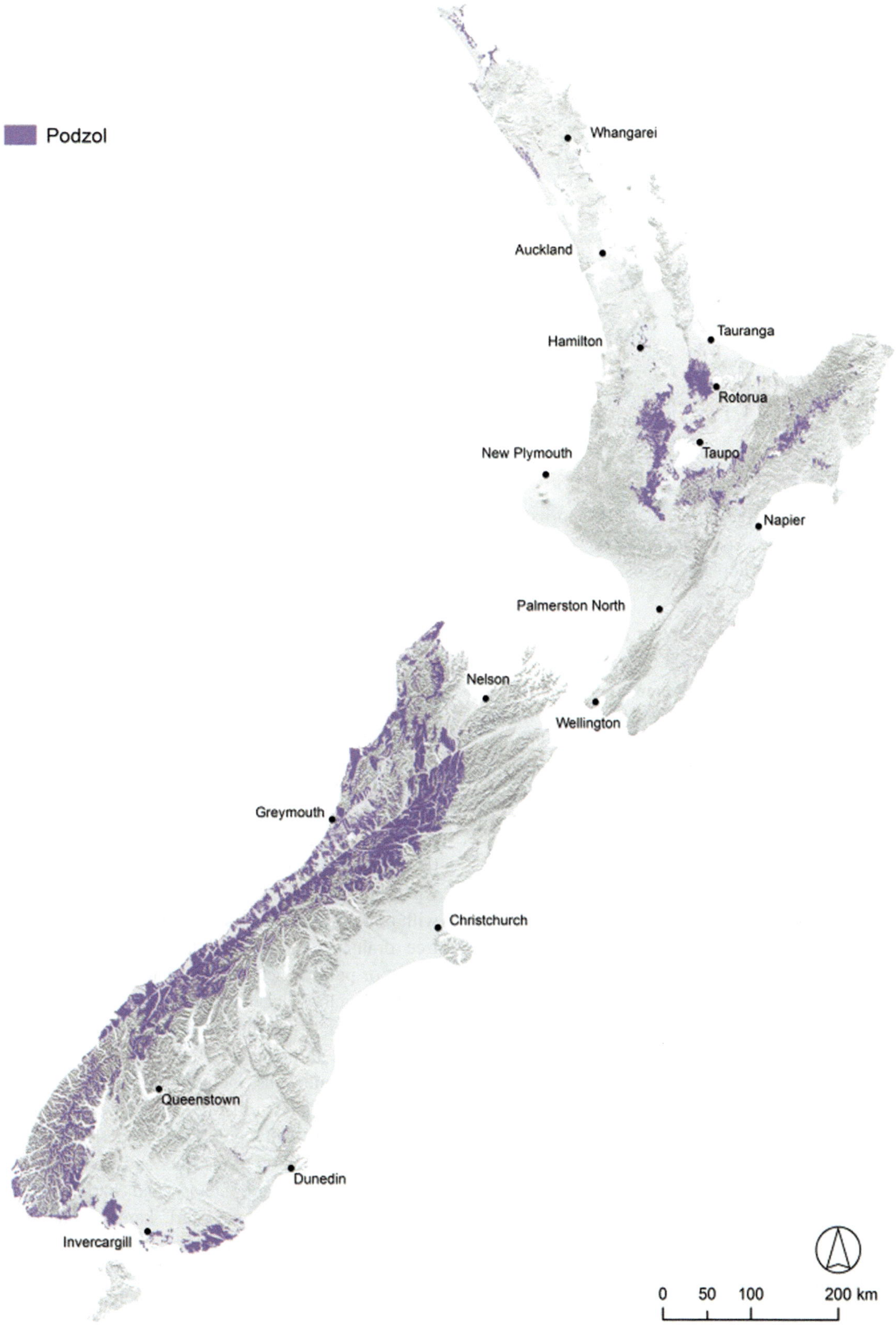

Fig. 11.2 Distribution of Podzol Soils

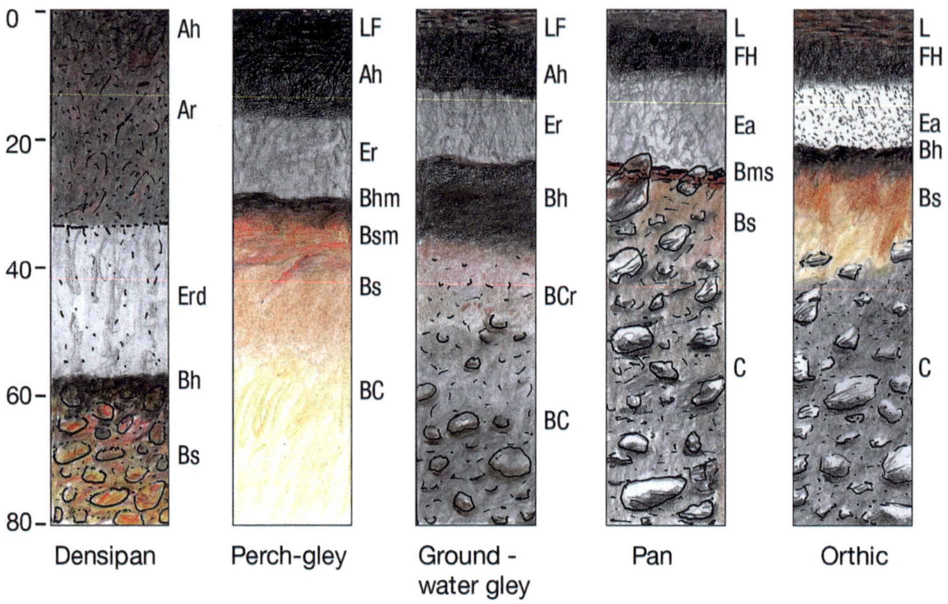

Fig. 11.3 Typical soil profiles of soil groups within the Podzol Soil order. Vertical axis is depth in cm

Allophanic Soils, but concentrated in Bs horizons, the Podzols key out in the *New Zealand Soil Classification* immediately before the Allophanic Soils.

11.1.4 Origin of the Soil Order Name

'Podzol' is derived from a Russian word meaning 'ashes underneath' (*pod*, under, *zola*, ash) from the way the ash-grey (bleached) E horizon resembles an underlying layer of burnt ash (from a fire) when turned over by a plough. The name 'podzol' is in wide use internationally for similar soils. In *Soil Taxonomy,* the podzol soils are named 'Spodosols' from a Greek word, *spodos*, wood ash.

11.2 Soil Profile Genesis

Vegetation that has acid-rich litter, along with a moderate to high annual rainfall (greater than about 1250 mm), and moderate or well-drained conditions, provides the acidic, strongly leaching environment that leads to the formation of Podzol Soils. The major acid-forming native trees that lead to podzol development are rimu (*Dacrydium cupressinum*) and kauri. Podzol Soils also form under kamahi (*Weinmannia racemosa*), mountain or Hall's totara (*Podocarpus laetus*), kaikawaka (*Libocedrus bidwillii*), monoao (*Halocarpus kirkii*), and four species of southern beech (*Nothofagus* sp.) forests in higher rainfall areas. Some introduced conifers, such as *Pinus radiata*, also have acidic litter under which there is potential, if they were to remain in place for centuries, to also contribute to Podzol Soil development.

In addition to the high leaching environment and acid-litter-forming (moroid) vegetation, a siliceous and free-draining parent material, and site stability, also help explain the distribution of Podzol Soils. High leaching is not easily defined just in terms of precipitation (rain and snowfall) because the effects of vegetation, parent material, topography, site stability, and temperature interact. Nevertheless, in the South Island most Podzol Soils occur in areas with more than 1250 mm mean annual precipitation, including extensive areas in Westland exceeding 2000 mm yr^{-1}. In the Northland region, where Podzol Soils occur on relatively old, stable, aeolian sand deposits (usually on terraces in the landscape), the mean annual rainfall may be around, or possibly a little less than, $\sim$1250 mm. Elsewhere a high annual rainfall ($\geq \sim$1800 mm yr^{-1}) together with podocarp- or beech-dominated vegetation, and typically free drainage in unconsolidated, silica-rich (quartzofeldspathic or pumice-rich) parent materials, can largely account for the development of most Podzols.

Rainwater is acidified as it moves from tree foliage and stems down through acid-litter-forming vegetation in the form of organic acids, including fulvic and humic acid. The mor (L, F, and H) horizons contain water-soluble organic molecules, especially polyphenols, derived from both the partial decomposition of the litter and from leachates formed on the tree's foliage, stems, and bark. Under confers, podocarps such as rimu, kauri, kamahi, and beeches, the polyphenols are coated in tannins that, together with strongly acid conditions, help protect organic material from microbial decomposition (in contrast to deciduous foliage which decomposes readily). Thus the organic materials are moved

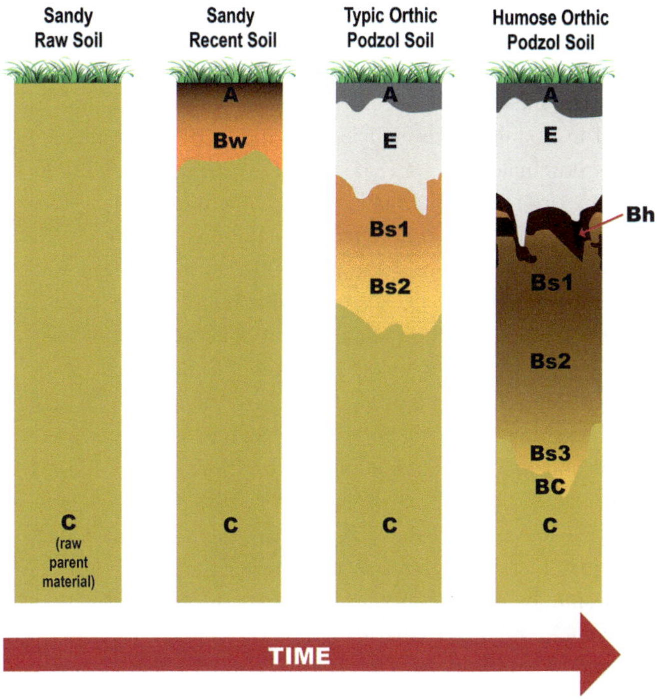

Fig. 11.4 Sequence of development of a Podzol Soil on well-drained sand showing the deepening of the pale E horizon, and simultaneous darkening and thickening of the complementary Bs and Bh horizons, over time. Modified after Schaetzl and Thompson (2015).

down into the underlying soil and topsoils may sometimes be black with concentrated organic matter.

The organic acids also combine with iron and aluminium to form metal–organic complexes (chelate complexes) that migrate through the soil. Alternatively, aluminium, iron, and silicon form inorganic Al–Fe-Si complexes (called proto-imogolite), which move in solution as 'sols' independently of the organic compounds before precipitation or adsorption in the lower profile. Fine-grained humus particles may move through soil profiles in suspension as well as solution. Thus, iron, aluminium, silicon, and organic matter complexes are removed from the upper subsoil leaving behind a pale-coloured, often sandy, quartz-rich, E horizon (Figs. 11.1, 11.3, and 11.4). The translocated materials are redeposited or precipitated in the underlying Bh and Bs horizons.

The degree of podzolization can be quantified by comparing the hues and values of the E horizon with the colours of the Bs and Bh horizons (becoming redder and darker with increasing podzolization), or by comparing the optical densities of acid oxalate extracts from the B with those of the E horizons. To provide a little more detail, two main mechanisms, or theories, are used to explain the podzolization process.

1. *Organic (chelate-complex) theory*. Under this mechanism, regarded as the classic process, the negatively charged organic compounds including polyphenols combine with acid-weathering-derived iron and aluminium cations, which are soluble at low pH, to form metal–organic complexes (chelates) that migrate from the acid (pH $\sim$ 4) upper horizons through the soil. The word chelate means 'claw' (Greek, *khele*) and denotes a chemical species that captures ions within its molecular structure. The chelates release the metal and organic compounds in the illuvial B horizon, the release being attained by microbial decomposition, a change in the chemical environment that destabilises the chelate (likely an increase in pH to ≥ 5), or charge saturation of the organic complexes by cations, or all of these. The loss of the metal cations, especially iron, leaves behind the white- or bleached eluvial E horizon that can be dominated by white grains of quartz if it is predominant in the parent material.
2. *Inorganic (or proto-imogolite) theory*. This mechanism was first proposed by Vic Farmer (of Scotland) and others in the early 1980s. Initially, cations of aluminium, iron, and silicon generated from strong weathering (hydrolysis) under acid conditions (pH < 5) form inorganic Al–Fe–Si complexes (referred to as proto-imogolite). These complexes move as proto-imogolite sols in solution at pH 4 *independently* from the organic compounds, precipitating at depth in the lower profile as nanocrystalline (allophane, ferrihydrite) and paracrystalline (imogolite) minerals, or as crystalline iron or aluminium oxides (sesquioxides), or both, where the pH rises to ≥ 5. The humic materials move separately (in solution and/or suspension) and accumulate or are adsorbed (forming a Bh horizon) on top of the nanomineral- and sesquioxide-rich Bs horizon.

Fig. 11.5 A Podzol Soil on Taupo Tephra at ~560 m altitude on Mamaku Plateau under ~2000 mm annual rainfall. Left: tall podocarp forest under which the soil is formed. Centre: Humose Orthic Podzol Soil (Tihoi soil, near Galaxy Road). Right: Photo of equivalent soil published in 1932. (Photos left and centre David Palmer, right from Grange and Taylor 1932)

A third mechanism, the so-called *rock-eating fungi theory*, has also been suggested. The rock-eating fungi theory is associated with mycorrhizal fungi found in many soils under pines and heaths commonly in parts of Europe, Scandinavia, Indonesia, and elsewhere. Dutch researchers, Antoine Jongmans, Nico van Breemen, and others, showed that the fungal hyphal tips, which produce organic acids (e.g. citric and oxalic acids), effectively 'drill' directly into mineral grains including feldspars and hornblendes in podzols, bypassing the external acid soil solution and enhancing weathering by forming complexes with aluminium. Thus, the E horizon in this model is regarded as the fungal 'eaten' part of the soil. The plant-ectomycorrhizal combination is seen by its proponents as a major driving force in podzolization that integrates the two mechanisms above. However, van Breemen and others, in the year 2000, suggested there was no evidence for mycorrhizal fungi facilitating weathering and podzolization under kauri in New Zealand.

Podzol Soils can form rapidly where conditions are favourable and the distinctive profile form can be seen within a few hundred years as shown by podzolized soil horizons on dated pumiceous tephras in the central North Island (e.g. Fig. 12.7).

Podzol Soils developed in the pumice of Taupo Tephra, (~1800 years old, erupted AD 232 ± 10, Fig. 11.5), were the first Podzols to be recognised in New Zealand (by Norman Taylor in 1930). Although the age of Taupo Tephra was not known in 1930, it was evident that the process of podzolization had been unusually rapid from a pedological viewpoint. The rate of podzolization was aided by the fragmental, fine-grained, silica-rich, well-drained character of the Taupo pumice along with the moroid (acid litter forming) forest vegetation cover. Before Norman Taylor recognised that podzolization had transformed the pumice, the grey Ea horizon was (erroneously) thought to represent a separate tephra layer on a buried soil, but after 1930 it had been recognised as an enleached (podzolised) E horizon (then labelled an A_2 horizon, Fig. 11.5).

The clay minerals and colloidal materials of podzolic-B horizons (allophane, ferrihydrite, and imogolite) also occur in the Bw horizons of Allophanic Soils and Allophanic Brown Soils. Allophanic and Brown Soils are frequently associated with Podzol Soils in the landscape. The processes of podzolization, and the formation of allophane, ferrihydrite, and imogolite in Podzol Soils, can be contrasted with processes involved in forming the same minerals in well-drained, tephra-derived, glass-shard-rich Allophanic Soils (Chap. 2). In Allophanic Soils allophane, ferrihydrite, and imogolite form in situ (even though the process requires desilication via leaching), but podzolization, in contrast, always involves the *translocation* of organic and/or inorganic complexes, from the very acidic horizons in the upper profile, and subsequent immobilisation and accumulation in less-acidic lower horizons. The synthesis of allophane and ferrihydrite via podzolization (also by strong leaching of silicon as occurs in Allophanic Brown Soils) can thus take place on parent materials that are devoid of any tephra-derived volcanic glass. However, podzolization, and the related processes that occur to form Allophanic Soils, can occur together on tephra deposits, for example, in central North Island where Podzol Soils are developed on Holocene tephras.

The Densipan Podzol Soil (Wharekohe soil) of Northland is formed on clays (formed from weathered banded sandstone and mudstone), which is globally unusual (although they may not quite meet the full criteria for Spodosols, the *Soil Taxonomy* equivalent to a Podzol Soil). Podzol Soils generally form on permeable, unconsolidated materials that support strong leaching, such as aeolian sands. The Podzol Soils on clay are a reflection of the potency of the kauri tree as a source of acids as well as the stability of the landscape, with long periods without disturbance under long-lived trees, during which the soil develops (see grey box). An unusual feature of the kauri-derived podzol, in many places, is the nature of the bleached horizon. Named a 'densipan' it has a very high density without cementation. The densipan has been recognised in Australia, the Caribbean, and England but the Wharekohe soil would seem to represent its most spectacular recorded example.

11.3 Soil-Landscape Relationships

11.3.1 Overview

At a national scale, Podzol Soils occur in areas of New Zealand with high rainfall, a history of acid-forming vegetation, usually on unconsolidated, silica-rich, well-drained parent material. Many Podzol Soils are on stable forested hill country. They are also often found on sand dunes, including in coastal Northland, Bay of Plenty, and Coromandel, and on the West Coast of the South Island. On Holocene coastal dune sequences, Podzol Soils are invariably formed on the older, innermost (landward) dunes. Podzol Soils also occur on tephra deposits at altitudes $>\sim 500$ m, and are especially common on the loose, highly silicic (rhyolitic) pumice deposits emplaced by the Taupo eruption in AD 232 $\pm$ 10 (Fig. 11.5). On stable slopes, a succession of different Podzol Soil groups and subgroups can be identified associated with podzolization and other processes operating according to landscape position. A catenary relationship may be identified with Orthic Podzol Soils on summits and backslopes giving way to Groundwater-gley Podzol Soils on foot slopes.

A unique feature of Podzol Soil's soil-landscape relationships in New Zealand is the 'egg-cup' podzol (where the pattern of E, Bh, and Bs horizons form a concave egg-cup shape) associated with individual kauri trees in Northland. Chronosequences (sequences of soils in similar landscape positions, with similar climate, vegetation, and parent materials but varying age) are of value in showing how soils, such as the Podzol Soils, develop over time, including their eventual retrogression. Chronosequences that include Podzol Soils have been studied throughout New Zealand and have contributed to the understanding of the genesis of both the soils and landscapes. Here we report the Franz Joseph chronosequence (Sect. 11.3.3).

11.3.2 The Egg-Cup Podzol Soils of Northland

A source of fascination, to both New Zealand and international soil aficionados, is the so-called 'egg-cup' Podzols of Northland. In such soils, the E horizon and underlying Bh and Bs horizons form a distinctive egg-cup shape (Fig. 11.6). The egg-cup shape is thought to be derived from greater leaching occurring immediately adjacent to a large, long-lived tree (most likely a kauri tree) that has, over a long period of time, provided plant-derived acids that have hastened the leaching near, and beneath, the tree. Because the kauri lives for hundreds of years, large amounts of litter and leachates are able to accumulate on a specific site. Egg-cup podzols have been observed adjacent to more usual Podzol Soils that have flat-lying soil horizons, or egg-cup podzols may form a Podzol Soil 'island', a few metres across, surrounded by Ultic Soils. Such a pattern may relate to a mosaic of trees evident in remnant kauri forests today (see grey box).

Fig. 11.6 The foot of a kauri tree and an egg-cup podzol in Northland. Left: the base of an undisturbed kauri is surrounded by a deep layer of kauri litter (and a few epiphytes) that have fallen from above. Near the tree, there is minimal competing vegetation growing, in contrast to the background area. Right: David Lowe investigates the prominent, egg-cup-shaped, E horizon and underlying Bh (dark) and Bs (reddish-yellow) horizons (by blade of spade) where once a kauri stood. The concave shape is approximately the mirror image of the convex cone of litter that would have been around the tree

On Podzol Soils and trees

At the time of early European exploration, it was often assumed that tall forest trees indicated high soil fertility. However, in New Zealand such ideas proved to be wrong and tall podocarps were found to be flourishing on low fertility Podzol Soils. So how can it be possible that big trees can be established in low fertility soils? There are two options: the soil developed first and the tree 'selected' the site, or the tree happened to grow at the site and caused the soil to develop towards a lower fertility state. In the case of the Podzol Soils, the evidence strongly suggests the latter option.

The 'egg-cup' Podzol Soils (Fig. 11.5) form directly beneath mature podocarps. Away from the tree, the soil may not be a Podzol. The egg-cup shape is intriguing and graphically demonstrates the ability of an individual tree to drive the podzolization processes to create its own soil. It is curious, though, that the tree would produce a very low fertility, extremely acid, soil for its home.

The most spectacular egg-cup Podzol Soils form under massive, slow-growing kauri where the acidic, low fertility environment helps the tree compete with its faster growing competitors. By supressing its competition, the kauri can out-compete them in the struggle for nitrogen. The kauri's competitive advantage was not just the production of acidity, which kauri can tolerate, but also its remarkably efficient recycling of nitrogen. A kauri tree produces nearly twice the amount of wood, bark, and cones for a given amount of nitrogen than most other forest trees. Dutch soil scientist Eric Verkaik and others showed that the nitrogen that accumulates in the litter below kauri trees is protected from N-mineralisation (breakdown) by kauri tannins that protect proteins. Thus, the availability of nitrogen, as well as phosphorus and calcium, is lower under kauri than in the surrounding vegetation. As kauri seedlings are more tolerant of low nutrient availability, and drought, than other tree seedlings, the conditions under mature kauri give kauri seedlings a competitive advantage. The low nutrient availability under mature kauri trees supports the idea of a positive soil–plant feedback driven by the poor 'decomposability' of kauri litter.

Soil scientists have observed that mature kauri exploit their great longevity—many live for 600–1200 years, some even longer. The dense, strong, root systems help prevent landslides, thus preventing land disturbance, and the litter buildup and acid leaching from low fertility Podzol Soils. The stable landscape and low soil fertility ensure that return to early stage forest regeneration (with competitors) is avoided or postponed as kauri exploits its ability to out-live other species. Thus the kauri could be described as a 'landscape engineer'.

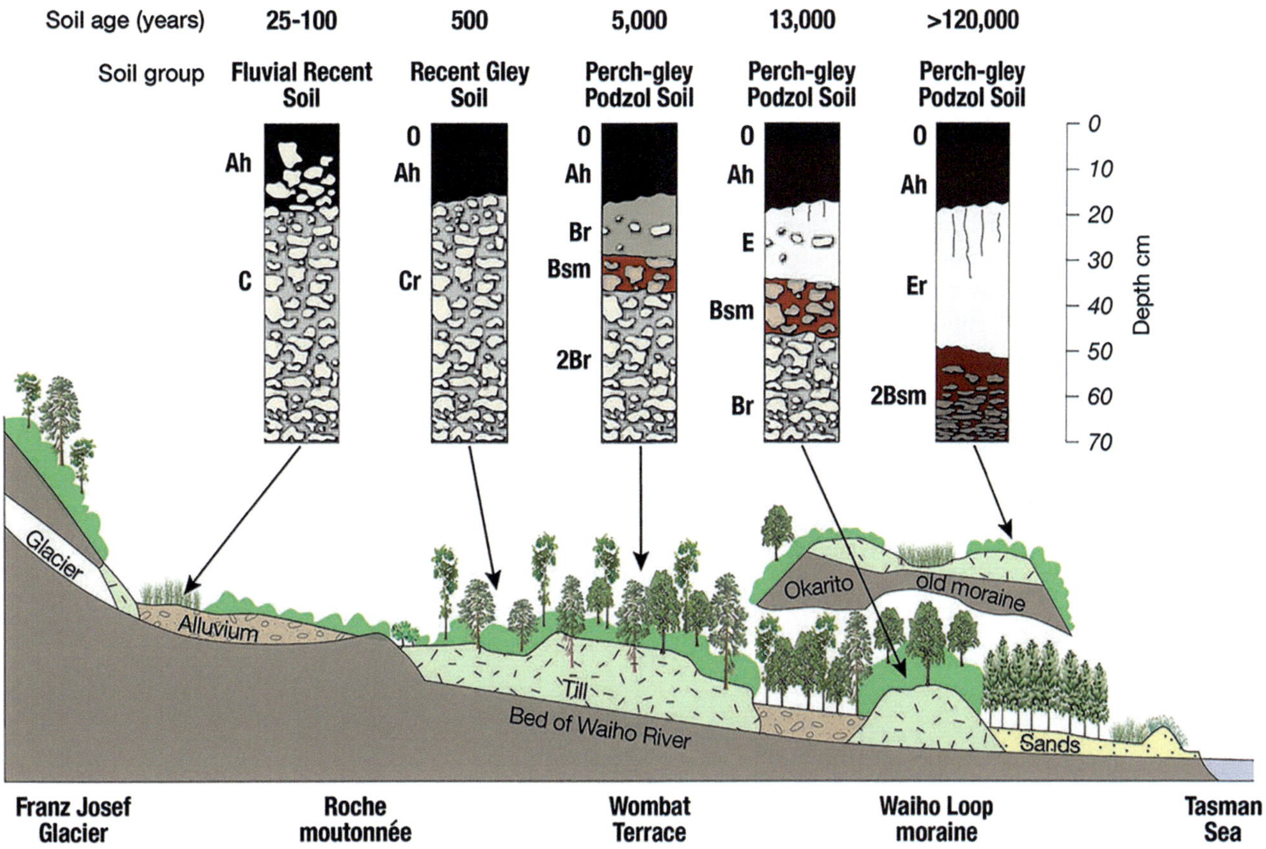

Fig. 11.7 The Franz Josef chronosequence provides an example of Podzol Soil development over the last ~120,000 years. All the soils are underlain by till (± alluvium) of differing ages. In the three oldest soils, thin deposits of loess have been added incrementally to the earlier land surface, giving rise to upbuilding pedogenesis in the upper parts of the soil. Glass shards of the 25,400 year old Kawakawa/Oruanui Tephra have been detected within the loess at about 0.3 m depth, providing a valuable dated marker (adapted from Molloy 1993 and Almond and Tonkin 1999; see also Peltzer et al. 2010)

11.3.3 Franz Josef Chronosequence

The chronosequence near Franz Josef glacier is a sequence of soil and vegetation sites carefully located across the landscape so that the age of the sites varies from a few years to tens of thousands of years. The sites have yielded much new understanding about soil and ecosystem development, especially the processes that release phosphorus by weathering (particularly through the work of Tom Walker, Keith Syers, and Peter Stephens).

The Waiho River drains the snout of the Franz Josef glacier in south Westland. Typical of glacial rivers it is choked with stony alluvium, generated by the grinding and conveying actions of the glacier during its advances and retreats in response to changes in temperature and precipitation. The river's broad alluvial fan is bounded by 400-m-high lateral moraines dating from the Last Glacial Maximum, approximately 30,000 to 18,000 years ago. The till making up the moraines is a chaotic mix of boulders, cobbles, gravels, sand, and silt. Compared to low terraces and moraines, the upper land surfaces, of high terraces and moraines, are older and the soils have developed over longer periods of time. Dating of the landforms, and particularly the more stable gently sloped land surfaces, including the Waiho Loop moraine, has provided a series of sites that form a time sequence (chronosequence) of landforms, ecosystems, and soils (Fig. 11.7).

Loess has accumulated on the older Last Glacial Maximum-related tills, hence the soils in these landscape positions are two-storeyed soils with a complex history that has been described by Peter Almond, Andre Eger, and Philip Tonkin. Initially topdown pedogenesis began transforming the glacial till into soil horizons and then as the loess slowly accumulated (at a mean rate of <10 mm per century) it was simultaneously transformed into a second upper soil via developmental upbuilding pedogenesis (as described in Chap. 10, but in a much wetter environment). The soils have developed under a rainfall of between ~2500 and 5000 mm per year.

Ecosystem changes through the chronosequence are shown by changes in soil profile development, soil fertility, forest biomass, and species dominance. Initial soil

Table 11.1 Sand, silt, and clay contents of a Humose Orthic Podzol Soil (Mamaku soil, SB09580)

Horizon	Depth	Sand (%)	Silt (%)	Clay (%)
Ah	0–8	42	55	2
Eu	8–15	35	63	1
2bBs1	15–25	43	55	2
2bBs2	25–43	48	45	7
3bBs	43–58	54	31	15
4bBs	58–100	37	44	19

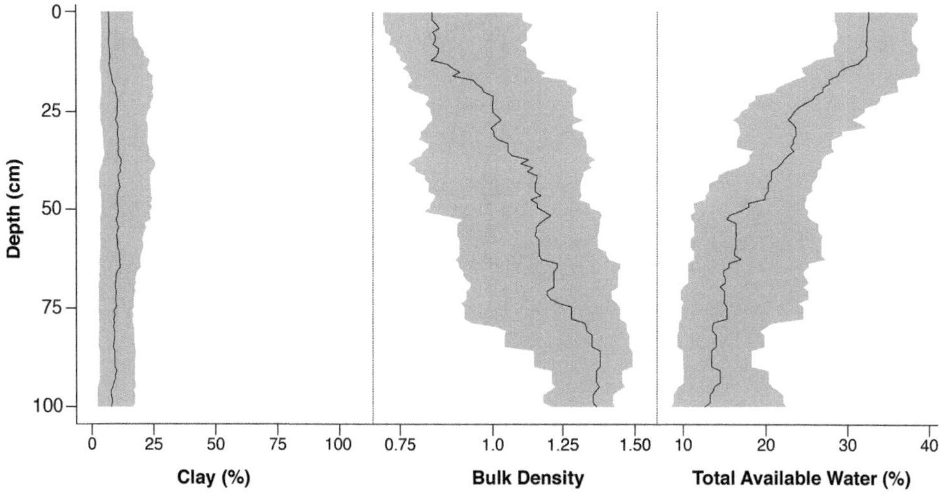

Fig. 11.8 Median and upper and lower quartiles of clay content (% in <2 mm fraction), soil dry bulk density (t m^{-3}), and total soil available water capacity for Podzol Soils in the NZ National Soil Data Repository

development from Raw Soils (not depicted in Fig. 11.7) to Recent Soils, Gley Soils (alternatively, acidic Brown Soils may occur depending on local drainage), and then Podzol Soils, reflects the progressive phase of ecosystem development. Fertility builds up and the soil profiles become differentiated into distinct horizons. Following this, a period of relative stability occurs, and then, as nutrients are gradually leached from the soil the ecosystem becomes nutrient limited so that soil fertility and associated plant growth gradually decline (retrogressive phase).

From soil initiation and through the progressive phases, phosphorus (P) is released by hydrolysis from minerals in the parent materials at rates sufficient to ensure that ecosystem development is not nutrient limited. Over time, the release of plant-available phosphorus slows because of P losses in leaching or runoff, and by sequestration (where an increasing proportion of total P becomes unavailable to plants by becoming locked up in clay minerals and organic compounds). The proportions of primary (rock) and secondary (clay) minerals control the main supply of nutrients for biological and ecosystem processes through the sequence. Eventually, P becomes the limiting nutrient, and if there is no disturbance and insufficient aeolian inputs to provide fresh parent material for weathering, the regressive stage of ecosystem change takes hold. Over the time span of the Franz Josef chronosequence (~120,000 years), soil phosphorus shows an eightfold decline. As P is a key nutrient in the control of plant biomass, the decline is demonstrated in many ecosystem features including reduced tree height, forest density, diversity, litter decomposition rates, and biomass.

11.4 Key Soil Properties

11.4.1 Soil Composition

Podzol Soils can occur in a wide range of parent materials, both rock and regolith, but the podzol form is most clearly expressed in sand derived from weathered or reworked quartzo-feldspathic sedimentary rocks (schist, greywacke, sandstones), and in silica-rich tephra deposits including fine-gained pumice. It has been shown that kauri do not cause podzolization on more iron-rich andesitic or basaltic deposits.

Complexes of iron and aluminium oxides, allophane (± ferrihydrite, imogolite), together with organic matter, are *strongly differentiated in the profile* and produce the distinct horizon colours. In A and E horizons, sand and silt grains are uncoated, and the clear sugar-like grains of quartz or feldspar

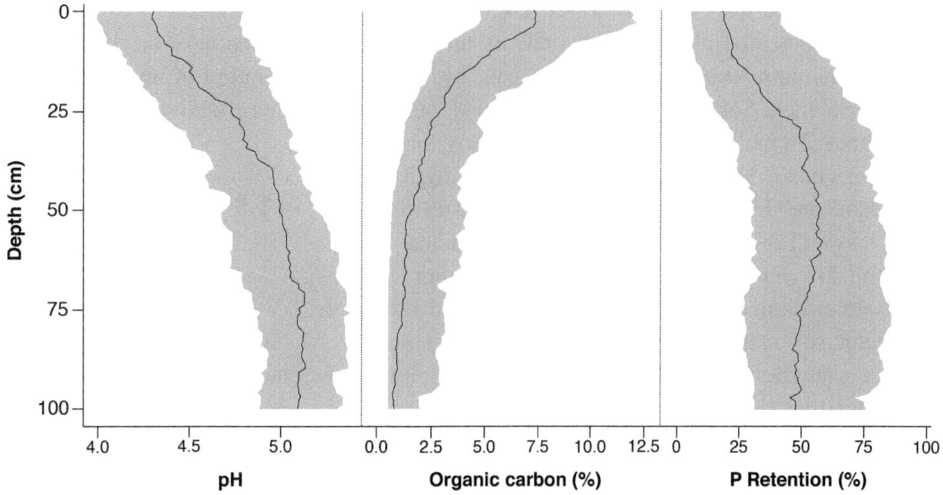

Fig. 11.9 Median and upper and lower quartiles of soil pH, organic carbon, and P retention for Podzol Soils in the New Zealand Soil Data Repository

Table 11.2 Soil chemical properties of a Humose Orthic Podzol Soil (Mamaku, SB09580)

Horizon	Depth	pH (in H_2O)	Carbon (%)	CEC[a] (cmol$^{(+)}$ kg^{-1})	Sum bases (cmol$^{(+)}$ kg^{-1})	P retention (%)
Ah	0–8	5.4	6.1	15.8	5.2	41
Eu	8–15	5.0	2.4	5	1.8	4
2bBs1	15–25	5.1	5.0	21.2	4.1	65
2bBs2	25–43	5.3	6.8	28.6	4.5	96
3bBs	43–58	5.6	5.1	26.1	2.3	98
4bBs	58–100	5.6	4.9	24	1.4	98

[a]CEC = cation exchange capacity

may be seen glinting in sunlight. The grains in B horizons are coated, and in some soils cemented, by the complex oxides, nanominerals, and organic matter.

Podzol Soil profiles have distinct mineralogy *with m*ica-smectite or smectite frequently occurring in A and E horizons (some probably derived from aeolian dust), and hydroxy-coated or interlayered minerals, and usually allophane, and possibly ferrihydrite or imogolite, in B horizons. In some Northland soils, secondary silica is dominant in B horizons.

11.4.2 Physical Properties

Physical properties of Podzol Soils are dominated by either the parent material or the manner in which the material has been modified by soil formation. Thus there is no one typical set of soil physical properties (Fig. 11.8, Table 11.1). However, Podzol Soils, especially those on (youthful) sand dunes and pumice materials are commonly sandy with low clay contents. Clay contents range from near 0 to 40% of the soil, with the majority of Podzol Soils having <25% clay (Fig. 11.8).

The wide range of soil dry bulk densities that may occur is evident in Fig. 11.8. In Densipan Podzols, the original material of the E horizon has been strongly enleached with apparent loss of stabilising secondary oxides and consequent reduction of porosity and increasing density. The densipan may be as dense as a fragipan. Podzols in sand (including pumice) are dominated by the permeability of sand but as B horizon pans are formed (humus- or ortstein-pans or placic horizons) the permeability and porosity (depending on the continuity of the pans) may be sharply reduced.

Barriers to root extension in Podzol Soils include pans, dense layers, acidity, aluminium toxicity, rock, and strong gleying causing limitations of available oxygen. The sandy nature of many Podzol Soils, along with leaching of potential aggregate-binding organic and oxide materials means there is often minimal soil aggregation, particularly in E horizons.

11.4.3 Chemical Properties

Podzol Soils are often strongly to extremely acid, particularly in the A and E horizons (Fig. 11.9, Table 11.2). Soil pH is used as an indicator for aluminum toxicity, and soils where values are less than 4.8 are at risk. A more precise measure of potential aluminium toxicity is the amount of labile aluminium extracted from the soil by potassium chloride solution. Soils with KCl-extractable Al values more than about 2 $cmol^{(+)}$ kg^{-1} are toxic to many plants and values in Podzol Soils are predominantly very high.

Topsoil and E horizon CEC levels in Podzol Soils are mainly low (Table 11.2) because they are limited by low clay and low humification of soil organic matter. Subsoil CECs are medium to high because of accumulated B horizon allophanic soil material and related humus complexes.

Base saturations in unfertilized soils are very low in A and E horizons.

Unexpectedly, total P values are commonly very high in B horizons in Podzol Soils because P anions are entangled in the leaching process and accumulate in B horizons. P retention values are generally low in the highly leached A and E horizons, and high or very high in B horizons where allophanic soil material and related metal-humus complexes accumulate. Thus the P available to plants may not be so high.

11.4.4 Biological Properties

Podzol Soils are associated with mor-forming forest trees with acid, low nutrient litter (as described above). They have low natural fertility including very low natural levels of nitrogen, phosphorus, potassium, and other nutrients in the topsoil. Low levels of faunal activity (due to the acidic and low nutrient conditions) occur with subsequent low rates of mineralization (conversion of mineral elements held within organic matter back to plant-available mineral forms). The lack of soil fauna means there is little soil mixing, hence horizon boundaries are typically distinct or sharp. Carbon/nitrogen ratios are very high which is also an indicator of the limitations to organic matter decomposition. The accumulation of weakly decomposed humus and minimal incorporation of humus into the upper mineral soil results in the mor (relatively undecomposed) humus form (e.g. Fig. 11.5).

11.5 Distinguishing Between Podzol Soils and Related Soil Orders

A typical Podzol Soil is distinctive in its profile morphology but in the landscape it can merge into associated soils, including Brown, Pumice, and Gley Soils, and to Ultic Soils which may also have an E horizon, making it difficult to recognise a clear spatial boundary. For this reason, the *New Zealand Soil Classification* defines the Podzol Soils using two alternative characteristic (diagnostic) features. The first is the presence of a dark-coloured podzolic-B horizon (Bh or Bs, or both), usually with an overlying pale E horizon, and a pH less than 5.5. The second feature is the occurrence of an ortstein-pan (Bsm) with pH < 5.5, or a dark-coloured humus-pan (Bhm) (both types of pans must exceed 10 mm in thickness). The Mamaku soil profile (Tables 11.1 and 11.2) has dark reddish brown Bs horizons (the reddish hue, 7.5YR or redder, being important). Although no humus-rich (Bh) horizons are evident, the presence of the overlying E horizon helps confirm the Mamaku soil as a Podzol Soil formed on tephras.

The Gley Soils and Brown Soils both have 'Acid' soil groups, and Ultic Soils have 'Albic' soil groups. Such groups incorporate soils that had previously been identified as 'podzolic' soils in the earlier New Zealand genetic soil classification. Subgroups within the Acid groups indicate soils that are currently considered marginal (i.e. intergrading) to the Podzol Soils—for instance, Placic Acid Brown Soils. Similarly, a few subgroups still use the adjective 'Podzolic' (e.g. Podzolic Yellow Ultic Soil and Podzolic

Table 11.3 Correlation[a] between Podzol Soil groups and equivalent classes of *Soil Taxonomy, World Reference Base,* and the New Zealand genetic soil classification

NZ Soil Classification	*Soil Taxonomy*	*World Reference Base*	NZ genetic soil classification
Densipan Podzol Soils	Haplorthods, Epiaquods	Stagnic Podzol	Gley podzol
Perch-gley Podzol Soils	Epiaquods, Placorthods	Stagnic Podzol	Gley podzol
Groundwater-gley Podzol Soils	Endoaquods	Gleyic Podzol	Gley podzol
Pan Podzol Soils	Placohumods, Durihumods	Placic Podzol Ortsteinic Podzol	Podzol
Orthic Podzol Soils	Haplorthods, Haplohumods	Haplic Podzol	Podzol

[a]The correlations given here are a guide only and for accurate classifications the relevant soil classification documents should be consulted. The two major international soil classification systems are *Soil Taxonomy,* which was developed in the USA, and *World Reference Base,* which was developed primarily in Europe. The NZ genetic soil classification was used in NZ prior to 1992.

Orthic Pumice Soil). The relationship between Brown, Gley, and Podzol Soils is often related to position in the landscape. While it may appear complex, at some sites, there may also be a catenary relationship on hillslopes, for instance, with Podzol Soils on stable summit and shoulder slopes, Gley Soils on poorly drained lower slopes and Recent or Organic Soils on valley floors.

11.6 Correlation with Other Classification Systems

The Podzol Soil groups mainly correlate with the Spodosols of *Soil Taxonomy* and were formerly classified as Podzols in New Zealand (Table 11.3).

11.7 Use and Management of Podzol Soils

The Podzol Soils are limited in natural fertility and require substantial fertiliser (N, P, K, Ca, and micronutrients) and lime additions to bring them into economic agricultural production. Where the pH is below 4.8 then many plant species will suffer from Al toxicity and other key nutrients will have limited availability. Thus the need for lime. Pan Podzol Soils have humus- or ortstein-pans (Bhm, Bsm) or placic horizons (Bfm) that may present abrupt rooting barriers as well as limiting downward moisture movement. Thus actions to break up pans such as ripping may be needed.

In the South Island, the largest areas of Podzol Soils are in the high rainfall zones within the Southern Alps and south Westland and so remain in native forest under conservation management. Native conservation ecosystems are valuable as they support biodiversity and tourism and prevent erosion. The forested catchments also have an important role in slowing floodwater runoff, and thus moderating flood peaks in the intense rainfall-prone region. In some areas, the forests have been milled and then allowed to regenerate or have been planted into *Pinus radiata* or Douglas fir production forestry.

Perhaps the most extreme conditions are in soils that are locally known as pākihi (meaning open or barren land, Fig. 11.10) where native forest trees do not establish. Instead, there is an impoverished vegetation that is tolerant of the wet, strongly acidic, low nutrient conditions that prevail on Perch-gley Podzol Soils. One soil of particular interest is the Okarito soil of the West Coast of the South Island, a Peaty-silt-mantled Perch-gley Podzol Soil. The classification encapsulates well the geological stratigraphy (layering) in the soil which, together with the soil's high rainfall (>2000 mm yr^{-1}) and high water table, has produced a striking soil. It is essentially a two-storeyed soil, the lower part formed on coarse fluvioglacial aluvium that has been subsequently mantled by the slow accession of loess during the past 30,000 years. Thus the upper profile has formed by developmental upbuilding.

The Okarito soil (Fig. 11.10) is highly leached and generally saturated with water. As a result, it has an organic-matter rich topsoil, a thick E horizon, and strongly

Fig. 11.10 Perch-gley Podzols in the wet West Coast (South Island) environment. Left: an area of pākihi (naturally devoid of forest trees) on the west coast of the South Island (photo: H.S. Gibbs). Right: the Okarito soil, a Peaty-silt-mantled Perch-gley Podzol Soil with humus - laden water weeping from the exposed profile. (Photo courtesy of NZ Society of Soil Science)

formed iron pans (ortstein-pans) that result in the overlying E horizon being strongly reduced (Er). In order to view or sample the saturated soil, it is necessary to pump water from the profile pit. While probably best left in its natural state some such areas have been developed. For successful plantation forestry, it is necessary to break up the pan so roots can penetrate and to provide some contour so water can runoff and escape, and to provide the trees with some soil above the water table. Fertiliser also needs to be added—usually slow-release at the time of planting.

On the flatter lowlands on the West Coast, many areas have been converted to pasture which is increasingly being used for dairy production. Areas in dairy production attest to the potential of these soils for agricultural use following substantial drainage and fertiliser and lime input. The soils occur mainly in areas with a precipitation of 1250 mm or more and they have a soil water surplus for most of the year (see Fig. 1.4).

Deep ripping, flipping, and humping and hollowing are applied to break up iron pans for deeper root penetration and to remediate poor drainage. Explosives have sometimes been used to break through the pans to make holes for fence posts in the Perch-gley Podzol Soils (Fig. 11.10). It has been assumed that the humus/ortstein-pans were responsible for the low subsurface hydraulic conductivity but measurements by Rick Jackson showed that the very slow permeability and extremely slow lateral drainage was caused by the massive unstructured silty soil materials in the upper part of the profile. The Groundwater-gley Podzol Soils are poorly drained because of prolonged high groundwater tables. Drainage should be preceded by examination for pans or other slowly permeable horizons in the soil that may frustrate attempts to drain them.

In the central North Island, pine trees (*Pinus radiata)* grow well on Orthic Podzol Soils formed on the fine-grained, pumiceous ignimbrite of Taupo Tephra on the Mamaku Plateau (Humic Orthose Podzol Soil, Fig. 11.5) which do not have the pans and slow permeability of other Podzol Soils, although they do have low pH and thus potential for aluminium toxicity. Pines may suffer from trace element deficiencies, particularly boron. The steeper slopes are prone to erosion of the low-density pumice materials if the vegetation is removed so care needs to be taken during forest harvest.

In Northland, the Densipan Podzol Soils, associated with existing or former kauri trees, support pasture and pine forest. The densipan is not cemented but has high strength, bulk density, and penetration resistance and can benefit from soil ripping to improve permeability, though the effect may not persist. As with all Podzol Soils, there is a high requirement for fertiliser and lime in order to provide effective pasture growth.

Further Reading

Alloway BV, Almond PC, Piet M, al, (2018) Mid-latitude trans-Pacific reconstructions and comparisons of coupled glacial/interglacial climate cycles based on soil stratigraphy of cover-beds. Quatern Sci Rev 189:57–75

Almond PC (1996) Loess, soil stratigraphy and Aokautere ash on late Pleistocene surfaces in south Westland: interpretation and correlation with the glacial stratigraphy. Quatern Int 32(33):163–176

Almond PC, Tonkin PJ (1999) Pedogenesis by upbuilding in an extreme leaching and weathering environment, and slow loess accretion, south Westland, New Zealand. Geoderma 92:1–36

Almond PC, Moar NT, Lian OB (2001) Reinterpretation of the glacial chronology of south Westland, New Zealand. NZ J Geol Geophys 44:1–15

Almond PC, Barrell D, Hyatt O et al (2007) Quaternary geomorphology, stratigraphy, and paleoclimate of the central Southern Alps, South Island, New Zealand. In: INQUA 2007 post conference field trip guide, p 71

Anderson B, Mackintosh A (2006) Temperature change is the major driver of late-glacial and Holocene glacier fluctuations in New Zealand. Geology 34(2):121–124

Anderson HA, Berrow ML, Farmer VC et al (1982) A reassessment of podzol formation processes. J Soil Sci 33:125–136

Barrell DJA (2011) Quaternary glaciers of New Zealand. In: Ehlers J, Gibbard PL, Hughes PD (eds) Developments in Quaternary science, vol 15. Elsevier, Amsterdam, pp 1047–1064

Barrell DJA, Almond PC, Vandergoes MJ et al (2013) A composite pollen-based stratotype for inter-regional evaluation of climatic events in New Zealand over the past 30,000 years (NZ-INTIMATE project). Quatern Sci Rev 74:4–20

Buurman P, Jongmans AG (2005) Podzolisation and soil organic matter dynamics. Geoderma 125:71–83

Campbell AS (1975) Chemical and mineralogical properties of a sequence of terrace soils near Reefton, New Zealand. PhD thesis Lincoln College, University of Canterbury, New Zealand

Churchman GJ, Lowe DJ (2012) Alteration, formation, and occurrence of minerals in soils. In: Huang PM, Li Y, Sumner ME (eds) Handbook of soil sciences, 2nd edn., vol 1: Properties and processes. CRC Press, Boca Raton, FL, pp 20.1–20.72

Claessens L, Verburg PH, Schoorl JM et al (2006) Contribution of topographically based landslide hazard modelling to the analysis of the spatial distribution and ecology of kauri (*Agathis australis*). Landscape Ecol 21:63–76

Claessens L, Veldkamp A, ten Broeke EM et al (2009) A Quaternary uplift record for the Auckland region, North Island, New Zealand, based on marine and fluvial terraces. Global Planet Change 68:383–394

Daly BK (1982) Identification of podzols and podzolised soils in New Zealand by relative absorbance of oxalate extracts of A and B horizons. Geoderma 28:29–38

Daly BK, Rijkse WC (1974) Chemistry: a typical sequence of soils derived from air-fall tephra. In: Read NE (ed) Soil groups of New Zealand. Part 1. Yellow-brown Pumice Soils. NZ Society of Soil Science, Wellington, pp 65–75

Eden DN, Hammond AP (2003) Dust accumulation in the New Zealand region since the last glacial maximum. Quatern Sci Rev 22:2037–2052

Farmer VC (1982) Significance of the presence of allophane and imogolite in podzol Bs horizons: a review. Soil Sci Plant Nutr 28:571–578

Farmer VC, Fraser AR (1982) Chemical and colloidal stability of sols in the Al_2O_3–Fe_2O_3–SiO_2–H_2O system: their role in podzolization. J Soil Sci 33:737–742

Farmer VC, Lumsdon DG (2001) Interactions of fulvic acid with aluminium and a proto-imogolite sol: the contribution of E-horizon eluates to podzolization. Eur J Soil Sci 52:177–188

Farmer VC, Russell JD, Berrow ML (1980) Imogolite and proto-imogolite allophane in spodic horizons: evidence for a mobile aluminium silicate complex in podzol formation. J Soil Sci 31:673–684

Farmer VC, Russell JD, Smith BFL (1983) Extraction of inorganic forms of translocated Al, Fe and Si from a podzol Bs horizon. J Soil Sci 34:571–576

Farmer VC, Mew B, Clayden B et al (1984) Hypothesis on the pedogenesis of the 'gley podzols' of the west coast of South Island, New Zealand. J Soil Sci 35:99–102

Farmer VC, McHardy WJ, Robertson L et al (1985) Micromorphology and sub-microscopy of allophane and imogolite in a podzol Bs horizon: evidence for translocation and origin. J Soil Sci 36:87–95

Hewitt AE (2010) New Zealand Soil Classification, 3rd edn. Landcare research science series no. 1, Manaaki Whenua Press, Lincoln, New Zealand

Jongkind AG, Buurman P (2006) The effect of kauri (*Agathis australis*) on grain size distribution and clay mineralogy of andesitic soils in the Waitakere Ranges, New Zealand. Geoderma 134:171–186

Jongkind AG, Velthorst E, Buurman P (2007) Soil chemical properties under kauri (*Agathis australis*) in the Waitakere Ranges, New Zealand. Geoderma 141:320–331

Jongmans AG, van Breemen N, Lundström US et al (1997) Rock-eating fungi. Nature 389:682–683

Kaplan MR, Schaefer JM, Denton GH et al (2010) Glacier retreat in New Zealand during the Younger Dryas stadial. Nature 467:194–197

Lee R (ed) (1980) Soil groups of New Zealand, Part 5, Podzols and Gley Podzols. New Zealand Soil Science Society, Wellington

Lundström US, van Breeman N, Bain D (2000) The podzolisation process. A review. Geoderma 94:91–107

Mew G (1983) Applications of the term 'pakihi' in New Zealand—a review. J Roy Soc New Zealand 13:175–198

Orwin J (2013) Kauri-forest. www.teara.govt.nz/en/kauri-forest

Palmer DJ, Lowe DJ, Payn TW et al (2005) Soil and foliar phosphorus as indicators of sustainability for *Pinus radiata* plantation forestry in New Zealand. For Ecol Manage 220:140–154

Peltzer DA, Wardle DA, Allison VJ et al (2010) Understanding ecosystem retrogression. Ecol Monogr 80:509–529

Richardson RJH (1985) Quaternary geology of the North Kaipara barrier, Northland, New Zealand. NZ J Geol Geophys 28(1):111–127

Richardson SJ, Peltzer DA, Allan RB et al (2004) Rapid development of phosphorus limitation in temperate rainforest along the Franz Josef soil chronosequence. Oecologia 267–276

Schaetzl RJ, Harris W (2012) Spodosols. In: Huang PM, Li Y, Summner ME (eds) Handbook of soil sciences, 2nd edn., vol 1. Properties and process, pp 33.113–33.127

Schaetzl RJ, Mokma DL (1988) A numerical index of podzol and podzolic soil development. Phys Geogr 9:232–246

Schaetzl RJ, Thompson ML (2015) Soils - genesis and morphology, 2nd ed. Cambridge Univ. Press, New York, 778 p

Soil Survey Staff (2014) Keys to Soil Taxonomy, 12th edn. USDA Natural Resources Conservation Service, 362 p https://www.nrcs.usda.gov/wps/PA_NRCSConsumption/download?cid=stelprdb1252094&ext=pdf

Stephens PR (1963) A chronosequence of soils and vegetation near the Franz Josef glacier. PhD thesis, Lincoln College, University of Canterbury, New Zealand

Suggate RP, Almond PC (2005) The Last Glacial Maximum (LGM) in western South Island, New Zealand: implications for the global LGM and MIS 2. Quatern Sci Rev 24:1923–1940

Tonkin PJ, Basher LR (1990) Soil stratigraphic techniques in the study of soil and landform evolution across the Southern Alps, New Zealand. Geomorphology 3:547–575

Turner BL, Condron LM (2013) Pedogenesis, nutrient dynamics, and ecosystem development: the legacy of T.W. Walker and J.K Syers. Plant Soil 367:1–10

Turney CSM, Roberts RG, de Jonge N et al (2007) Redating the advance of the New Zealand Franz Josef Glacier during the last termination: evidence for asynchronous climate change. Quatern Sci Rev 26:3037–3042

Ugolini FC, Dahlgren RA (1991) Weathering environments and occurrence of imogolite/allophane in selected Andisols and Spodosols. Soil Sci Soc Am J 55:1166–1171

Ugolini FC, Dawson H, Zachara J (1977) Direct evidence of particle migration in the soil solution of a Podzol. Science 198:603–605

Ugolini FC, Dahlgren RA, Shoji S et al (1988) An example of andosolization and podzolization as revealed by soil solution studies, southern Hakkoda, northeastern Japan. Soil Sci 145:111–125

Ugolini FC, Dahlgren R, LaManna J et al (1991) Mineralogy and weathering processes in Recent and Holocene tephra deposits of the Pacific Northwest, USA. Geoderma 51:277–299

van Breemen N, Lundström US, Jongmans AG (2000) Do plants drive podzolization via rock-eating mycorrhizal fungi? Geoderma 94:163–171

Vandergoes MJ, Dieffenbacher-Krall AC, Newnham RM (2008) Cooling and changing seasonality in the Southern Alps, New Zealand, during the Antarctic cold reversal. Quatern Sci Rev 27:589–601

Vandergoes MJ, Hogg AG, Lowe DJ et al (2013) A revised age for the Kawakawa/Oruanui tephra, a key marker for the Last Glacial Maximum in New Zealand. Quatern Sci Rev 74:195–201

Vandergoes MJ, Newnham RM, Denton GH et al (2013) The anatomy of Last Glacial Maximum climate change in the southern mid-latitudes derived from pollen records in south Westland, New Zealand. Quatern Sci Rev 74:170–194

Verkaik E, Braakhekke WG (2007) Kauri trees (*Agathis australis*) affect nutrient, water and light availability for their seedlings. New Zealand J Ecol 31:39–46

Verkaik E, Jongkind AG, Berendes F (2006) Short-term and long-term effects of tannins on nitrogen mineralisation and litter decomposition in kauri (*Agathis australis*) forests. Plant Soil 287:337–345

Vucetich CG (1981) Grange and New Zealand soil science. Geol Soc New Zealand Newsl 54:46–49

Walker TW, Syres JK (1976) The fate of phosphorus during pedogenesis. Geoderma 15:1–19

Wang C, McKeague JA, Kodama H (1986) Pedogenic imogolite and soil environments: case study of spodosols in Quebec. Canada Soil Sci Soc Am J 50:711–718

Wells N, Northy RD (1985) Strengths of a densipan, humus-pan, and clay-pan in a spodosol developed under kauri (*Agathis australis*) and the implications for soil classification. Geoderma 35:1–13

Williams PA, Courtney S, Glenny D et al (1990) Pakihi and surrounding vegetation in north Westland, South Island. J Roy Soc New Zealand 20:179–203

Wood V (2003) Appraising soil fertility in early colonial New Zealand: the 'biometric fallacy' and beyond. Environ Hist 9:393–405

World Reference Base (2015 update) https://www.fao.org/3/a-i3794e.pdf

Young AW (1988) A catena of soils on Bealey Spur, Canterbury, New Zealand. PhD thesis. Lincoln University, NZ

12 Pumice Soils

Born of pyroclastic eruption
Pumice Soils host wood production
Pine roots reach to paleosols
nutrients and water are their goals.

Abstract

Pumice Soils cover about 16% of North Island, formed mainly in unconsolidated rhyolitic (silica-rich) pumice deposited by the Taupo eruption (AD 232 ± 10) from the huge caldera now occupied by Lake Taupo, and the Kaharoa eruption (AD 1314 ± 12) from Mt Tarawera. Other areas of Pumice Soils are formed in 3500-year-old Waimihia pumice deposits and on fluvially reworked pumice alongside rivers. Pumice Soils comprise sandy to gravelly pumice materials. They are weakly weathered with a low clay content and have A, and weakly developed Bw, horizons within pumice deposits that are at least 25 cm thick. Pumice Soils have low reserves of major nutrients and are likely to be deficient in trace elements including cobalt, copper, molybdenum, boron, iodine, and selenium. For agricultural development, it is necessary to add fertiliser and trace elements, particularly cobalt as its deficiency causes 'bush sickness' (vitamin B_{12} deficiency) which leads to death of sheep and cattle. *Pinus radiata* trees have been successfully grown, supporting a major plantation forestry industry. However, in recent decades, many plantations have been converted to dairying. Pumice Soils are readily eroded by water, and if not protected by vegetation, severe gully erosion can form in a single rainstorm.

12.1 Important Features of Pumice Soils

12.1.1 Concept and Key Features of the Soil Order

Pumice Soils are formed in unconsolidated, weakly or non-weathered coarse-textured rhyolitic pumice fragments (clasts) deposited largely within the last 1800 years from the Taupo and Kaharoa eruptions (Sect. 12.2). Consequently, the soil properties relate strongly to those of the (geologically young) parent materials (Sect. 12.4). Pumice is a low-density, glass-rich, vesicular (i.e. containing vesicles or holes) material derived from pyroclastic eruptions of viscous ('sticky') magma. The term 'pyroclastic' describes fragmental material (as opposed to lava) of all grain sizes erupted explosively in a volcanic eruption. Rhyolitic pumice is pale coloured and enriched in silica (70–78 wt% SiO_2) along with potassium, sodium, and other elements. The vesicularity of pumice clasts means that soil horizons have a low bulk density (usually <1 t m^{-3}), and pumiceous sand grains, unlike equivalent grains of quartz or feldspar, contain vesicles filled with air and/or water. Pumice clasts of the Kaharoa deposits are harder, whiter, and less vesicular than those of the Taupo deposits, which are typically pale cream coloured. Taupo lapilli (clasts 2–64 mm in diameter) are often crushable between thumb-nails, whereas Kaharoa lapilli are not.

A. E. Hewitt et al., *The Soils of Aotearoa New Zealand*, World Soils Book Series,
https://doi.org/10.1007/978-3-030-64763-6_12

Fig. 12.1 Landscape and Pumice Soil profile. Left: Pumice Soil grazed by dairy cattle near Mangakino following conversion from pine forest. In background is the eroded andesitic cone of Titiraupenga. Right: an Orthic Pumice Soil that has been ploughed giving a sharp boundary between the Ap and Bw horizons. A diffuse boundary is evident between the Bw and the unweathered, pumice-rich, tephra of the C horizon. Profile is about 80 cm deep

Pumice Soils are sandy to gravelly and dominated by weakly weathered pumice, and glassy siliceous sand and silt (also referred to as ash, which comprises particles <2 mm in diameter). Pumice clasts >64 mm in diameter (described as blocks or bombs volcanologically or as very coarse gravels to boulders pedologically) can be common depending on the depositional setting and location.

A Pumice Soil profile comprises an A horizon over a weathered-B horizon (usually Bw) that is at least 5 cm thick with visible weathering and alteration of the pumice (Fig. 12.1). The minimum thickness of pumice needed to qualify as a Pumice Soil is 25 cm (or the pumice layer is 35 cm or more thick within the upper 60 cm of the profile). The C horizon usually comprises unconsolidated gravel-sized pumice lapilli and sand (ash) particles. The clay content is low and dominated by allophane.

12.1.2 Areas of Occurrence

Pumice Soils occur predominantly in the central North Island, particularly on the central Volcanic Plateau and Bay of Plenty (Fig. 12.2). They cover about 7% of New Zealand or 16% of the North Island. The Pumice Soils provide almost continuous cover on the Volcanic Plateau in a nearly circular pattern, about 160 km across, around Taupo (formed from Taupo eruptives) and in the Bay of Plenty (formed mainly from Kaharoa eruptives). Some of the 'gaps' on the map comprise Podzol Soils formed on Taupo pumice at high elevations under high rainfall, and a zone around Mt Tarawera comprising basaltic scoria and ash, and Rotomahana 'mud', deposited on 10 June 1886 on top of the Kaharoa pumice. Pockets of Pumice Soils also occur in northern Hawke's Bay on older pumice deposits.

12.1.3 Variation within the Pumice Soil order

The following soil groups are defined in the New Zealand Soil Classification (Fig. 12.3):

Perch-gley Pumice Soils—periodic wetness caused by a water table perching on a slowly permeable horizon, usually a paleosol,

Impeded Pumice Soils—have a hard, dense, or tightly packed, compact layer, which may be cemented with redeposited silica in places, potentially impeding roots and water,

Orthic Pumice Soils—other Pumice Soils.

Intergrading subgroups recognise transitions from the Pumice Soil order to other related soil orders or soil groups: Mottled (to Gley Soils), Podzolic (to Podzol Soils),

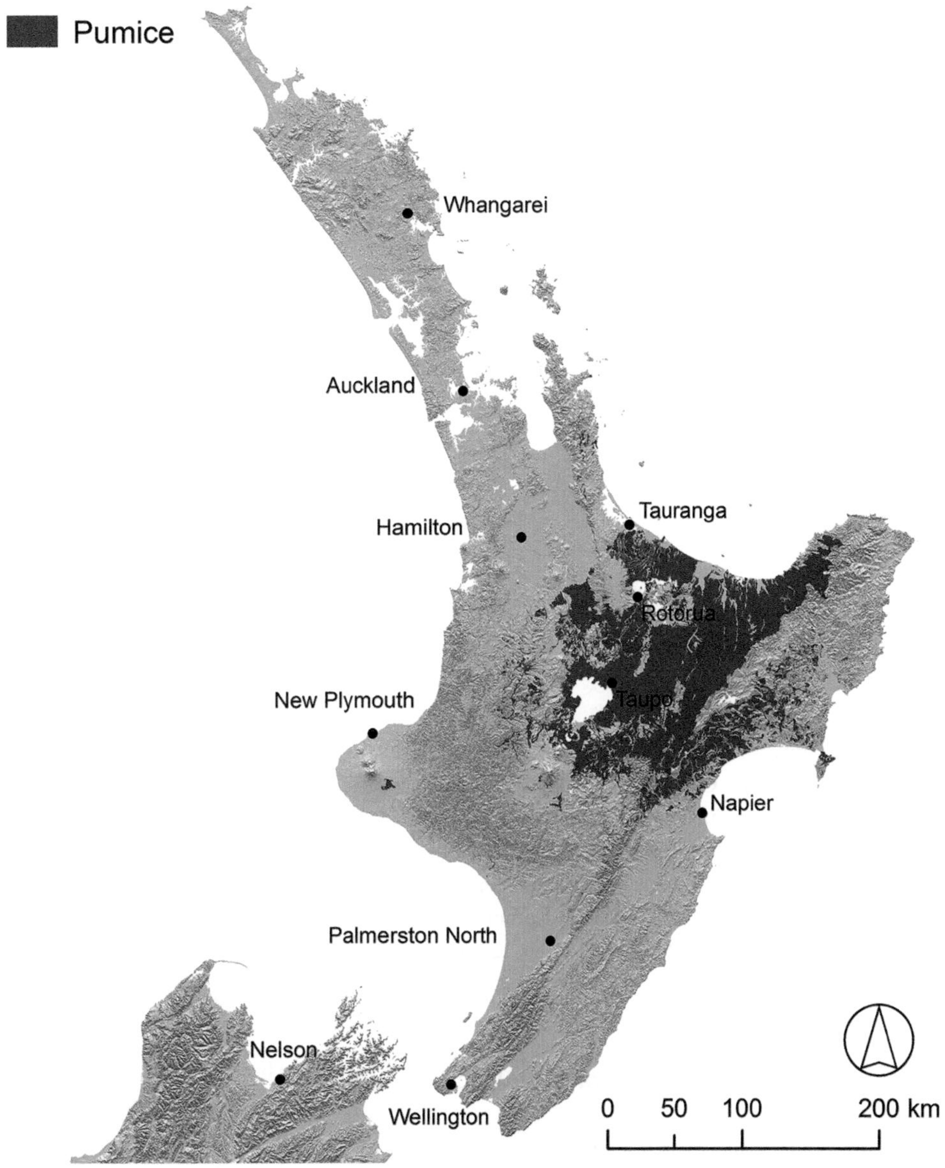

Fig. 12.2 Distribution of Pumice Soils

Allophanic (to Allophanic Soils), and Immature (to Recent Soils). Extragrading subgroups depart in some specific way from the soil group (Welded, and Buried-allophanic). The subgroup closest to the central concept of the group is identified as Typic.

12.1.4 Origin of Soil Order Name

The Pumice Soils are named for the pumice which dominates the parent material and the soil's properties.

12.2 Soil Profile Genesis

12.2.1 Introduction

Most Pumice Soils are derived largely from unconsolidated or loose, silica-rich pumice deposited by two major rhyolitic volcanic eruptions: (1) the Taupo eruption in AD 232 ± 10 (i.e. about 1800 years ago), from vents in the 25,400-yr-old Oruanui caldera now occupied by Lake Taupo, and (2) the Kaharoa eruption in AD 1314 ± 12 (about 700 years ago)

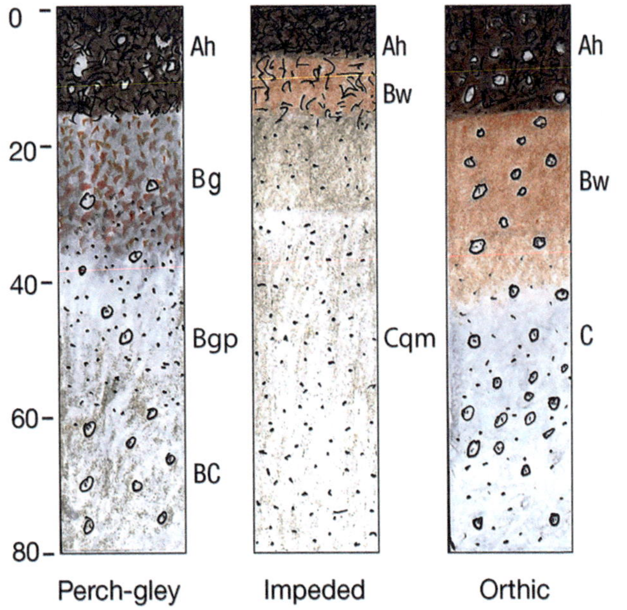

Fig. 12.3 Groups within the Pumice Soil order. Vertical axis is depth (cm)

from Mt Tarawera (which forms the southern part of the Okataina Volcanic Centre) (Fig. 12.4). Small areas of Pumice Soils are formed in older rhyolitic pumice deposits of the Waimihia eruption (erupted from Taupo volcano about 3500 years ago) where they are preserved at the land surface in northern Hawke's Bay, and others are formed on fluvially reworked pumice deposits on terraces alongside rivers including the Waikato and Whanganui.

12.2.2 Magmatic Origins of Rhyolitic Pumice

Both the Taupo and Kaharoa eruptions, the most recent rhyolitic events in New Zealand's geological history, were explosive, large-scale, events. Their sources were the Taupo and Okataina Volcanic Centres, respectively, which are the two most frequently-erupting and productive rhyolite volcanoes in the world. As the Pacific tectonic plate is subducted beneath the North Island, which is located on the Australian plate, it is heated and water and other fluids are boiled off and stream into the mantle rocks beneath the island. The fluids cause chemical changes that enable the otherwise solid rock of the mantle to melt, forming basaltic magma. The magma rises until it gets trapped beneath the continental crust of the North Island. The very hot basalt magma acts like a huge blow-torch, melting the crust and mixing with it to form andesite magma (with SiO_2 ~50–70 wt%), which is then erupted as cone-shaped stratovolcanoes, such as Mt Ruapehu and Taranaki Maunga. However, where further melting of the continental crust occurs, extremely viscous rhyolite magma forms in a magma chamber that may be only a few kilometres beneath the ground. As magma rises to the surface, the drop in pressure causes constituent gases (volatiles) to expand violently, generating eruption columns of shattered magma and rock fragments (lithics) and gases high into the upper atmosphere, and even the stratosphere, before falling back to the ground. The higher the pyroclastic materials (tephra) are ejected, the farther from the volcano they will be carried by the wind. In some cases, the explosive emptying of the magma chamber results in the ground becoming unsupported and so the land collapses, dropping like a piston, to form a large depression or caldera. Nine large caldera volcanoes including Taupo and Okataina (Fig. 12.4, inset) are concentrated in the central part of the Taupo Volcanic Zone. The rhyolitic volcanoes erupt much greater volumes of magma (both pyroclastics and lava), far less frequently, than the eruptions from the andesitic volcanoes.

Taupo is regarded as the archetypal 'inverse' super-volcano in that most eruptions have been so explosive, and dispersed material so widely, where the accumulation of material around the vents has not compensated for collapse associated with caldera formation. Consequently, the lake bed is the lowest point in the landscape for over 40 km in any direction. The Okataina Volcanic Centre, in which Mt Tarawera forms the southern part, comprises a caldera that has been partly infilled with numerous lava flows and domes and so has a greater relief than the Taupo Volcanic Centre.

12.2.3 Taupo Eruption and Its Products

The Taupo eruption took place in AD 232 ± 10 (see grey box for age details) and was the 27th eruption to have taken place in the Taupo Volcanic Centre (which includes Lake Taupo) since the Oruanui super-eruption occurred about 25,400 years ago. The Oruanui (also known as Kawakawa) eruption was extremely voluminous with a total magma volume of ~530 km^3 (equivalent to 1170 km^3 as bulk pyroclastic material). Caldera collapse associated with the Oruanui eruption generated the wide basin that forms the northern half of Lake Taupo today (Fig. 12.4).

The Taupo eruption in ~AD 232 (unit Y in the post-Oruanui eruption sequence, Fig. 12.5) was one of the most violent and complex rhyolite eruptions in the world in the past 5000 years. The eruption was centred on at least three vents near the Horomatangi Reefs under the waters of eastern Lake Taupo. It consisted of five explosive phases, both 'wet' (involving magma-water interactions, i.e. phreatomagmatic) and 'dry' (driven by the expansion of internal volatiles, i.e. magmatic), which deposited pyroclastic fall deposits (subunits Y1–Y5, Fig. 12.5), followed by the sixth and climactic final

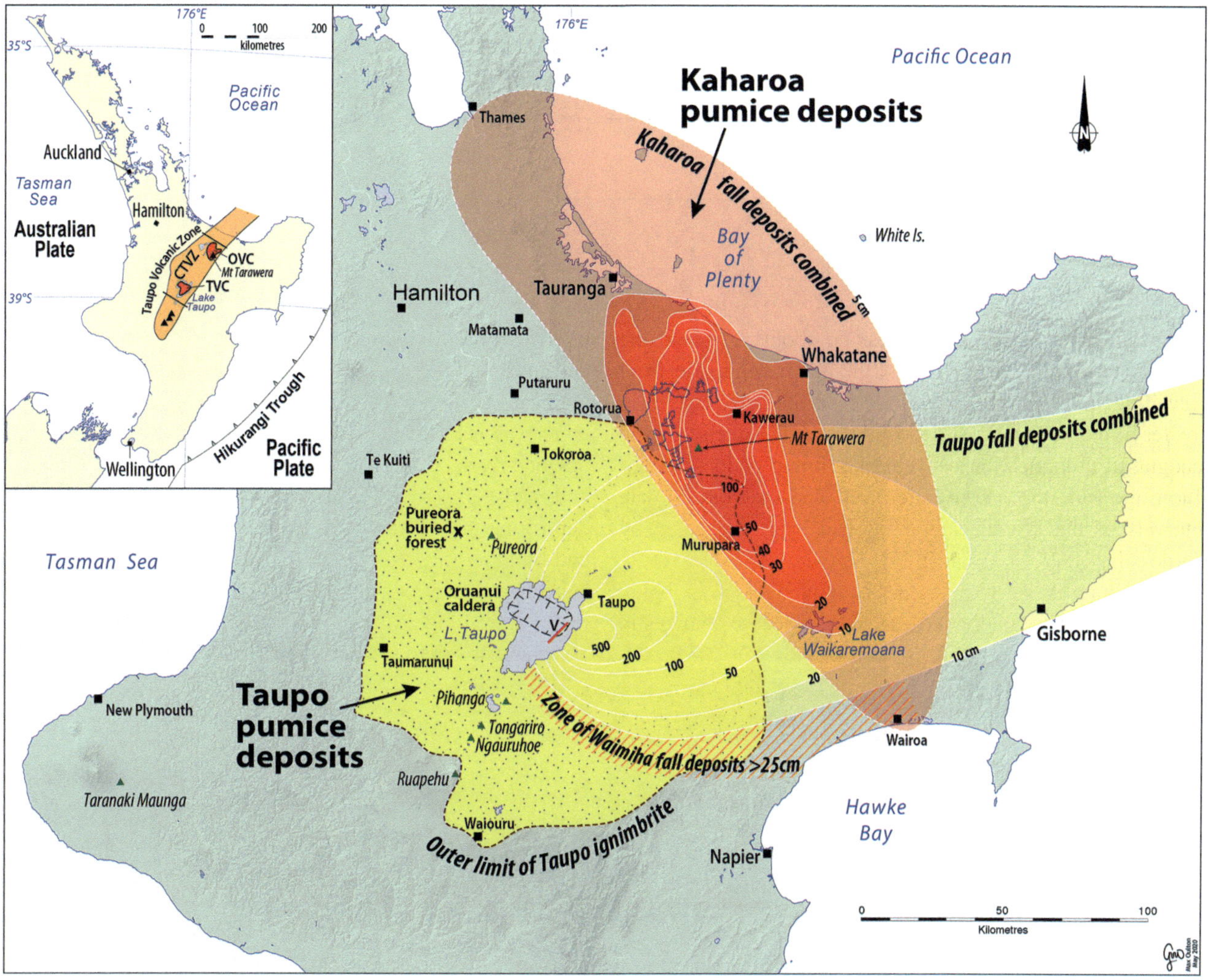

Fig. 12.4 Distribution of the products of the Taupo (~AD 232) and Kaharoa (~AD 1314) eruptions in central North Island on which almost all of the Pumice Soils are formed. The Taupo eruptives (yellow) include the area covered by (i) ashfall to a depth of 10 cm or more shown with thickness contours (isopachs) in centimetres (directed strongly eastward), and (ii) the non-welded Taupo ignimbrite that was emplaced radially around Lake Taupo by an extremely energetic pyroclastic flow, the deposit covering all peaks except Ruapehu. Vents for the eruption occurred on a fissure along the eastern side of the lake marked with the symbol V. The Kaharoa ash-fall eruptives (red) are depicted by isopachs (in cm) with only a small area comprising pyroclastic flow deposits within about 5 km of source Mt Tarawera (not shown). Inset map shows the plate tectonic setting of North Island. (CTVZ = Central Taupo Volcanic Zone, OVC = Okataina Volcanic Centre, and TVC = Taupo Volcanic Centre. Redrawn after Wilson and Walker (1985), Walker (1981), and Sahetapy-Engel et al. (2014)

phase that formed the non-welded (loose) Taupo ignimbrite (subunit Y6). The total volume of the Taupo eruptives is about 35 km^3 as magma, equivalent to about 105 km^3 of bulk pyroclastic material.

The fall deposits rained out over the landscape while being blown more or less eastward towards Hawke's Bay and beyond into the Pacific Ocean (Fig. 12.4). The fall beds are quite variable in character but are everywhere loose and fragmental, and so are easily reworked. During the eruption, a break occurred, possibly a few days in duration, during which intense rain, or ejected water from Lake Taupo, or both, resulted in the formation of rills (shallow gullies) in the loose deposits. The rills deepen closer to Lake Taupo. The renewal of the eruption saw the rills being infilled with a dark grey, fine-grained deposit forming a distinctive V-shaped layer in the sequence (known as Rotongaio ash or subunit Y4).

The sixth phase of the eruption was the catastrophic emplacement of the Taupo ignimbrite as the product of a hot pyroclastic flow. Pyroclastic flows commonly originate when a towering eruption column collapses and falls to the ground, causing the mass of hot pumice, ash (mainly glass shards and crystals), lithics, and gas to race outwards across the landscape in the form of a dense (but air-supported) ground-hugging, gravity-controlled pyroclastic flow (also

Fig. 12.5 Plinian eruption of Taupo volcano, resulting deposits and soil. Top: reconstruction of the final phase of the ~232 AD Taupo eruption when the high Plinian column began collapsing generating 'clouds' of hot gas and solid particles racing radially outwards across the land as a violent ground-hugging pyroclastic flow, forming a non-welded ignimbrite deposit in about 10 min. Image by Mark Garlick. Bottom: the result of the Taupo eruption: an Immature Orthic Pumice Soil (on Te Toke Rd about 30 km north of the vent) has developed on both fall beds and pyroclastic flow deposits (Taupo ignimbrite) that amount to about 2 m in thickness. The difference between layers 1 and 2 of the ignimbrite is evident

called a pyroclastic density current) that looks like grey, incandescent billowing clouds.

The Plinian eruption column (of phase 5) that collapsed to generate the Taupo ignimbrite attained a height of about 35–40 km (Fig. 12.5). The amount of magma being erupted (mass discharge rate) to sustain the very high Plinian column, and the ensuing pyroclastic flow, was extremely high (between 10^6 and 10^{10} kg s^{-1}). The hot pyroclastic flow was so energetic and powerful that it travelled outwards from Taupo at speeds at or exceeding about 200 to 300 m s^{-1} (600–900 km h^{-1}) for about 80 km in all directions, and the flow stopped only when the material ran out. The flow covered an area of about 20,000 km^2 in around 7 to 15 min. The high speed and fluidity gave the flow sufficient momentum to allow it to easily surmount mountains >1500 m high, including Tongariro, with only Ruapehu high enough to block it (Fig. 12.4). Recent large-scale laboratory simulations and modelling at Massey University have shown that pyroclastic flows generate their own air lubrication, forming a near-frictionless basal region displacing particles upwards, thereby allowing huge masses (thousands to millions of tonnes) of material to be transported over uneven and upsloping terrains. Two layers are recognised in the Taupo ignimbrite: layer 1 comprises

Fig. 12.6 Carbonised logs and charcoal fragments scattered through pyroclastic flow deposit in cutting in Palmer Mill Road north of Wairakei. Cutting tool 30 cm long

pumice-rich and dense, lithic- and crystal-rich zones that underlie layer 2, which is dominated by pumiceous materials consisting of coarse clasts (lapilli, blocks) and abundant fine material (ash) (Fig. 12.5). In places, the flow tore up or overturned parts of the underlying soil and incorporated them into the ignimbrite as rip-up clasts.

The violent flow flattened and engulfed an entire forest, turning it to charcoal and ingesting about 1 km^3 of timber in total. Charred logs and charcoal are ubiquitous in the ignimbrite (Fig. 12.6). Until the charcoal-bearing unit was identified as a pyroclastic flow deposit, it was a puzzle as to how the deposit had remained so hot throughout its 160-km-wide distribution. Leslie Grange estimated, in 1931, that the charcoal-bearing unit must have been at least 250 °C. More recent estimates (from charring experiments and thermal remnant magnetic work) indicate the ignimbrite's temperature was 400–500 °C at sites beyond 40 km from vent and, surprisingly, cooler (150–300 °C) at sites within 30–40 km of the vent.

The extreme violence and fluidity of the Taupo pyroclastic flow caused the ignimbrite to be spread as an essentially continuous, but thin, sheet over the entire landscape (its average thickness is about 1.5 m). Unlike more conventional ignimbrites, the Taupo ignimbrite's powerful emplacement was not controlled by the existing topography, or only barely. The Taupo ignimbrite covered a near-circular area and reached about the same distance after crossing several substantial mountain ridges as it did after travelling in other sectors over nearly level ground. The thickness varies according to the underlying topography, however, and the resultant Pumice Soils also vary somewhat in their properties (Sect. 12.3).

George Walker and Colin Wilson, who have studied Taupo volcano and its products intensely, commented that the widespread nature of the ignimbrite reflected an extremely powerful and dangerous event despite its thinness. During emplacement, the flows infilled valleys as well as surmounting slopes as steep as 30°, spilling over ridges into neighbouring catchments, and swinging around valley bends, as a speedway-car would take a speedway curve, and splaying out to cover plains. Taupo ignimbrite is thus thicker in valleys, forming deposits about 5–70 m deep, typically >5 m (known as valley-ponded deposits). As the flow crossed topographic highs, it emplaced thin deposits 0.25–8 m deep, typically 0.5–2 m (known as ignimbrite-veneer deposits).

Unlike a number of older ignimbrites, the Taupo ignimbrite is everywhere loose and unconsolidated, a state referred to as *non-welded.* In volcanology, the term 'welded' means transformed by the sintering together of hot pumiceous fragments and glass shards under a compactional load to form hard rock, typically where the deposits exceed 550 °C and are thick. Welded ignimbrites tend to have prominent cliffs, often with vertical joints, such as that occur in the Hinuera Valley. Originally the term 'ignimbrite' (meaning 'fiery storm-clouds'), introduced in 1932, was associated with hard (welded) rocks and it took many decades for unconsolidated pyroclastic flow deposits to be recognised also as ignimbrites. The term is different, however, in meaning to the subgroup term 'welded' used in the *New Zealand Soil Classification* which applies to compact, tightly packed but nevertheless unconsolidated (and usually non-cemented) pumice clasts.

Because of its extreme violence and energy release (150 ± 50 megaton TNT explosive yield, cf. Hiroshima bomb 0.015 Mt), and by analogy with the 1883 Krakatau event, it is likely that the Taupo ignimbrite-emplacement phase generated a volcano-meteorological tsunami (essentially the collapsing eruption column displaced a large volume of air laterally, forming pressure waves) that may have reached coastal areas worldwide.

12.2.4 Kaharoa Eruption and Its Products

The Kaharoa eruption was the most recent rhyolite eruption in New Zealand and forms the dominant parent material for Pumice Soils in the Bay of Plenty (Fig. 12.7). The Kaharoa Tephra was erupted in AD 1314 ± 12 (see grey box for age details) from Mt Tarawera (which forms the southern half of the Okataina Volcanic centre). The Kaharoa eruption followed nine previous rhyolite eruptions in the Okataina centre since 25,400 years ago. The most recent eruption from Tarawera, on 10 June 1886, was initially of basaltic pyroclastic material, depositing Tarawera Scoria. Then hydrothermally altered (weathered) rhyolitic material, along

Fig. 12.7 Kaharoa tephra on Ash Pit Road about 10 km southeast of Mt Tarawera. Top: road cutting showing the white layers of Kaharoa tephra mantling the landscape. Mt Tarawera in the background. Bottom: Wim Rijkse views the Allophanic Orthic Pumice Soil developed mainly in the layered Kaharoa tephra fall beds overlying buried soils in older eruptives. Here the Rotomahana Mud is not thick enough to qualify the soil as a Raw or Recent Soil because the Kaharoa pumice layers (vitric soil material) are within 60 cm of the soil surface

with lake sediment, was erupted, depositing Rotomahana Mud (Fig. 12.7).

The Kaharoa eruption was a complex, multi-episode rhyolitic event comprising up to 13 separate and very explosive (Plinian-style) eruption phases in two stages. The early stage generated seven pumice lapilli fall units dispersed solely to the southeast of the volcano, and the later stage produced six fall units that were dispersed almost entirely to the north and northwest. The elongated NW–SE shape of the Kaharoa isopachs (Fig. 12.4) thus reflects the 180°-change in wind direction during the eruption episode. Within about 5 km of Tarawera, fine-textured ash beds occur interbedded with the fallout deposits and represent deposition from pyroclastic flows. The explosive eruptive activity during the Kaharoa eruption lasted about two weeks in total (subsequent lava dome formation, and block-and-ash flows, continued for another five years or so), with the total magma volume being about 9 km^3 (equivalent to about 15 km^3 as bulk pyroclastic material). Break-out flood deposits were also formed during the Kaharoa eruption.

The Kaharoa tephra deposits are everywhere unconsolidated, and Pumice Soils, formed entirely on the fall beds, occur in the central and eastern parts of the Bay of Plenty where pumice layers are >25 cm thick. Many soils are multi-layered because the Kaharoa tephra thins relatively quickly away from the source and so the pumice layers are only between about 25 and 75 cm thick in most locations.

12.2.5 The Development of Pumice Soils

The short durations of the explosive phases of the Taupo and Kaharoa eruptions means that large volumes of pyroclastic material were deposited rapidly over large swathes of land in central and eastern North Island. Pre-existing soils were buried in an instant, turning them into paleosols. The emplacement of the Taupo ignimbrite within about 10 min provided a pumice-rich parent material that is exactly the same age wherever it occurs. It means that studies, such as those looking at the effects of different land use and management, can be undertaken on Taupo soils with the surety that the age of the parent material is not a compounding factor. In addition, soil development under differing conditions can also be studied with the time factor constant. The Pumice Soils on the Taupo and Kaharoa eruptives thus provide an exceptional example of retardant upbuilding pedogenesis (Fig. 12.8). Soil formation in the newly buried soil, instantly isolated from surface processes, was stopped in its tracks, and pedogenesis was forced to start again from time zero in the freshly arrived pumice at the land surface. The pre-existing soils (now paleosols) were sterilised by a combination of heating and anoxia. Where pumice deposits are less than 1 m thick, the profiles contain identifiable older tephra layers and buried paleosols that together form multi-layered soils (Sect. 12.3, Fig. 12.10). Figure 12.8 shows developmental and retardant upbuilding pedogenesis,

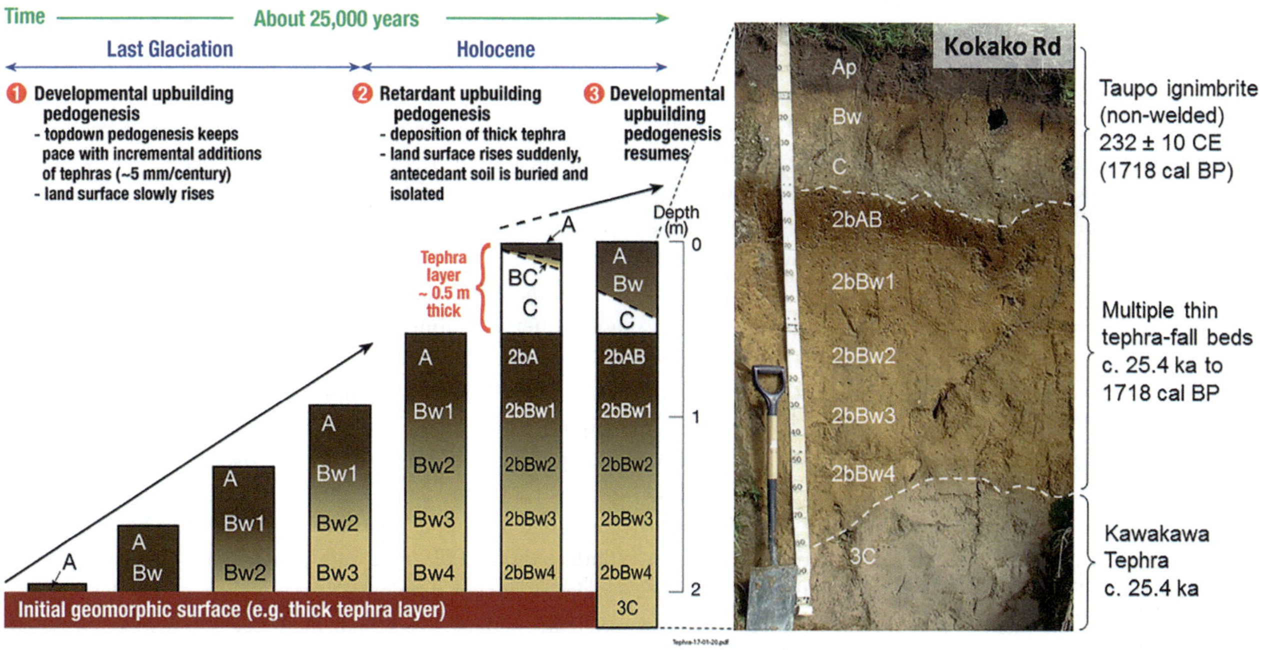

Fig. 12.8 Developmental and retardant upbuilding pedogenesis, and a resultant multi-layered Buried-allophanic Orthic Pumice Soil on Kokako Rd near Lichfield (cal BP = calendar years before 1950, ka = 1000 years ago) After McDaniel et al (2012)

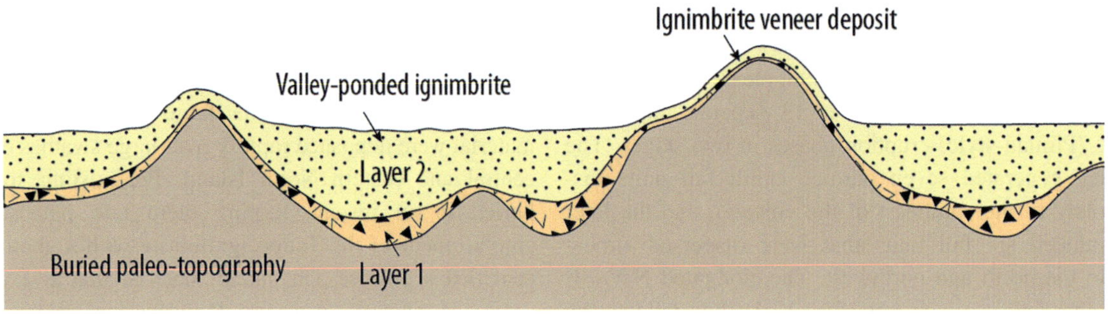

Fig. 12.9 Relationship between modern and past landscape and the two types of non-welded ignimbrite that were deposited by the pyroclastic flow emplaced at the climax of the Taupo eruption. The Taupo ignimbrite itself comprises two main units, layers 1 and 2 (Fig. 12.5). Pumice Soils formed on the valley-ponded deposits differ somewhat in their physical properties from those on the veneer deposits. After Wilson (1985)

and a resultant multi-layered Buried-allophanic Orthic Pumice Soil.

Soon after emplacement, the pumiceous deposits are subjected to the onset of weathering and soil-forming processes operating from the land surface downwards (topdown pedogenesis). The Pumice Soil develops from an initial Raw Soil, through Recent Soil, and eventually to Pumice Soil. Given further time, and depending on environmental conditions, the Pumice Soils may eventually develop into Allophanic, and ultimately Granular Soils. Some have already developed into Podzol Soils where elevation and rainfall are high (Chap. 11).

The pumiceous parent material, like the typically finer grained rhyolitic and andesitic tephra deposits in which Allophanic Soils are formed, is dominated by volcanic glass (Sect. 12.4). Hence in most cases, the processes that form the nanocrystalline clays, allophane and ferrihydrite, are followed (Chap. 2). Gradually, vegetation becomes established on the new materials, starting with tough pioneer species of lichens and plants such as bracken that are tolerant of low-nutrient, low-moisture environments. The pioneer plants stabilise the Raw Soil surface and plant materials break down to provide organic matter, especially nitrogen, that supports growth of a wider range of plants. Over time the plant community strengthens and organic matter starts to accumulate in the soil, leading to the development of an A horizon (the process of melanisation) and thus formation of a Recent Soil.

Chemical and physical weathering processes are aided by weakly acidified water leaching through the soil. Carbon dioxide, from the atmosphere and from soil microbial respiration, dissolves in water to form carbonic acid (H_2CO_3) which, along with organic acids from plants, acidifies the soil water. The weakly acidic soil solution weathers volcanic glass (as well as mafic and felsic minerals, such as feldspar) in the tephra by hydrolysis. Hydrolysis occurs because hydrogen ions (H^+) are continuously released from the acids, and attack the easily weathered glass fragments and mineral grains to release silicon, aluminium, and various cations into the soil solution. The silicon solutes form silicic acid (H_4SiO_4) and the aluminium forms an Al-rich gibbsite-like product, and these precipitate together as allophane and imogolite. Ferrihydrite is formed in small amounts as a nanocrystalline, iron-rich, oxide clay equivalent to allophane.

Clay formation is rapid in Pumice Soils, because volcanic glass is thermodynamically unstable and vesicular pumice fragments have a high surface area and are porous. The relatively warm, humid, climate in central North Island also supports relatively rapid chemical weathering processes.

A weathered-B horizon (Bw) will, therefore, form relatively quickly with the small amounts of clays coating the particle surfaces and imparting yellowish-brown colours (a process known as brunification). To qualify as a Pumice Soil, a Bw horizon at least 5 cm thick must have formed. That such a soil horizon can form in barely 700 years (the age of the Kaharoa tephra deposits) attests to the rapidity of pedogenesis in the pumice materials in the North Island environment.

On the 1800-year-old Taupo deposits, profile morphologies differ according to increasing elevation and rainfall, and with change in forest-type cover, so that leaching increases markedly. Wim Rijkse and Brian Daly showed that on a sequence of soils on Taupo pumice materials, the B horizons changed from a thin, yellowish-brown (10YR) Bw horizon under 1200 mm annual rainfall (at 150–450 m elevation) though to a thick, reddish-brown (7.5YR to 5YR) Bs horizon under 1500–2000 mm rainfall (at >550 m elevation and under podocarp-dominated forest, such as on the Mamaku Plateau—Chap. 11). Thus, the three soils, identical in age and parent material, have developed into an Orthic Pumice Soil, a Podzolic Pumice Soil, and an Orthic Podzol Soil as leaching and podzolisation increased.

Dating the Taupo and Kaharoa eruptions

Taupo

The Taupo ignimbrite is full of charcoal (Fig. 12.6) and so was an obvious target when radiocarbon (^{14}C) dating was first undertaken in New Zealand. NZ-1 (New Zealand's first carbon date published in 1953) gave a date of 1820 ± 150 ^{14}C yr BP (68% probability). Since then, dozens of samples from Taupo eruptives have been dated but various problems, including in-built age and stratigraphic considerations, and imprecision in calibrating ages from ^{14}C years to calendar years, meant that the date of the eruption could have been at any time between about AD 130 and AD 320. A solution to deriving a much more precise date arose from a quirk of the eruption itself. During its emplacement, the Taupo ignimbrite flattened a remnant of forest at Pureora (Fig. 12.4) which was then perfectly preserved, as a buried forest, by peat development because of a fortuitous alteration in drainage. After developing a high-precision kauri (*Agathis australis*)-based ^{14}C calibration curve with the assistance of dendrochronologists, Alan Hogg (Waikato Radiocarbon Dating Laboratory) and colleagues used a large tanekaha (celery pine, *Phyllocladus trichomanoides*) log from Pureora, and a technique called ^{14}C wiggle-matching, to derive a new calendar date of AD 232 ± 10 (95% probability). That date put paid to various dates obtained previously including a date based on ancient Roman and Chinese records as well as erroneous ice-core-based dates. The Taupo eruption took place in late summer to early autumn (typically late March to early April) on the basis of fruit and seeds preserved at Pureora and the lack of an outer latewood ring on preserved logs. Preserved insect assemblages suggested the ignimbrite was deposited in the late afternoon.

Kaharoa

As was the case for Taupo, ^{14}C dates obtained for the Kaharoa eruption were not able to be calibrated with much precision. However, after a fruitless search for suitable wood on Mt Tarawera, a partially carbonised tanekaha log was discovered in Kaharoa pyroclastic flow and surge deposits at the foot of the mountain. With the help of dendrochronologist Jonathan Palmer, Alan Hogg used the ^{14}C wiggle-matching method to obtain a date of AD 1314 ± 12 (95% probability). The outermost ring of the log included both early and late wood, meaning the tree was killed in the period between growth cessation—because of the onset of winter—but before the start of spring growth. The exact month of the eruption is uncertain, but contemporary tree-ring studies of the same species indicate that the wintery period from May to September is likely.

12.3 Soil-Landscape Relationships

12.3.1 Forming and Modifying Landscapes and Changes in Pumice Soils Over Distance

The deposition of tephra-fall beds from the Taupo and Kaharoa eruptions, and the emplacement of the Taupo ignimbrite, resulted in sudden and marked changes to the landscape. The fall beds draped the pre-existing land, specially making it smoother (a landscape-modifying event). The emplacement of the Taupo ignimbrite, in contrast, was mainly a landscape-forming event in which pre-existing valleys became plains upon infilling. The deposition of the ignimbrite veneer deposits on high points in the landscape mimicked the effect of the fall beds in generally rounding the landscape contours (Fig. 12.9).

The soils on the valley-ponded ignimbrite deposits are typically much more compact and dense (Impeded Pumice Soils) in contrast with those on the 'soft' and loose ignimbrite veneer deposits (Orthic Pumice Soils).

The patterns of deposition of fall and flow deposits, and the inherent properties of the deposits, especially layering and thickness, have led to differences in the Pumice Soils depending on their location. Despite an initial period of landscape instability and reworking the young Pumice Soils were generally well preserved by vegetation cover once forests became established. Hence at sites relatively close to Lake Taupo, especially immediately east where the thickest deposits occur, the soils are formed entirely in pyroclastic fall and flow deposits >2 m deep (e.g. Fig. 12.5). Beyond about 50 km, the Taupo eruptives make up profiles of about 1 m or less in thickness. Near Mt Tarawera, the proximal Kaharoa fall beds about 8 km southeast of the mountain are nearly 2 m thick (Fig. 12.7) but the thickness drops off markedly to only about 0.5 m or less within a few kilometres (Fig. 12.4).

At distal sites where the Taupo or Kaharoa eruptives are only about 0.5 m thick or less, the soils are multi-layered and complex, with Taupo or Kaharoa deposits forming just the upper part of the soil profile. Such soils occur on the margins of the distribution of Taupo ignimbrite, towards Taumarunui, Tokoroa, Waiouru, and Murupara, for example (Fig. 12.10).

12.3.2 Tephrostratigraphy and Paleopedology

In the many multi-layered profiles, such as that at Murupara (Fig. 12.10) the various tephra deposits are able to be identified using their positions in the sequence together with their physical and (if necessary) mineralogical and glass compositional properties (derived from laboratory analysis). Alan Pullar and Colin Vucetich identified and mapped tephra layers, mainly in the 1950s to 1970s, in the central North

Fig. 12.10 Distinct distal tephra layers exposed near Muraupara. Left: river terrace covered with a stack of tephra beds dating back to 14,000 years ago (just below the floor of the quarry) with buried (dark-coloured) paleosols between each tephra deposit. Kaingaroa Plateau forms the distant skyline. Right: closer view of the Immature Orthic Pumice Soil formed at the current soil surface. The top three layers are the Kaharoa Tephra AD 1314 ± 12; Taupo Tephra AD 232 ± 10; and Whakatane Tephra 5526 ± 145 calendar years old

Island and Bay of Plenty-Gisborne areas using field characteristics and stratigraphy.

In cases such as at Murupara, layers are usually readily identifiable (Fig. 12.10), with paleosols often preserved where the former soil profile or part thereof has been buried by materials from subsequent eruptions (an example of retardant upbuilding pedogenesis). Paleosols (buried soils) represent the passage of time between eruptions when there was an opportunity for soil formation and horizon development to take place on the deposit at the land surface where topdown pedogenic processes occur. Burial by the next thick tephra deposit effectively isolates that soil, turning it into a buried paleosol which, because it is now at depth, is no longer affected (or only weakly influenced) by surface-based pedogenic processes (e.g. Figs. 12.8 and 12.10). The properties of paleosols (including the analysis of the remains of plant cells (phytoliths), or the extraction of ancient DNA) can be used to infer the nature of the environment at the time the soil was exposed at the land surface.

12.3.3 Tephra Reworking

During the Taupo eruption, much of Lake Taupo was expelled, evaporated, or drained into a caldera-collapse structure beneath the current lake floor. Afterwards, the lake

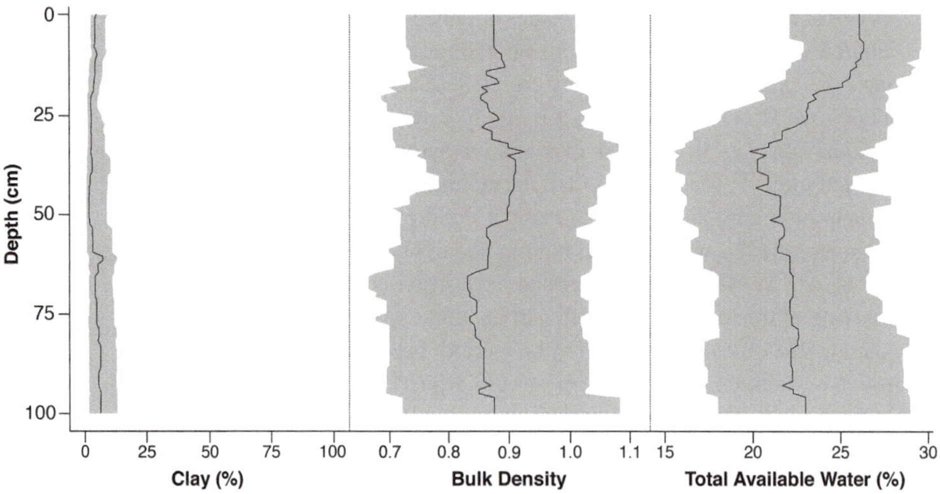

Fig. 12.11 Median and upper and lower quartiles of soil clay (%), soil dry bulk density (t m^{-3}), and total available water (%) for Pumice Soils in the New Zealand Soil Data Repository

refilled over about 15–20 years, reaching a height of 400 m asl, about 30–40 m above its present level (357 m asl) to form a terrace on which modern-day Taupo town is located. Catastrophic failure of a pumiceous dam led to the re-establishment of the Waikato River and the release of about 20 km^3 of water, the peak discharge being 20,000–40,000 m^3sec^{-1}, equivalent to the Mississippi River in flood. Vern Manville and others traced the break-out flood deposits (called Taupo Pumice Alluvium) about 220 km downstream of Lake Taupo. Parts of Hamilton city are built on Pumice Soils formed on Taupo Pumice Alluvium on low terraces alongside the Waikato River. Other North Island rivers, including the Whanganui, Rangitaiki, Mohaka, and Ngaruroro, were choked with fall deposits and Taupo ignimbrite materials. The immediate aftermath of the eruption would have seen erosion and reworking on a massive scale in the river catchments (especially where the vegetation had been entirely buried) involving rilling and debris flows, fan generation, and the development of overbank flood plains and terraces underlain by Taupo Pumice Alluvium. The process is still ongoing and it is still possible to find pumice, carried down by the rivers and washed up on beaches on both the east and west coasts of North Island.

12.4 Key Soil Properties

12.4.1 Soil Composition

Unconsolidated rhyolitic pumice is the dominant parent material of Pumice Soils. The soils may also contain minor amounts of 'country rock' (lithics) caught up in the eruption at the vent, or picked up during the ignimbrite emplacement as a pyroclastic flow, together with clasts of obsidian and small crystals of felsic (feldspar and quartz) and mafic (iron, magnesium, and titanium) minerals, and glass shards. Volcanic glass is an amorphous solid with poorly ordered internal structure (hence is a mineraloid) comprising loosely linked SiO_4 tetrahedra around which cations such as sodium occur in the intermolecular spaces.

Charcoal fragments are ubiquitous in the uppermost pumiceous ignimbrite layer throughout its distribution (Fig. 12.6). In outcrops near Taupo and in road cuttings along state highways between Turangi and Waiouru (e.g. along the 'Desert Road') carbonised trees that were flattened and burnt within the pyroclastic flow are abundant.

Clay minerals are dominantly allophane, with some ferrihydrite and imogolite, and occur as coatings around glass or pumice particles. However, halloysite has also been identified (by Peter McIntosh) 2 m below the land surface, attributed to synthesis from a soil solution enriched in silicic acid because of strong silicon migration through the highly siliceous pumice materials above. Determining that halloysite, as well as allophane, can form directly in such young materials was a notable advance in the understanding of the formation of clays from the products of the dissolution and hydrolysis of glass.

12.4.2 Physical Properties

Pumice Soils are sandy or gravelly soils dominated by pumice or pumice-derived sand and silt (ash) with a high content of volcanic glass. Many of the large pumice clasts are notably rounded because of transport in a pyroclastic flow. The uppermost layer in many profiles on the Taupo eruptives is typically fine-grained and pumiceous (Taupo ignimbrite), and it overlies coarser (pre-ignimbrite) fallout beds.

In the fall deposits, the size of pumice clasts generally decreases with distance from the eruption source while the proportion of sand and silt size particles typically increases with distance from source. However, pumice clasts in the pyroclastic flow deposits show a complex relationship of size with distance from source that depends on the specific origin of the deposit. Clasts in the ignimbrite veneer deposits (on ridge tops) decrease in size away from source but those in the valley-ponded deposits remain uniform or increase slightly in size away from source to about 40 km from vent until decreasing beyond 60 km from vent.

Clay contents are low, generally less than about 5% or so (Fig. 12.11, Table 12.1). Like the pumice from which it is derived, soil dry bulk densities are generally less than 1 g cm^{-3} but may be higher in massive or firm, tightly packed and compact horizons.

Pumice Soils are usually apedal earthy in the topsoil and single-grain beneath, although some soils may have extremely fine spheroidal structures. The Pumice Soils have weak to very weak soil strength, except in tightly packed layers which are massive and may have higher strength. Such compact materials, sometimes described as brittle, feature in some of the Impeded Pumice Soils or Perch-gley Pumice Soils. Pumice Soils are non-plastic and sensitive meaning that they have low soil strength when disturbed.

Very high macroporosity enables rapid profile drainage at low soil–water tensions and many Pumice Soils, despite lacking much clay, are capable of storing large amounts of water (high available water capacity) for plants. Partly, the very fine vesicularity of the glass and pumice fragments provides a higher (than expected for sand-sized grains) water storage capacity. However, some of the shallower pumice soils with pumice lapilli in the near-surface horizons drain very rapidly and readily become droughty during dry spells. Subsurface paleosols, with finer allophanic soil materials,

Table 12.1 Typical example of physical properties of an Immature Orthic Pumice Soil (Taupo series, SB09577)

Horizon	Depth (cm)	Sand (%)	Silt (%)	Clay (%)	Dry bulk density (t m^{-3})
Ap	0–9	46	50	3	0.75
Bw1	9–24	35	58	6	0.82
Bw2	24–34	63	33	3	0.90
BCu	34–64	99	0	0	0.94
Cu1	64–95	87	8	4	0.64
Cu2	95–102	52	45	1	1.30

are important sources of water and nutrients for plants (particularly pine trees) whose roots can penetrate through the coarse lapilli horizons to the underlying paleosols.

Pumice Soils generally have deep rooting potential, although plant roots may be limited in the Impeded Pumice Soils group where the root barrier is a massive, compact layer with high undisturbed soil strength. In Pumice Soils that have coarse lapilli horizons near the soil surface, the rapidly draining and drying lapilli can be a barrier for grass root penetration. There is often a proliferation of roots at the top of such horizons as the plants seek access to moisture.

Near Taupo, the lithic- and crystal-rich layers (the basal part of layer 1 of the Taupo ignimbrite) underlying the uppermost pumice-rich ignimbrite (layer 2, Fig. 12.5) are often loose. The looseness can cause problems with stability of built structures (e.g. power poles). In places, thin deposits of mainly andesitic tephra, from more recent volcanic eruptions, overlie the pumice deposits including the area around the Tongariro Volcanic Centre (notably visible in cuttings on the Desert Road, i.e. State Highway 1 and on State Highway 47).

12.4.3 Chemical Properties

The pumice, mineral, and lithic grains in Pumice Soils vary from unweathered to weakly weathered. Because the glassy rhyolitic pumices are very high in SiO_2 (c. 75–78 wt%), the deposits have *low reserves of* major nutrient elements and trace elements. Sulphur, potassium, nitrogen, phosphorus, and magnesium are usually required for agricultural or horticultural crops, and heavy initial fertiliser applications (especially for phosphorus) are essential. Reserve potassium is low and exchangeable magnesium is very low, particularly in subsoils. Trace elements including cobalt, copper, molybdenum, boron, iodine, and selenium are likely to be deficient. Phosphate-retention is medium and soil carbon is medium in topsoils and very low in the subsoil (Fig. 12.12, Table 12.2). Soil pH is moderately acid in the topsoil, slightly acid beneath and so lime applications rarely increase pasture growth. *Pinus radiata* forestry does increase acidity in topsoils over time (compared with soils under manuka scrub). Soil pH levels may be maintained by small annual lime applications to counter the acidification that occurs under pastoral management.

12.4.4 Biological Properties

Soil animal populations are low with most species concentrated in the topsoil. Generally, high porosity should provide a good range of soil habitats but seasonal drought, low soil organic matter contents, and low content of weathering products are significant limitations. The coarse nature of subsoil materials is a limitation for earthworm survival in Pumice Soils. Trace element deficiencies (including copper, molybdenum, boron, iodine, and selenium), left uncorrected, may strongly affect grazing animals and may also affect organisms in the soil. *P. radiata* has been successfully grown on Pumice Soils. Where the pine roots can reach into subsurface paleosols, such as that on the Whakatane tephra (Figs. 12.7 and 12.10), then the tree has access to a good store of water and nutrients which help it to thrive. On soils with a deep layer of Taupo pumice that prevent roots from reaching the underlying paleosols, deficiencies of a range of nutrients, most notably boron, may restrict pine productivity.

12.5 Distinguishing Between Pumice Soils and Related Soil Orders

Wim Rijkse characterised the Pumice Soils as an 'intermediate' state between Recent Soils in recently deposited tephra and Allophanic Soils in older tephra (strictly, the age and degree of weathering of the ~700-year-old Pumice Soils match those of many Recent Soils—the key difference, though, is their preponderance of glass-dominated pumice). Pumice Soils are easy to recognise due to the abundance of pumice clasts which distinguishes them from Allophanic Soils. Towards the outer margins of its depositional range, the thinning pumice typically overlies pre-existing allophanic soil materials in the profile (see Sect. 12.2). The resultant multi-layered soils are Buried-allophanic Orthic Pumice Soils if the depth of pumice from the land surface is

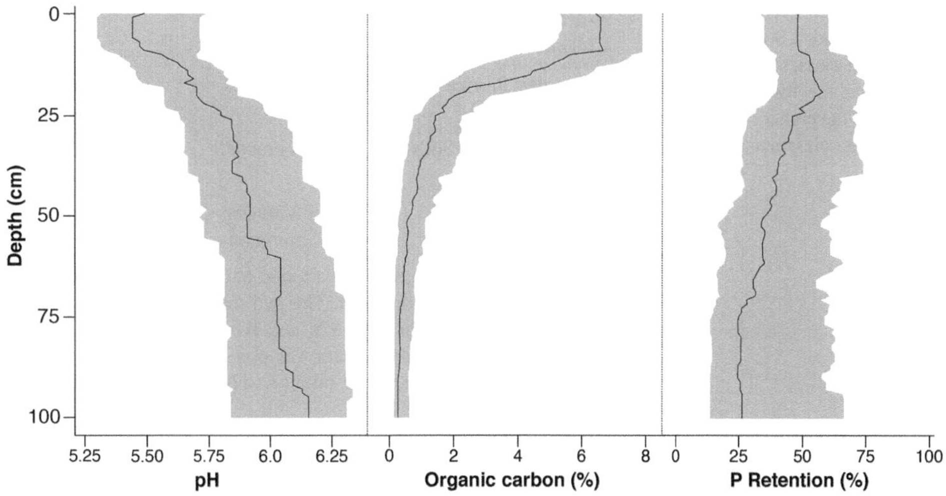

Fig. 12.12 Median and upper and lower quartiles of soil pH, organic carbon (%) and P retention (%) in Pumice Soils in the New Zealand Soil Data Repository

Table 12.2 Typical example of chemical properties of an Immature Orthic Pumice Soil (Taupo series, SB09577)

Horizon	Depth	pH (in H_2O)	Carbon (%)	Nitrogen (%)	CEC[a] (cmol$^{(+)}$ kg^{-1})	Sum bases (cmol$^{(+)}$ kg^{-1})	P retention (%)
Ap	0–9	5	5.3	0.35	17.3	4.19	44
Bw1	9–24	5.5	1.4	0.08	6.3	1.76	44
Bw2	24–34	5.9	0.4	0.03	3.5	1.70	29
BCu	34–64	6.4	0.3	0.01	4.6	4.11	5
Cu1	64–95	6.7	0.2	0.01	2.9	2.48	11
Cu2	95–102	6.3	0.1	0.01	3.7	3.05	11

[a]CEC = cation exchange capacity

at least 25 cm (or the pumice layer is ≥ 35 cm in thickness within the upper 60 cm of the profile).

Where pumice is not obvious, and allophanic soil material is not positively identified, then the glass (vitric material) content of the sand fraction needs to be estimated. The presence of a weathered Bw horizon at least 5 cm thick distinguishes a Pumice Soil from a Recent Soil. The Tephric Recent Soil group is distinguished from the Pumice Soils by specifically lacking pumice, instead comprising tephric soil material ≥ 30 cm thick in the upper profile. Fluvially redeposited pumice materials that immediately post-date the Taupo eruption occur extensively on terraces and other landforms along the Waikato, Rangitaiki, Waipunga, and Whanganui rivers. Where the soils formed in pumiceous deposits (Taupo Pumice Alluvium) lack the ≥ 5-cm-thick Bw horizon needed for Pumice Soils they are instead classed as Recent Soils.

Other soils, including Allophanic Soils and Brown Soils, may incorporate pumiceous tephra that was deposited in small amounts on the periphery of eruption fall-out or reworked as tephric loess or alluvium. In such soils, the amount of glassy pumiceous material is too low to dominate the soil properties.

12.6 Correlation with Other Classification Systems

Both *Soil Taxonomy* and the *World Reference Base* combine what in New Zealand are called Allophanic Soils and Pumice Soils, into one soil order—Andisols (in *Soil Taxonomy*) and Andosols (in *World Reference Base*). New Zealand has recognised the Pumice and Allophanic Soils as two separate soil orders because they are largely geographically distinct, have contrasting properties that require different management, and between them make up about 28% of the North Island. The Pumice Soils key out as 'vitric' in the international systems reflecting the high content of volcanic glass (Table 12.3).

12.7 Use and Management of Pumice Soils

12.7.1 Introduction

Pollen evidence shows that the forest was re-established within about 200 years after the ~232 AD Taupo eruption. However, deforestation began about 1280 AD, which is

Table 12.3 Correlation[a] between Pumice Soils and equivalent classes of *Soil Taxonomy*, the *World Reference Base*, and the earlier New Zealand genetic soil classification

New Zealand Soil Classification	*Soil Taxonomy*	*World Reference Base*	NZ genetic soil classification
Perch-gley Pumice Soils	Vitraquands	Gleyic Vitric Andosols	Yellow–brown pumice soils
Impeded Pumice Soils	Udivitrands	Vitric Andosols	Yellow–brown pumice soils
Orthic Pumice Soils	Udivitrands, Vitricryands	Vitric Andosol	Yellow-brown pumice soils[b]

[a]The correlations given here are a guide only and for accurate classifications the relevant soil classification documents should be consulted. The two major international soil classification systems are *Soil Taxonomy*, which was developed in the USA, and *World Reference Base*, which was developed primarily in Europe. The NZ genetic soil classification was used in NZ prior to 1992

[b]Also composite yellow-brown pumice soils on yellow-brown loams

attributed to rapid and continuing use of fire following the arrival of Polynesians. When Europeans arrived at the central Volcanic Plateau, the vegetation was predominantly fern-land, tussock grassland, and scrubland with a little forest. European settlers brought sheep and cattle to the region hoping they would prosper as had the sheep grazing farmers in the South Island tussock grasslands. Initial growth of clover was encouraging, but sheep and cattle grazing on the Pumice Soils suffered continuing ill-thrift or a 'wasting' livestock disease first reported in 1893. Agricultural chemist, Bernard Aston, described the condition in 1911. The stock became thin and sickly, suffered anaemia, and offspring born in the area could not be reared, and many animals (especially lambs) died. The illness, called 'bush sickness', was described by Aston as 'occurring in ruminants pastured on land which has been recently reclaimed from forest, or in open country … laid down in English pasture for over twenty years'. As well, the depredations of wild dogs, rampant weeds, and high transport costs further inhibited development. Thus, many farming operations on Pumice Soils were abandoned in the 1900s to 1920s and plantation forests were planted as trees proved to grow successfully, culminating in the establishment of the Kaingaroa Forest, the second-largest plantation forest in the Southern Hemisphere.

12.7.2 Chemical Limitations Including Cobalt Deficiency

The condition of 'bush sickness' in ruminants, so prevalent on the Pumice Soils, was also recognised elsewhere including in the King Country ('Mairoa dopiness'), the Nelson area ('Glenhope ailment'), and Southland ('Morton Mains disease'). However, the most extensive occurrences were on the Pumice Soils. Many possible causes were advanced. Pasture plants were tested for toxins, along with animal tissues, but with no identification of a clear cause. Meantime in the quest for a cure to bush sickness, thinking shifted onto possible deficiencies because the characteristic anaemia suggested iron deficiency, and a possible link with soil type was later considered as well.

Veterinarian C.J. Reakes (Director of Agriculture) suggested (in 1912) that bush sickness was caused by an iron deficiency. Leslie Grange in 1929 was the first to name and map soil-forming tephra deposits in the Rotorua area, and he suggested that some deposits related more to the incidence of bush sickness than others. A relationship between 'geology' (parent material) and bush sickness was also implied by Norman Taylor in 1930. Grange and Taylor (in 1932) confirmed the connection between the Pumice Soils (on Taupo and Kaharoa eruptives) and bush sickness. They noted that areas that had tephra inputs of iron-rich basaltic material from the 1886 Tarawera eruption were no longer prone to bush sickness.

An encouraging animal health response was found in 1931 by administering an iron oxide that has been known as 'limonite' (which mainly comprises the mineral goethite). It seemed the case was solved. For some time up until 1934, the research focussed on better ways of delivering iron to animals. But variation in the effectiveness of limonite mined from different sources suggested there may be some aspect of the ore, such as its solubility or grain size, or that a 'contaminating element' might be responsible for the curative effectiveness. In a study that got tantalisingly close by identifying trace elements including cobalt, but failed to recognise its importance, Dick Grimmett and Brian Shorland (1934) reported the analyses—by Fred Seelye–of four limonite ores (all derived from strongly weathered Oxidic Soils developed on basalt in Northland). They discovered trace amounts of cobalt in three of them along with copper, chromium, arsenic, antimony, and nickel. Various forms of the ores were trialled on bush-sick farms (all on Pumice Soils) at Atiamuri, Tokoroa, and Putaruru. In 1934 Grimmett and Shorland concluded that 'The differences in chemical composition between the various deposits would appear, however, totally inadequate to account for the variations observed in the feeding value of the ores'.

Meanwhile, Australian researchers, Eric Underwood and John Filmer, were working on the equivalent problem to bush sickness (known as 'enzootic marasmus') in Western Australia. The breakthrough came in June 1935 when from systematic trials, involving different groups of elements, Underwood and Filmer concluded that 'enzootic marasmus is due to a deficiency of cobalt in the food', and that 'cobalt must be regarded as an essential element in animal nutrition'. In South Australia, Ted Lines reported almost simultaneously (in May, 1935) that tiny quantities of cobalt had successfully treated the disease in sheep.

The South Australian success was reported in the journal *Nature* on 28 September 1935. It was picked up by Hugh Ramage in the UK, who had developed a new method for measuring trace elements including cobalt. In New Zealand, D.F.S. (Sandys) Wunsch (also known as Wunsh), who managed a dairy factory at Edendale in Southland, and local farmers, had begun trying to find a cure for Morton Mains disease. They used 'Whangarei' limonite (unsuccessfully) along with soils from Nelson and Bluff (successfully). In 1935, they obtained analyses from Ramage who found that the Nelson and Bluff soils both contained nickel and cobalt, but their local Southland soil did not. Wunsch and the farmers, armed with the South Australian findings and Ramage's analyses, set about testing cobalt's efficacy in 1935–36. Wunsch reported that 'the results obtained have been so spectacular that almost magical properties have been ascribed to it [cobalt]' and that it was 'difficult to believe… that traces of cobalt so minute' were so effective.

In 1948, vitamin B_{12} (cobalamin) was discovered and the crucial role of cobalt in its formation was identified in 1950. Vitamin B_{12} is essential in ruminants (and humans) for various functions including red blood cell production, the development and functioning of brain and nerve cells, and DNA synthesis. Thus cobalt deficiency may be more accurately described as a vitamin B_{12} deficiency.

Soil surveyors related the tephra layers to cobalt levels and so soil maps could be used to help identify areas of cobalt deficiency, providing confidence in farm development in the central North Island. On the pumice deposits (and similar high-silica materials), cobalt was evidently an *inherited* deficiency. The amount of cobalt in Pumice Soils is around 1 to 5 ppm, whereas similarly young soils on basaltic tephras (Rangitoto Tephra and Tarawera Scoria) have about 100 ppm. In other (non-pumice-mantled) areas that had displayed the symptoms of bush sickness, strong leaching had removed most cobalt that had been present, and thus the soils developed an *acquired* deficiency of cobalt. For example, an Allophanic Soil at Waihi (with annual rainfall $\sim$2000 mm yr^{-1}) contained only about 3 ppm of cobalt. Ken McNaught developed a new method in 1938 for assessing the cobalt content of pasture and found that herbage on Pumice Soils contained on average <0.04 ppm, whereas that from healthy areas averaged about 0.11 ppm.

Besides cobalt, the Pumice Soils are generally also deficient in boron, copper, molybdenum, iodine, and selenium. There are low reserves of nutrient elements, with sulphur, potassium, nitrogen, phosphorus, and magnesium usually required for agricultural or horticulture crops. Reserve potassium is low and exchangeable magnesium is very low, particularly in subsoils.

12.7.3 Establishment of Plantation Forestry Including *Pinus radiata*

Prior to the cause of bush sickness being established, with livestock farming unsuccessful, exotic plantation trees such as larch, Douglas fir, and eucalypts were planted to supplement the native-tree based forestry industry. *Pinus radiata*, the Monterey pine from California, was observed to grow well. Initial plantings were made from 1898 at Whakarewarewa (3250 ha, 17 million trees) and then near Waiotapu from 1901 (3000 ha). A Royal Commission on Forestry was established in 1913, and a proposal to plant trees on the Kaingaroa Plains was developed in 1915. World War I prevented any planting until 1923 when an extensive planting campaign of *P. radiata* ('radiata pine') on the Pumice Soils of the tussock-dominated Kaingaroa Plains began, with 42 500 ha planted by the late 1920s.

Government-funded planting continued into the depression years from 1928 to 1933, partly to create work for people desperate for employment. Smaller scale, but nevertheless extensive, pine forests were planted in the Tokoroa plateau area from 1923 by private companies, and by 1930 included 47 700 ha of *P. radiata* (11.8 million trees) and 230 660 ha of *P. ponderosa.*

The radiata-dominated pine forestry became a success story on the Pumice Soils with radiata able to be harvested at about 25 years old and so many areas are now into fourth and fifth rotations (crops) of trees. *Pinus radiata* is largely tolerant of low fertility soils (boron deficiency aside) and it has a long growing season. The low bulk densities of the pumice deposits permit tree roots to penetrate deeply (up to 3–6 m) into buried paleosols to access extra water and nutrients. Radiata pine now makes up about 92% of the plantation forests in central North Island, with Douglas fir comprising 4%.

12.7.4 Soil Erosion

Under native forest, the infiltration rate in Pumice Soils is very high and generally exceeds rainfall rates. This means that water added to the soil during rainfall all seeps into the soil with no surface runoff (and consequently few surface streams). However, topsoils under pasture have a lower content of large pores than topsoils under forest, making the soils more vulnerable to runoff, especially where vehicle or stock trampling has increased soil compaction, leading to potential for erosion and thus sediment and nutrient losses.

In the 1970s, Michael Selby examined runoff from Pumice Soils under pasture grass, un-grazed grass, and scrub. Climatic, soil, vegetation, and slope variables were evaluated. Surface water runoff, caused by intense rainfall, coupled with low soil moisture content, was found to be greater on developed pasture than on un-grazed grasslands or scrub. The runoff was exacerbated where soils had been compacted by animals or vehicles. In spite of the greater inherent erodibility of soil under scrub, due to higher runoff, the actual erosion was found to be greater on pasture lands. Factor analyses showed that the high porosity and low density of the pumice, the soil's degree of aggregation, and the soil organic matter content together accounted for 83% of the erosion variability.

One heavy rainstorm, which causes runoff on exposed soil, can result in large volumes of soil removal and rapid formation of steep-sided gullies. Gully erosion is attributable largely to the low density and easy entrainment of pumice particles. Thus the erosion of Pumice Soils can be rapid if vegetation is removed and the soil is exposed to running water in which pumice floats. Hence, it is recommended that vegetation cover, preferably using plants with strong, dense, root systems, is maintained to protect surface soils. If a pasture renewal programme is to be undertaken it is strongly recommended to use no tillage. Instead, direct drill to protect the soil from erosion. Topsoils are relatively resistant to compaction, but treading damage can occur under grazing with reduction of macroporosity and thus increased risk of water runoff and potential for erosion if vegetation cover is not strong. Vulnerable areas such as valley floors and steep valley sides can be protected as water-absorption areas (planted in trees for example) with animals and vehicles excluded to prevent soil compaction. Extra care has to be taken when managing construction sites or road margins to prevent channel development and runoff water from causing severe damage.

12.7.5 Developments on Pumice Soils in the Twenty-First Century

Land use change on Pumice Soils has occurred at pace since the late 1990s through intensification of current pastoral systems (e.g. drystock to dairying) and the conversion of established plantation forests to pasture, most commonly for dairying. Conversion was driven by the perceived better economic returns from farming compared with forestry. Large-scale conversion of established plantation forest to pasture occurred, primarily in the Waikato River catchment, between Taupo and Lake Karapiro.

The main environmental impact is from the increased losses of nitrogen and, to a lesser extent, phosphorus, into ground and surface waters. At the same time, the change in land use from plantation forest to dairy pasture on Pumice Soils has resulted in a moderate increase or no change in soil storage of carbon. Regional council staff and others have been working with farmers, iwi, and other stakeholders to develop new ways of managing farms to mitigate environmental degradation.

Treated wastewater from Taupo town is successfully irrigated onto Pumice Soils. The phosphate (or anion) retention capacity assists in preventing leaching of phosphates and other anions from effluents. In well-drained Pumice Soils, nitrate may readily leach from the soil profile. However, Glen Trewick and Erin Telfer showed that at the Taupo town wastewater irrigation site, pasture harvest and export removed at least 80% of applied nitrogen from the site with pasture production two to three times higher under effluent irrigation than unirrigated controls. The mean concentration of nitrate in water reaching the subsoil was 1.3 g N m^{-3}, considerably below the Ministry of Health guideline limit of 11.3 g N m^{-3}. Perch-gley Pumice Soils will have seasonally anaerobic conditions which support denitrification and thus further reduce risk of nitrate leaching to groundwater.

Further Reading

Anon (1935) Cobalt and 'coast disease' of sheep in Australia. Nature 135:518

Askew HO, Dixon JK (1937) Influence of cobalt top-dressing on the cobalt status of pasture plants. New Zealand J Sci Technol 18:688–693

Aston BC (1911) The chemistry of bush sickness. Trans Proc New Zealand Inst 44:288–298

Aston BC (1924) The chemistry of bush sickness, or iron starvation, in ruminants. Trans Proc New Zealand Inst 55:720–723

Barker SJ, Wilson CJN, Illsley-Kemp F et al (2021) Taupō: an overview of New Zealand's youngest supervolcano. NZ J Geol Geophys. https://doi.org/10.1080/00288306.2020.1792515

Brink NG, Kuehl FA Jr, Folkers K (1950) Vitamin B_{12}: the identification of vitamin B_{12} as a cyano-cobalt coordination complex. Science 112:354

Dixon JK (1936) Investigation of a sheep ailment in Southland. New Zealand J Sci Technol 17:600–619

Dixon JK (1936–37) Mineral content of pastures. Report on Morton Mains investigations. New Zealand Department of Scientific and Industrial Research Annual Report 11:47–49

Eden DN (1992) A standard method for determining volcanic glass content in Andisols. New Zealand Department of Scientific and Industrial Research Land Resources Scientific Report 2

Filmer JR, Underwood EJ (1934) Enzootic marasmus. Treatment with limonite fractions. Aust Vet J 10:83–87

Froggatt PC, Wilson CJN, Walker GPL (1981) Orientation of logs in the Taupo Ignimbrite as an indicator of flow direction and vent position. Geology 9:109–111

Gibbs HS (1968) Volcanic-ash soils in New Zealand. New Zealand Department of Scientific and Industrial Research Information Series 65, 39 p

Gibbs HS, Cowie JD, Pullar WA (1968) Soils of North Island. In: "Soils of New Zealand Part 1". New Zealand Soil Bureau Bull 26 (1):48–67

Grange LI (1929) A classification of soils of Rotorua county. New Zealand J Sci Technol 11:219–228

Grange LI (1931) Volcanic ash showers. New Zealand J Sci Techno 12:228–240

Grange LI, Taylor NH (1932) The occurrence of bush sickness on the volcanic soils of the North Island. A. The distribution and field characteristics of bush-sick soils. New Zealand Department of Scientific and Industrial Research Bulletin 32:21–35

Grimmett RER, Shorland FB (1934) Some characteristics of "limonite" used in the cure and prevention of bush sickness. Trans Roy Soc New Zealand 64:191–213

Grimmett RER, Shorland FB (1935) Use of limonite in bush sickness—influence of cobalt content. New Zealand J Agric 50:367

Hewitt AE (2010) New Zealand Soil Classification, 3rd edn., Landcare Research Science Series No.1, Manaaki Whenua Press, Lincoln, New Zealand, 136 p

Hogg AG, Lowe DJ, Palmer JG et al (2012) Revised calendar date for the Taupo eruption derived by ^{14}C wiggle-matching using a New Zealand kauri ^{14}C calibration data set. The Holocene 22:439–449

Hogg AG, Wilson CJN, Lowe DJ et al (2019) Wiggle-match radiocarbon dating of the Taupo eruption. Nat Commun 10:4669

Kidson EB (1937) Cobalt status of New Zealand soils. New Zealand J Sci Technol 18:694–707

Lines EW (1935) The effect of the ingestion of minute quantities of cobalt by sheep affected with "Coast disease": a preliminary note. J Counc Sci Ind Res (Australia) 8:117–119

Lowe DJ, de Lange WP (2000) Volcano-meteorological tsunamis, the c. AD 200 Taupo eruption (New Zealand) and the possibility of a global tsunami. The Holocene 10:401–407

Lowe DJ, Green BE (1992) A hydrofluoric acid dissolution method for determining volcanic glass content of tephra-derived soils (Andisols). Aust J Soil Res 30:573–581

Lowe DJ, Pittari A (2014) An ashy septingentenarian: the Kaharoa tephra turns 700 (with notes on its volcanological, archaeological, and historical importance). Geosci Soc New Zealand Newsl 13:35–46

Lowe DJ, Alloway BV, Shane PAR (2015) Far-flown markers. In: Graham I (ed) A continent on the move: New Zealand geoscience revealed, 2nd edn. Geoscience Society of New Zealand with GNS Science, Wellington. GSNZ Miscellaneous Publication 141, pp 172–175

Lowe DJ, Pittari A (2021) The Taupo eruption sequence of AD 232 +/- 10 in Aotearoa New Zealand - a retrospection. Journal of Geography (in press)

Manville V, White JDL, Houghton BF et al (1999) Paleohydrology and sedimentology of a post-1.8 ka breakout flood from intracaldera Lake Taupo, North Island, New Zealand. Geol Soc Am Bull 111:1435–1447

Manville V, Newton EH, White JDL (2005) Fluvial responses to volcanism: resedimentation of the 1800a Taupo ignimbrite eruption in the Rangitaiki River catchment, North Island, New Zealand. Geomorphology 65:49–70

Marston HR (1935) Problems associated with "Coast disease" in South Australia. J Aust Counc Sci Ind Res J (Australia) 8:111–116

Maslin MJ (1998) 'Wunsch, Donald Frederick Sandys', Dictionary of New Zealand Biography, first published in 1998. Te Ara—The Encyclopaedia of New Zealand. https://teara.govt.nz/en/biographies/4w28/wunsch-donald-frederick-sandys. Accessed 27 Apr 2020

McDaniel PA, Lowe DJ et al (2012) Andisols. In: Huang PM, Li Y, Sumner ME (eds) Handbook of soil sciences, 2nd edn., vol 1: Properties and processes. CRC Press, Boca Raton, FL, pp 33.29–33.48

McIntosh PD (1980) Weathering products in Vitrandept profiles under pine and manuka, New Zealand. Geoderma 24:225–239

McIntosh PD (1980) Soil changes under radiata pine in Kaingaroa Forest, central North Island, New Zealand. NZ J Sci 23:83–92

McNaught KJ (1937) Cobalt content of limonites used in the treatment of bush sickness. New Zealand J Sci Technol 18:655–661

McNaught KJ (1938) The cobalt content of North Island pastures. New Zealand J Sci Technol 20:14–30

Milne JDG, Clayden B, Singleton PL et al (1995) Soil description handbook. Revised Edition. Manaaki Whenua Press, Lincoln, New Zealand. ISBN 0-478-04549-2. 157p

Ministry for the Environment and Stats NZ 2019. New Zealand's Environmental Reporting Series: Environment Aotearoa 2019. Available from www.mfe.govt.nz and www.stats.govt.nz

Neall WM, Ahmann CF (1937) Cobalt as an essential element in animal nutrition. Science 86(2227):225–226

Palmer JG, Ogden J, Patel RN (1988) A 426-year floating tree-ring chronology from *Phyllocladus trichomanoides* buried by the Taupo eruption at Pureora, central North Island, New Zealand. J Roy Soc New Zealand 18:407–415

Perry GLW, Wilmshurst JM, McGlone MS et al (2012) Explaining fire-driven landscape transformation during the initial burning period of New Zealand's prehistory. Glob Change Biol 18:1609–1621

Ramage H (1936) Biological distribution of metals. Nature 137:67

Ramage H (1936) Analysis of tissues for metallic content. Nature 138:762–763

Read NE (ed) (1974) Soil groups of New Zealand. Part 1: Yellow-brown pumice soils. New Zealand Society of Soil Science, Wellington. 251 p

Reakes CJ (1912) Bush sickness. New Zealand J Agric 5:118–120

Rijkse WC (1974) Definition: boundaries between yellow-brown pumice soils and related or associated soils, p. 22–27; Ranges and limits of morphological properties, p. 37–40; Diagnostic criteria and classification of mapping units, p. 40–47. In Read NE (ed), Soil groups of New Zealand. Part 1: Yellow-brown pumice soils. New Zealand Society of Soil Science, Wellington

Rijkse WC, Vucetich CG (1980) Soils of Wairakei Research Station, North Island. New Zealand, New Zealand Soil Survey Report, 57 p

Roberts A (2014) Soil fertility on Pumice Soils. In: Lowe DJ, Balks MR, Laubscher N (eds) Guidebook for field trip "Hot volcanic soils", New Zealand Society of Soil Science conference, Hamilton (1–4 December), pp 25–27

Sahetapy-Engel S, Self S, Carey RJ et al (2014) Deposition and generation of multiple widespread fall units from the c. AD 1314 Kaharoa rhyolitic eruption, Tarawera, New Zealand. Bulletin of Volcanology 76, article 836

Selby MJ (1972) The relationships between land use and erosion in the central North Island, New Zealand. J Hydrol (NZ) 11:73–87

Selby MJ, Hosking PJ (1973) The erodibility of pumice soils of the North Island, New Zealand. J Hydrol (NZ) 12:32–56

Soil Survey Staff (2014) Keys to Soil Taxonomy twelfth edition. USDA Natural Resources Conservation Service. 360p. https://www.nrcs.usda.gov/wps/PA_NRCSConsumption/download?cid=stelprdb1252094&ext=pdf. Accessed 4 March 2020

Sparling GP, Lewis R, Schipper LA et al (2014) Changes in soil total C and N contents at three chronosequences after conversion from plantation pine forest to dairy pasture on a New Zealand Pumice Soil. Soil Res 52:38–45

Taylor NH (1930) The relation of geology to sheep sickness in Mairoa district. New Zealand J Sci Technol 12:1–10
Treweek G, Balks MR, Telfer E (2014) Nitrogen leaching from effluent irrigation: Taupo District Council View Road effluent treatment site. In: Lowe DJ, Balks MR, Laubscher N (eds) Guidebook for field trip "Hot volcanic soils", New Zealand Society of Soil Science conference, Hamilton (1–4 December), pp 57–59
Underwood EJ, Filmer JF (1935) The determination of the biologically potent element (cobalt) in limonite. The Aust Vet J 11:84–92
Walker GPL (1980) The Taupo plinian pumice: product of the most powerful known (ultraplinian) eruption? J Volcanol Geoth Res 8:69–94
Walker GPL (1981) The Waimihia and Hatepe plinian deposits from the rhyolitic Taupo Volcanic Centre. NZ J Geol Geophys 24:305–324
Walker GPL, Wilson CJN (1983) Lateral variations in the Taupo ignimbrite. J Volcanol Geoth Res 18:117–133
Wall EM (1937) Trace elements in relation to bush sickness. New Zealand J Sci Technol 18:642–650
Wells N (1968) Element composition of soils and plants. In: "Soils of New Zealand Part 2". New Zealand Soil Bureau Bull 26(2):115–131
Wilson CJN (1985) The Taupo eruption, New Zealand II. The Taupo ignimbrite. Philos Trans Roy Soc, Lond A314:229–310
Wilson CJN (1993) Stratigraphy, chronology, styles, and dynamics of late Quaternary eruptions from Taupo volcano, New Zealand. Philos Trans Roy Soc, Lond A343:205–306
Wilson CJN, Walker GPL (1981) Violence in pyroclastic flow eruptions. In: Self S, Sparks RSJ (eds) Tephra Studies. Reidel, Dordrecht, pp 441–448
Wilson CJN, Walker GPL (1985) The Taupo eruption, New Zealand. I. General aspects. Philos Trans Roy Soc, Lond A314:199–228
World Reference Base (2015) https://www.fao.org/3/i3794en/I3794en.pdf. Accessed 2 March 2020
Wuns[c]h DFS, (1937) Tracking down a deficiency disease. Chem Ind 56:855–859
Wunsch DFS (1939) Trace elements in livestock diseases. Chem Ind 58:531–533

Raw Soils

13

Youngest of young, barely soil at all
where lavas flow or tephras fall,
mudflats, sand-dunes, or a landslide scar
That's where you'll find the Raw Soils are.

Abstract

Raw Soils have minimal evidence of soil development, usually because of a short time since the parent materials were deposited or exposed at the land surface. The Raw Soils are associated with some of New Zealand's most spectacular scenic landscapes. Raw Soils are scattered throughout New Zealand, particularly in association with high mountains (eroding alpine rock areas and active screes), braided rivers, beaches and tidal estuaries, non-stabilised sand dunes, recently emplaced lavas or tephras, and active geothermal areas. They cover about 3% of New Zealand (about 700,000 ha). Raw Soils lack a distinct topsoil or the topsoil is very thin (<5 cm) and they have no B horizon. The soil profile properties, therefore, correspond largely with those of the parent materials.

13.1 Important Features of Raw Soils

13.1.1 Concept and Key Features of the Soil Order

Raw Soils are the youngest of soils with minimal evidence of soil development, usually due to a short time since the parent materials were deposited or exposed at the land surface. Raw Soils lack a distinct topsoil or the topsoil is very thin (less than 5 cm) and there are no B horizons formed as a result of pedogenic (soil forming) processes. Some Raw Soils are fluid at a shallow depth (e.g. on tidal mudflats). Raw Soils occur in environments where the development of topsoils (and other horizons) is prevented by rockiness, active erosion, or active deposition including beach sands and active sand dunes, actively eroding alpine rock areas, active screes, lagoons, tidal estuaries, braided river beds, areas of recently deposited tephra or alluvium (flood deposits), landslide erosion scars or runout deposits, and active hydrothermal areas.

Raw Soils were introduced as a soil order in New Zealand in 1992 when the *New Zealand Soil Classification* was first published, mainly because they are regarded as soil by engineers. Some Raw Soils also have potential to be protected from flooding or erosion and undergo development for agriculture and so it is important to recognise and understand their limitations and potential (Fig. 13.1). However, the occurrence of Raw Soils indicates a dynamic environment that is likely to present hazards to most land uses.

The properties of Raw Soils generally coincide with those of the soil parent material and so may vary widely from loose unconsolidated sand to hard in-situ bedrock. Thus there is no one 'typical' Raw Soil. While Raw Soils generally do host some life, and may have small amounts of accumulated soil organic matter, many lack the soil organic matter needed to store and supply nitrogen, phosphorus, sulphur, and other nutrients required to support a substantial plant and animal biomass. Due to short time frames and low nutrient status of most Raw Soils, the natural vegetation cover, if present, is generally sparse, consisting of colonising species including ephemeral herbaceous plants, mosses, and

A. E. Hewitt et al., *The Soils of Aotearoa New Zealand*, World Soils Book Series,
https://doi.org/10.1007/978-3-030-64763-6_13

Fig. 13.1 A Fluvial Raw Soil formed in alluvium deposited by Cyclone Bola on the Waipaoa River flood plain in April 1988, photographed 25 years later. The site has been protected by a new stop bank and the fertile silty alluvium has been developed into a vineyard. The soil profile is about 1 m deep. Inset is a nearby site with a vineyard buried in mud deposited by Cyclone Bola photographed in August 1988

lichens. However, human intervention, including addition of fertilisers and plant establishment, may lead to rapid revegetation on some Raw Soils (eg. Fig. 13.1).

13.1.2 Areas of Occurrence

The Raw Soils are associated with some of New Zealand's most spectacular and scenic landscapes. Raw Soils are scattered throughout New Zealand, particularly in association with high mountains (alpine rock areas and active screes), braided rivers, beaches, very recent lava flows, tephra deposits (including scoria), or flood deposits, recently active landslides, and tidal estuaries. They cover about 3% of New Zealand (about 700,000 ha). The most extensive areas of Raw Soils are in the high alpine areas of the Southern Alps (Fig. 13.2). Raw Soils are also found in the central North Island on recently emplaced or exposed tephras and lavas on Mounts Ruapehu, Ngauruhoe, and Tongariro, and on Taranaki Maunga. Some of the exposed and wind-eroded sandy areas of relatively sparse vegetation, formed by frequent additions of tephra adjacent to the Desert Road (State Highway 1), qualify as Raw Soils where topsoils are lacking and where any buried A horizons are below 60 cm depth. Rocky Raw Soils occur on ~700-year-old lavas on Rangitoto Island and on the most recent lavas (or erosion-exposed lavas) on Mounts Tongariro, Ngauruhoe, and Ruapehu.

Small areas of Raw Soils are common throughout New Zealand, such as the beach sands and dunes along Ninety Mile Beach in Northland and Farewell Spit in the Nelson area. However, many areas are too small to show on a national-extent map.

13.1.3 Variation Within the Raw Soil Order

The following soil groups are defined in the New Zealand Soil Classification (Fig. 13.3):

Gley Raw Soils—periodically or permanently flooded,
Hydrothermal Raw Soils—in active hydrothermal areas affected by hot water,
Rocky Raw Soils—on rock outcrops subject to erosion or on recently emplaced, unmodified lavas. The soil volume available to roots is severely restricted. The soil may only consist of lichen adhered to a rock surface by fungal hyphae,
Sandy Raw Soils—in areas of active sand erosion, transportation, and deposition, most commonly as dune sands,

Fig. 13.2 Map of distribution of Raw Soils

Fluvial Raw Soils—in sediments deposited by flowing water, on land surfaces with a high risk of flooding,
Tephric Raw Soils—in unconsolidated deposits of volcanic origin including ash, scoria, lapilli, pumice, and other pyroclastic materials, and reworked tephric materials including lahar and lake deposits,
Orthic Raw Soils—on eroded land including landslide scars, on rock outcrops, or on redeposited products of erosion such as landslide runout deposits.

The Gley Raw Soil group is the only Raw Soil group to have subgroups. Acidic, Sulphidic, Fluid, Saline, Sandy, and

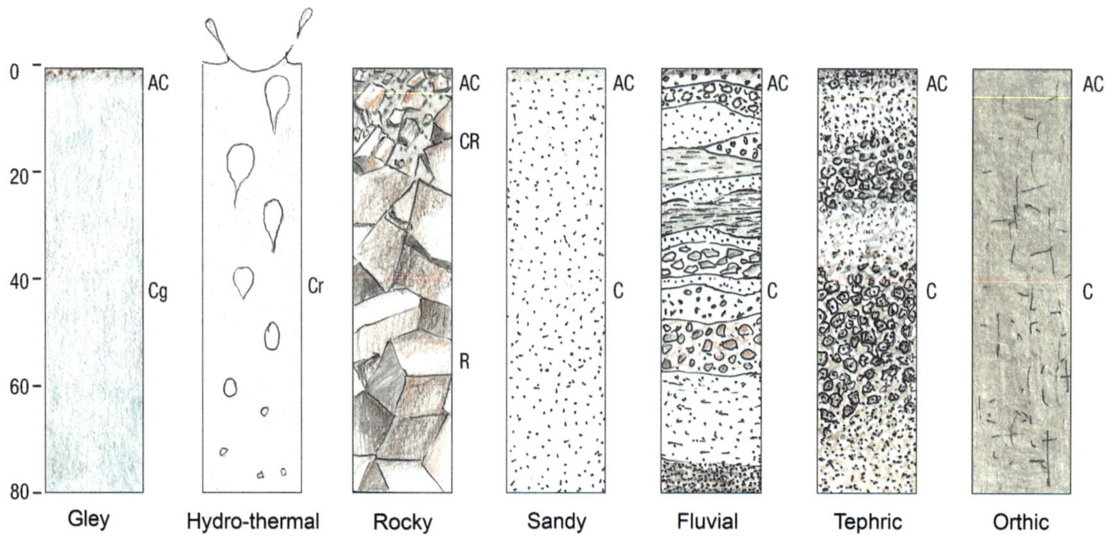

Fig. 13.3 Soil groups within the Raw Soils order. The vertical axis is depth (cm)

Typic, Gley Raw Soils are recognised. No subgroups are currently recognised for the other Raw Soil groups in the *New Zealand Soil Classification*.

13.1.4 Origin of Soil Order Name

Prior to the *New Zealand Soil Classification*, Raw Soils were either not recognised or included with Recent Soils. The 'Raw Soils' name was selected to clearly express the raw and unformed nature of the soils.

13.2 Soil Profile Genesis

Raw Soils are the result of natural processes that expose 'new' materials at the land surface where weathering processes and soil formation may commence. The exposed earth materials are diverse, ranging from hard rock, exposed by erosion, to soft mud in a geothermal hot pool. Because of the diversity of soil parent materials, Raw Soils are divided into geologically based soil groups, related to rock types or deposits, landforms, tectonics, and geomorphic landscape evolution. Thus the soil profile genesis (such as it is) is different for each of the groups within the Raw Soil order. The names and brief outlines of the Raw Soil groups (Sect. 13.1.3) give a clear description of the major materials and methods of emplacement, and hence will not be discussed further here.

Raw Soils provide a fascinating opportunity to study the initiation of soil formation, and the interplay of incipient soil processes in dynamic landscapes. Many Raw Soils may be considered to have an on-again/off-again lifestyle that relates in part to broader themes of 'classical' topdown pedogenesis (on a stable site) versus upbuilding pedogenesis (where new material is added to the land surface at varying rates). We can recognise progressive (on-again) stable phases when the soil is forming through topdown processes, generating soil horizons, and regressive (off-again) phases when the soil is disrupted, either by the removal of soil by erosion (as may occur on steep slopes during a rainstorm or earthquake) or by burial. The associated ecosystems are dynamic as pioneering flora and fauna become established, or die out, in response to changing conditions.

The basaltic lavas and scoria of Rangitoto Island, erupted 600–700 years ago, support the largest single tract of pohutukawa (*Metrosideros excelsa*) forest in New Zealand. Pohutukawa have played a critical role in colonising bare lava as described particularly in studies by Bruce Clarkson and Peter de Lange. As well as pohutukawa, lichens, mosses, and liverworts on the lavas play an important role in converting lava to 'soil' thereby allowing other plants to become established. Pohutukawa facilitate the establishment of a succession of plant species on the lava by influencing the microclimate and light regimes.

Some Raw Soils will 'never grow old' because the geological or climatic event that established the original Raw Soil is likely to recur and inundate or destroy the pioneer soil, unless something changes to enable its preservation. For example, the dynamics of shifting sediment in a braided river may rapidly change when the river channels change course. At some point earlier profiles will be either eroded or buried, re-starting the soil formation process. However, when the river has a major change of course, permanently abandoning its previous channel, some Raw Soils may be able to progress to become a Recent Soil because the profile is on a stable site. Thus, once soil weathering and profile formation become more advanced the Raw Soils develop and transform into soils of the other soil orders.

A number of key stages of soil initiation from a newly exposed substrate, or emplaced deposit, have been recognised, supported by a range of physical and chemical weathering processes.

1. *Surface disintegration* of rock caused by a range of processes including stress release due to removal of overlying materials and thermal expansion and contraction through successive cycles of heat or cold. The effectiveness of disintegration may be strongly controlled by the nature of the rock such as pre-weathering, pre-existing discontinuities, jointing, or rock fabric. Unconsolidated material, such as alluvial or pyroclastic deposits, have a 'head start' over hard, continuous rock because they are fragmental and so have greater surface areas at the outset. The mineralogical composition of the rock or deposit also impacts the rate of disintegration as some minerals are more susceptible to chemical weathering than others.
2. *Penetration of air and water* into the earth materials provides for chemical weathering processes, including oxidation and reduction, hydration, hydrolysis, and dissolution of minerals and the formation of clay materials, to commence. Water and air also provide sustenance for colonising roots and wide-ranging life forms.
3. *Establishment of life* is facilitated by accumulation of seed, spores, dust, and microbes which may be dispersed by wind, birds, animals, water, and lateral root spread. Invasion of roots, and addition of leaf litter, from trees growing nearby, introduce not only plant roots but also passengers including bacterial and fungal/rhizobial flora, and higher trophic level organisms. Some plants will be nitrogen fixers and will build vital nitrogen fertility. Plants also introduce organic materials, for example, acids and chelating organic biochemicals, which further facilitate weathering of the geologic substrate.
4. *Further weathering* follows the introduction of life, along with water and air. Weathering drives the increasing pace of the transformation of soil parent material to clay, which, along with decomposing organic matter, builds the essential soil fertility by providing increasing cation and anion exchange for nutrient storage, pH buffering capacity, and water-holding capacity.

The rate and timing of ongoing soil development will depend on the climate as well as ongoing landform stability, parent materials, and other factors.

13.3 Soil-Landscape Relationships

The geomorphic processes that drive the formation of Raw Soils are often obvious—for example, Sandy Raw Soils in aeolian dune lands and Rocky Raw Soils on high and steep mountain tops. The dynamism of the New Zealand landscape is ultimately controlled by the geological setting of New Zealand on the boundary of the Pacific and Australian tectonic plates and in the 'roaring forties' latitude (around 40°S) which means the New Zealand archipelago lies in a belt of strong prevailing westerlies and is prone to high winds and intense rainfall. Thus the combination of rapid uplift, active erosion and deposition, volcanic activity, and (most commonly) periodic storm events cause landscape changes that lead to Raw Soil establishment.

Rocky Raw Soils are mainly formed in mountainous areas, particularly the Southern Alps, where rapid and ongoing uplift, to high altitudes, leads to the land being exposed to cold temperatures, high precipitation, and gale-force winds that promote erosion and make plant establishment difficult. A complex pattern of Raw and Recent Soils often occurs on a local scale as plants become established and pockets of Recent Soil form in areas with more sheltered microclimates, while Raw Soils remain on the most exposed faces. Some of the Raw Soils in high altitude alpine areas, particularly adjacent to glaciers and on shaded southerly faces, may contain permafrost, and many exhibit some periglacial features such as down-slope movement due to freeze-thaw processes. Small areas of Rocky Raw Soil may also be observed where river down-cutting or coastal erosion has exposed rocky cliff faces.

Younger lava flows, such as those on Rangitoto Island, or in Mangatepopo valley from eruptions of Ngauruhoe, where there has been insufficient time for soil development to progress, also support Rocky Raw Soils (Fig. 13.4). There is a progression of weathering with age of lava flows from Raw to Recent, then to Granular or Podzol, then Oxidic Soils. However, many of the older lava flows in New Zealand are subject to later additions of tephra that, if thick, reset the soil development process back to Tephric Raw Soils that may then progress through Recent to Pumice or Allophanic Soils.

Hydrothermal Raw Soils are scattered in small areas through the central North Island where geothermal activity results in areas of relatively inhospitable soils that often have very low pHs ($\sim$3.4) and temperatures as high as 100 °C (Fig. 13.5). As well as Ketetahi Springs in the northern flanks of Tongariro, many of the more extensive sites, including Whakarewarewa, Waiotapu, Craters of the Moon, hydrothermal areas on the margins of Lake Rotorua, and Orakei Korako, are important tourist attractions. Vegetation if present is typically stunted and shallow rooted.

Fig. 13.4 Soil landscape of Rocky Raw Soil on lavas from Ngauruhoe erupted in 1870. The track runs through a 'mounded' levee landform adjacent to a channel at right

Fig. 13.5 Hydrothermal Raw Soils at Whakarewarewa in Rotorua. The visible yellow sulphur deposits indicate the low pH, and boiling mud, the high temperature, of these largely inhospitable soils. Pohutu geyser is playing in the background

Hydrothermal Raw Soils (Fig. 13.5) include two quite distinct geothermal situations and marked differences in the resulting soil properties. In areas associated with alkali-chloride features, such as Pohutu geyser at Whakarewarewa and the Champaign pool at the Waiotapu geothermal area, the water that is erupting has an alkaline pH. As the water flows out over the land surface and cools some of the ions dissolved in the water, predominantly Si^{4+}, precipitate out and gradually build up the land surface with a hard silica sinter coating. The famous Pink and White terraces (that were largely destroyed in the Tarawera eruption of 1886, although surveys show probable remnants of both terraces at the bottom of Lake Rotomahana) were formed of silica sinter. However, as is evident in the photo of Pohutu (Fig. 13.5), sulphur may also be deposited and accumulate. Where the yellow sulphur deposit interacts with water, sulphuric acid will form, imparting a very low pH. In contrast, areas of boiling mud are defined as acid-steam features. Here the pH is likely to be extremely low and the acidic, boiling, mud provides one of the most hostile environments for most living organisms.

Sandy Raw Soils occur predominantly on beaches and coastal sand dunes. Ninety-mile beach in Northland and Farewell Spit at the north-western tip of the South Island are two outstanding areas that have formed from sand being carried along the west coast of New Zealand. The Manawatu sand country, which extends from about Levin northward to the outlet of the Rangitikei River, is the largest dune field in New Zealand.

The prevailing strong westerly winds in the Manawatu have carried sand inland, forming a distinct series of dunes ranging in age from active dunes to some over 5000 years old. Raw Soils occur on the current beaches and active dunes adjacent to the coast. Pockets of Raw Soils also occur further inland where dunes have been reactivated and the sand has been eroded and redeposited. The rate of weathering and soil formation has been widely studied in sand in the Manawatu environment with pioneering work by Des Cowie. Raw Soils are associated with the Waiterere dune phase which are dunes formed in the last 150 years or so. Recent Soils (with A horizons at least 5 cm thick) are recognised on the Motuiti dune phase which has been dated at about 500–1000 years old. In the Foxton dune phase (1600–6500 years old) the Recent Soils have yellow-brown B horizons in addition to topsoils.

Within the Manawatu sand country there is also a clear soil-landscape pattern between the droughty dunes and the wetland areas where drainage is cut off by dunes, with Sandy Raw Soils on the youngest dunes and Sandy Gley Raw Soils in the adjacent low-lying areas of high water table between the dunes (Fig. 13.6). Near the coast, in the low-lying coastal wetlands the water table is controlled by the tide, rising and falling over the course of a day.

Fluvial Raw Soils occur on river flood plains (Fig. 13.1) where the sediment deposited in flood events is deep enough, and/or frequent enough, to prevent a Recent Soil from developing. Thus the Fluvial Raw Soils tend to occur in thin strips on river banks, river levees, and riparian margins. They are associated with Recent Soils on slightly higher ground where flood deposits are thinner, and with Gley Raw Soils where water tables are high. The largest areas of Fluvial Raw Soils are associated with rivers that flow from the erosion-prone mudstone hill country in the Rangitikei/Whanganui area and on the east coast of the North Island from the Wairarapa northward to East Cape. Such rivers periodically carry extremely high sediment loads of material eroded from the mudstone hills, and deposit it on the river flood plains during storm/flood events. Murray Hicks and others calculated the suspended sediment yields from the East Cape area (from East Cape to Mahia Peninsula) to be ~69 Mt/yr, with the two main rivers, Waiapu and Waipaoa, delivering 35 Mt/yr and 15 Mt/yr, respectively. Their combined yield represents 42% of the yield from the North Island and 24% of the total yield to the entire New Zealand coast. The yield from the southwest coast of North Island (taking in the Rangitikei/Whanganui hinterland) is also relatively high at ~12 Mt/yr.

Gley Raw Soils often occur in association with Fluvial Raw Soils in the lower lying areas on river flood plains. The more unusual, but increasingly recognised, Gley Raw Soils are those of tidal mud flats where the material was deposited in water and has never dried out and thus, has never been exposed to oxygen (Fig. 13.7). Such sites often comprise Fluid Gley Raw Soils which have low weight bearing capacity so that a person may sink into the 'mud' if attempting to walk across the area. The Fluid Gley Raw Soils merge into the sea/estuary floor on their seaward boundary (i.e. non-soils), and often into Sandy or Saline Gley Raw Soils on the margins where the soil material has built to a depth above sea level on the coastal margin.

Tephric Raw Soils occur in unconsolidated deposits of explosive volcanic origin, collectively termed tephra, or in reworked (secondary) tephric material. The tephra materials include ash, lapilli, blocks and bombs, and pumice and scoria (cinders), all of which originated via explosive activity that generated (most commonly) fall deposits or ground-hugging pyroclastic flow deposits. Hence Tephric Raw Soils are prominent on very young, andesitic, tephra-mantled surfaces or on older tephra deposits freshly exhumed by recent erosion. Tephric Raw Soils are found:

- on Mounts Ruapehu, Tongariro, and Ngauruhoe in central North Island (Fig. 13.8);
- on exposed andesitic tephras around the steep peak of Taranaki Maunga;
- on the summit and shoulder slopes of Mt Tarawera (comprising a combination mainly of basaltic scoria erupted on 10 June 1886 mixed with re-erupted clasts of ~1314 AD Kaharoa eruptives);
- on andesitic ash and other materials on Whakaari (White Island); and
- on some of the steeper slopes of the 600-year-old basaltic scoria cones near the summit of Rangitoto Island (amidst sparse Tephric Recent Soils).

Fig. 13.6 Sandy Raw Soils on sand dunes on the Manawatu coast. Top: foredune with active sand movement. Bottom left: landscape, with salt tolerant vegetation. Bottom right: soil profile on sandflat behind (inland of) the dune with water table that rises and falls with the tide (at about 50 cm depth in the photo). Photos: Tanya O'Neill

Narrow strips of Tephric Raw Soils also occur on secondary lacustrine tephric materials around the southern and eastern shores of Lake Taupo, and on recently deposited lahars on Mounts Ruapehu and Taranaki. Lahars are rapidly moving, water-laden flows, other than normal stream flow, that carry poorly sorted pyroclastic and volcanic debris. A good example is the young laharic deposits in the Whangaehu River valley that were emplaced following the breaching of the tephra-impounded Crater Lake of Mt Ruapehu in March 2007. Such secondary materials typically contain at least some tephra material and are, therefore, appropriately classed as Tephric Raw Soils (in the widest sense of 'tephric'), but they potentially cross over to Fluvial Raw Soils given their fluvial-like origins.

Orthic Raw Soils are scattered throughout New Zealand on erosion scars, and the debris (runout materials) deposited in the lower reaches of areas of hillslope mass movement (Fig. 13.9). Within the New Zealand hill country, particularly

Fig. 13.7 Fluid Gley Raw Soils. Left: the margin of a mangrove forest on the Northland coast where Fluid Gley Raw Soils gradually merge to Saline Gley Raw Soils with increasing distance from the water's edge. Right: auger samples from the Waihola-Waipori wetland in Otago. While brown colours occur on the oxidised soil surface, beneath is anaerobic dark blue grey material. Within a few minutes of exposure to air, the anaerobic material changes colour to brown as the iron-bearing minerals are oxidised

on the easily eroded mudstone substrates in eastern North Island, there is a mosaic soil pattern comprising:

- young scars or landslide deposits that have not been weathered greatly and so are Orthic Raw Soils;
- older landslide scars and deposits that have a distinct topsoil formed and so are Recent Soils; and
- un-eroded areas that comprise the 'climax' soil of the area, e.g. Brown, Pallic, Podzol, Granular, or Ultic Soils.

13.4 Key Soil Properties

For most of the Raw Soil groups, the normal soil profile morphology, comprising pedogenically modified material in the form of soil horizons, is not applicable, and so the usual suite of physical, chemical, and biological analyses have not been undertaken. Nevertheless, the properties of many of the earth materials which make up Raw Soils have been studied by geotechnical scientists, engineers, soil conservators, and pedologists seeking 'time zero' soils for studies on soil evolution. Such knowledge contributes to understanding the genesis, behaviour, and management requirements of Raw Soils.

The soil mineralogy is inherited from the generally relatively unweathered parent material. Coarse fragments and sand particles are typically without coatings, although clasts in scree may have weathering rinds and hydrothermally altered material may include secondary products.

Fluid Raw Soils have been studied by soil scientists and engineers particularly in The Netherlands where land above the water table is at a premium. Fluid Raw Soils occur in young sediments that have a continuously high water table, such as in tidal mudflats, and exist in an 'unripened' state. Ripening is the process of consolidation of sediment due to dewatering and consequent settlement of the material with an increase in bulk density and development of strength and cohesion. The simplest method of estimating the state of 'ripening' of a sample is to squeeze a sample in the hand and judge its fluidity. A fist-full of soil is squeezed firmly in the hand and a ripened soil will not flow out of the hand, or between the fingers, under this pressure.

A quantitative method for determining a state of ripening (n) was determined in The Netherlands by Pons and Zonneveld in 1965 as follows:

$$n = (A - 0.2R)/(L + bH)$$

where

n = degree of ripening (or the water factor) in g of water adsorbed by 1 g of the clay fraction,
A = is gravimetric water content (g/100 g dry soil),

Fig. 13.8 Formation of Tephric Raw Soils. Top: erupting wind-borne plume of tephra from Mt Ruapehu at ~1230 h on 18 June 1996. Only at sites with thick deposits (≥ 60 cm) would the soils become Tephric Raw Soils. Bottom left: profile, about 1.5 m high, of a Tephric Raw Soil on young (<1800 years old) bedded tephra deposits near the Emerald Lakes (their source vents) on Mt Tongariro. Bottom right: variously coloured scoriaceous tephra beds in exposed scoria cone surrounding Red Crater on Mt Tongariro (note figures on skyline for scale), from which issued the most recent lavas flows from Tongariro (1855 and probably 1926). The thick basaltic andesite scoria deposits, suppporting Tephric Raw Soils, were deposited by explosive eruptions within the last ~365 years

R = is mass of sand + silt (g/100 g dry soil),
L = is mass of clay (g/100 g dry soil), and
H = is mass of organic matter (g/100 g dry soil).
$b = 3$ where organic matter in the soil is highly decomposed, and 4 if clearly visible plant material remains.

A soil is considered ripe (i.e. consolidated) if $n < 0.7$. The n value is helpful in predicting whether a soil can be grazed by livestock or can support other loads, and in predicting what degree of subsidence would occur after drainage.

Fig. 13.9 Orthic Raw Soils on erosion scars and redeposited material near Gisborne following Cyclone Bola in 1988. A pine forest was successfully established on the 'Y-Wurri Station' slip shown on the photo at right. The mosaic effect of disturbed and stable soils is evident in these photographs

Without human intervention, soil fertility is generally low in Raw Soils, with nitrogen deficiency usually the primary limitation. Nitrogen accumulation relies on establishment of nitrogen-fixing plant species. The accumulation of soil organic matter, and mineral weathering, gradually build nutrient and water storage to support a wider range of living organisms. Over time a Raw Soil will gradually build the capacity to entrap and store the suite of nutrients needed to support viable ecosystems (nitrogen, phosphorus, potassium, sulphur, calcium, magnesium, potassium, sodium, and trace elements). However, in some Fluvial Raw Soils, the alluvial materials, if eroded from pre-existing fertile soils, may have sufficient nutrients, and a texture (containing clays) conducive to moisture storage, so that plants may readily become established and flourish.

Most definitions of soil require or assume the presence of soil organic matter or life, but do Raw Soils contain life? We know that most Raw Soils host at least some microbial life, and at all but the most extreme sites nematodes, insects, mosses, lichens, and other plants will quickly become established in newly exposed materials. However, vegetation cover in Raw Soils is generally either absent, sparse, or patchy, and often consists of ephemeral herbaceous plants, mosses, or lichens. A number of plant, and other, species that specialise in colonising newly exposed materials depend on periodic changes in the environment to provide them with an ecological niche to survive.

Hydrothermal Raw Soils host a suite of microbes that are adapted to the hot, often acidic, conditions. Such thermophilic microbes have been closely studied as they have potential for a number of technical applications, particularly in work led by Roy Daniels and Hugh Morgan.

The mangrove tree is specially adapted to the Fluid Raw Soils of estuaries in the northern half of the North Island. Mangrove roots are adapted to the oxygen-depleted substrate with pneumatophores that extend upwards to reach air, at least for a few hours during low tide (Fig. 13.6). Mangrove seeds will float long distances in the sea to eventually wash up and become established on a tidal mudflat area. Once established, mangroves tend to slow the rate of water flow, thus encouraging further settling of 'mud' from the water column. Over time the mangroves build up and entrap sediment and extend out into the tidal zone, with the inland edge becoming ever more sheltered from waves and erosion, and gradually building up to a Saline, rather than a Fluid, Raw Soil.

Along the southern margin of the Firth of Thames the accumulation of sediment, in collaboration with mangrove establishment, has extended the shore line out into the former marine area by 800 m between the 1960s and about 2010 (Fig. 13.10). The margins of the mangroves, which became established outside a stop bank that was built along the coastline in the early 1960s, is now 2 m higher than the land on the inside of the stop bank, which provides a good buffer against storm surges. About 60% of the sediment trapped by the mangroves is considered to have come from sediment that was within the Firth of Thames by the 1960s with the balance being carried down the rivers. Sediment can be re-suspended by storms. Construction of stop-banks along the margins of the Waihou and Piako Rivers prevents flooding of adjacent land and means that sediment that was formerly deposited on the flood plains is now all carried to the sea. Between 1963 and 2007, the mangrove area expanded by about 19 m yr^{-1}. However, the rate is now slowing and was about 8 m yr^{-1} from 2002 to 2007, with no marked expansion from 2007 to 2017.

When is a soil, a soil?

It might seem strange that something as common as soil (from Latin, *solum*, ground) does not have a universally agreed definition. Partly this is because of the complexity and diversity of soil (in terms of origin, character, functioning) and partly because it has different meanings to different people or disciplines. Prior to 1999, the soil science definitions of soil, such as that in *Soil Taxonomy*, generally required a 'soil' to include an accumulation of organic matter in the topsoil and to support plants. Thus many of the Raw Soils defined in New Zealand, along with surface materials in ice-free areas of Antarctica, the Moon, and planets such as Mars, were not considered to qualify as 'soil'. However, engineers use the word 'soil' to mean any disaggregated rock material that is not a solid or coherent mass, and recognise three groups: sands and gravels, silts and clays, and *peat*.

Scientists studying surface materials in Antarctica, and on the Moon and Mars, felt strongly that such materials should be considered soils and clearly come within the interests of soil science. As a result, in 1999, the definition of a soil, in *Soil Taxonomy*, was changed. Thus soil is now defined as '*a natural three-dimensional (Some define soil as four-dimensional, the 4th dimension being time.) body (pedon) typically ~1–2 m thick comprised of solids (minerals and organic matter), liquid, and gases, that occurs on the land surface, occupies space, and is characterised by one or both of the following: horizons, or layers, that are distinguishable from the initial material as a result of additions, losses, transfers, and transformations of energy and matter or the ability to support rooted plants in a natural*

Fig. 13.10 Fluid Raw Soils and Saline Raw Soils under mangroves that have become established on the southern margin of the Firth of Thames since building of the coastal stop bank was completed in 1963. Photo courtesy of Waikato Regional Council

environment'. Therefore the ground surface materials in Antarctica, and on planets, moons, and asteroids in space, could be recognised as 'soils'. Back in New Zealand, the updated definition meant we can now formally identify all our Raw Soils as 'soils'.

Finally, it is certain that 'soil' is not 'dirt'. This may seem a subtle distinction for most people but not for soil scientists. The association derives from Old French term, *soillier/suillier* (from Latin *sus*, a sow) meaning to 'soil' in the sense of 'defile'. Dirt refers to unhygienic or unwanted material. The term 'dirt' derives from the Old Norse (and still used in modern Norway) word, *drit*, meaning excrement. By contrast, soil, rather than being 'unclean', paradoxically is a universal purifier, cleaning and regulating water flow everywhere it occurs. Many of the most widely used antibiotics have come from the soil. As well, soil is essential for supporting life on Earth, providing the foundation of diverse ecosystems and the basis of all civilisations. The relationship of soil to dirt is akin to that of plants to weeds, or animals to pests. This may be a nicety, but the important thing is that we should never, ever, treat soils like dirt!

13.5 Distinguishing Between Raw Soils and Related Soil Orders

'Raw' means that the soil is undeveloped, and is dominated by the properties of the geologic substrate materials of which it is formed. Raw Soils lack distinct topsoil development (meaning that they have no topsoil, or the topsoil, including organic horizons if present, is less than 5 cm thick) or they are fluid and wet at a shallow depth. At older more stable sites, Raw Soils merge into Recent Soils, with formation of distinct topsoil at least 5 cm thick. Organic soils can be fluid at a shallow depth but their high organic matter content excludes them from the Raw Soils.

If a soil has been eroded then the remaining 'not-eroded' soil can still be allocated to the original soil order if sufficient soil profile remains to allocate it to a soil order in the key to the *New Zealand Soil Classification*. Because the Raw Soils are placed right at the end of the key, any soil that is not allocated to an order earlier will be caught in the Raw Soils.

13.6 Correlation with Other Classification Systems

Raw Soils have often been regarded as geological materials and so are not usually included in soil classification systems (Table 13.1). The Gelisols of *Soil Taxonomy* would equate to a 'frozen' or 'cryic' Raw Soil at sites where there is permafrost, although at the time of writing no such soil group was defined in the *New Zealand Soil Classification*. Consequently, the soils of Antarctica and possibly some high mountain areas would be classified as Rocky or Orthic Raw Soils.

13.7 Use and Management of Raw Soils

While most Raw Soils are generally not a suitable substrate for food production, they are important for a variety of uses and have some special management requirements. Wherever Raw Soils are identified, they invoke an inherent warning: they may not be safe places to build or establish costly

Table 13.1 Correlation[a] between Raw Soils and equivalent classes of *Soil Taxonomy*, the *World Reference Base*, and the New Zealand genetic soil classification

New Zealand Soil Classification	*Soil Taxonomy*	*World Reference Base*	NZ genetic soil classification
Gley Raw Soils	Not recognised[b] or Aquents or Wassents	Regosols	Not recognised
Hydrothermal Raw Soils	Not recognised or Orthents	Not recognised	Not recognised
Rocky Raw Soils	Not recognised or Lithic Orthels	Not recognised	Not recognised
Sandy Raw Soils	Psamments	Arenosols	Not recognised
Fluvial Raw Soils	Not recognised or Fluvents	Regosols	Not recognised
Tephric Raw Soils	Not recognised or Orthents	Regosols	Not recognised
Orthic Raw Soils	Not recognised or Orthents or Orthels	Regosols	Not recognised

[a]The correlations given here are a guide only and for accurate classifications the relevant soil classification documents should be consulted. The two major international soil classification systems are *Soil Taxonomy,* which was developed in the USA, and *World Reference Base,* which was developed primarily in Europe. The NZ genetic soil classification was used in NZ prior to 1992

[b]Sometimes referred to as 'not soil'

infrastructure, as they tend to be sites of geological instability, subject to erosion or deposition. Where development is proposed, careful regard may need to be taken to the soil properties and position and specific engineering solutions may need to be considered.

The ecosystems that develop on Raw Soils are often unique as they provide an environment where pioneer, early succession plants have a particular niche. There are many introduced invasive weed species that will readily colonise Raw Soils, preventing the natural ecological succession of New Zealand native plants. Invasive plants that will readily colonise Raw Soils include pampas (*Cortaderia selloana*), wildling pines (often *Pinus contorta*), reed sweet grass (*Glyceria maxima,* on water margins), and *Hieracium pilosella* (mouse-eared hawkweed, a plant that inhibits other plants and invades bare ground in the South Island high country). Constant vigilance is needed to prevent introduction of weeds, and rapid response to small invasions can prevent a need for much larger efforts later. In some regions major efforts are being, or have been, undertaken to try to control the spread of such weeds.

Because of the large variety of Raw Soils, and their potential uses, specific management requirement, and hazards, each soil group is considered separately in the following discussion.

Rocky Raw Soils are an important component of the scenic landscapes that attract tourists to New Zealand. Many are in parks or protected areas managed by the Department of Conservation. Some underlie ski fields while others attract hikers/trampers and mountain climbers. The main objectives in managing Rocky Raw Soils are often to maintain the natural environment and to protect users from hazards, for example, potential rockfalls at sites where steep cliffs are exposed.

At some sites, such as the Tongariro Alpine Crossing Track, in the Tongariro National Park, trampling by humans can greatly damage sensitive Raw Soils (Fig. 13.4). Up to 3000 tourists per day walk the Tongariro Crossing with a total of about 125,000 in 2016. Human trampling of the soils and associated flora and fauna (sparse as it is) can lead to damage to rare ecosystems, compaction, and concentration of runoff, and thus increased risk of rill and gully erosion. Much effort is put into encouraging walkers to stay on marked tracks, and to form and manage tracks to accommodate tourists and minimise damage to the wider environment.

Both Fluvial and Gley Raw Soils are often in low-lying areas and thus are at risk of flooding. With development of flood protection works and drainage, such soils may be developed for pastoral land use but expensive infrastructure should generally be avoided unless special measures are in place to provide protection from inundation. Gley Raw Soils may also have very low bearing strength. Consequently, heavy vehicles or buildings may risk sinking into the soil. In areas such as tidal mudflats, with Fluid Gley Raw Soils, the potential to sink into and be trapped in the soil extends to people and animals trying to walk across the area. However, Fluid Gley Raw Soils may be drained and reclaimed for building infrastructure or agriculture development, and even nature herself can make such a change. During the Napier earthquake of 3 February 1931, an area including the brackish Ahuriri Lagoon was raised about 1.5 m above its previous level. Part of the Ahuriri Lagoon, supporting Fluid-saline Gley Raw Soils, became dry. The salinity of the uplifted soils varies over time, decreasing with leaching of salt during wet seasons and increasing due to capillary rise of saline groundwater during dry spells. The uplifted, formerly, Fluid Gley Soils are now Saline Recent Gley Soils.

When Fluid Gley Raw Soils are drained there is a risk that they will become extremely acidic, forming soils known globally as 'acid sulphate soils' (or Sulfaquents in *Soil Taxonomy*). When the soil is drained, oxygen enters, and if there is sulphur present (which is usual in brackish lakes, coastal marine environments, landscapes comprising mine tailings, or soils containing pyrite, FeS_2), then the oxygen reacts with reduced forms of the sulphur (sulphate ions are reduced to sulphide ions by bacteria) to form sulphuric acid:

$$FeS_2 + 2H_2O + 3O_2 \rightarrow FeSO_4 + H_2SO_4 + 1440\ \text{kJ}$$

The soil pH is likely to drop to below 4, and can be as low as 2, possibly the lowest pH for any mineral soil. Such an acidic environment is capable of dissolving concrete in structures such as bridge abutments, and rapidly corroding steel pipelines and other infrastructure. Thus special engineering precautions are needed in such environments. The low pH also allows toxic concentrations of soluble Al and Fe to be released, together with heavy metals that may include arsenic. The marked acidity inhibits most plant growth and large inputs of lime are needed to provide for agricultural development. Keeping the water table as high as possible helps minimise oxygen infiltration and thus acid formation. Jarrosite, a distinctive, yellow, iron-sulphate mineral often forms in drained Fluid Gley Soils in coastal areas, and an orange iron-sulphate mineral, schwertmannite, forms typically in wetland situations or on acid mine tailings between pH 2 and 4.

There is potential for acid sulphate soils to develop in any former tidal mudflat environments that are drained. Such areas are particularly common adjacent to the harbours in the northern North Island.

Many Hydrothermal Raw Soils, such as those at Whakarewarewa in Rotorua and Craters of the Moon at Wairakei, are managed primarily as tourist attractions and sites of unique ecological and geological value. The main considerations are to find means to provide interesting experiences for tourists while preventing injuries from contact with the hot, and often acidic, environment.

There is a need to protect geothermal fields from too much water and energy extraction to protect the geothermal features. In Rotorua City, a large number of houses and businesses took up using geothermal waters for heating, spa pools, and the like. The first hot water wells were drilled in the 1920s and by 1985 around 450 wells were recorded. As a result, the natural heat flow in the geothermal water table was reduced by 30% and the geothermal activity at Whakarewarewa and Kuirau thermal park declined, and some spring activity ceased. The central Government launched the fraught 'Bore War' in 1987 by closing down many of the private bores within 1.5 km of the Pohutu geyser in 1987, with subsequent recovery of the geothermal field.

Sandy Raw Soils in coastal dune systems may be subject to wind erosion and sand deposition. Encouraging vegetation establishment, and preventing disturbance by foot and vehicle traffic, can help stabilise dunes and prevent major sand 'blowout'. Mobilised sand may cause dunes to move inland, burying everything in their path including roads, buildings, forests, and farms. While plants such as lupin (which fixes nitrogen) and marram have been successfully used to stabilise sand, they tend to out-compete native plants, and therefore encouragement of plants such as pingao (*Ficinia spiralis*) and kowhangatara (*Spinifex sericeus*) is favoured in many areas. Planting is often undertaken by local 'beachcare' groups.

On the west coast of the North Island large areas of Raw Soils occur on sands that are rich in iron and titanium. The iron sands are mined at Taharoa, for export to Japan, and near the mouth of the Waikato River where the constituents are made into steel at the Glenbrook steel mill. The iron-rich sand is concentrated with the remaining sand grains redeposited as tailings and stabilised by revegetation. Quartz-rich Sandy Raw Soils at Parengarenga Harbour in Northland are mined for use in glass making.

Tephric Raw Soils may occur in unconsolidated volcanic ejecta on unstable slopes, and when volcanically active, there are associated hazards of pyroclastic flows and tephra fallout, together with the potentially dangerous emplacement of lahars in valleys. As recently as August 2012, a phreatic (steam-driven) eruption occurred in a new vent below the Upper Te Māri crater, on the northern slopes of Mt Tongariro, near the Alpine Crossing Track, and blocks as large as 1 m in diameter were deposited up to 2 km from the vent.

The eruption also caused a landslide which blocked streams. Eventually the blockage failed and released a lahar that damaged part of the track, flowing as far as State Highway 46 (Fig. 13.11). The lahar was sufficient to change the track walking conditions with lahar hazard signs now installed.

The Tephric Raw Soils occur mainly in reserves and public conservation land in areas such as Taranaki Maunga, Mt

Fig. 13.11 Tephric Raw Soil on debris from a lahar emplaced in forest flanking a stream near the SH 46 end of the Tongariro Alpine Crossing after an eruption from Te Maari Crater in August 2012

Ruapehu, and Mt Tongariro where generally they are protected as natural features and natural processes are allowed to play out. There is an issue that such areas are fragile and easily damaged by foot traffic, particularly on the popular Tongariro Alpine Crossing. Efforts have been made to restrict people to formed paths and in some areas wooden walkways have been constructed to prevent damage to the fragile soils.

Orthic Raw Soils often occur on slopes where material has been removed or deposited as a result of hillslope mass movement. Such sites are particularly common on the soft rock Tertiary mudstone (papa) that underlies the steep hilly terrain on the east coast of the North Island. A range of measures have been shown to help prevent such erosion events, including spaced tree planting and gully dams. Where accelerated erosion has occurred, spreading seed and fertiliser speeds up re-establishment of vegetation cover, thereby helping rehabilitate such sites. Work in the Wairarapa hill country has shown that pasture production on landslide scars was about 20% lower than on adjacent non-eroded sites 80 years after the erosion event. Forest productivity has been shown to be about 10% lower on eroded, than non-eroded, areas in east coast hill country.

Further Reading

Basher LR (2013) Erosion processes and their control in New Zealand. In: Dymond JR (ed) Ecosystem services in New Zealand—conditions and trends. Manaaki Whenua Press, Lincoln, New Zealand, pp 363–374

Briggs RM, Laurent JG, Hume et al (2009) Provenance of black sands on the west coast, North Island, New Zealand. Proceedings, AusIMM New Zealand Branch Annual Conference, Queenstown (25-28 August), pp 41-50

Certini G, Ugolini FC (2013) An updated, expanded, universal definition of soil. Geoderma 192:378–379

Clarkson BD, Clarkson BR, Juvik JO (2015) Pattern and process of vegetation change (succession) on two northern New Zealand island volcanoes. Surtsey Res 13:45–48

Clement AJH, Sloss CR, Fuller IC (2010) Late Quaternary geomorphology of the Manawatu coastal plain, North Island, New Zealand. Quatern Int 221:36–45

Cody AD (2003) Geology, history and stratigraphy of hydrothermal eruptions in the Rotorua Geothermal Field. Unpublished MSc thesis, University of Waikato, Hamilton

Cowie JD (1967) Soils of the Manawatu—Rangitikei sand country. Soil Bureau Bulletin 29. New Zealand Department of Scientific and Industrial Research, 58 p

Cowie JD (1968) Pedology of soils from wind-blown sand in the Manawatu district. NZ J Sci 11:459–487

Cowie JD, Rijkse WC (1977) Soils of Manawatu county, North Island. New Zealand, NZ Soil Survey Report 30, 34 p

Crozier MJ (2005) Multiple-occurrence regional landslide events in New Zealand: hazard management issues. Landslides 2:247–256

Davies TRH (2017) Mountain process geomorphology: conceptual progress in the Southern Alps. In: Shulmeister J (ed) Landscape and Quaternary environmental change in New Zealand, Springer/Atlantis Press, pp 205–233

de Ronde CEJ, Fornarib DJ, Ferrinic VL et al (2016) The Pink and White terraces of Lake Rotomahana: what was their fate after the 1886 Tarawera Rift eruption? J Volcanol Geoth Res 314:126–141

Glade T (1998) Establishing the frequency and magnitude of landslide-triggering rainstorm events in New Zealand. Environ Geol 35:160–174

Grab SW, Dickenson KJM, Mark AF et al (2008) Ploughing boulders on the Rock and Pillar Range, south-central New Zealand: their geomorphology and alpine plant associations. J Roy Soc New Zealand 38(1):51–70

Hartemink AE (2016) The definition of soil since the early 1800s. Adv Agron 137:73–126. https://doi.org/10.1016/bs.agron.2015.12.001

Heaphy MJ, Lowe DJ, Palmer DJ et al (2014) Assessing drivers of plantation forest productivity on eroded and non-eroded soils in hilly land, eastern North Island, New Zealand. New Zealand J For Sci44:24 (10p)

Hewitt AE (2010) New Zealand Soil Classification 3rd edition, Landcare Research Science Series No.1, Manaaki Whenua Press, Lincoln, New Zealand, 136p

Hicks M, Shankar U, McKerchar AI et al (2011) Suspended sediment yields from New Zealand rivers. J Hydrol (NZ) 50(1):81–142

Ling LL, Schneider T, Peoples AJ et al (2015) A new antibiotic kills pathogens without detectable resistance. Nature 517:455–459

Lowe DJ, Shane PAR, de Lange PJ et al (2017) Rangitoto Island field trip, Auckland (FT3). Geosci Soc New Zealand Miscellaneous Publ 147B:56p

Lynn IH, Manderson AK, Page MJ et al (2009) Land use capability survey handbook, 3rd edn. AgResearch/Landcare Research/GNS Science, 163 p

Metson AJ, Janica GE, Cox TE et al (1977) The problem of acid-sulphate soils with examples from North Auckland, New Zealand. N.Z. J Sci 20:371–394

Pons LJ, Zonneveld IS (1965) Soil ripening and soil classification. Initial soil formation in alluvial deposits and a classification of the resulting soils. International Institute for land reclamation and improvement. Publication 13. H Veenman and Zonnen NV, Wageningin, The Netherlands, 128 p

Roberts RC, McConchie J (2017) Preliminary assessment of the acid sulphate soils hazard in the Auckland region. In: Alexander GJ, Chin CY (eds) Proceedings of the 20th NZGS Geotechnical Symposium, Napier

Sattler K, Anderson B, Macintosh A et al (2016) Estimating permafrost distribution in the Maritime Southern Alps, New Zealand, based on climatic conditions at rock glacier sites. Front Earth Sci 4:4. https://doi.org/10.3389/feart.2016.00004

Shepherd MJ, Lees CM (1987) Holocene alluviation and transgressive dune activity in the lower Manawatu Valley, New Zealand. NZ J Geol Geophys 30:175–187

Soil Survey Staff (2014) Keys to Soil Taxonomy twelfth edition. USDA Natural Resources Conservation Service. 360p https://www.nrcs.usda.gov/wps/PA_NRCSConsumption/download?cid=stelprdb1252094&ext=pdf

Townsend DB, Leonard GS, Conway CE et al (compilers) (2017) Geology of the Tongariro National Park area, New Zealand. GNS Science geological map 4. 1 sheet + 109 p, scale 1:60,000, GNS Science, Lower Hutt

Vucetich CG, Wells N (1978) Soils, agriculture, and forestry of Waiotapu region, central North Island, New Zealand. NZ Soil Bur Bull 31:100p

Williams PW (2017) New Zealand landscape—behind the scene. Elsevier, Amsterdam, 470 p

World Reference Base (2015) https://www.fao.org/3/i3794en/I3794en.pdf. Accessed 2 March 2020

14 Recent Soils

Recent Soils are in their prime
the key thing is enough time
with topsoil and a weak B
plants will grow productively

Abstract

Recent Soils cover about 6% of New Zealand and are developed where material has been deposited, or eroded, and soil development has recommenced on new parent materials. Recent Soils occur where repeated deposits are thin enough, or there has been enough time, for a topsoil ≥ 5 cm thick to form. They may also have a weakly developed B horizon. Many Recent Soils occur on the flat alluvial flood plains of our major river systems. Such Fluvial Recent Soils are productive and widely used for dairy, and other pastoral, farming as well as for cropping and horticulture. However, many Recent Soils remain at risk of flooding, and extensive flood protection works have been installed to safeguard Recent Soils in both urban and rural areas in many regions. A number of towns, including Whangarei, Paeroa, Edgecumbe, Whakatane, Gisborne, Wairoa, Palmerston North, Whanganui, Westport, Hokitika, and Mosgiel, are built, in part, on Recent Soils, and have been subject to flooding. Recent Soils also occur on eroded hill and mountain country, older relatively stable sand dunes, and surfaces mantled with young tephra deposits. Sandy Recent Soils are particularly prone to wind erosion, hence maintenance of vegetation cover to prevent damage is important.

14.1 Important Features of Recent Soils

14.1.1 Concept and Key Features of the Soil Order

Because of its geologically active plate tectonic setting, and hence geomorphic instability and generally rapid rates of natural erosion and deposition, New Zealand has extensive areas, and a wide variety, of Recent Soils. Recent Soils are weakly developed, showing limited signs of soil-forming processes because of youthfulness as a consequence of soil erosion, or deposition of parent materials, or, occasionally, because the soil parent material is especially resistant to weathering. Soil formation is sufficient to develop a distinct topsoil but a B horizon is either absent or only weakly expressed as a 'colour' B (namely, Bw) horizon. A continuous cover of plants is normally well established. Soil development in a Recent Soil, albeit limited, is by definition more advanced than in Raw Soils.

The most economically important and extensive areas of Recent Soils in New Zealand are on the highly productive river flood plains where typically thin additions of material during flooding episodes have allowed the soil to gradually build up (Fig. 14.1). Such incremental deposition and the

A. E. Hewitt et al., *The Soils of Aotearoa New Zealand*, World Soils Book Series,
https://doi.org/10.1007/978-3-030-64763-6_14

Fig. 14.1 Recent Soil and landscape. Left: productive land uses on Recent Soils adjacent to a river on the Canterbury Plains. Right: a Fluvial Recent Soil on the Taieri Plains shows thin layering from the addition of material in repeated flood events and evidence of earthworm mixing of material at the base of the topsoil

subsequent modification, by pedogenesis, of the freshly added sediment is an example of developmental upbuilding (i.e. the ongoing formation of soil via topdown soil-forming processes while new materials are periodically added to the land surface by geological processes). In contrast, where a thicker layer of alluvium is deposited that deeply buries the previous soil (turning it into a paleosol), the soil's evolution is reset to a new time zero forming a Raw Soil.

Layers in the subsoil of Fluvial Recent Soils record the history of alternating inputs of: gravels, sand, or mud (silt and clay), deposited by floodwaters; organic matter accumulated as shallow peats in wet swales or back swamps; and organic matter accumulating in topsoil between flood events. In older soils, the layers may be obliterated by bioturbation—mixing by roots, worms, and other burrowing organisms. In some locations in North Island, such as on the extensive Rangitaiki Plains in eastern Bay of Plenty, tephra deposition has provided additional parent materials. Tephras have been deposited both through fallout and as material reworked by rivers. Thus, many soils on the plains are complex combinations of layered alluvium and tephra-derived materials interspersed with organic and gleyed materials.

Recent Soils also form on sand dunes that have been stable for long enough develop a topsoil. Such soils have loose, relatively unweathered sand in the subsoil. The Recent Soils on stable but previously eroded surfaces, such as older landslide scars, often comprise just topsoils over the rocky or unconsolidated parent materials. In softer rocks, the progress from a Raw to a Recent Soil can be relatively rapid (decades). However, on harder rocks, it may take some centuries to form a Recent Soil and considerably longer for the Recent Soil to develop into a soil of a more mature soil order.

The age of Recent Soils varies depending on the environment and soil parent materials, but most are probably less than 1000–2000 years old. Some Recent Soils have been formed on gradually accumulating material (e.g. from repeated floods or eruptions) throughout the Holocene (last 11700 years). The ongoing additions do not allow time for more than thin, rudimentary, Bw horizons to form and so further development is prevented and a Recent Soil remains. In such cases, the soil age spans some hundreds of years rather than having a specific age. A Recent Soil can also be formed on a very old rock material, such as greywacke or mudstone, that has been eroded so that the soil formation process has started over to form a young soil that may comprise only an A horizon over rock.

On sand dunes in the Manawatu region, a Recent Soil is recognised on dunes that are about 500 years old, with younger dunes supporting Raw Soils (Chap. 13), and older dunes supporting Brown Soils. Similarly dated dunes on which Recent and Raw Soils are formed occur along the Coromandel and Bay of Plenty coasts, including in the Papamoa area where the dunes and soils have been dated using tephrochronology by Glenn Wigley and David Lowe.

14.1.2 Areas of Occurrence

Recent Soils are widespread throughout New Zealand (Fig. 14.2), covering about 6% of New Zealand (about 1.6

Fig. 14.2 Distribution of Recent Soils in New Zealand. The main areas are mainly associated with river flood plains

million hectares). Recent Soils are not restricted to particular parent materials or climate zones, but occur on young land surfaces, including alluvial flood plains, fans, sand dunes, unstable hill and mountain slopes, and surfaces mantled by young tephra deposits. Recent Soils occupy sites where erosion or deposition occurs at rates low enough, or with only thin deposits, to allow time for well-developed topsoils to form. Recent Soils may also occur on sites with shallow rock where the slow rate of rock weathering retards soil profile development.

14.1.3 Variation Within the Recent Soil Order

Recent Soils comprise six soil groups (Fig. 14.3) with 15 subgroups. The six soil groups are;

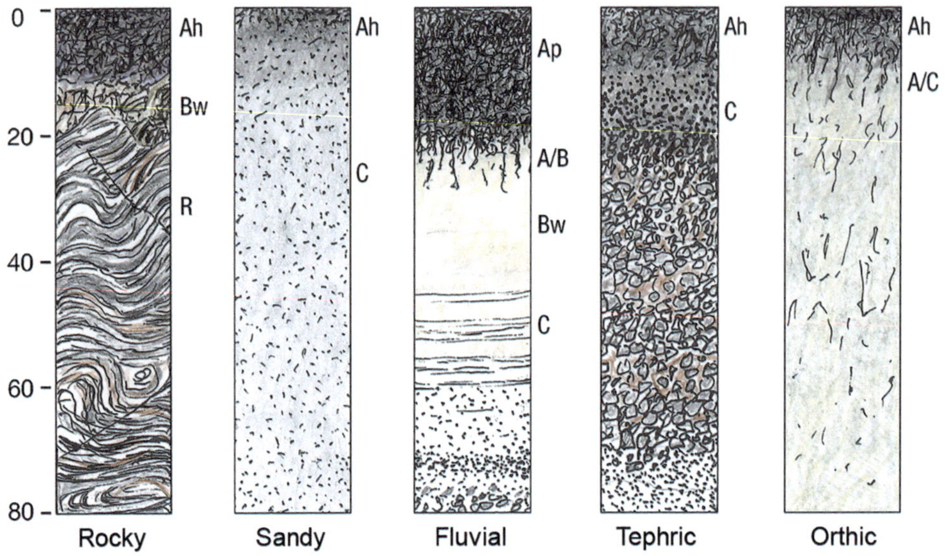

Fig. 14.3 Soil groups of the Recent Soil order. Hydrothermal Recent Soils not shown

Hydrothermal Recent Soils—naturally warmed by current, previously active, or extremely acidic geothermal activity, usually in areas adjacent to Hydrothermal Raw Soils,
Rocky Recent Soils—have a lithic (hard rock) or paralithic (sufficiently hard to resist cutting with a spade) contact at shallow depth, usually directly beneath an A horizon, often formed as a result of erosion,
Sandy Recent Soils—dominated by sand or loamy sand textures to depths of 60 cm,
Fluvial Recent Soils—in sediments with fluvial features, i.e. as deposited by flowing water, and with not more than 30 cm of non-fluvial material,
Tephric Recent Soils—in tephric soil materials at least 30 cm thick (originating as unconsolidated, fragmental volcanic ejecta, namely tephra), and
Orthic Recent Soils—other Recent Soils, most commonly on eroded hill-slopes.

The subgroups are mainly of two types: those that recognise variation within the group (Peaty, Saline, Weathered, Acidic); and those that recognise features that vary towards the character of other groups or orders, i.e. intergrades (Mottled, Acidic, Tephric, Pallic, Buried-pumice, Buried-allophanic, Weathered). All of the subgroups depart in some properties from the central concept of the soil group, but the simplest expression of the soil group is identified as the Typic subgroup with 'no aberrant properties'. Definitions of the soil groups and subgroups are provided in the *New Zealand Soil Classification*.

14.1.4 Origin of the Soil Order Name

The name *Recent* derives from the Latin *recens*, meaning 'fresh, young or lately made or done' and hence means that the soil has been recently formed, and is likely to be still forming by addition of sediment (including alluvium and dust), tephra, landslide debris, or by weathering of rock fragments. 'Recent' means young in age but the term is used in the context of the geological time scale. It means the soils are only a few hundreds to a few thousands of years old. Recent Soils are recognised globally, and named 'Entisols' in *Soil Taxonomy*.

14.2 Soil Profile Genesis

The genesis of Recent Soils varies depending on the geomorphic or geologic processes that have led to the generation or exposure of a new parent material, and hence the initiation of soil formation. Such processes include:

- the deposition of alluvial sediments by swollen rivers;
- gravity-induced mass movement (e.g. landsliding) on hill or mountain slopes generating erosion scars, and the exposure of underlying rock or regolith, as well as mass movement deposits on lower slopes;
- the aeolian deposition of sand along a coast or river bank; or
- tephra deposition following a volcanic eruption.

The repetition of geomorphic/geologic events, such as the regular deposition of sediment by flooding, or tephra-fall, every few decades or centuries, means that many Recent Soils are formed on multi-layered deposits through upbuilding pedogenesis, with the land surface slowly rising at a rate that depends on the frequency and thickness of the fresh geologic additions to the profile. Because soil weathering in Recent Soils is always in an early stage, and horizon transformation is restricted to the uppermost part of the profile, the character of the unaltered parent material has a strong influence on the soil's overall properties.

Regardless of the parent material and its origins, the key process in Recent Soil development is the formation of an A horizon by the input, and continuing accumulation, of soil organic matter (a process known as melanisation). In more developed Recent Soils, along with most other soil orders, organic matter accumulation is balanced by decomposition (humification) providing a quasi-equilibrium in soil organic matter content.

Organic matter is accumulated in all mineral soils by four paths:

1. growth and eventual death and decay of plant roots, which are especially strong under grasslands;
2. fall of plant (leaves, branches, and bark), and/or animal (urine, dung, and carcase) remains, to the soil surface with consequent decay and incorporation by soil animals;
3. chemicals released from roots as 'root exudates', for example, organic compounds that inhibit competing organisms; and
4. soluble organic chemicals, such as polyphenols, washed by rainwater from leaves and bark and channelled into the soil (as leachate) in the leaching water.

Paths 3 and 4 may only contribute small quantities of organic matter to the soil, but the compounds may be highly active as soil-forming agents. For example, polyphenols, washed from leaves and bark, act as potent chelating agents carrying iron, aluminium, and phosphorus ions from near-surface horizons, and releasing them deeper in the soil (see Podzol Soils, Chap. 11).

Many Recent Soils have a weakly developed B horizon where oxidation of iron-bearing minerals has led to the accumulation of enough rust-coloured iron oxide material, often goethite, to change the soil material from the colour of the parent material to a warmer brown colour (a process sometimes called brunification or rubification). Often the iron oxides just form a thin coat over the rock particles or mineral grains.

Fluvial Recent Soils are formed on river floodplains by deposition of sediment from floods. The soil profile often provides clues about the nature and frequency of past floods that have delivered the sediment. Understanding the sediment history is useful to support the management of flood hazards to life, crops, and infrastructure. Excellent examples are provided in the detailed soil survey reports of Alan Pullar in the Rangitaiki Plains (eastern Bay of Plenty) and in the Gisborne Plains.

Where alluvial sediments have been deposited by repeated floods, then the silt and sand layers may be visible as fine strata in the subsoil (e.g. Fig. 14.1). Thick layers of material indicate floods with higher sediment loads. Layers that include gravel-sized (or coarser) particles indicate floods, or former river beds, that had high energy (high water velocity and turbulence) to carry the larger material. On the positive side, sediments can add fresh soil material to build soil fertility.

Observing the thickness of the topsoil and any subsoil layers darkened by humus can give insights into the frequency, and amount, of material laid down in, flood or tephra deposition events (Fig. 14.4). Dark layers in the profile are usually the topsoil of a buried soil. Generally, the more strongly developed (thicker and darker coloured) the buried soil horizons, the longer they must have had to develop before the addition of new material cut them off from near-surface soil processes.

The Rotomahana Mud, on which the uppermost horizons are formed to a depth of 35 cm (in Fig. 14.4), provides an example of how rapidly a soil may develop from a Raw Soil to a Recent Soil. The A horizon has a clay content of up to 30%, which would normally indicate a period of considerable weathering and clay synthesis. However, in this case, the clay was 'pre-weathered' hydrothermal material, derived from land near (pre-1886) Lake Rotomahana that was violently engulfed and redeposited during the Tarawera eruption. Vegetation was rapidly established on the fine material and under the relatively cool, moist, climate organic matter was able to accumulate to form an A horizon in about 100 years.

Soil surveyors have recognised three kinds of Fluvial or Tephric Recent Soils: (a) non-accumulating Recent Soils with no evidence of younger sediments or buried soil horizons at or near the soil surface; (b) rapidly accumulating Recent Soils with evidence of buried topsoils or fresh sediment or thicker than normal topsoils at the soil surface; and (c) an intermediate category of moderately accumulating Recent Soils. The last two cases (b and c) exemplify retardant and developmental upbuilding pedogenesis (depending on thickness and frequency of deposits).

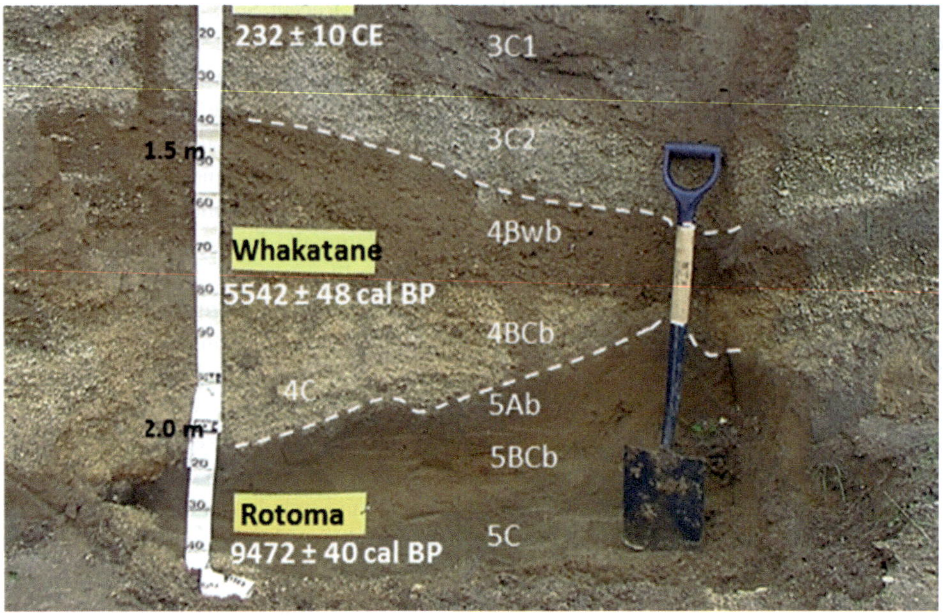

Fig. 14.4 A Buried-pumice Tephric Recent Soil near Mt Tarawera, showing a series of five buried soils stacked on top of one another. Periods of quiescence (pauses between eruptions) are marked by the development of (now) buried soil horizons which range in colour from black to dark brown. The names and ages or dates relate to the eruptions that generated the tephra deposits. The uppermost material, the Rotomahana Mud (with subordinate Tarawera Scoria at the base), was deposited over a period of about four hours on 10 June 1886. After Hartemink et al (2020)

14.3 Soil-Landscape Relationships

14.3.1 High Soil Variability in Alluvial Flood Plain Soils

Soils on flood plains are commonly highly variable, with soil properties changing at the scale of a paddock, or even over a few metres (Fig. 14.5). Depending partly on catchment lithologies, coarse sediments ranging from boulders, cobbles, gravels, and sands tend to be deposited within the river bed, according to the turbulence and flow in the river, and finer sand, silt, and clay materials in overbank flood deposits. As a river gradually migrates back and forth over a landscape, different layers build up over one another. In back swamp areas, peat deposits may form, only to be later buried by gravelly sands or finer sediments deposited by a later shift in the main river bed. The complexity that can result from such processes is evident in a braided river bed (Fig. 14.1). The pattern in soil parent materials, and thus the soils that form, is often visible from the air, particularly in dry summer periods (Fig. 14.5) where the stonier (coarser) soils dry out faster and show up as browner than neighbouring areas of finer materials that hold moisture for longer.

When less sediment is available, or there has been tectonic uplift, or a change in climate, the rivers gradually cut down through the sediments, becoming entrenched. The nature of the sediment, its texture, mode of deposition, stoniness, post-deposition erosion, wetness gradients, and management effects, all contribute to complex patterns of soil variability. It is desirable to map soil variability within areas that can be practically managed so that land management can be matched to soil properties. However, the soil variability is often so complex that a very large-scale map is needed. Precision agriculture is developing to address such issues of spatial variability.

The base level, the lowest level to which a stream can erode, influences the landforms and sedimentation patterns that control the development of Fluvial Recent Soils. The base level is set ultimately by sea level, with lakes, and bedrock highs, providing sub-base levels within a catchment. Where land is above the base level, a river will have a reasonable slope, and so flows swiftly and has potential energy to cut into its bed, erode sediment, and entrench its self into the landscape. Where a river is close to base level, slopes are low and in such a near-flat situation a river has no power to erode its bed but is able to transport and deposit sediments carried from upstream erosion. Many of the highly productive alluvial basins of New Zealand, such as the lower Manawatu, are formed by material deposited near a base level. The Rangitaiki Plains in eastern Bay of Plenty are subsiding tectonically and the land is remaining above sea level (maintaining freeboard) mainly because of recurring volcanic events and flooding that have deposited new material during the Holocene. Tectonism, volcanic eruptions, storminess, an easily eroded catchment rock (mudstone), and several large meandering rivers have similarly combined to control the development of the Gisborne Plain (Poverty Bay Flats) and its detailed fluvial history and predominant, fertile, Fluvial Recent Soils.

Fig. 14.5 Soil variability, reflecting former river flow paths and consequent particle size variability and associated soil moisture holding capacities, in a landscape dominated by Fluvial Recent Soils on the Canterbury Plains, near the Waimakariri River, is evident in this aerial photograph

The Taieri Plain is another example of an alluvial basin close to the base level. Lake Waihola at the southern end is at sea level and controls the base level of the plain and, thus, the sedimentation regime. The Taieri River, and the smaller Silverstream, flow across the basin, through Lake Waihola, and then drain to the Pacific Ocean via an antecedent gorge. Where there is no appreciable down cutting by streams and rivers, into the surface of the plain, the soil materials are dominantly constructed by accumulated flood sedimentation (Fig. 14.6). However, in the higher reaches of the Silverstream the alluvial sediment has been weakly eroded. It is interpreted to be a relict lake floor, possibly a former extension of Lake Waihola. It is possible that the main erosion event occurred following a lowering of base level, by either tectonic uplift or lowering of sea level, with consequent dewatering of the former lake.

In the town of Mosgiel, sediment from the Silverstream enters in the northeast corner and splays out as a series of gentle fine sandy loam ridges with a channel along the crest in-filled with sand or sandy loam. The channels have the form of a distributary pattern because they distribute drainage of water outward to the former lake. The pattern is that of a delta that remains from pre-human times. There is a range of sedimentary features associated with flood plain rivers, such as those of the Taieri River. First, is the substantial natural levee of the river that meanders across the plain. Floods have frequently over-topped its bank and formed a large-scale levee/back swamp system. Second, is a fingered array of sediment called a crevasse splay. It is a point where the river levee was at one time breached and sediment rapidly spewed out and down the levee slope. The sediments are often wide-ranging in texture and with gravels indicating a disordered event. Third, below the crevasse splay feature (though not necessarily related) are two characteristics typical of levee systems: (a) a texture gradient from relatively coarse texture near the crest (where water

Fig. 14.6 Flood event on the Taieri River in the mid-1980s. This is an area where the stop bank is designed to overflow to remove water from the river and thus protect downstream areas from flooding. The flooded area still receives new inputs of sediment, whereas the protected areas do not usually flood, and so new flood deposits are prevented

velocities are highest) to fine textures in the back swamp where water velocities have slackened off; and (b) a drainage gradient from well drained soils on the crest, to poorly drained soils in the back swamp. Thus, the soils have a wide range of textures and drainage stages with Recent Soils on micro-topographical highs and Gley Soils or Organic Soils in the back swamp areas.

Flood events in New Zealand

In the early days of European settlement in New Zealand, drowning was described as 'the New Zealand death' such was the frequency of people being caught up in flood waters when trying to cross swollen rivers. Major flood events that caused catastrophic damage in New Zealand include the 1858 flood in the Wellington region (14 deaths), 1863 in Central Otago (about 200 deaths), 1897 in Hawkes Bay (12 deaths), 1938 in Hawkes Bay (21 deaths), 1988 Cyclone Bola on the east coast of North Island (3 deaths), 2014 in Northland (2 deaths), and 2020 in Southland and Otago (2 deaths). Many other floods have caused widespread damage to houses and farmland throughout New Zealand.

All regions of New Zealand are periodically subject to prolonged, high-intensity, rainfall events that lead to flooding. During floods, rivers overtop their banks, depositing alluvium on flood plains on which Recent Soils develop. The regular fronts that pass across the South Island drop large volumes of water in the Southern Alps, and on the West Coast, as the moisture-laden air is forced up over the mountain range. Rivers in steeply sloping catchments can rise rapidly, overtopping their banks both to the east and west of the Southern Alps. In the northern half of North Island, subtropical cyclones occasionally venture south far enough to bring extreme, prolonged, heavy rainfall.

Many flood plains are now protected from regular flooding by stop banks and flood control programmes, but periodically such defences are breached in exceptionally large storms or where weakness is exploited. Increased surface sealing by impermeable surfaces such as roads and roofs leads to lowered infiltration and more rapid runoff, exacerbating flooding in strongly urbanised catchments.

14.3.2 Soil Variability in Soft Rock Hill Country

Soft rock (papa or mudstone) hill country occurs predominantly in the extensive, productive, pastoral hilly land of the east coast from East Cape to the Wairarapa, and in the Rangitikei/Whanganui areas of southern North Island. The blueish grey papa/mudstones are sedimentary rocks derived from clays, silts, and sands that were deposited in the sea mainly during the Miocene period (between about 24 and 5 million years ago) which have been rapidly uplifted as a result of the collision of the Pacific and Australia tectonic plates. Because the rocks are relatively young and were not deeply buried they are not particularly indurated (cemented into hard rock). Thus, the papa has low strength compared with New Zealand's much older basement greywacke or schist rocks. Whereas basement rocks will breakdown by physical processes to form stony colluvium or alluvium, the soft rock papa readily disaggregates into the primary texture (sand, silt, or clay) particles. The papa in many subsoils is sufficiently soft that it can be cut with a spade.

Due to the rapid uplift and erodibility of the papa, the landscape tends to be hilly with steep gullies cut into the soft rock by streams and rivers. The soft rock hill country is susceptible to mass movement, especially landslides (slips), and mud- or earthflows. A soil slip is initiated when water accumulates at the contact between the soil and the underlying slowly permeable rock. Soil pores fill with water, and the weight of water-saturated soil, and consequent lowering of soil strength, opens crevasses in the soil cover. The mass of soil may crack across the slope and part company from the relatively stable ridge material. Once cracks have formed they provide a conduit for water to enter the interface between the soil and underlying papa.

During periods of prolonged heavy rain, the soil material becomes saturated and heavy. On steeper slopes, the slip material loses its coherency and a slide forms with a water-lubricated shear plane on the steep rock interface (e.g. see Fig. 4.7). Movement of the wet soil mass rapidly accelerates downslope and what is initially a landslide on a shear plane in the upper slope may transform into a mud-slide (earthflow) with material flowing on down the hill. Some material will be deposited on foot slopes where the flow loses momentum, as the slope declines, and runout deposits from mud/earth-slides can extend well down valleys. Some material may be carried by floodwaters into rivers where it may be deposited on river flood plains or carried out to sea to form the next generation of sedimentary rocks. A scar is often left with a steep exposure of raw blue papa at the head of the landslide. Deposits of displaced material, some still coherent with topsoils preserved, and some, just piles of saturated subsoil, are left on the lower slopes.

In the Taranaki-Whanganui region, Mike Crozier and Brad Pillans found that following recent deforestation, and associated slope destabilization, erosion rates increased to three times their previous rate. The forces involved in the erosion include earthquakes but the main culprit is rainstorms with a return period of about 10 years which, during a single event, can remove soil and underlying regolith from 5 to 15% of hillslopes over areas of up to 100 km^2. Michael Selby had earlier suggested that intense rainfalls are responsible for most of the slope erosion in the greywacke ranges of the lower and middle Waikato Basins. Crozier, in 2005, concluded that landsliding in New Zealand most commonly occurs in the form of multiple-occurrence landslide events, simultaneously involving thousands to tens of thousands of landslides over areas extending up to 20,000 km^2. Agricultural production loss through loss of soils, damage to road and rail infrastructure, and increased flooding, are the main consequences of these events.

Over time, especially in the years since the native forest was removed, continuing erosion and recovery in many parts of New Zealand has left a legacy of high soil variability. There are some discernible order to the soil variability on soft rock hill slopes, with a soil mosaic pattern, including Brown Soils (Chap. 2), Recent Soils, and Raw Soils (Chap. 13), related to landscape position, erosion slip processes, and time since erosion occurred. A number of different zones are often apparent.

1. *Ridges.* The ridges are generally relatively stable with erosion scars rarely extending to the ridge crest. Soils on the ridges are usually the least variable, often with strongly developed Brown, Pallic, Ultic, Podzol, or Allophanic Soils.
2. *Stable hillslopes.* Areas of the hillside that have not been subject to landslide activity for a considerable time (centuries or millennia) will usually have mature soils that relate to the climate of the area, ranging from Pallic through Brown to Podzol and Ultic Soils. In areas, marginal to the central North Island volcanoes, Allophanic Soils may predominate on the less-steep slopes. However, on steeper slopes ($>\sim 26°$), the tephras may have been largely removed by surface erosion and so Granular, Ultic, or Brown Soils will be formed in the previously underlying, more strongly weathered, clay-rich materials.
3. *Erosion head-scarps.* Head-scarps are exposed shear planes and usually occur in the lower shoulder slope or upper back slope of a hillside. Head-scarps are generally

areas of newly exposed, relatively unweathered, papa. Initially, such areas would be classified as Raw Soils. Because of the lack of soil material, they have low water holding capacity and steep slopes so will be relatively slow to re-establish vegetation and develop from Raw to Recent Soils.

4. *Translation zones.* Translation zones are the mid-slopes of the hillside over which landslide material has travelled. In the translation zone, the slip material reaches maximum velocity, often forming a slurry that is channelled downwards. Some landslides may deepen by stripping soil at the base of the slide down to a resistant layer and removing regolith material. The turbulent slough of material may over-flow the channel margins leaving levee-like deposits. The force of soil and water has great competency, sufficient to move large rocks and trees. Within the translation zone, the landslide will leave uneven surface erosion surfaces and deposited materials from which Raw Soils are formed. Where sand- or silt-dominated material is deposited the soil may be relatively fertile and pasture may be fairly quickly re-established. Soils on such sites will usually develop from Raw to Recent Soils within a few decades. On eroded areas, the rate of soil development will depend on the nature of the underlying material, but it may be many decades before the topsoil becomes sufficiently established for soil to qualify as a Recent Soil.
5. *Arrest zones.* Movement of the slip may be arrested, at least in part, where the land slope flattens, the slurry loses water, or where it is captured by a stream. In many sites, the base comprises disorderly debris of sediment, rock, or regolith material and conveyed vegetation. In some places, large blocks may remain intact and so the deposited soil may retain its original properties, which could range from those of a Recent Soil to the predominant mature soil in the region such as a Brown, Pallic, Allophanic, or Ultic Soil. However, most materials in the arrest zone will probably be highly disturbed and hence will likely initially be classified as Raw Soils. In contrast in the transition zone, topsoils may develop relatively quickly and thus progress to Recent Soils. Often arrest zones are hummocky and pockets of cut-off drainage will occur where Gley Soils may develop.
6. *Foot slopes.* The foot slopes of the papa hill country are usually fan-shaped, gently sloping areas. They may be cut by steep gullies where streams come through. The foot slopes are predominantly formed from colluvium, some carried down and deposited in large mass movement events, some just as a result of minor surface erosion material being carried by overland flow. Often foot slopes are marked by springs where groundwater seeps from the adjacent hillside and they also receive runoff water in heavy rainfall. Hence, the soils are often wet for prolonged periods, as evidenced by rushes and the presence of Gley Soils.

Thus, the soils of landscapes where soil slip erosion is common are highly variable with a mosaic of soils formed in a complex, but generally predictable, pattern depending on landscape position and the time since the last erosion or depositional event at any given site. One soil surveyor who worked intensively in alluvial floodplain soils sought experience in eroded soft rock hill country. He reported that, though alluvial floodplain soils had set a standard for complex variability, the soft rock terrain variability greatly exceeded the soil complexity of alluvial terrain. For example, in a sample of 11 farms in the Pohangina Valley of the Manawatu, a soil survey reconnaissance revealed at least 21 soil subgroups.

14.4 Key Soil Properties

14.4.1 Soil Composition

Because Recent Soils occur throughout New Zealand, in a wide range of environments, their properties may vary considerably from region to region. For example, soils in a high rainfall climate may have higher soil organic matter and greater acidity, than soils of comparable age with similar parent materials in a semiarid climate.

The soil clay mineralogy varies with the inherited mineralogy of the parent material. For instance, in a landscape dominated by Ultic Soils, the associated Recent Soils may contain substantial kaolin subgroup clays. In less weathered terrain, illite group minerals will be more likely, and in tephric terrain, the primary materials will often be dominated by volcanic glass. Soil development is mostly confined to organic matter accumulation and humification in topsoils. The B horizons, if present, may be expressed by a difference in colour in comparison with that of the lower subsoil, indicative of some oxidation of iron-bearing minerals, but the limited time for the development of Recent Soils precludes much formation of secondary minerals, thus the mineralogy remains predominantly that of the parent materials.

14.4.2 Physical Properties

Recent Soils have high spatial variability in soil texture, reflecting the texture of the parent materials. It is common to have a stratification of contrasting materials in subsoils. However, as a consequence of the limited time for weathering, clay contents in Recent Soils are generally less than about 20% with silt- and sand-dominating soil textures (Fig. 14.7, Table 14.1). An exception is the Rotomahana soil

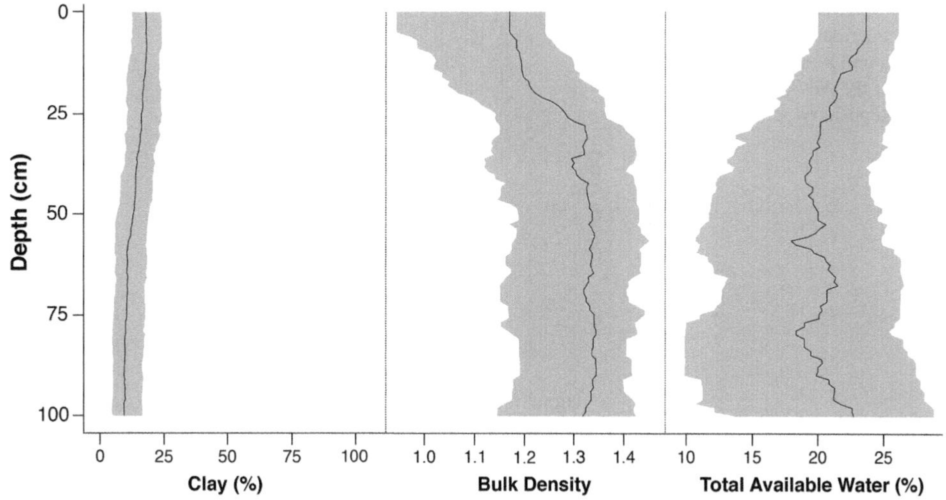

Fig. 14.7 Median and upper and lower quartiles of clay content, soil dry bulk density (t m^{-3}), and total available water holding capacity for Recent Soils in the New Zealand Soil Data Repository

Table 14.1 Typical example of physical properties for a Weathered Fluvial Recent Soil (Waimakariri series, SB10105)

Horizon	Depth (cm)	Sand (%)	Silt (%)	Clay (%)	Dry bulk density (t m^{-3})	RAW (%)	TAW[a] (%)
Ap1	0–18	33	49	17	1.2	12.7	26.0
Ap2	18–27	33	49	17			
BA	27–38	41	46	12	1.4	11.5	18.4
Bw	38–46	48	43	8	1.3	13.3	20.3
Cu1	46–54	58	37	4			
Cu2	54–80	76	19	4	1.4	9.2	17.0
Cu3	80–113	59	34	6	1.4	9.2	17.0
Cu4	113–124	60	32	7			
2bBw1	124–134	14	63	22			
2bBw2	134–160	24	57	18			

[a]RAW = readily available water, TAW = total available water, both in % by volume

near Mt Tarawera (Fig. 14.4) with a tephra fall-derived parent material, the Rotomahana Mud erupted during the Tarawera eruption of 10 June 1886, containing ~30% clay from weathering prior to its eruption. Soil dry bulk densities tend to be relatively high in subsoils but somewhat lower in topsoils. Recent Soils are usually deep rooting unless rock or dense clay is present. Consequently, except for rocky or stony soils, Recent Soils typically have high-profile plant-available water capacity. Good drainage is common except where there are shallow water tables.

14.4.3 Chemical Properties

Recent Soils generally have slightly acid to moderately acid pHs, low organic carbon, and low P retention (Fig. 14.8, Table 14.2). In Fluvial Recent Soils, organic carbon values may fluctuate in the subsoil with the presence of buried soils. Recent Soils are usually naturally fertile and base saturations are usually high (meaning the cation exchange complex of the soil is well supplied with the cations calcium, magnesium, potassium, and sodium) except in high rainfall areas where soils are likely to be more leached and acidic.

14.4.4 Biological Properties

A continuous cover of vascular plants is normally well established. Recent Soils are among New Zealand's most productive and versatile soils, and if soil productivity is used as an indicator of soil biological health, then under good management the soils are expected to favourably support soil biodiversity.

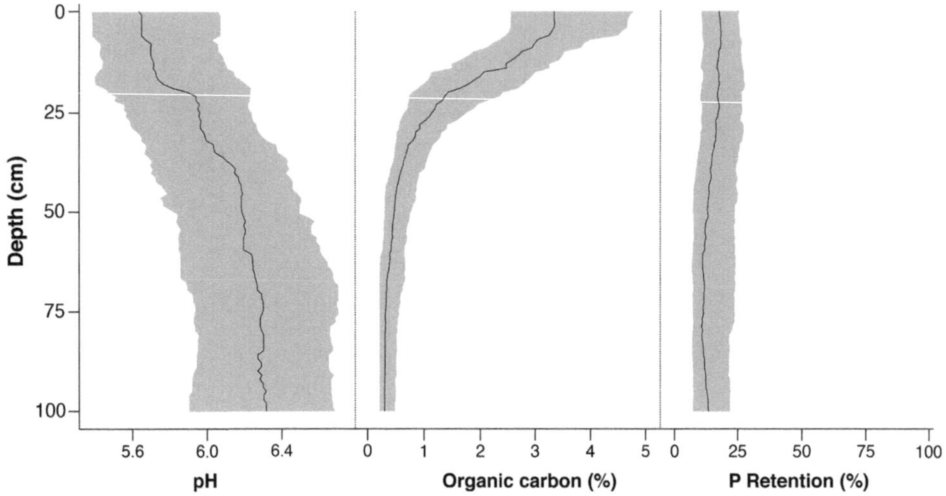

Fig. 14.8 Median and upper and lower quartiles of soil pH, organic carbon, and P retention for Recent Soils in the New Zealand Soil Data Repository

Table 14.2 Typical example of chemical properties for a Weathered Fluvial Recent Soil (Waimakariri series, SB10105)

Horizon	Depth	pH (in H_2O)	Carbon (%)	Nitrogen (%)	CEC[a] (cmol$^{(+)}$/kg)	Sum bases (cmol$^{(+)}$kg)	P retention (%)
Ap1	0–18	5.5	2.7	0.23	11.9	8.50	15
Ap2	18–27	5.4	2	0.17	10	6.76	16
BA	27–38	5.4	1.2	0.11	7.4	5.69	15
Bw	38–46	5.5	0.63	0.06	6.4	4.00	12
Cu1	46–54	5.5	0.37	0.04	4.3	3.18	10
Cu2	54–80	5.9	0.2	0.03	3.2	2.58	6
Cu2	80–113	6.1	0.16	0.03	3.3	2.93	5
Cu3	113–124	6.4	0.11	0.02	2.8	2.60	4
2bBw1	124–134	6.7	0.25	0.04	9.1	9.02	13
2bBw2	134–160	6.7	0.28	0.04	8	7.92	12

[a]CEC = cation exchange capacity

14.5 Distinguishing Between Recent Soils and Related Soil Orders

In the *New Zealand Soil Classification* key to soil orders, Recent Soils are almost at the end of the key (being followed only by Raw Soils). The key acts as a logic filter and Recent Soils are identified by lacking the features that distinguish other soil orders that have priority earlier in the key.

The Recent Soils are frequently associated with landscapes where active, or recently active, erosion and sedimentation processes are prevalent. Common landscape associates are Raw Soils that differ by having no B horizon, no topsoil, or a topsoil less than 5 cm thick. Gley Soils occur where high ground water tables are prolonged and Organic Soils occur where water tables are permanently high and there has been an accumulation of humified plant material.

14.6 Correlation with Other Classification Systems

The Recent Soils mainly correlate with the Entisols of *Soil Taxonomy* and with a variety of soil orders in the *World Reference Base* (Table 14.3). They were formerly classified as Recent soils in New Zealand.

14.7 Use and Management of Recent Soils

Recent Soils on flat or gently sloping land are some of the most versatile and productive dairying, grazing, and cropping land in New Zealand. However, there are a number of potential hazards that must be considered, particularly the risk of flooding, hillslope mass movement and deposition,

Table 14.3 Generalised correlation[a] between Recent Soils and *Soil Taxonomy, World Reference Base,* and the New Zealand genetic soil classification

New Zealand Soil Classification	*Soil Taxonomy*	*World Reference Base*	NZ genetic soil classification
Hydrothermal Recent Soils	Orthents	not recognised	not recognised
Rocky Recent Soils	Lithic subgroups of Cryorthents Ustorthents Udorthents	Leptosols	Skeletal soils
Sandy Recent Soils	Psamments	Arenosols	Recent soils
Fluvial Recent Soils	Fluvents,Inceptisols	Fluvisols, Umbrisols, Cambisols or Regosols	Recent soils
Tephric Recent Soils	Vitrandic subgroups of Orthents	Regosols, Leptosols, or Cambisols	Recent soils
Orthic Recent Soils	Typic, Aquic, or Cryic subgroups of Orthents	Leptosols	Recent soils

[a]The correlations given here are a guide only and, for accurate classifications, the relevant soil classification documents should be consulted. The two major international soil classification systems are *Soil Taxonomy,* which was developed in the USA, and *World Reference Base,* which was developed primarily in Europe. The NZ genetic soil classification was used in NZ prior to 1992

streambank erosion, and reactivation of wind-blown sand dunes.

Most Fluvial Recent Soils are naturally fertile because they have received regular inputs of alluvial (flood deposited) material and, because of limited time, leaching has not yet removed cations such as calcium, magnesium, and potassium. The soil consistence is usually weak and friable, enabling roots to explore all of the soil volume. Thus, plants grow well and soil organic matter accumulates which provides good storage of phosphorus and nitrogen. Therefore, the Fluvial Recent Soils are highly productive flat land and so are some of the most intensively used soils (including farming and horticulture, Fig. 14.9), such as on the Gisborne, Heretaunga, and Rangitaiki Plains, on the flats near Dargaville, around Otaki and Te Horo on the Kapiti Coast, and in the Marlborough and Tasman (Nelson) regions, including Waimea Plains. However, if products are harvested then, to be sustainable, it is necessary to add fertilisers to replace the nutrients removed with the crops.

Many New Zealand towns are situated, at least in part, on Fluvial Recent Soils. Flat sites adjacent to river mouths, or upstream where rivers were navigable, were important areas for human settlements. Such towns include Hokitika, Westport, and Greymouth on the West Coast of the South Island where high rainfall in the Southern Alps makes them particularly prone to intense flood events. Invercargill, Mosgiel, Blenheim, Palmerston North, Whanganui, Wairoa, Gisborne, Opotiki, Whakatane, Kawerau, Edgecumbe, Paeroa, Thames, and Whangarei are also towns where serious flooding has occurred.

In many regions, Fluvial Recent Soils have been protected from flooding by the construction of river stop banks that prevent flooding, and thus further additions of alluvium (which also has the consequence of increasing the amount of flood water and sediment carried directly to estuaries and the marine environment). It is easy to become complacent about flood risk in areas protected by stop banks. However, occasional large floods may overtop or breach any flood protection, inundate the land, and deposit fresh alluvium, such as occurred in Kawerau and Edgecumbe in 2017, in Whanganui in 2015 (described as the worst flood on record), and in Whangarei and Northland in 2020. Thus any infrastructure development on flood plains needs to be carefully considered.

Buildings on alluvial plains may be constructed on stilts, or with raised floor levels, to increase resilience to flood events. Roads may be constructed on raised platforms, and other infrastructure may best be installed on higher land. New Zealand has invested a great deal in flood protection works, including stop banks, and designed overflow mechanisms (eg. Fig. 14.6) to protect our productive, and often intensively settled, alluvial plains. Ongoing diligence in the maintenance and protection of flood protection works is important. Such work sometimes meets resistance, due to the cost, as the flood protection has been so effective that people forget there is a hazard.

Given that flood plains contain a river or stream, there is also a risk of stream bank erosion, as rivers naturally tend to migrate across their flood plains, cutting into the outside of bends and depositing material on the inside. There have been recent incidences of stream bank erosion causing problems with cutting into old, abandoned, rubbish dumps on the Haast River on the West Coast of the South Island, and on the Manawatu River in Foxton. Thus, a long-term vision needs to be applied when siting infrastructure on flood plains.

Fluvial or Sandy Recent soils with silt or sandy substrates and high groundwater tables are particularly susceptible to

Fig. 14.9 Intensive orchard and vegetable production compete with lifestyle blocks and expanding suburbs on Fluvial Recent Soils near Christchurch

liquefaction in earthquakes. While liquefaction events are relatively infrequent, the damage to infrastructure may be significant as experienced in the Christchurch area after the 2011 and 2012 earthquakes. In areas of risk, buildings can be designed for resilience to liquefaction. Farmland generally recovers fairly readily from liquefaction damage.

A few New Zealand towns, including Te Aroha (in the Waikato region), Matatā (in the Bay of Plenty), and Mosgiel (in Otago), are built on gently sloping alluvial fans. Such fans have formed from material, often including large boulders, carried in debris flows from adjacent hill and mountain streams. Occasional major storm events have brought great damage to both Te Aroha and Matatā. At the time of writing, it was proposed that all houses be removed from the parts of Matatā that are at risk, with the Whakatāne District Council (supported financially by Bay of Plenty Regional Council and central Government) offering to buy homes at risk from further debris flows at market value as part of a managed retreat. Several homes have already been demolished under this scheme, which was seen to be more cost-effective than attempting to remediate and mitigate against further debris flow events.

There are a number of areas of Sandy Recent Soils, particularly on the west coasts of both the North and South Islands, formed on sand dunes that have been stable for a few hundred years (Chap. 13). Sandy Recent Soils have been developed for pastoral farming or forestry. Such soils should not be ploughed, and grazing animals, or forest harvesting, need to be managed carefully, as any loss of vegetation cover may lead to wind erosion and reactivation of the dunes. Large moving dunes are difficult to stabilise and wind-blown sand moving across the landscape has potential to bury anything in its path. Some areas, such as those in South Westland, are under protection for conservation as unique ecosystems.

Orthic Raw Soils are common in the soft-rock hill country. After erosion has stripped the upper soil profile leaving a Raw Soil, how quickly can a Recent Soil develop and the former productivity recover? Is recovery under the pastoral management sufficiently rapid that erosion is not of significant concern for sustainable production in this landscape? The effect of soil slip erosion on pasture production in soft rock hill country, at Te Whanga in the Wairarapa, was investigated by Keith Vincent and Derek Milne in 1990 and followed up 20 years later by Brenda Rosser and Craig Ross. The soils on soil slip scars, initiated during storms in 1941, 1961, and 1977, were compared with soils on adjacent uneroded sites to evaluate rates of recovery. The work showed that production had returned to about 80% of uneroded levels 70 years after the landslide event, prompting the conclusion that it is unlikely that pasture production on the slip scars will return to pre-erosion production levels in human time scales. This finding has implications for sustainability. The slow recovery was attributed to the loss of topsoil, well as the immediately underlying subsoil, meaning that recovery had to occur within the remaining deeper, poorly structured, less weathered, and more compacted, parent materials.

In plantation forestry on hilly to steepland, including Raw and Recent Soils, north of Napier, Marie Heaphy and others showed that soil total nitrogen, total carbon, total phosphorus, and organic matter were lower, and tree volume was 10% lower, in eroded plots than non-eroded plots. Based on an assessment of log quality, trees in the eroded plots were forecast to produce 16% less volume from high-quality pruned logs than trees in non-eroded plots. In contrast, Mark Smale, Malcolm McLeod and others reported no difference in growth rates of kanuka/manuka (*Kunzia ericoides/Leptospermum scoparium*) stands on landslide scars compared to stands on uneroded areas in the East Cape region. There is potential for more work to be undertaken to determine the length of time needed to develop from a Raw to a Recent Soil in a range of different soil parent materials and climates.

Further Reading

Almond PC, Tonkin PJ (1999) Pedogenesis by upbuilding in an extreme leaching and weathering environment, and slow loess accretion, south Westland, New Zealand. Geoderma 92:1–36

Beecroft FG, Hewitt AE, Smith SM (1991) Soils of the Taieri Plain, northeast Taieri River, Otago, New Zealand. DSIR Land Resour Sci Rep 17:65p

Blaschke PM, Trustrum NA, DeRose RC (1992) Ecosystem processes and sustainable land use in New Zealand steeplands. Agr Ecosyst Environ 41:153–178

Briggs RM, Lowe DJ, Esler WR et al (2006) Geology of the Maketu area, Bay of Plenty, North Island, New Zealand—Sheet V14 1:50 000. Department of Earth Sciences, University of Waikato, Occasional Report 26. 44 p + map

Chittenden ET, Hodgson L, Dodson KH (1966) Soils and agriculture of Waimea County. NZ, NZ Soil Bureau Bull 30. 66 p

Crozier MJ (2005) Multiple-occurrence regional landslide events in New Zealand: hazard management issues. Landslides 2:247–256

Crozier MJ, Pillans BJ (1991) Geomorphic events and landform response in south-eastern Taranaki, New Zealand. Catena 18:471–487

Crozier MJ, Gage M, Pettinga JR et al (1992) The stability of hillslopes. In: Soons JM, Selby MJ (eds) Landforms of New Zealand, 2nd edn. Longman Paul, Auckland pp 63–90

Fransen P, Brownlie R (1995) Historical slip erosion in catchments underpasture and radiata pine forest, Hawke's Bay hill country. New Zealand Forestry (November), pp 9–33

Glade T (1998) Establishing the frequency and magnitude of landslide-triggering rainstorm events in New Zealand. Environ Geol 35(2–3):160–174

Hartemink AE, Zhang Y, Bockheim JG et al (2020) Soil horizon variation: a review. Advances in Agronomy 160:125-185

Heaphy MJ, Lowe DJ, Palmer DJ et al (2014) Assessing drivers of plantation forest productivity on eroded and non-eroded soils in hilly land, eastern North Island, New Zealand. New Zealand J For Sci 44:24 (10p)

Hicks M, Shankar U, McKerchar AI et al (2011) Suspended sediment yields from New Zealand rivers. J Hydrol (NZ) 50(1):81–142

Lambert MG, Trustrum NA, Costall DA (1984) Effect of soil slip erosion on seasonally dry Wairarapa hill pastures. New Zealand J Agric Res 27:57–64

Lynn IH, Manderson AK, Page MJ et al (2009) Land use capability survey handbook, 3rd edn. AgResearch/Landcare Research/GNS Science, 163 p

Marsaglia KM, DeVaughn AM, James DE (2010) Provenance of fluvial terrace sediments within the Waipaoa sedimentary system and their importance to New Zealand source-to-sink studies. Mar Geol 270:84–93

Mazengarb C, Speden IG (compilers) (2000) Geology of the Raukumara area. Institute of Geological & Nuclear Sciences 1:250 000 geological map 6 (1 sheet + 60 p)

Page MJ, Reid LM, Lynn IH (1999) Sediment production from Cyclone Bola landslides, Waipaoa catchment. J Hydrol NZ 38 (2):289–308

Pillans BJ, Pullar WA, Selby MJ et al (1992) The age and development of the New Zealand landscape. In: Soons JM, Selby MJ (eds) Landforms of New Zealand, 2nd edn. Longman Paul, Auckland, pp 31–62

Pullar WA 1962. Soils and agriculture of Gisborne Plains. NZ Soil Bureau Bull 20. 90 p

Pullar WA (1985) Soils and land use of Rangitaiki Plains, North Island, New Zealand. New Zealand Soil Survey Report 86 (75 p + map 1: 31680)

Pullar WA, Penhale HR (1970) Periods of recent infilling of the Gisborne Plains basin. Associated marker beds and changes in shoreline. NZ J Sci 13:410–434

Pullar WA, Selby MJ (1971) Coastal progradation of Rangitaiki Plains, New Zealand. N.Z J Sci 14:419–434

Pullar WA, Hewitt SR, Heine JC (1978) Soils and land use of Whakatane Borough and environs, Bay of Plenty, New Zealand. New Zealand Soil Bureau Bulletin 38. 100 p + multiple maps

Rijkse WC (ed) (1985) Soil groups of New Zealand, Part 8. New Zealand Society of Soil Science, Recent Soils, 44 p

Smale MC, McLeod M, Smale PN (1997) Vegetation and soil recovery on shallow landslide scars in Tertiary hill country, East Cape region, New Zealand. NZ J Ecol 21(1):31–41

Sparling G, Ross D, Trustrum N et al (2003) Recovery of topsoil characteristics after landslip erosion in dry hill country of New Zealand, and a test of the space-for-time hypothesis. Soil Biol Biochem 35:1575–1586

Trustrum NA, DeRose RC (1987) Soil depth: age relationship of landslides on deforested hillslopes, Taranaki, New Zealand. Geomorphology 1:143–160

Vincent KW, Milne JDG (1990) Soil pattern or Te Whanga trail site, Wairarapa: soil landscapes, effects of erosion, and soil amelioration in steep eroded siltstone hill country. Unpublished DSIR Land Resources report

Wigley GNA (1990) Holocene tephrochronology and evolution of the Te Puke lowlands, Bay of Plenty, New Zealand. Unpublished MSc thesis, lodged in the Library, University of Waikato, Hamilton

Semiarid Soils

15

In the drylands of Central Otago
on Semiarid Soils thyme doth grow.
Through summer sun and winter
frost all moisture to the air is lost.

Abstract

Semiarid Soils form in the inland basins of Central Otago and South Canterbury under rainfalls of less than about 500 mm yr^{-1}. The soils have an annual moisture deficit with hot summers and cold winters. Parent materials are predominantly alluvium, colluvium, and loess derived from greywacke and schist. Semiarid Soils range from immature gravelly sandy soils, on low terraces, to moderately weathered soils with clay accumulation on older, higher, terraces, fans, and hills. Accumulations of salts and illuvial clays can occur in subsoils. Pockets of Solonetzic Semiarid Soils occur with high pH (>8.5) and sufficient sodium to preclude the growth of plants, except for a few specially adapted salt-tolerant plant species. With irrigation, Semiarid Soils can be developed for productive agriculture and horticulture. Stone fruit, including apricots and cherries targeting the Northern Hemisphere Christmas market, and grape growing for wine production are important industries in the region, along with merino fine wool. Although there has been concern about the potential for irrigation to cause salinisation, such fears have proven largely unfounded as the plentiful use of high-quality water, from the major rivers that flow through the region, along with good drainage, has tended to wash any salts from the soil.

15.1 Important Features of Semiarid Soils

15.1.1 Concept and Key Features of the Soil Order

Semiarid Soils (Fig. 15.1) form in the driest regions of New Zealand with rainfall ranging from about 300 to 500 mm per year. Semiarid Soils generally lack a strongly developed topsoil. However, many profiles show evidence of salt and/or clay accumulation. The presence of visible salts (e.g. Fig. 15.1 centre) is an indicator of the semiarid character of the soils. The Semiarid Soils (with some minor exceptions in north and south Canterbury) are the only soils in New Zealand where there is potential for significant natural salt accumulation. On some higher terraces, the soil is likely to be very old, possibly of the order of c. 350,000 years or so (e.g. Fig. 15.1 right). The clay in such soils is considered to be a very old, relict feature, and the marked dark reddish brown clay-enriched argillic horizon between 35 and 60 cm depth is a paleosol. Stoniness is common but there are areas of deep soils with few or no stones that have a high capacity to store water and so respond well to irrigation.

Much of the early soil science investigation in Central Otago was undertaken, particularly by John McCraw and Michael Leamy, to ensure that the development of irrigation

A. E. Hewitt et al., *The Soils of Aotearoa New Zealand*, World Soils Book Series,
https://doi.org/10.1007/978-3-030-64763-6_15

Fig. 15.1 Semiarid Soils and associated landscape. Left: landscape near Alexandra where wild flood irrigation provides some green pasture growth. Centre: Typic Immature Semiarid Soil on an intermediate terrace (Molyneux shallow stony sandy loam). White calcium carbonate is visible at the base of the B horizon. Right: Aged-argillic Semiarid Soil on a high terrace (Lowburn stony sandy loam). (Soil profile photos reproduced with permission of NZ Society of Soil Science.)

Fig. 15.2 Distribution of Semiarid Soils

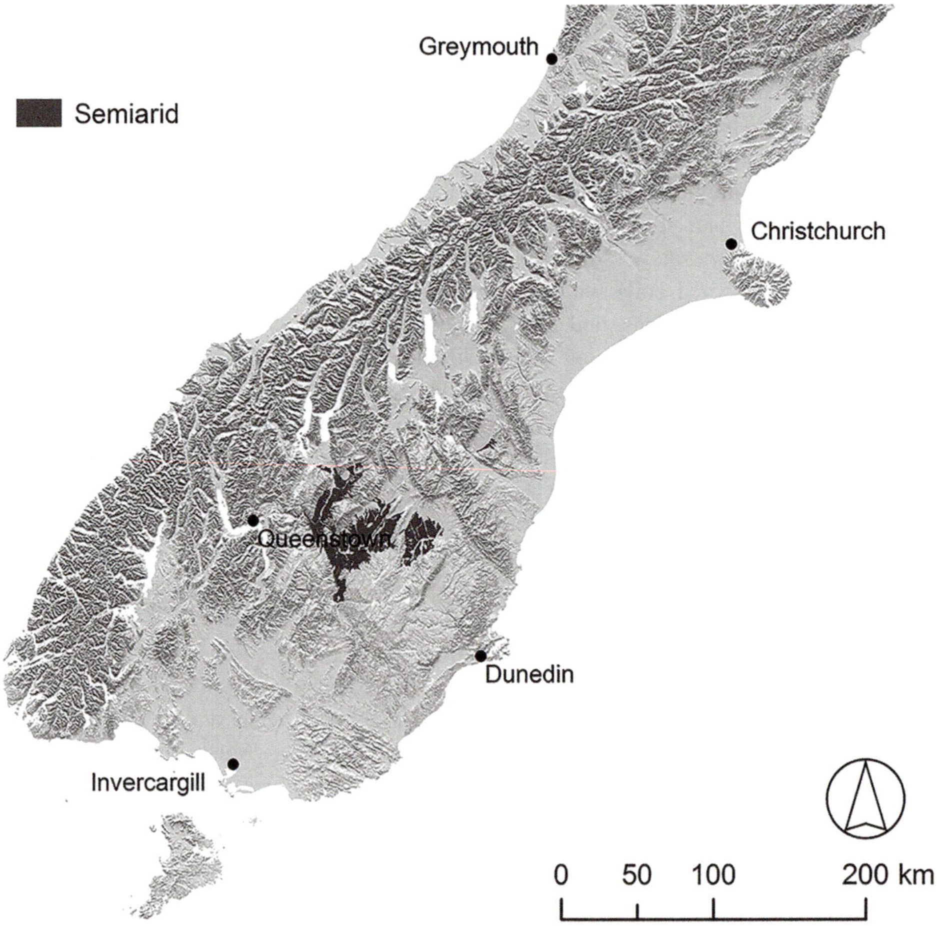

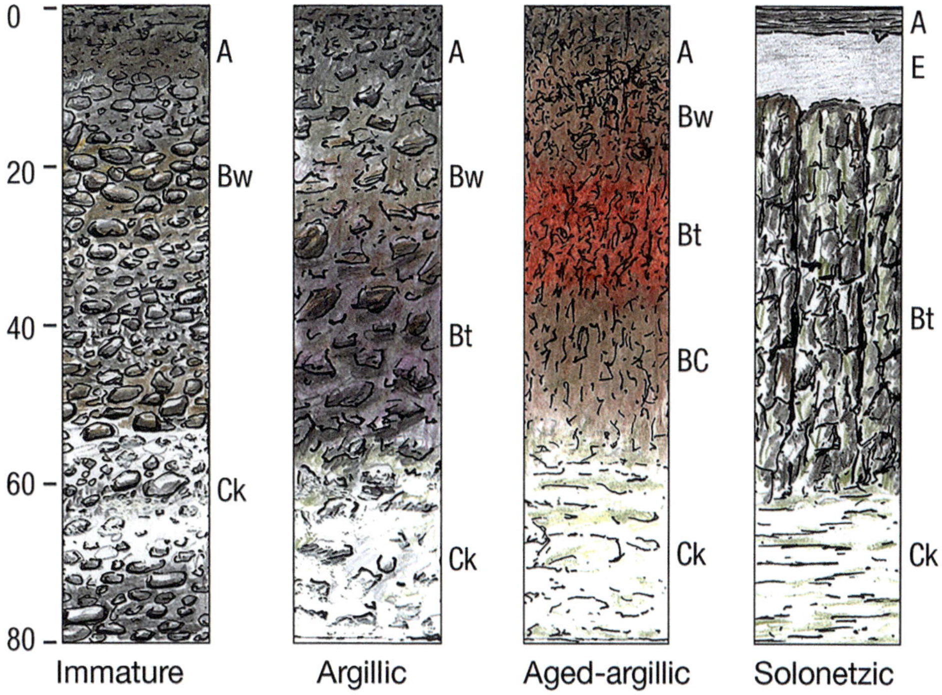

Fig. 15.3 Groups within the Semiarid Soil order

did not lead to adverse effects of salt accumulation or redistribution. The earliest soil-related mapping in Otago was undertaken by Hartley Ferrar in the 1920s.

Central Otago has the lowest national recorded mean annual rainfall, 212 mm (in Alexandra in 1964), the lowest national recorded air temperature of −25.6 °C (in Ranfurly in 1903), and some of the highest recorded temperatures in the country. Central Otago is the region farthest from oceanic influence, experiencing a near continental climate, explaining the temperature extremes. The surrounding mountain ranges provide a rain shadow situation which explains the low annual rainfall in the dry inland basins. Semiarid Soils are dry for most of the growing season as, although the highest rainfalls occur in the summer months, the moisture is rapidly lost because of high rates of evapotranspiration at that time of year. Rain is not sufficient in most years to leach through the soil (Fig. 1.4), and so salts can accumulate in the lower subsoil (Fig. 15.1). Thus, the soils have a high base saturation (rich in ions of Na, K, Ca, and Mg) although the cation exchange capacity may be limited because of the coarse textures and low clay and organic matter content of many Semiarid Soils. The lack of moisture limits plant growth and hence drought-tolerant plants predominate unless the land is irrigated.

Prominent ancient rocky outcrops of schist (tors) are an important and unique feature of the Central Otago landscape. The tors tower 10–20 m high above the surrounding surface and were formed by a combination of physical weathering, notably freeze-thaw cycles in periglacial conditions, during glacials, and differential chemical weathering during interglacials. Debris has been removed by wind and water erosion as well as by solifluction (the slow, downslope movement of water-saturated unconsolidated material, instigated by freeze-thaw processes).

15.1.2 Areas of Occurrence

Semiarid Soils occur in the inland basins of Otago and southern Canterbury, and in the upper Waitaki and Hakataramea valleys, where annual precipitation is less than about 500 mm (Fig. 15.2). They cover about 1% of New Zealand. Although not yet tested by fieldwork, it has been suggested by John Leathwick, from findings of the 'Land Environments of New Zealand' project, that some of the inland basins of Marlborough were climatically similar to those of Central Otago basins and therefore Semiarid Soils may also be present in them.

15.1.3 Variation Within the Semiarid Soil Order

Four soil groups are defined within the Semiarid Soils in the *New Zealand Soil Classification* (Fig. 15.3):

Immature Semiarid Soils—features are weakly expressed,
Argillic Semiarid Soils—clay accumulation in subsoils as thin coatings on peds or in pores,
Aged-argillic Semiarid Soils—reddish coloured clay accumulation in subsoils, and

Solonetzic Semiarid Soils—strongly developed argillic horizons and high proportion of sodium on clay surfaces in subsoils.

Soil subgroups, defined by the *New Zealand Soil Classification*, recognise soils that are related to other soil orders or soil groups (Mottled), or soils that depart in some specific way from the soil group or soil order (Weathered, Alkaline, Thick, Saline, or Laminar). The central concept of the soil group is the Typic subgroup.

The *New Zealand Soil Classification* expresses the varying levels of salinity at the subgroup level by identifying saline soils (with an electrical conductivity >0.8 mS cm^{-1}), and alkaline soils (pH $\geq$ 8.6). Sodic soils (where sodium ions dominate the cation exchange), with subsoil argillic horizons that have dark-coloured coatings (dispersed soil organic matter with clay), are recognised in the Solonetzic group.

15.1.4 Origin of the Soil Order Name

'Semiarid Soils' are named for the semiarid climate which is the dominant soil-forming factor that determines their occurrence. However, Semiarid Soils are not directly identified by climate but by a number of soil properties that are climate-related, including subsoil salt accumulation and low organic matter contents in topsoils.

15.2 Soil Profile Genesis

15.2.1 Introduction

The formation of Semiarid Soils is dominated by the effects of the climate, with a water deficit in most months of the year (Fig. 1.4), hot summers, and cool winters. The lack of water influences pedogenic processes (slowing weathering and supporting salt accumulation) and restricts plant growth (thus limiting the organic matter available to accumulate in the soil). The major soil-forming processes that influence the form of Semiarid Soils are the movement and accumulation of salts and clay within the soil. With increasing age, the soils form sequences that show increases in both salt and clay accumulation. Many of the land surfaces in the dry inland basins are old and so there has been time for advanced soil development to occur. The parent materials are predominantly quartzo-feldspathic in origin, and include alluvium derived from local schist deposits and from greywacke gravels carried from the mountain ranges and deposited in the river terraces by the Clutha, Manuherikia, Taieri, and other rivers. Locally derived colluvium, and loess blown off river flood plains, also form soil parent materials. Thus, it is likely that many of the Semiarid Soils reflect upbuilding pedogenesis whereby new geological materials, such as loess, are added to stable land surfaces, while at the same time pedogenic processes are actively modifying the parent materials. McCraw recognised that soils on high terraces and flatter parts of hills in the area typically included a mantle of loess which had accumulated incrementally, at a very low rate, over thousands of years. Thus, pedogenesis would continue (albeit at slow rates in the semiarid climate) unabated by the ongoing loess inputs.

15.2.2 Argillic Horizon Development

A characteristic of Semiarid Soils is the accumulation of clay in the subsoil (forming argillic, or illuvial clay-enriched, horizons) and the eventual reddening of the argillic horizons over time. In the Clutha and Kawarau river valleys, the age of river terraces and fan surfaces increases with height above the river level. In the Clutha River valley, Michael Leamy used reasoning similar to that outlined for the Rangitikei River sequence (Chapter 10) to describe a chronosequence of Semiarid Soils. The chronosequence displays the major pedogenic changes over time, and the range of soils evident in the Central Otago landscape. The chronosequence ranges from an Immature Semiarid Soil through a Saline Argillic Semiarid Soil and a Typic Argillic Semiarid Soil to a Weathered Aged-argillic Semiarid Soil (Fig. 15.4).

The Immature Semiarid Soil, the youngest in the sequence (although there are younger, Raw and Recent Soils, on lowermost terraces and modern floodplains), has no argillic horizon but subtle dark staining of the gravels described (misleadingly) as an "incipient clay pan". The staining is probably organic matter mobilised under the influence of weak sodium accumulation (with little or no illuvial clay accumulated at this stage of soil development). A salt-rich calcareous horizon (Ck) lies beneath the Bw.

The Saline Argillic Semiarid Soil, second in the sequence, has a weakly developed argillic horizon stained by soil organic matter, described as a brown clay pan, and underlain by a calcareous horizon. The Typic Argillic Semiarid Soil is third in the sequence with a red argillic horizon described as a thin red clay pan, and underlain by a calcareous horizon.

The Weathered Aged-argillic Semiarid Soil is fourth in the sequence with a red argillic horizon identified by Leamy as a thick red clay pan. It is inferred that the Weathered Aged-argillic Semiarid Soil has developed through each of the preceding three stages. If true, then the calcareous horizon must have formed in the earlier stages and then been subsequently removed by leaching. A feature of the materials of the first three stages of soils is that the spaces between the stones and gravel are infilled with permeable

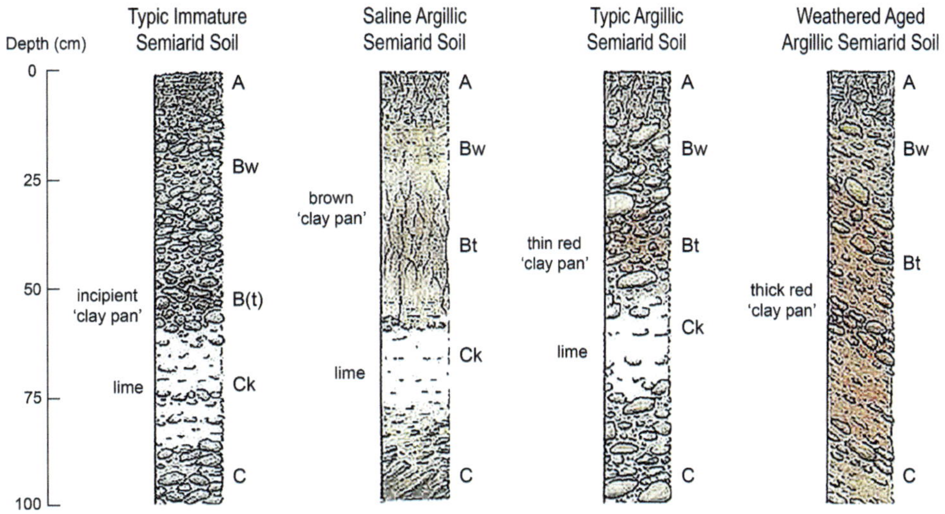

Fig. 15.4 Four stages in the development of Semiarid Soils were recognised by Leamy in the Lower Manuherikia Valley on land surfaces (river terraces and fans) that increase in age from left to right

sands or loamy sands. However, in the Weathered Aged-Argillic Semiarid Soil the gravel interstices are infilled with sandy clay. The clay-bound gravels are common on the older, higher, land surfaces and it is understood that the clay has been moved in suspension and 'filtered' amidst the gravels over the long period since the gravel deposition.

Leamy suggested that the oldest soils with very thick, red clay pans represented the accumulation of clay for at least four glacial-interglacial cycles inferred to date back to a period described geologically as the Porikan stage, which is now dated at c. 350,000 years ago. One hypothesis regarding the red colouration of the argillic horizons is that the red colour is derived from additions of red-weathered clay-sized dust particles blown across the Tasman Sea from Australia. Such material (much finer than local, New Zealand-derived loess) is common in southern and eastern Australia and termed 'parna'. The wide distribution of red dust is documented in southern Europe, the Middle East, and oceanic islands with sources mainly in North Africa. Red dust has been observed falling in New Zealand following Australian dust storms, thus the hypothesis is plausible. In addition to the historical evidence of Australian-derived dust fall, analyses by Sam Marx and others of pre-historic dust accumulation in a peat bog on Old Man Range showed that the average deposition rate of Australian dust since c. 8000 years ago was 0.6 g m^{-2} yr^{-1}, with deposition rates as high as 1.4 g m^{-2} yr^{-1} during the mid-Holocene, implying that Australian dust has contributed >5 kg m^{-2} of dust to Otago (and other parts of New Zealand) during the Holocene period (since c. 11,700 years ago). Greater amounts would have accumulated during glacial periods as shown from studies of sediment in cores from the Tasman Sea.

An alternative hypothesis is that the red-weathered material is derived from clay that was originally red-weathered during the Cretaceous and which was then preserved on some old land surfaces in Otago (Chap. 16), and as far north as Wellington.

15.2.3 Salts, Sodium, and the Origin of Salinity and Sodicity

Most soils generate salts as a product of chemical weathering. Under higher rainfall the salts are leached from the soil into the groundwater and eventually pass through the hydrological network to the ocean. However, in Semiarid Soils, in most years, there is too little rain to wash the salts through the soil. Instead, the salts precipitate and accumulate in the soil. In Central Otago, the salts probably have three main sources: (1) from the chemical dissolution and hydrolysis of minerals in rocks and soil; (2) the deposition, with rainfall, of salts carried with water evaporated from the ocean (marine aerosol flux); and (3) aeolian dust carried across the Tasman Sea from Australia. Deposition with rain occurs in miniscule amounts, and the Australia aeolian dust influx is also relatively small, but accumulations nevertheless build up over thousands of years, in a dry, stable, landscape.

In Otago, the most saline (sodium salt-rich) areas occur where the weathered schist is almost totally transformed to kaolinite. Such areas act as reservoirs that slowly leak salt, via water tables, into the adjacent land. The highly weathered schist, and associated weakly saline mudstones (Manuherikia group sediments), are the common basement rocks beneath the fans and terraces. The area where salt accumulation has been most obvious is in the Maniototo Basin where saline patches have been observed from the time of the early settlers and presumably have formed naturally.

The white Ck soil horizons (designated as calcareous horizons in the *New Zealand Soil Classification*), which form in some Semiarid Soils, contain accumulations of salts, dominated by calcium carbonate (lime). No significant natural geologic limestone rocks occur in the Central Otago region and isotope analyses by Leamy and Athol Rafter established that the calcium carbonates were of pedogenic origin. The calcium carbonate is formed from the release (by hydrolysis) of calcium and other ions from silicate minerals in non-calcareous parent materials, typically gravelly and sandy alluvium, and carbon dioxide contributed from the respiration of microbes living in the pores of the soil.

Over time, sodium may displace calcium and accumulate in the soil. Sodium is of particular interest and concern as sodium accumulation may impact on soil structure and permeability as well as plant growth and survival. Sodium in soils occurs mainly in two chemical states. First, it exists as salts commonly as sodium chloride (common table salt) or sodium sulphate, together with calcium salts. In dry soil, the salts are crystalline whereas in wet soil they are dissolved in the soil water. The second state occurs when sodium cations become attached to the negatively charged cation exchange complex of the clay minerals and soil organic matter. These two chemical states of sodium give rise to three soil forms:

- *saline* soils dominated by sodium salts—e.g. Saline Immature Semiarid Soils;
- *sodic* soils dominated by exchangeable sodium—Solonetzic Semiarid Soils; and
- *saline-sodic* soils where sodium is present both as salt and in the exchangeable state—Saline Solonetzic Semiarid Soil.

The three soil forms three soil forms underpin a soil development sequence. The initial soil has a cation exchange complex dominated by calcium and magnesium bivalent cations, for example, the Typic Immature Semiarid Soil (Figs. 15.1, 15.4) which contains free calcium carbonate. With the incursion of a shallow saline water table, monovalent sodium cations will gradually displace the bivalent cations and, with gradual dominance of sodium, a saline-sodic soil forms. While the soil solution salinity remains high, the soil structure will remain flocculated, and the soil will remain permeable. If the water table is lowered, rain water (or irrigation water) of low ionic strength will leach out the saline soil solution, but sodium cations that are relatively firmly bonded to cation exchange sites will remain in the exchange complex, and a sodic soil is formed. Soil aggregates may then disperse, fine clay platelets will migrate in suspension down the soil profile to settle out on ped faces or void walls (which act as filters), or flocculate (where charges and pH are favourable) to form a clay-enriched (argillic) horizon.

In the *New Zealand Soil Classification*, the three soil forms are defined by the electrical conductivity (EC), which estimates the total salt content, and the exchangeable sodium percentage (ESP), which estimates the percentage of cation exchange capacity occupied by exchangeable sodium. Alkalinity is associated with sodicity and so a pH below 8.5 indicates saline soils, and pH above 8.5 indicates sodic soils. Saline sodic soils have pH values near or above 8.5.

The soil/salt development sequence is consistent with the distribution of the soil types in the Central Otago landscape, where saline and saline-sodic soils occur on younger, or lower, parts of the landscape where water tables remain high. Sodic soils with characteristic argillic horizons are on higher parts of the landscape (Fig. 15.5), on relatively uneroded hill slopes or terraces, or where water tables have been lowered (over geological time-scales) by natural down-cutting of streams into the landscape.

Fig. 15.5 Sodium-rich soils. Left: salt pans on foot slopes in the 1950s (photo: J.D. McCraw). Right: profile of a Saline Solonetzic Semiarid Soil (Manorburn soil) with a columnar structure that is associated with such sodic soils. (Photo reproduced with permission of NZ Society of Soil Science.)

15.3 Soil-Landscape Relationships

There is a strong soil-landscape pattern that is repeated across the inland basins of Central Otago and inland Canterbury where Semiarid Soils predominate (Fig. 15.6, see also Fig. 4.4). The inland basins generally comprise wide river valleys with a series of flat terraces that merge into the hills of the adjacent, uplifted, ranges. Alluvial fans and colluvial deposits form along the margins of the hill/steeplands and on some terrace scarps.

Recent Soils occur on active river flood plains and some of the lowest terraces, with Anthropic Soils in areas that have been subject to gold mining by sluicing or dredging. Above the contemporary river flood plains is a series of terraces, generally divided into low, intermediate, and high terraces. The low terrace formed during, and/or following, the last stadial of the last glaciation (since about 20 000 years ago) with progressively older terraces at higher altitudes. The Semiarid Soils in the weathering sequence, described in Sect. 15.2.2, each occur on progressively older, higher land surfaces. Immature and Argillic Semiarid Soils occur on intermediate terraces with Argillic and Aged-argillic Soils on high terraces and the lower slopes of the adjacent hills. Solonetzic Semiarid Soils form in topographic hollows and foot- or toe-slopes where there is strong salt accumulation. Some of the Solonetzic Semiarid Soils are adjacent to and receive drainage from, Tertiary sedimentary deposits that have high salt contents.

The high terraces merge into fans on the margins of the ranges, and into hills and steeplands ascending to high altitudes. Within the hills and steeplands, some erosion has occurred, but where surfaces have been stable the soils are predominantly Aged-argillic Semiarid Soils.

With increasing rainfall—which occurs as one moves away from the floors of the inland basins, be it east, west, north, south, or up (in altitude) onto the ranges—the Semiarid Soils systematically give way to Pallic and then Brown Soils (Fig. 15.6).

Fans at the margins of South Island basins provide a gallery of classic landforms and soils. The soils on the fans are frequently comprised of schist gravels (stones), but their compensating attribute is that the inter-stone spaces are infilled with a range of loamy soil material with good water holding capacity, and provision of good rooting depths.

Fans are built from sediments that are carried swiftly in mountain streams and deposited where flow rates decrease as the water exits from steep gorges onto the flatter terraces. The resulting sediment is a jumbled mix of boulders, cobbles, gravels, and fine material. Deposition may take the form of stratified layers or, if a dense debris-flow develops, which carries particles of all sizes, then the resulting deposits are mixed with an assortment of boulders, cobbles, gravel, and sand. Successive floods or debris flows do not build up in the same place as the last one but slip off sideways to marginally lower land. In this way, over the course of time, sediment distribution snakes out across upbuilding fan surface and constructs the fan shape of the deposit. Fans although providing usually useful soils are also potentially hazardous because of sudden, and usually spatially unpredictable, flooding or debris-flow events.

Although fan deposits can be highly variable there are guidelines that help to predict the nature of the soil pattern. The apex of the fan, where the sediment issues from the mountain stream, is where the soils are mostly stony with sizes ranging up to boulders. Further down, towards the fan toe, stone sizes tend to be smaller and textures of the inter-stone infill finer (for example, silt loams and sandy clay loams). The down-fan particle size trend is also reflected in the drainage pattern with well drained permeable soils at the fan apex and increasing wetness and lower permeability down the fan, and with Gley Soils likely to occur at fan toes.

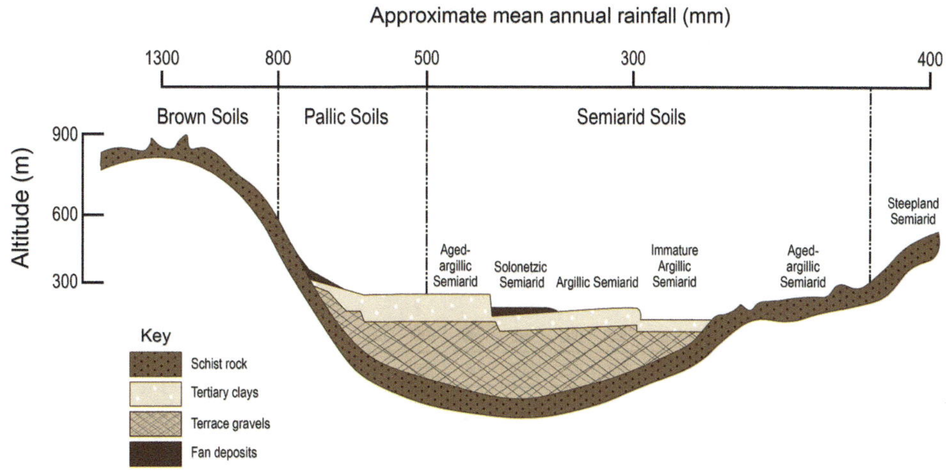

Fig. 15.6 Diagram of typical cross section of a Central Otago basin showing general geological structure, rainfall, and pattern of soils in relation to landforms and lithologies (adapted from Leamy 1966)

Other factors that complicate the variability of fan sediments and soils include the weathering of the slope mantle of rock and soils and their slope stability, the length of time since the slope mantle was last eroded, and the proportion of the slope mantle carried from the slopes and deposited onto the fan. The processes also depend on the volume of water flowing from the catchment and the turbulence of the sediment-charged streams.

The processes generating such variability result in complex fan systems, driven by degrees of chaotic behaviour. One mitigating factor that smooths out soil textural variability is the deposition of loess on older fans which provides a relatively homogeneous blanket of varying thicknesses of silt and very fine sand. If the direction of loess transport can be inferred, then it is usually thicker on the downwind side of the fan. Such soils are formed by upbuilding pedogenesis involving some complexity: episodic deposition of alluvium and the slow accretion of loess during glacials takes place at the same time as weak soil formation by topdown pedogenesis.

15.4 Key Soil Properties

15.4.1 Soil Composition

Semiarid Soils are mainly formed in coarse alluvial sediments that include boulders, cobbles, gravels, and sand, as well as loess, derived from the non-calcareous greywacke, schist, and mudstone rocks that dominate the semiarid Otago and inland Canterbury regions. The parent materials are dominated by quartz and feldspar, and by mica (which is derived from schist). Weathering of the parent material is generally limited due to the cool winters and generally dry conditions, and in many cases, but not all, limited time. Thus many of the soils tend to have the pale grey colours of the parent materials (lithochromic).

Many Semiarid Soils are formed in the material deposited from about 2000 to at least 30,000 years ago, and others on high terraces (as noted earlier) may be an order of magnitude older. The clay mineralogy reflects the limited soil weathering with clays dominated by hydrous mica or illite, plus chlorites, mostly little altered from the original mica of the parent material. Semiarid Soils generally have low concentrations of secondary oxides. However, on some high, older, surfaces, and in areas of salt concentration, in Aged-argillic and Solonetzic Semiarid Soils, kaolinite is common.

Some 'amorphous' (potentially nanocrystalline) hydrous oxides of aluminium and iron also occur. Peter McIntosh suggested in 1994 that small quantities of allophane (which is a nanocrystalline mineral) were possibly present, coating sand grains, in the old soils on high terraces at Pisa Flats near Cromwell. This inference was based on reactions, recorded as moderate to strong, for lower horizons using the NaF (allophane) field test. Despite the low rainfall, which would normally seem to preclude any possibility of silicon being sufficiently enleached to enable allophane to form, McIntosh argued instead that the likely old age of the soil (hence long time available), coupled with the subsoil's highly permeable character, generating strong acidity (pHs of 4.9 and 5.4), would make up for this limitation. NaF does react with calcium carbonate but this was absent from the soil examined by McIntosh (and the low pHs support such an absence).

15.4.2 Physical Properties

While most of the Semiarid Soils are formed on gravelly sand materials, on alluvial terraces, some are formed on weathered rock materials and many contain loess (silty but also sandy close to river bed sources). Thus, there is a wide range of soil physical properties within the Semiarid Soil order (Fig. 15.7, Table 15.1). Due to the slow rates of

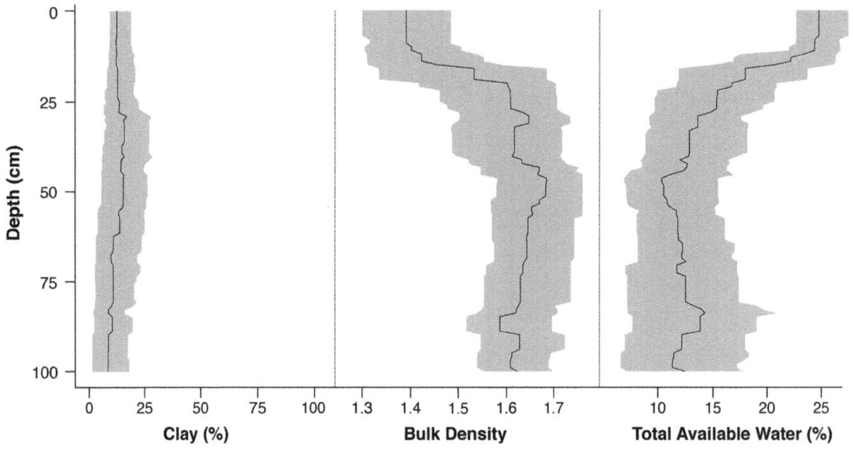

Fig. 15.7 Median and upper and lower quartiles of clay content, soil dry bulk density (t m^{-3}), and total available water holding capacity for Semiarid Soils in the New Zealand Soil Data Repository

Table 15.1 Example of the physical properties of a Mottled Argillic Semiarid Soil (Ranfurly series, SB09891)

Horizon	Depth	Sand (%)	Silt (%)	Clay (%)	Dry bulk density (t m^{-3})	RAW[a] (%)	TAW[a] (%)
Ap	0–15	18	53	27	1.40	5	23
BA	15–24	16	53	29	1.68	5	10
Bt1	24–40	15	48	36	1.70	5	7
Bt2	40–52	15	47	37	1.67	5	10
Bw	52–73	19	46	34	1.60	8	13
BCk	73–89	25	47	26	1.57	5	15
Cg	89–107	24	57	17	1.72	5	15

[a]RAW = readily available water, TAW = total available water

weathering, limited by lack of water and cool average temperatures, clay contents are usually relatively low. However, the older soils with the well-developed relict argillic horizons can contain nearly 40% clay (Table 15.1). Soil dry bulk densities are high to very high and total available water-holding capacities range from low to very high, depending on the gravel contents of the soils and the amount of finer interstitial fill.

Soil structure is frequently weakly developed because there is relatively little organic matter or clay to bind soil aggregates. The sodic forms of the soils with sodium-enriched argillic horizons can have well developed prismatic to columnar structures (Fig. 15.5). Permeability (saturated hydraulic conductivity) is commonly slow especially in soils with argillic horizons, or horizons with silty or fine sandy textures. However, permeability may be rapid or very rapid in gravelly and/or sandy Immature Semiarid Soils. Infiltration may be reduced by traffic or treading when soils are saturated by irrigation water. Slow permeability may induce a perched water table under irrigation management. There are no naturally poorly, or very poorly, drained Semiarid Soils. Soils with prolonged high water tables, and grey colours indicative of gleying, are classified as Gley Soils. However, soils with slow permeability, particularly towards the toes of fans where water seepage may occur, may have some mottling in subsoils (e.g. Mottled Argillic Semiarid Soils).

Root exploration is limited in Semiarid Soils over rock and also in soils with very low available water-holding capacity, especially in stony (gravel-rich) soils. The quality of stony soils depends greatly on the nature of the fine-earth material that fills the gaps between stones. Stony soils may have limited fine material between stones, but where the stone gaps are filled with loamy soil materials the available water capacity may be a little higher.

15.4.3 Chemical Properties

Topsoil and upper subsoil pH values are generally slightly acid to near neutral (Fig. 15.8, Table 15.2), although some Solonetzic groups may have surface salt accumulation and alkaline topsoils. Deeper subsoils are more likely to be moderately to strongly alkaline, with a pH as high as 9 in some soils. Stony (gravelly) Immature Semiarid Soils often have low clay and organic matter contents and thus low pH buffering capacity, and hence pH may change markedly in response to management activities such as fertiliser addition.

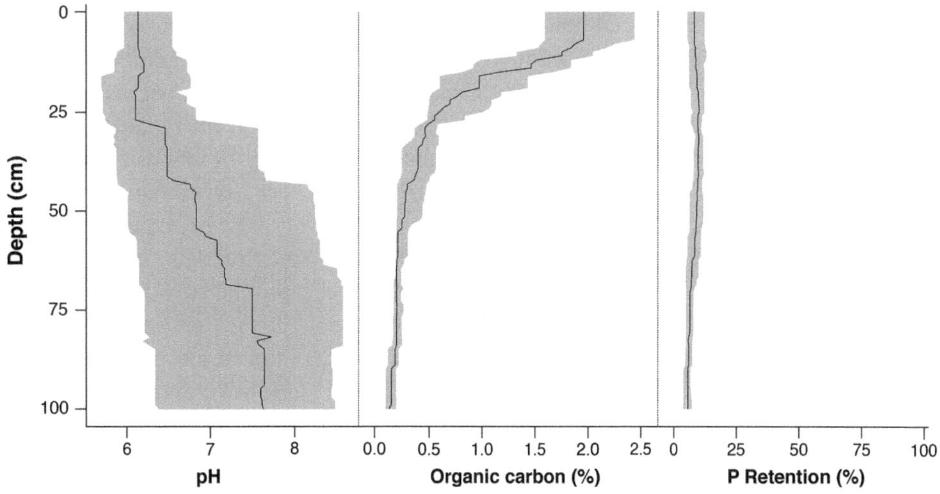

Fig. 15.8 Median and upper and lower quartiles of soil pH, organic carbon (%), and P retention for Semiarid soils in the New Zealand Soil Data Repository

Table 15.2 Example of chemical properties of a Mottled Argillic Semiarid soil (Ranfurly series, SB09891 SWAMP)

Horizon	Depth (cm)	pH (in H_2O)	Carbon (%)	Nitrogen (%)	CEC[a] (cmol(+)/kg)	Sum bases (cmol(+) kg)	P retention (%)
Ap	0–15	7.0	1.8	0.18	10.1	11.3	8
BA	15–24	7.0	0.9	0.10	9.8	9.8	9
Bt1	24–40	8.2	0.5	0.07	11.7	13.5	10
Bt2	40–52	8.8	0.4	0.05	13.9	19.7	11
Bw	52–73	9.1	0.2	0.03	13.9	21.3	10
BCk	73–89	9.3	0.3	0.02	10.9	42.8	9
Cg	89–107	9.2	0.1	0.01	8.9	13.0	6

[a]CEC = cation exchange capacity

Organic carbon contents, phosphate retention, and cation exchange capacity are all low to very low, reflecting the lack of clay, the lack of oxide clays, and limited organic matter in most Semiarid Soils. As a consequence of the low CEC, and lack of leaching of salts, the subsoil base saturations are high to very high. Many soils have base saturations >100% which means that the cation exchange complex comprises the normal Ca, Mg, K, and Na ions, and the soil contains free salts, chiefly sodium chloride, calcium sulphate (gypsum), sodium sulphate, and calcium carbonate.

Total P values are generally very high indicating low weathering. A high proportion of the inorganic phosphorus is non-occluded (i.e. it is available for plants) but a relatively high proportion of total phosphorus is in an organic form. Because of the relatively high pHs and relative lack of weathering, the exchangeable aluminium values are very low and so aluminium toxicity is unlikely to occur.

15.4.4 Biological Properties

Semiarid Soils have relatively low biological activity, in their natural state, because the soils are usually water-deficient for most months of the year (apart from late winter or spring) and because of marked winter chill. Thus, only plants adapted to such conditions thrive.

Prior to initial human arrival about 750 years ago (around 1250 AD or soon after), most of the dry inland basins of Otago were forested, but by the time of European settlement, tussock grasslands and native shrublands dominated the region. Matt McGlone and others, using pollen records derived from peat bogs in the area, determined that in the early Holocene, tussock grassland and shrubland predominated before full forest cover was established by c. 8500 years ago, several millennia after podocarp-dominated forest began to occupy coastal regions in Otago (reaf-forestation was delayed inland because of the drier climate). Repeated fires by Polynesian settlers destroyed both lowland and upland forest and tall scrub communities. Initially, bracken followed by grassland replaced the burnt forest. Following European settlement, the introduction of sheep, along with rabbits, and repeated use of fire to stimulate new, palatable, growth on the tussocks, led to severe depletion of the remaining native vegetation.

A notable feature of the vegetation on less intensively managed Semiarid Soils is the very thorny, small-leaved native matagouri (*Discaria toumatou*), along with introduced thyme (*Thymus vulagris*) which gives a distinct aroma to many of the dryland hill areas, and the sweet briar rose (*Rosa rubiginosa*). *Hieracium pilosella* is another invasive weed, well adapted to the environment, that out-competes both the native vegetation and improved pasture. The relatively low rate of plant growth, along with the warm summer conditions that favour rapid biodegradation of plant remains, means that the soils have low organic matter contents.

There are a few species of native plants, most notably *Lepidium kirkii*, that are especially adapted to grow on the small patches of extremely salty Saline Solonetzic Semiarid Soils where very little other vegetation can survive. The Central Otago native salt-tolerant plants are found nowhere else on Earth and, because irrigation with high-quality water has removed salts from many of the formerly salty areas, their habitat is endangered. Efforts have been made to preserve some of the last remaining salt-affected soils and their associated flora.

The Semiarid Soil environment provides excellent habitat for rabbits because dry burrows provide an ideal nursery for rabbit litters and the herb/grass dryland pastures are attractive for rabbits to browse. Densities of more than 50 rabbits per hectare have been estimated during plagues.

The dry conditions make survival for soil species, such as earthworms, difficult and Semiarid Soils do not show the prominent worm mixing prevalent at the base of the topsoil in the seasonally moist Pallic Soils. Microbial activity is limited by both organic matter and soil moisture availability. However, under irrigation, the soils are productive and with good management soil organic matter content and associated soil biomass are able to increase.

15.5 Distinguishing Between Semiarid Soils and Related Soil Orders

Subsoil lime ($CaCO_3$) deposits (forming Bk or calcareous horizons) are one of the key properties used to identify Semiarid Soils. Where the calcareous horizon is lacking, an alkaline pH (greater than 7.5) is indicative of a Semiarid Soil.

The transition of Semiarid Soils to Pallic Soils is controlled by increasing rainfall. There is a zone of intergrades where it is sometimes difficult to distinguish between Semiarid and Pallic Soils. The key determinants that distinguish the Semiarid Soils are the lack of a fragipan, the presence of salts, and generally high pH. Gley Soils have gley profile features in the form of dominant pale grey colours often accompanied by redox segregations (orange or reddish mottles). Recent Soils lack an argillic horizon and do not have calcareous nor alkalinity features. Raw Soils lack a topsoil, or if present it will be less than 5 cm thick.

The grinding of schist

It seems puzzling that although Central Otago is dominated by schist rock, many of the soils on the extensive river alluvial terraces are formed in hard rounded greywacke gravel. The nearest greywacke outcrops are in the headwaters of the major rivers. However, unlike the terraces, most of the fans are formed from schist alluvium. Philip Tonkin explained this conundrum in terms of the contrasting vulnerability of greywacke and schist to impacts during fluvial transport, and the distance travelled in high-energy river systems.

When observed in a rock outcrop, both greywacke and schist are hard rocks. But in comparison with greywacke, the schist has a weakness: its foliation. When greywacke is reduced, by erosion, to cobbles or gravel, they are usually structurally homogeneous and strong. In contrast, the foliations (alternate bands of light-coloured quartz-rich, and dark-coloured, platy, mica-rich layers) in schist form planes of weakness in the rock. When schist is eroded and discharged into large swift rivers, the stones are cast into the violent rock-against-rock society of the turbulent river bed-load. Here, greywacke stones fare well. The bed-load collisions finish sculpting the greywacke to rounded shapes. The same process serves, however, to fracture the schist and pulverise the fragments.

The dominance of schist materials in fans issuing from schist hill country, and the greywacke materials in the large river terraces, reflects the transport distance. In fans, fluvial transport is short, there is no 'competition' from greywacke, and hence schist survives the journey. However, in the large, long, rivers distance and time favour the survival of greywacke as boulders, cobbles, gravel, and sand.

In contrast, much of New Zealand's loess consists of the minerals quartz, feldspar, and mica that are derived predominantly from the abrasion and pulverisation of schist and siltstone, rather than greywacke. The high-energy river beds are the 'dust engines' creating loess deposits in New Zealand. In flood plains, fine sediments along with coarse gravels accumulate. The fines are the source of loess, especially during glaciations. Small amounts of fines are also contributed by glacial grinding, freeze–thaw processes, and as aeolian dust from Australia.

15.6 Correlation with Other Classification Systems

All New Zealand Semiarid Soils fit within the Aridisols of *Soil Taxonomy* as they have an aridic soil moisture regime and other criteria required for the Aridisols order. A key feature of the aridic soil moisture regime, apart from water balance, is the occurrence of secondary salts in soil profiles. The different Semiarid Soils, recognised within the *New Zealand Soil Classification*, key out with different names at

Table 15.3 Correlation[a] between Semiarid Soils and equivalent classes of *Soil Taxonomy*, *World Reference Base*, and the earlier NZ genetic soil classification

New Zealand Soil Classification	*Soil Taxonomy*	*World Reference Base*	NZ genetic soil classification
Aged-argillic semiarid soils	Paleargids	Luvisols chromic	Brown-grey earth
Solonetzic semiarid soils	Natrargids	Solonetz	Solonetzic soil
Argillic semiarid soils	Calciargids, Haplargids	Haplic Luvisols	Brown-grey earth
Immature semiarid soils	Haplocalcids, Haplocambids, Aquicambids	Cambisols eutric	Brown-grey earth

[a]The correlations given here are a guide only and, for accurate classifications, the relevant soil classification documents should be consulted. The two major international soil classification systems are *Soil Taxonomy,* which was developed in the USA, and *World Reference Base,* which was developed primarily in Europe. The NZ genetic soil classification was used in NZ prior to 1992

the suborder and great group levels of *Soil Taxonomy* (Table 15.3). The *World Reference Base* (WRB) classification does not have a soil order equivalent to the Semiarid Soils, which therefore correlate with a range of WRB soil classifications, depending mainly on their degree of weathering. The brown-grey earths in the New Zealand genetic soil classification all fall within the Semiarid Soils. The solonetzic soils of the New Zealand genetic soil classification are also included in Semiarid Soils, with the solonetzic features being recognised at soil group level.

15.7 Use and Management of Semiarid Soils

15.7.1 Introduction

There is evidence from preserved plant and bird remains in the Earnscleugh Cave, and from the pollen record, that immediately prior to human arrival the Central Otago landscape was largely dominated by coniferous forest. Since humans entered the inland basins, the soils and their environment have been subjected to a number of impacts. The first was the burning of forest by Polynesians to enable easier penetration inland to herd moa and other game. When Europeans arrived in the early nineteenth century, the landscape comprised mainly tussock grasslands and shrublands with most forest vegetation lost to repeated fires, except in small refuges. Sheep farming followed with further burning to promote grass growth. The gold rush in the 1860s intensified the impacts with large numbers of new settlers arriving in the region. The effects of sheep grazing, along with the introduction of rabbits, which reached plague proportions, resulted in the removal of vegetation (including the litter, hence exposing bare soil to frost-heave and wind erosion) and set off a period of accelerated soil erosion.

The unique climate and plentiful water supplies, carried by the Clutha River from the high rainfall regions to the west and from snow melt, support irrigated pasture and intensive stone fruit and wine production in the region. Some Central Otago fruit, such as cherries, have a unique global niche, ripening in time for the Northern Hemisphere Christmas market. Land management issues for the Semiarid Soils are different from those in other regions of New Zealand, with irrigation and erosion management both important. The spectacular landscapes are an attraction for tourists. Hence, there is a concern to protect landscape heritage values including the remarkable, and globally extremely rare, tor-dominated landscapes and other periglacial features including patterned ground, both relict and active, such as relict stone stripes and active plough blocks in alpine zones.

15.7.2 Soil Management Considerations

On the hill and steeplands, merino sheep farming (Fig. 15.9), primarily for high-value fine wool, has long been a mainstay of dryland farming on the Semiarid Soils. The merino sheep are well adapted to the dry climate and steep hill country and do not thrive in wetter regions of New Zealand. Dryland sheep farming is undertaken on an extensive basis with relatively low stocking rates. In the past, fire was used to stimulate fresh, palatable plant growth, but is now largely abandoned as repeated fire gradually depletes the vegetation, making the soil vulnerable to erosion. Care needs to be taken to avoid overgrazing which, like fire, leaves the area vulnerable to erosion and weed invasion.

The Central Otago alluvial fans are particularly important economically because they have some of the region's most versatile soils, and are used for a wide range of crops. The Aged-argillic Semiarid Soils are on the oldest land surfaces,

Fig. 15.9 Land uses on Semiarid Soils. Left: grapes growing at Bannockburn, near Cromwell. The eroded landscape in the background was caused by gold sluicing in the nineteenth century. Right: merino sheep, grown primarily for their high-value fine wool, thrive in the dryland environment provided stocking rates are low.

which are usually on terraces or fans elevated high in the landscape. The age is probably not significant in terms of fertility but it is relevant to the root environment. With age, the clay is reddened and, in many sites, it clogs the subsoil soil pores. This may limit the potential for roots to explore the deeper soil for water and nutrients. Under irrigation, there may be a risk of waterlogging in the subsoil.

Argillic Semiarid Soils have accumulations of clay in the subsoil but, unlike Aged-argillic soils, the clay may not affect the potential root depth and may boost the water holding capacity. They are valued soils, especially on fan landforms where deep schist alluvium provides good nutrients, rooting conditions, and high water-holding capacity that facilitates irrigation management. On fan soils, it is important to avoid over-watering as drainage will run laterally, channelled by the soil layers down the fan where it may cause waterlogging or salt accumulation at the foot or toe of the fan.

The Immature Semiarid Soils have simple profiles but have a wide range of soil materials and textures. Rapidly permeable gravelly soils, and some strongly structured (prismatic) soils, pose a risk of nutrient losses to water tables, with consequences for water quality in lakes and rivers. As for the other soils, droughty conditions usually extend from spring through summer and autumn and so irrigation is required for most crops. The stony profile is poor for most crops not only because of the low rainfall but also because the abundant gravels reduce the amount of fine soil available to trap any rain that does fall. There is little stored water for roots to tap into.

Stone fruit orcharding thrives on the Immature Semiarid Soils of the Earnscleugh flats and Cromwell area (Molyneux soils). The soils are gravelly sands and have not been regarded as versatile soils because the stones/gravels limit the profile available water capacity. However, the gaps between the stones are important. They are filled with sand material that enables the soil to drain rapidly. Stone fruit blossom can be severely damaged by heavy frosts that spoil the crops. An efficient method of counter-acting frost is to spray the blossoming trees with water that coats the flowers with ice, which provides insulation and the latent heat released by water freezing is enough to prevent frost damage (Fig. 15.11). On many soils, the large volume of frost-fighting water accumulates and the soil becomes waterlogged with consequent increased risk of *Phytophthora* infection. However, the rapidly draining Molyneux soils circumvent this risk. The Molyneux soils may not be highly versatile but they are recognised as good specialist soils for stone fruit production in frost-prone land where well monitored and managed micro-irrigation can keep the trees with near-ideal soil moisture conditions.

The Solonetzic Semiarid Soils occur only as small patches and are infrequent enough that it is generally worth avoiding developing them for productive use in order to protect their rare native ecosystems (unique to the inland basins of New Zealand). The Solonetzic Semiarid Soils could, however, be rehabilitated for productive use with the installation of drainage, application of calcium (gypsum) to displace the sodium, and careful use of quality irrigation water to displace the salts.

There has been a rapid expansion in growing grapes for wine on Semiarid Soils since the 1990s (Fig. 15.9). A range of soils are used and the dry climate means that it is relatively easy to control the water available to the crop to ensure maximum quality grape and wine production.

The risk of bypass flow is high where there are permeable or well-structured prismatic or columnar subsoils where drainage water is channelled down the fissures, largely avoiding the mass of the soil within structural units. Bypass flow is a significant transport mechanism for water-borne contaminants or nutrients. However, in this dry environment, bypass flow can be prevented by avoiding soil saturation by ensuring that irrigation is at rates slower than the saturated hydraulic conductivity, and also by making sure water does not pond on the soil surface. Thus, the old flood irrigation systems should be avoided to prevent water and associated contaminant loss to the groundwater.

Semiarid Soils have high slaking and dispersion potential. Topsoil structures may break down under prolonged impact by heavy machinery or by continuous tillage and, like most soils, such risks are higher if they are wet. Weak soil structures render the soils susceptible to erosion, especially when exposed to water runoff or wind erosion aggravated after particles are separated by frost-heave. Managing soils to increase soil organic matter can help improve soil structural stability.

Many of the most intensively used soils, on the fans and terraces, have low contents of clay and organic matter, and thus, very low cation exchange and pH buffering capacities, along with low oxalate-extractable iron and aluminium, phosphate retention, absorbed sulphate, and aggregate stabilities. Gary Beecroft and Don Brash documented some of the effects of low buffering capacity to alert land managers to potential risks in managing low-buffering soils. They included (1) lowering of soil pH due to application of ammonium sulphate; (2) rapid topsoil compaction where water is used to counter spring frosts in stone fruit orchards; (3) herbicide penetration of topsoils in orchards affecting tree roots; (4) serious wind erosion in topsoils under cultivation; and (5) the development of Gley Soil features (such as mottles and low-chroma colours indicative of prolonged saturation) reported in Semiarid Soils under 50 years of irrigation in Bannockburn.

When transferring management techniques from other regions, it is important to be sure that soils at the place of origin are comparable with those of the transfer site. Although low buffering is a feature of Semiarid Soils, it is by no means a feature of Aridisols throughout the world. Knowledge transfer from Aridisols elsewhere in the world requires particular attention to cation exchange capacity, anion characteristics, and soil particle size, as well as other factors including topography and land management history.

15.7.3 Irrigation History and Management

In Semiarid Soils, soil water deficits limit plant growth for most of the year. Dryland pasture supports only a low stocking rate. For horticultural production, or more intense farming, irrigation is necessary. When the gold rush brought people to the region, it created a need for water for both irrigation and gold-mining operations. Much of the gold mining was undertaken by 'sluicing' (washing material down off slopes and through a sluice or trough where the heavier gold could be extracted from the remaining soil/subsoil material). A series of water races were built along contours to conduct water, under gravity, from upstream areas to the places where it was needed for irrigation or for gold extraction.

Much of the early irrigation was carried out using wild flooding (blocking a water race to encourage water to overflow across the paddock). An improvement on wild flooding was provided in the twentieth century with the introduction of border dyke irrigation where the land was recontoured into a series of "borders" that provided a more even spread of the water flow. Although flood irrigation provides inefficient use of water (and promotes leaching of nutrients and contaminants), it is inexpensive and, where water is plentiful, some flood irrigation systems are still in use in the twenty-first century (Fig. 15.10).

Now many of the old water races have been converted to piped water supplies, which prevent water loss to seepage and evaporation. For pastures, spray irrigation systems provide more efficient water use, and a better spread of water across the landscape. Within orchards, combinations of overhead sprinklers, micro-sprinklers, and drip irrigation systems are used (Fig. 15.11). Overhead sprinklers can also be used for frost fighting in the spring. Micro-irrigation systems allow water to be targeted at tree roots, with minimal losses. Fertilisers may also be added to the irrigation water to deliver carefully controlled tree nutrition.

Secondary salinisation occurs where salt is redistributed in the landscape as a result of irrigation, or other management that impacts the soil water regime, causing a concentration of salt in previously productive areas. Salt accumulation is a major cause of soil degradation in many dry regions of the world, including in our neighbouring Australia. In some areas in Central Otago, salinity did increase when water carrying dissolved salts moved downslope, emerging at the toes of fans, or where the water table reached near to the ground surface. Water evaporated and salts were left behind to accumulate at the ground surface. Past wild flooding and border dyke irrigation have probably contributed to some secondary salinisation. Much of the early soil science work in Central Otago was undertaken due to concern that irrigation could cause wide-spread soil salinisation . However, while there have been some small local salt incursions, widespread salt accumulation, as a result of irrigation, has not occurred.

The irrigation water in the inland basins is predominantly derived from the large rivers, such as the Clutha and Waitaki, that are fed from an abundant high country or alpine sources, and have miniscule salt contents. The use of abundant, high-quality water, on permeable soils, or with drainage installed if needed, means that the salts have tended to be leached from the soil. The irony is that there is concern that the last remaining areas of Solonetzic Semiarid Soils will be lost as habitats for rare, endangered, salt tolerant, flora and flora. In some areas, efforts have been made to

Fig. 15.10 Flood irrigation in Central Otago. Left: Installation of a border dyke irrigation system in the 1960s (photo J.D. McCraw). Right: water flowing from a water race across a paddock using wild flooding in 2018, as practised in Central Otago since the mid-nineteenth century

Fig. 15.11 Irrigation in orchards. Left: stone fruit trees with overhead sprinkler irrigation. Centre: frost fighting using overhead sprinklers. Right: drip irrigation delivering water to individual trees with tensiometers to monitor the soil moisture content

preserve the remaining Solonetzic Semiarid Soils with cut off drains to prevent up-slope irrigation water from reaching such areas. Thus, it is not likely that widespread secondary salinisation will become an issue for New Zealand.

15.7.4 Soil Erosion

It is evident that following the arrival of humans, first with fire, then with grazing animals (including sheep, deer, and thar, and notably rabbits that rapidly increased to plague proportions), vegetation was rapidly depleted in many areas of the Semiarid Soils. The loss of vegetation cover, including through spring burning to encourage new shoots but at the expense of protective litter cover, led to a period of extensive accelerated erosion. For instance, Leamy reported that a period of aggradation of alluvial fans on the Hawea flats commenced in the mid-eighteenth century but had subsided by the mid-twentieth century. Leamy estimated that a minimum volume of about 6 million cubic metres of debris had been eroded from the mountains to the east of Hawea Flat and deposited as alluvial fans, lowering the surface of the eroded area by about 30 cm on average.

Semiarid Soils have a significant wind and water erosion risk when not protected by good vegetation cover. Soil has often been exposed, particularly during rabbit plagues when areas of bare ground can be substantial, and also following periods of prolonged drought and consequent overgrazing by sheep. Loss of vegetation may be greater where soil water storage and root depth are limited by stoniness, bedrock, or very firm substrates. For example, in a prolonged drought in the Hakataramea Valley in the early 1980s, severe wind erosion of exposed soil left some paddocks with a desert pavement-like stony surface cover. However, with careful land management, vegetation was re-established and the land appears to be recovered. Care needs to be taken to destock or provide supplementary feeding during droughts to prevent surface vegetation removal and reoccurrence of erosion events. Given the potential for droughts, and long periods of soil exposure if vegetation is removed, no-till methods (and avoiding burning) are strongly recommended for establishing arable crops or pasture improvement.

Surface erosion can also be slow and insidious with small quantities of soil lost at any one time. In a study of soil erosion over 40 years, using caesium-137 (^{137}Cs, which was deposited as a result of atmospheric nuclear weapon testing in the 1950s and early 1960s) as a tracer, Allan Hewitt estimated that about 34 mm of soil depth was lost over the 40 years. This rate seems small but when it is projected over the remaining (shallow) soil depth, the soil at the study site had a life expectancy of only 44–72 years before it is exhausted. Conversely, in a study in the Mackenzie Basin in 2015, Hannah Leckie and Peter Almond showed, using both ^{137}Cs and cryptotephrochronology (analysis of fine tephra-derived glass shards), that the soils had undergone negligible wind erosion since 1953. *They identified* tiny concentrations of glass shards of the Kawakawa Tephra, blown to Otago during the Oruanui super-eruption of Taupo volcano c. 25,400 years ago, distributed in soil profiles as a non-visible cryptotephra (i.e. only as sparse glass shards, not as a layer), which helped in dating the soils.

The effects of past erosion would appear to be borne out by the apparently denuded appearance of many parts of the semiarid landscape. However, the overall removal of sediment exported from the region by large rivers is low compared to the sediment volumes removed by large rivers draining the heavily bush-clad, hill and mountain slopes of catchments in South Westland. The West Coast rivers deliver about two and a half times more sediment than the Clutha River, which drains the Central Otago catchments. The higher sediment loads in South Westland are attributed to the high, and often intense rainfall (>7,000 mm yr^{-1} in many places), steep slopes, and rapid tectonic uplift in the Westland catchments.

Many soils in the Semiarid Soil landscapes of Central Otago have thin, poorly aggregated topsoils. Soil organic matter is low, especially where not irrigated or well vegetated. Are these a fair representation of the soils and landscapes prior to human occupation? Probably not, given the unequivocal evidence from prehistoric pollen records that show essentially continuous forest cover (apart from montane or alpine areas) since c. 8500 years ago. Soil quality is partly a function of healthy vigorous vegetation. Loss of plant vigour impacts on soil organic matter levels and the protective, insulating effects of ground surface litter. Large accumulations of moa bones have been discovered in some of the basins, prompting a question: what was the carrying capacity of the land to support a sizable grazing population of these large birds? The questions remain: are the soils in a generally depleted state, and could land management and ecosystem restoration return the soils to a higher level of soil quality and ecosystem functioning? Soil organic carbon modelling by Aroon Parshotam shows that, given good vegetation cover, soil organic carbon is capable of improvement, and suggests that, given ideal conditions, the soil organic matter levels may have once been higher and could potentially be returned to that state.

Further Reading

Allen RB, McIntosh PD (1994) Soils and plant communities and their management, Pisa Flats. Central Otago. Landcare Research Contract Report, Landcare Research New Zealand Ltd, Dunedin

Anderson A (2013) A fragile plenty: pre-European Maori and the New Zealand environment. In: Pawson E, Brooking T (eds) Making a new land—environmental histories of New Zealand, new edn. Otago University Press, Dunedin, pp 35–51

Balks MR, Hewitt AE (1987) The prospects for maintaining saline soils for *Lepidium kirkii*.N.Z. Soil Bureau Contract Report 87/197

Balks MR, Beecroft FG (1985) Soil, irrigation suitability and drainage information for selected irrigation schemes in Otago, 1985. Bound soil data file No. 5. DSIR Soil Bureau, Dunedin

Barrell DJA, Cox SC, Greene S, Townsend DB (2009) Otago Alluvial Fans Project: Supplementary maps and information on fans in selected areas of Otago. GNS Science Consultancy Report 2009/052. Prepared for Otago Regional Council

Basher LR (1998) Surface erosion assessment in New Zealand using caesium-137. In: Proceedings of SPERA98, environmental radioactivity and its application in environmental studies, 5th biennial conference of the South Pacific environmental radioactivity association, Christchurch, 16-19 February, pp 31–38

Basher LR (2000) Surface erosion assessment using ^{137}Cs: examples from New Zealand. Acta Geologica Hispanica 35:219–228

Basher LR, Webb TH (1997) Wind erosion rates on terraces in the Mackenzie Basin. J Roy Soc New Zealand 27:499–512

Basher LR, Meurk CD, Tate KR (1990) The effects of burning on soil properties and vegetation: a review of the scientific evidence relating to the sustainability of ecosystems and land use in the eastern South Island hill and high country. DSIR Land Resources Technical Record 18, 93 p

Beecroft FG, Balks MR (1986) The drainage of land suitable for orcharding, Blackmans Road area, Earnscleugh irrigation scheme, Central Otago. N.Z. Soil Bureau Contract Report 86/05

Beecroft FG, Balks MR (1985) Examination of the Kohine and Galloway soils for orchard suitability, Earnscleugh irrigation scheme. Central Otago, Soil Bureau District Office Report DN, 28 p

Beecroft FG (1978) Gleying induced by irrigation in two Central Otago soils. In: Rijske WC (ed) Soil groups of New Zealand: part 3. New Zealand Society of Soil Science, Gley soils, pp 42–43

Brash DW, Beecroft FG (1987) Soil resources of Central Otago. Proc New Zealand Grassland Assoc 48:23–30

Brook MS, Ferrar S (2019) Hartley Travers Ferrar (1879–1932) and his geological legacy in Antarctica, Egypt and New Zealand. Earth Sci Hist 38:43–58

Chen L, Arimoto R, Duce RA (1985) The sources and forms of phosphorus in marine aerosol particles and rain from northern New Zealand. Atmos Environ 19:779–787

Cook FJ, Beecroft FG, Joe EN, Balks MR (1986) Flow of water in a gravelly sandy loam soil from a point source. N.Z. Soil Bureau Contract Report 86/06

Fahey BD (1981) Origin and age of upland schist tors in Central Otago, New Zealand. New Zealand J Geol Geophys 24(3):399–413

Fieldes M (1968) Clay mineralogy. In: Soils of New Zealand (Part 2). NZ Soil Bureau Bull 26(2):22–39

Fieldes M (1971) Clay mineralogy of Brown-grey Earths. In: Miller RB, Willoughby J (compilers) Soil Groups of New Zealand (second printing). NZ Society of Soil Science, Lower Hutt, pp 17–18

Fieldes M, Perrott KW (1966) The nature of allophane in soils. Part 3. Rapid field and laboratory test for allophane. NZ J Sci 4:623–629

Grab SW, Dickinson KJM, Mark AF et al (2008) Ploughing boulders on the Rock and Pillar Range, south-central New Zealand: their geomorphology and alpine plant associations. J Roy Soc New Zealand 38:51–70

Hewitt AE (1982) Soils of Waipori farm settlement, east Otago. New Zealand, NZ Soil Survey Report, 65 p

Hewitt AE (1984) Review of the brown-grey earths. NZ Soil Resources Report SR2. Landcare Research

Hewitt AE (1994) A soil-landscape model for the Conroy land system. In: Webb TH (ed) Soil-landscape modelling in New Zealand, Landcare Research Science Series 5:80–86

Hewitt AE (1995) Soils of the Conroy land system, Central Otago. Landcare Research Science Series 13, 56 p

Hewitt AE (1996) Estimating surface erosion using ^{137}Cs at a semiarid site in Central Otago, New Zealand. J Roy Soc New Zealand 26:107–118

Hewitt AE, Balks MR (1988) Review of some halophyte habitats, Central Otago. N.Z.Soil Bureau Contract Report 88/07

Hewitt AE, Beecroft FG (1989) Aridisols of New Zealand. In: Proceedings of the sixth international soil correlation meeting (VI ISCOM): characterisation, classification and utilization of cold Aridisols and Vertisols, Saskatchewan to Utah, USDA, pp 67–72

Leamy ML (1966) The soils of Central Otago. In: Proceedings of the NZ Grassland Association, vol 28, pp 7–18. https://www.grassland.org.nz/publications/nzgrassland_publication_1713.pdf

Leamy ML (1969) Recent aggradation in the Hawea Flat District, Otago. New Zealand J Sci 12(2):373–379

Leamy ML (1973) Subsoil claypans as Quaternary markers in semi-arid Central Otago. J Geol Geophys 16(3):611–622

Leamy ML, Saunders WMH (1967) Soils and land use in the Upper Clutha Valley, Otago. NZ Soil Bureau Bulletin 28. DSIR, Wellington, p 110

Leamy ML, Rafter TA (1972) Isotope ratios preserved in pedogenic carbonate and their application in paleopedology. In: 8th international conference on radio carbon dating, Wellington, New Zealand

Leamy ML, Wilde RH (1971) Soils of Roxburgh District, Central Otago. Soil Bureau Publication, New Zealand. 478 p

Leamy ML, Wilde RH (1971) Soil map of Roxburgh District, Central Otago, New Zealand. Scale 1:63360. N.Z, Soil Bureau Map 80

Leathwick J, Wilson G, Rutledge D et al (2003) 'Land Environments of New Zealand'. Bateman, Auckland. 184pp + 2 maps

Leckie HD, Almond PC (2015) Evidence of prehistoric wind erosion of the Mackenzie Basin, South Island, New Zealand: an assessment based on ^{137}Cs and Kawakawa-Oruanui tephra. Soil Res 53:56–66

Lowe DJ, Tonkin PJ, Palmer J et al (2015) Dusty horizons. In: Graham I (ed) A continent on the move: New Zealand geoscience revealed, 2nd edn. Geoscience Society of New Zealand with GNS science, Wellington, GSNZ Miscellaneous Publication 141, pp 286–289

Marx SK, Kamber BS, McGowan HA (2005) Provenance of long travelled dust determined with ultra-trace-element composition: a pilot study with samples from New Zealand glaciers. Earth Surf Proc Land 30:699–716

Marx SK, McGowan HA, Kamber BS (2009) Long-range dust transport from eastern Australia: a proxy for Holocene aridity and ENSO type climate variability. Earth Planet Sci Lett 282:167–177

McCraw JD (1964) Soils of Alexandra District. Soil Bureau Bulletin 24. DSIR, Wellington, 91 p

McCraw JD (1968) The soil pattern of some New Zealand alluvial fans. Trans 9th Int Congress Soil Sci 4:631–640

McCraw JD (1971). Brown-grey Earths. In: Miller RB, Willoughby J (compilers) (eds) Soil groups of New Zealand (second printing), NZ Society of Soil Science, Lower Hutt, pp 15–17

McGlone MS, Mark AF, Bell D (1995) Late Pleistocene and Holocene vegetation history, Central Otago, South Island, New Zealand. J Roy Soc New Zealand 25:1–22

McIntosh P (1994) Allophane in Semiarid Soils. New Zealand Soil News 42:85–86

McWethy DB, Whitlock C, Wilmshurst JM et al (2009) Rapid deforestation of South Island, New Zealand, by early Polynesian fires. The Holocene 19(6):883–897

Newnham RM, Lowe DJ, Williams PW (1999) Quaternary environmental change in New Zealand: a review. Prog Phys Geogr 23 (4):567–610

Parshotam A, Hewitt AE (1995) Application of the Rothamstead carbon turnover model to soils in degraded semiarid land in New Zealand. Environ Int 21:693–697

Raeside JD, Cutler EJB, Miller RB (1966) Soils and related irrigation problems of part of the Maniototo Plains, Otago. NZ Soil Bureau Bulletin 23, DSIR, Wellington, New Zealand

Raeside JD, Vucetich G, Cox JE et al (1968) Soils of South Island. In: Soils of New Zealand (Part 1). NZ Soil Bureau Bull 26(1):67–88

Rogers G, Hewitt AE, Wilson B (2000) Ecosystem-based conservation strategy for Central Otago's saline patches. Science for Conservation 166. Department of Conservation, Wellington, 38 p

Ross DJ, Speir TW, Tate KR et al (1997) Burning in a New Zealand snow-tussock grassland: effects on soil microbial biomass and nitrogen and phosphorus availability. New Zealand J Ecol 21 (1):63–71

Tonkin PJ, Basher LR (1990) Soil-stratigraphic techniques in the study of soil and landform evolution across the Southern Alps, New Zealand. Geomorphology 3:547–575

Woods R (2011) Otago alluvial fans: high hazard fan investigation. Report for Otago Regional Council, Dunedin, 79 p https://www.orc.govt.nz/media/2948/otago-alluvial-fans-high-hazard-fan-investigation.pdf

Ultic Soils

16

Yellow and old, weathered and worn
an Ultic Soil, with time is born
Rich in clay and poor in vigour
farming's tough and needs some rigour.

Abstract

Ultic Soils are strongly weathered, clay-rich, soils that form on stable land surfaces, mainly on quartz- and feldspar-rich parent materials, predominantly in the northern half of the North Island. Scattered occurrences also occur on extremely old surfaces in other regions of New Zealand including Wellington, Marlborough/Nelson, Otago, and Southland. Some Ultic Soils developed on iron-rich volcanogenic, calcite-bearing sandstones of Tertiary age in the Auckland-Northland region have distinct relict reddening in lower subsoils. One subgroup of the Yellow Ultic Soils in the northern Waikato region is formed on halloysitic tephra deposits over a buried clay-rich paleosol. Ultic Soils generally have slow permeability and are prone to wetness in winter with attendant risks of compaction or pugging, and are droughty in summer. They are strongly leached and naturally acidic (pHs <5.5) with illuvial clay-rich subsoils and pale-coloured upper subsoils. They have low soil fertility. Thus substantial additions of lime and fertiliser are needed for productive use. Ultic Soils are best suited to pastoral agriculture and forestry. They are an option for urban development that avoids the more productive soils, provided the engineering design takes into account the potential for mud-sliding on the Tertiary sedimentary rocks (Waitemata Group) in the Auckland-Northland regions.

16.1 Important Features of Ultic Soils

16.1.1 Concept and Key Features of the Soil Order

Ultic Soils are strongly weathered, clay-rich, and strongly leached soils that occur predominantly in the warm, wet, relatively stable, hill country in northern North Island (Fig. 16.1) with some occurrences in central and southern New Zealand. Ultic Soils are generally weathered in place from quartzofeldspathic basement rocks (usually greywacke/argillite or schist or Tertiary sedimentary rocks) as well as from weathered halloysitic tephra in one special case in the Waikato (Kainui soil). Ultic Soils are the final soil in a broad climate-dominated weathering and leaching sequence that proceeds from Semiarid Soils through Pallic Soils and then Brown Soils to Ultic Soils. However, soils in some other orders, including (old) Podzol Soils, Granular Soils, Oxidic Soils, and Aged-argillic Semiarid Soils, may be of a similar age or similar stage of weathering and/or leaching (see Chap. 18, Fig. 18.2).

Ultic Soils have a strongly weathered, humus-darkened, but shallow, topsoil that has a lower clay content than the underlying well-structured, illuvial clay-enriched, lower subsoil Bt horizon. Upper subsoils comprise either highly leached E horizons or pale B horizons with colour values ≥ 5 in the transition to the lower clay-rich subsoil. Some E

A. E. Hewitt et al., *The Soils of Aotearoa New Zealand*, World Soils Book Series,
https://doi.org/10.1007/978-3-030-64763-6_16

horizons may be very dense, forming densipans (Ed horizons). The E or pale B horizons have paler colours than the lower subsoils, which are enriched with translocated clay, usually with clay coatings (argillans) and possibly humus coatings, forming Bt (argillic) horizons that also contain iron and aluminium ions leached from the upper profile.

The E horizons are usually much less prominent, or absent, in the Yellow Ultic or Sandy Ultic groups that have pale B horizons instead. The Bt horizons tend to have pale, often yellowish, colours. In some soils, rust-coloured mottles and other redox features such as Fe–Mn concretions point to restricted drainage and periodic soil saturation, and perched gleying. However, in the Auckland and Northland regions, some lower subsoils may be distinctly reddened as a result of differential weathering of a volcanogenic (Fe-rich) parent material rather than redox processes, and such 'red-weathered' horizons are therefore denoted by the suffix 'v', forming Btv horizons. The subsoil dry bulk density is usually high, forming a root restriction for many plants (Fig. 16.1). The vertical fissures that define the shapes of large polyhedral or blocky and prismatic peds in B horizons are usually easily observed in the profile. The fissures act as channels for drainage and translocation of clay and/or humus, and provide a habitat for soil fauna as well as playing a role in mass movement (especially mudsliding—Sect. 16.7).

In the lowermost subsoil, yellow brown colours are common. With increasing depth, bulk density increases, fissures are less prominent, and the soil mass is more tightly packed than above. Thus, porosity diminishes, potential rooting volume decreases, and permeability becomes slower with increasing depth.

Ultic Soils are strongly leached with limited natural fertility, low calcium, magnesium, potassium, and sodium cation contents, and acidic pH values that are <5.5 (some pHs are sufficiently low for Al-sensitivity or toxicity to occur). The challenges of the Auckland gardener are shared by the farmer because the soils become hard, as they dry, in summer and sodden in winter or spring. It can be a challenge to establish plants, provide drainage, and maintain soil fertility in the root zone.

16.1.2 Areas of Occurrence

Ultic Soils are most common in the northern North Island broadly from Hamilton northwards (roughly coincident with

Fig. 16.1 Ultic Soils and landscape. Left: hilly landscape, dominated by Ultic Soils on stable ridgetops and backslopes, between Warkworth and Orewa in Northland. Right: strongly weathered Mottled Yellow Ultic Soil (reproduced with permission of NZ Society of Soil Science)

Fig. 16.2 Distribution of Ultic Soils

the (mainly climate-controlled) distribution of kauri (*Agathis australis*). Limited areas of Ultic Soils occur in the Wellington, Marlborough, Nelson, Otago, and Southland regions. Ultic Soils cover 3% of New Zealand (Fig. 16.2). Beyond the north Waikato and Auckland-Northland regions, Ultic Soils occur on generally stable hills or old terraces in landscapes where the weathering surfaces have survived for at least hundreds of thousands, possibly millions, of years.

16.1.3 Variation Within the Ultic Soil Order

Five soil groups are defined within the Ultic Soil order in the *New Zealand Soil Classification* (Fig. 16.3) as follows.

Densipan Ultic Soils—have a densipan (an Ed horizon) which is usually just below the topsoil (upper subsoil). The densipan has a high soil dry bulk density, very high soil strength, and high penetration resistance although it is not

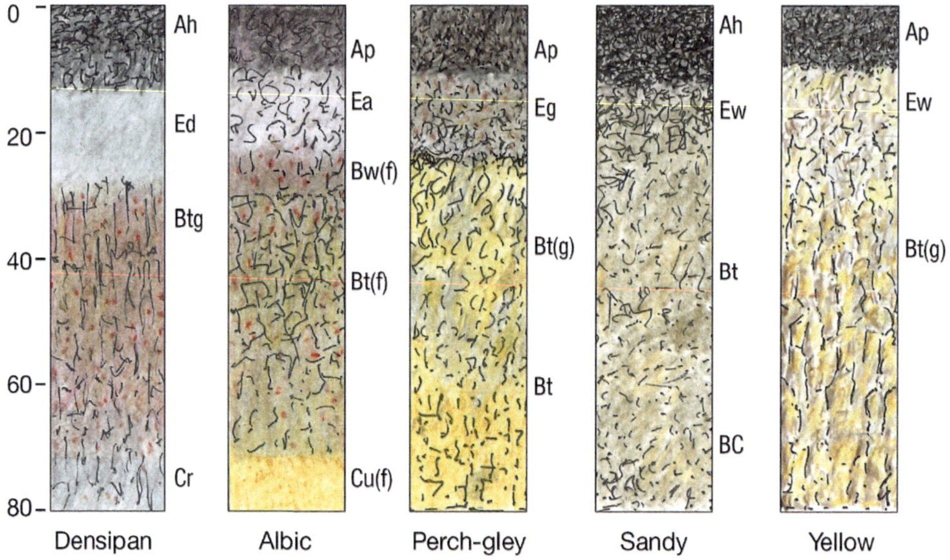

Fig. 16.3 Groups within the Ultic Soil order

cemented and so samples slake in water. Subsoils are usually pale grey or yellowish in colour but may also include rust-coloured mottles where saturated conditions occur for some time during winter/spring. The subsoils have many of the properties of a fragipan except for the prismatic structure. The Densipan Ultic Soils are transitional to Densipan Podzol Soils (the latter requiring a podzolic-B horizon or acidic ortstein-pan or humus-pan).

Albic Ultic Soils—have a pale-coloured (Ea) horizon just beneath the topsoil, in which sand and silt grains are clean and not covered with films of weathered material (an indication of acid leaching). Others may have Eg or Er horizons on firm, clayey Bt horizons. Albic Ultic Soils are also transitional to Podzol Soils, but do not have a densipan.

Perch-gley Ultic Soils—have prolonged, periodic wetness in the upper part of the profile caused by a water table perched on the slowly permeable subsoil (argillic horizon), technically a gley profile form.

Sandy Ultic Soils—occur in strongly weathered aeolian sands, or sandstones, with >60% sand in the B horizon. They may have higher permeability than other Ultic Soils.

Yellow Ultic Soils—the most common or 'default' group of soils that do not meet the criteria for those above, and which have yellow or yellow-brown colours in at least part of the subsoil. This group also includes soils with reddened lower subsoils and the special case, the Kainui soil formed on weathered tephra deposits including a buried paleosol.

Subgroups express variations based on gleying (Perch-gley, and Mottled), a lack of gleying or other diagnostic features (Yellow), sandy texture (Sandy), presence of an albic horizon (Albic), presence of ultramafic rock influence in the parent material (Magnesic), horizons with Granular Soil properties at depth (Buried-granular), and soils typical of the Ultic Soil group without any of the more specific properties (Typic).

16.1.4 Origin of Soil Order Name

The term 'Ultic' comes from the Latin *ultimus*, meaning 'final, last' and was chosen by Allan Hewitt to match the Ultisols order of *Soil Taxonomy* (but with a narrower range for New Zealand). Although it is now well appreciated that soils evolve by ebb and flow (based on the dynamic-rate model of pedogenesis) rather than developing in a linear fashion towards some ultimate point in soil development, the name 'Ultic' does invoke a highly weathered and strongly enleached character and thus an advanced stage of soil development.

16.2 Soil Profile Genesis

The vast majority of Ultic Soils have formed through top-down pedogenesis, with soil formation starting at the land surface on freshly exposed rock material. The soil profile gradually deepens as soil horizons are formed in a downward-moving front driven by the organic and hydrologic cycles. Ultic Soils are formed in saprolite on strongly weathered parent rocks, including sandstones and greywacke, that are rich in quartz, feldspar, and various other minerals that may include mafic crystals. The well-developed soils, with advanced weathering, attest to the long periods of time that the soils have been exposed on largely stable land surfaces. Many are probably several hundred thousand years old, or older. As well, the northern

North Island, where most Ultic Soils occur, has the warmest climate in New Zealand and hence rates of soil-forming processes are relatively fast compared with those elsewhere. The northern Ultic Soils form under generally humid conditions, although droughts are not uncommon. Because of the warm climate and high solar radiation, evapotranspiration is high. Thus, the generally low monthly water balances (rainfall/potential evaporation) tend to make the region susceptible to drought when rainfall is only a little lower than usual.

In terms of vegetation, pollen studies by Rewi Newnham and others have shown that the Northland Peninsula remained forested throughout the Last Glaciation, although the forest was dominated by beech and 'cool' tree species (rather than kauri and 'warm' species that characterise interglacials). During the Last Glaciation, mean annual temperatures in New Zealand were depressed by ~6 °C. Consequently, the Ultic Soils north of Auckland have always been under forest, whereas the rest of New Zealand was dominated mainly by shrubland-grassland vegetation during the Last Glaciation (the Waikato region occupied an intermediate position, comprising shrubland-grassland with some beech forest and rare conifers).

Putting these factors all together means that the Ultic Soils (in northern North Island) have been strongly weathered and hence clay formation (argillisation or clay synthesis) is well advanced with clay contents in lower subsoils often around 50–60% or more.

Plant growth is rapid and melanisation, humification, mineralisation, and nutrient recycling in Ultic Soils are also rapid processes in warm, humid conditions. Soil mixing by bioturbation, such as tree overturn and worm activity, have been important and, as a result, the horizon boundaries in many Ultic Soils are occluded or irregular, and indistinct to diffuse.

Leaching has been strong despite many soils being imperfectly to poorly drained (presumably because rainfall has always been at least moderate to high, and land surfaces are old, thus allowing plentiful time), and marked acidity is therefore a key feature for the order. Ultic Soils are defined as having a pH <5.5 for the major part of the profile to a depth of 60 cm. Hand-in-hand with the leaching-induced acidity is a low nutrient status.

Clay movement down the profile (illuviation or illimerisation) has occurred, shown by the common occurrence of either an E horizon or a pale B horizon beneath the A horizon and a complementary clay-enriched Bt (argillic) horizon in the lower subsoil. Dark humus material is also commonly translocated down cracks in the profile, coating, and staining ped faces (which may be prismatic, coarse blocky, or polyhedral) in the subsoil. The argillic horizon slows permeability and therefore horizons above it typically have redox and reductimorphic features, ranging from orange mottles, Fe–Mn concretions, to low-chroma colours sufficiently extensive to form Eg or Er horizons.

Although the Ultic Soils in Northland are formed by topdown pedogenesis on a range of parent materials, mineralogical studies of sand fractions in the soils show that minor amounts of volcanic glass also occur in the profiles. Such glass shards were probably largely derived from tephra blown directly from eruptions of volcanoes in far-off central North Island. This inference is supported by the occurrence of thin rhyolitic (silica-rich) tephra (volcanic ash) beds, mainly comprising glass, preserved unaltered in lake sediments and peat bogs in Northland. Such tephras include a ~10-cm-thick layer of Rotoehu Ash (c. 45,000 years old) in Lake Omapere near Kaikohe, along with dustings of Kaharoa, Taupo, and other tephras.

In only one known case in North Island has an Ultic Soil been formed by upbuilding (rather than topdown) pedogenesis, namely the tephra-derived Kainui soil (Sect. 16.3.3). In Southland, an Ultic Soil formed on strongly weathered alluvium also has a capping of loess and so it originated mainly via topdown pedogenesis but with some upbuilding pedogenesis in more recent times represented by incremental loess accumulation.

16.3 Soil-Landscape Relationships

16.3.1 Introduction

The strongly weathered, strongly leached, clay-rich, and acidic character of Ultic Soils is attributed to long periods of land stability that provide sufficient time for soil-forming process to operate. The land surfaces where Ultic Soils occur are survivors of the major perturbations (volcanic activity, tectonic uplift, and impacts related to glacial-interglacial cycles including the Last Glaciation) that have strongly influenced the soils and their parent materials in other regions.

As well as their Northland-Auckland-Waikato homeland, the Ultic Soils also occur in small isolated areas throughout the rest of New Zealand where remnants of land surfaces have remained stable. Thus, scattered Ultic Soils also occur in:

- Wellington, where remnants of old land surfaces occur as relatively stable ridge crests in hilly terrain.
- Marlborough Sounds, on spurs of low relief land near sea level with slopes less than 25° that protrude into the inlets and which have escaped burial by the active colluvium on steeper slopes closer to the main ridges, specifically in Port Underwood, and Tennyson Inlet. Beneath these soils, the schist and schistose-greywacke saprolite is deeply red weathered.

Fig. 16.4 The Otago plateau appears as a series of low-angle ramps leading up from valleys to the gentle summits of mountain ranges

- Tasman District (Nelson area), where there are extensive areas of weathered granite on rolling and gentle hills, close to the sea, where slopes have been long protected from colluvial modification.
- Inland Otago on stable broad, flattish ridge crests (Fig. 16.4).
- Southland, where on the southern margin of the upper Waimea valley, Yellow Ultic Soils are formed on the highest (and therefore oldest) alluvial terrace in highly weathered, stony, and clayey gravel with a thin surface layer of loess. The high terraces are isolated from colluvium on the Hokonui Hills (dominated by Mafic Melanic soils) and are exposed to winds that have prevented deposition of deep accumulations of loess.

16.3.2 Ultic Soils in Auckland and Northland

The Ultic Soils, along with Oxidic and Granular Soils, are the prominent soils of Northland and are also important in the Auckland and western and northern Waikato regions where younger tephra deposits are thin or absent.

Ultic Soils, developed on sedimentary rocks, in northern North Island have not undergone the rapid tectonic uplift that has elsewhere led to erosion and deposition of eroded materials, providing a new start to soil development. Because the Ultic Soils in Northland, along with Granular and Oxidic Soils, have always been under forest cover they have been 'insulated' sufficiently from many of the effects of the cold climate periods of the Last Glaciation (including erosion and deposition of loess, alluvium, and colluvium) to prevent initiation of new soil development processes.

Exotic plantation forestry was established in the 1970s, in the Warkworth area in Northland, mainly on Ultic Soils, when pasture was replaced with *Pinus radiata* trees, on blocks of steeper less productive farmland, to form Mahurangi Forest. To help support sustainable forest management Wim Rijkse and Haydon Jones developed a new understanding of soil-landscape relationships in the hill country of the southern Mahurangi Forest. Jones' soil-landscape model showed that three soil orders were represented (Ultic, Gley, and Raw Soils) with Ultic Soils predominating and Mottled Yellow Ultic Soils the most common subgroup.

In the landscape (Fig. 16.5), narrow ridge summits are dominated by Typic Yellow Ultic Soils (Warkworth series), broad ridge summits, and backslopes are dominated by Mottled Yellow Ultic Soils (Whangaripo and Puhoi series), and foot slopes are overlain by Typic Perch-gley Ultic Soils

Fig. 16.5 Hilly landscape dominated by Yellow Ultic Soils near Warkworth in Northland. The first rotation of plantation radiata pine has been harvested and since been replanted. Photo: Andrew Hosken

Table 16.1 Relationships between soil classes and landscape features in Mahurangi Forest, Northland (after Jones et al. 2000)

Landscape features	Dominant soil class	Subdominant soil class
Flat/narrow ridge summit	Typic Yellow Ultic	Mottled Yellow Ultic (Whangaripo series)
Broad/sloping ridge summit	Mottled Yellow Ultic (Whangaripo series)	Mottled Yellow Ultic (Puhoi series)
Ridge hillock	Typic Yellow Ultic	Mottled Yellow Ultic (Whangaripo series)
Ridge saddle	Mottled Yellow Ultic (Puhoi series)	Mottled Yellow Ultic (Whangaripo series)
Backslope	Mottled Yellow Ultic (Whangaripo series)	Mottled Yellow Ultic (Puhoi series), Typic Yellow Ultic, Orthic Raw Soil
Foot slope	Typic Perch-gley Ultic	
Narrow valley floor	Acidic Orthic Gley Soil	
Broad valley floor	Fluvial Raw Soil	

(Pohuehue series). Narrow valley floors support Acidic Orthic Gley Soils (Table 16.1).

Jones found that the soil-landscape relationships related mainly to a natural drainage sequence ranging from the well or moderately well drained Typic Yellow Ultic Soils in upper landscape positions to poorly drained Gley Soils in low lying areas. The drainage of the soil appears to be controlled mainly by a combination of topographic position and the nature of the underlying parent material. The parent material is saprolitic clay derived from strongly weathered Tertiary sandstone and siltstone (Waitemata Group), which is red weathered in places (Sect. 16.3.4). The drainage of the soil can have a marked influence on tree growth and can impact on harvesting operations and so the soil-landscape relationships established are important for forest management despite the limited taxonomic diversity.

16.3.3 Ultic Soils in the Northern Waikato Region Including the Special Case of the Kainui Soil

On steeper hill slopes in northern and western Waikato Ultic Soils occur on exhumed greywacke where tephra cover is not preserved. A relationship between slope angle and the occurrence of Ultic Soils was developed by Michael Selby and Robin Palmer in the 1970s for hilly land underlain by greywacke in the western Waikato. They found that on slopes of ~20°, only the topsoil horizons of the soils are derived from tephra deposits, the subsoil horizons being derived from weathered greywacke. Where slope angles exceeded ~26°, the soils are formed more-or-less entirely from weathered greywacke alone. Thus, in general, in this terrain, Ultic Soils occur on slopes ≥20° and Granular or Allophanic Soils are found on slopes <20°.

On the pervasive tephra mantle, in the western and northern Waikato region, Granular Soils (formed on old, strongly weathered Hamilton Ash beds) are common on rolling landscapes with Allophanic Soils on flat surfaces, as described by John Bruce in the Raglan County survey. However, there is one special case, the Kainui soils, where the tephra is sufficiently thick, and halloysitic in character so that an affinity to Ultic rather than Granular Soils is preferred.

The Kainui soil (Fig. 16.6) is a Buried-granular Yellow Ultic Soil, effectively an intergrade between Granular Soils and Ultic Soils. It occurs on low rolling hills of middle-Quaternary age mainly in the Hamilton lowlands in and to the north of Hamilton city. The Kainui Soil is a composite, multi-layered tephra-derived soil consisting of two distinct parts, upper and lower. The upper part is a cover-bed typically ~40–70 cm in thickness (~60 cm on average) comprising numerous thin late Quaternary rhyolitic and andesitic tephras that have been accumulating incrementally since c. 45,000 years ago (the age of Rotoehu Ash at the cover-bed's base). While the tephras were accumulating the

Fig. 16.6 Kainui soil near Hamilton comprising two parts: an upper composite cover-bed and lower dark reddish brown buried clay-rich paleosol (effectively a buried Granular Soil). The soil is classified as a Buried-granular Yellow Ultic Soil (ka = 1000 years) (after Lowe 2019)

soil was gradually developing (being pedogenically altered via developmental upbuilding pedogenesis) forming soil horizons. Any original depositional (tephra fall) bedding has been almost entirely masked by pedogenic alteration including clay formation (average clay content is ~30%) and soil mixing. The lower horizons of the cover-bed always show redox features including manganese oxide concretions and coatings on peds (mangans) with soil horizon notation Bw(f).

The moderately well-drained upper profile is dominated by halloysite (not allophane) in the clay fraction because of limited desilication as the result of impeded drainage and seasonal perching on a slowly permeable, buried, clay-rich paleosol (average clay content is ~85%) on upper Hamilton Ash beds (>c. 45,000 years old) which makes up the lower part of the two-storeyed Kainui soil. The cover-bed-paleosol boundary is a lithologic discontinuity (unconformity) marking a tree-overturn paleosurface that David Lowe reported (in 2019) may be c. 74,000 years old. The buried paleosol is strongly weathered and halloysitic with relict clay skins (forming paleo-argillic and/or paleo-kandic horizons) and redox features towards the base of the profile. The paleosol is inferred to have formed (via developmental upbuilding pedogenesis) during the Last Interglacial.

The classification of the tephra-derived Kainui soil has been problematic. Being non-allophanic, it clearly is not an Allophanic Soil, nor did it fit well in the Granular Soils (formed on weathered Hamilton Ash beds) because the 'granular' character (distinctive polyhedral pedality or soil fabric) associated with this order (Chap. 6) was at depths typically 60 cm or more below the land surface. Instead, the distinctive 'two storeyed' profile and its twin 'ultic' and subsurface 'granular' characters are reflected by its classification uniquely as a Buried-granular Yellow Ultic Soil.

16.3.4 Origin of Red Weathering

a. Red weathering in Northland

In the Mahurangi Forest area near Warkworth, Haydon Jones identified the main parent materials of the Yellow Ultic Soils as saprolite derived from alternating, volcanic-rich sandstones and siltstones of the Pakiri Formation (within the Waitemata Group). Jones identified two types of saprolite, one reddish (red-weathered) and the other yellowish (non-red-weathered, Fig. 16.7, centre). In soil profiles, the red saprolite is always in the lower subsoil, never at the surface.

Jones worked with micro-morphologist Antoine Jongmans to help explain the origin of the two types of saprolite. Jongmans examined a Mottled Yellow Ultic Soil underlain by yellow saprolite (Puhoi soil, Fig. 16.7, left) and also a Typic Yellow Ultic Soil underlain by red saprolite

Fig. 16.7 Red weathering. Left: Mottled Yellow Ultic Soil (Puhoi soil with yellow saprolite below 70 cm). Centre: Road cutting through saprolite showing layers of red and yellow-weathered material. Right: Typic Yellow Ultic Soil (Warkworth soil with red saprolite below 50 cm)

(Warkworth soil, Fig. 16.7, right) and the fresh, unweathered, parent rocks associated with each soil. Jongmans found that the different saprolites reflected a difference in the parent materials and thus how they weathered. The red saprolite is formed on a coarse sandstone enriched in mafic minerals (e.g. pyroxenes) and volcanic rock fragments, together, critically, with calcite (Pakiri Formation containing volcanogenic Parnell Grit). This parent material becomes porous as the calcite dissolves amidst the coarse Fe-rich mineral grains, favouring the synthesis of red haematite in a 'box-like' microfabric, provided the soil is well drained (and therefore likely to be in an upper landscape position). In contrast, the yellow saprolite is formed on fine-grained mudstone with little or no mafic material and no calcite (Pakiri Formation without Parnell Grit). It therefore remains massive and non-porous during weathering, lacking the right conditions to form red Fe oxides (probably it contains yellow goethite together with lepidocrocite in orange mottles).

The red weathering has a patchy, erratic distribution in the landscape as the bedded volcanogenic sandstone and siltstone parent materials are often highly folded and convoluted, and discontinuous because of erosion. The red-weathered saprolite is thus considered to be a relict feature because it has been overprinted by topdown pedogenesis in upper profiles, or because the contemporary soils are saturated (gleyed), or both.

b. Red weathering in other parts of New Zealand

Although the red weathering puzzle in Northland has been solved, it is not clear when or how the red weathering seen in other parts of New Zealand, commonly associated with Ultic Soils, was developed. Pockets of deep, red-weathered materials occur on greywacke surfaces throughout New Zealand as well as on granodiorites in northern Stewart Island and throughout eastern Otago and Canterbury. Red weathered soils are the norm in northeastern Marlborough and western Wellington. Kaolinitic red-weathered material in New Zealand appears to be commonly found in areas of relatively low tectonic uplift rates where the stability may have helped preserve the highly weathered material. An alternative possibility is that the red-coloured kaolinitic material is aeolian dust derived from Australia that has been deposited over long time periods on relatively stable surfaces in New Zealand (Chap. 15).

Another hypothesis is that the reddening occurred in a notably warmer interval in the distant past, either in the Quaternary (since 2.6 million years ago), in the Pliocene (5.3–2.6 million years ago), or Miocene (23–5.3 million years ago), or even in the Cretaceous (145–66 million years ago) (see grey box). It has been observed that the sedimentary deposits forming the earliest part of the Cenozoic (last 66 million years, such as the Broken River Formation) are quartz- and kaolin-clay rich. Such deposits came from a strongly, and presumably deeply, weathered greywacke regolith which some have attributed to weathering on a "Cretaceous peneplain" surface (about 80 million years ago) prior to the greywacke basement subsiding beneath the ocean.

Could this be the ultimate Ultic Soil?

Today, Central Otago has a low level of tectonic activity similar to that of the Northland region. We might expect to find Ultic Soils in Otago but, generally, their formation has been prevented by erosion related to periods of severe cold as glaciers advanced and retreated in the Southern Alps further west. Nevertheless, very old Semiarid Soils do occur (Chap. 15).

However, near Alexandra, in Central Otago, there is perhaps the oldest soil in New Zealand (Fig. 16.8). It is a paleosol, which is a fossil soil formed in an earlier time presumably under warmer, wetter, conditions than those of the present. The paleosol has been exposed in an old gold mining area, near the shore of Butchers Dam.

The white and brilliant red colours of the paleosol contrast with the subdued grey-brown colours of the overlying Late Quaternary ($\sim$20,000 years old) fluvioglacial gravels and pebbles underlying the present-day surface soil. The top horizon of the paleosol is the white clay that is thought to correspond to the ancient land surface. Both the white and red horizons have 55% clay with clay minerals dominated by kaolinite. The irregular, sharp nature of the boundary suggests a pedologic origin. The red is likely to be the Fe-oxide mineral haematite. Not shown in the figure is underlying white silt with 22% clay, again dominated by kaolinite, with the ghost rock structure of highly weathered schist.

The Butchers Dam soil is an important historical marker. It supports the occurrence of a lengthy period of stability on the erosion surface where soil weathering rates have exceeded those of erosion.

16.4 Key Soil Properties

16.4.1 Soil Composition

Ultic Soils comprise the clayey weathering products of mainly siliceous sedimentary rock of varying compositions and textures. The types of clays formed are wide-ranging but typically are dominated by kaolinite and halloysite, along with abundant smectite including montmorillonite in wetter soils, and minor micaceous clays including illite and

Fig. 16.8 The Butchers Dam paleosol is a remnant of an extensive ancient soil, which is at least 20 million years old, formed on weathered (kaolinised) schist

clay-vermiculite. That montmorillonite and halloysite are swelling clays has implications for slope stability. Small amounts of crystalline iron oxides, including lepidocrocite (in orange mottles) and goethite (in yellowish-coloured horizons), occur and reddish horizons contain haematite.

16.4.2 Physical Properties

Ultic Soils are have some of the highest clay contents of any soils in New Zealand (Fig. 16.9, Table 16.2). The effect of clay having been moved down through the soil profile is evident with clay contents of about 25–50% in topsoils and about 50–75% in subsoils. Much of the translocated clay is likely to be fine clay (<0.2 μm) because the fine clay/total clay ratio increases in Bt horizons (e.g. the ratio essentially doubles from 0.2 to 0.4 in the Waikare soil, a Perch-gleyed Albic Ultic Soil). The soil dry bulk density peaks just below the topsoil where a densipan occurs in many Ultic Soils. However, soil dry bulk densities remain high (1.2–1.4 t m^{-3}) throughout the subsoil. Topsoils have a greater porosity and this is reflected in the lower soil dry bulk density (about 1.0 t m^{-3}), and higher water holding capacity (about 30%), in the topsoil than in subsoils. The heavy clay subsoils have relatively low total available water-holding capacity as there is a lack of pore space and the clay strongly adsorbs any water that is in the subsoil (Fig. 16.9).

A layer with slower permeability (Bt horizons) occurs in most Ultic Soil profiles, thus the majority of soils are imperfectly to poorly drained, although a few are well drained. Soil water movement is mainly along planar voids which open up as the soil dries in summer, but in winter as the soil wets up, the common smectite clays and halloysite swell and voids are closed, leaving the soil with only very low hydraulic conductivity. Surface horizons, especially silty ones, are dispersible and are thus particularly prone to water erosion where the surface vegetation has been removed.

16.4.3 Chemical Properties

The pH of topsoil is, like base saturation, strongly affected by fertiliser and lime use. Subsoil pH values tend to be strongly, or, extremely acid (pH <5.5). Subsoils with pH values less than about 4.8 are likely to be affected by toxic levels of plant-available aluminium which may inhibit root function and contribute to shallow rooting habits in aluminium-sensitive

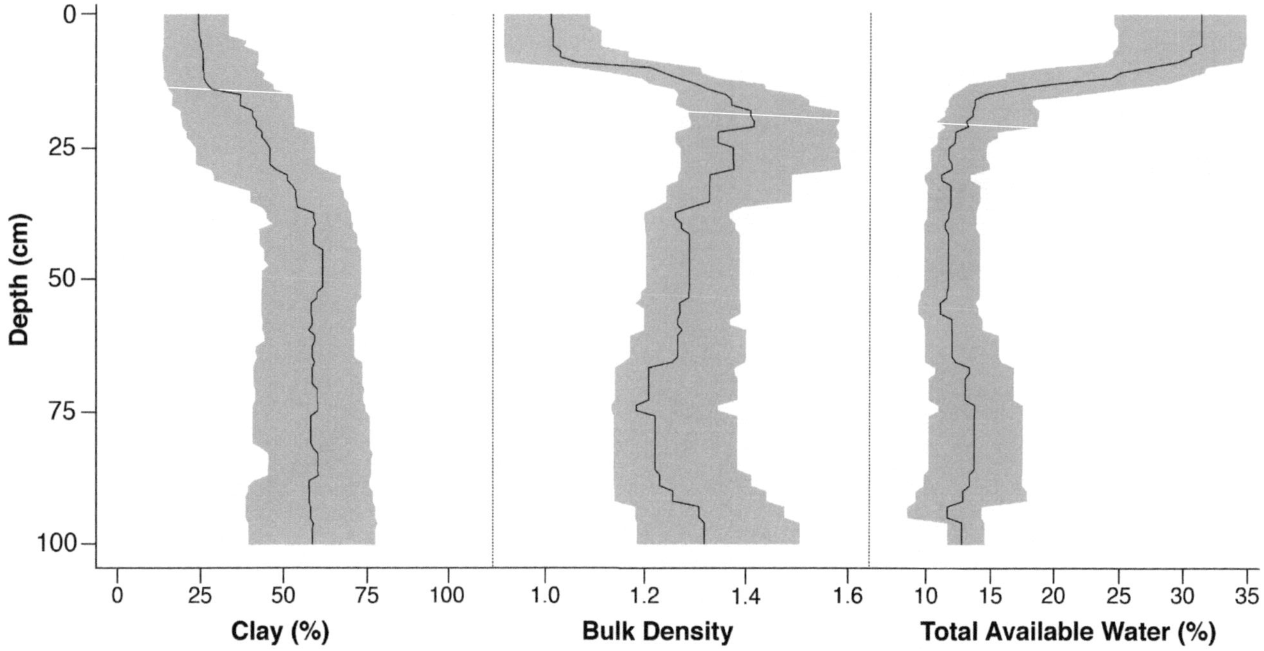

Fig. 16.9 Median and upper and lower quartiles of clay content, soil dry bulk density (t m^{-3}), and total available water-holding capacity for Ultic Soils in the New Zealand Soil Data Repository

Table 16.2 Example of the sand, silt and clay content of a Mottled Albic Ultic Soil (Mapua series, SB09292)

Horizon	Depth	Sand (%)	Silt (%)	Clay (%)
Ah	0–14	19	56	24
Bw	14–20	15	50	33
Bt(f)1	20–37	14	42	43
Bt(f)2	37–65	6	27	66
Bt(g)	65–104	9	23	66
C(g)	104–120	7	40	52

plants. Organic carbon is generally medium in topsoils and very low in subsoils (Fig. 16.10, Table 16.3).

Ultic Soils are strongly weathered and leached with low nutrient reserves. Phosphorus reserves are low with total phosphorus characteristically less than 20 mg/100 g. However, total P values may be very high in topsoils as a result of fertiliser inputs. Subsoil P values are usually less influenced by fertilisation and are frequently low indicating the marked weathering of the soil parent material. Low levels of 0.5M H_2SO_4-extractable phosphorus (usually less than 3 mg/l00g) are a characteristic feature. Strong weathering is indicated by low ratios of 0.5M H_2SO_4-extractable phosphorus to inorganic phosphorus, usually 0.25 or less. Potassium, magnesium, and molybdenum reserves are also low, again reflecting the strong weathering and leaching. Although P retention is usually medium in B horizons, it may be low in A and E horizons consistent with generally low contents of secondary iron oxides. Thus P added as fertiliser remains available for plant uptake. Subsoil base saturations are low or very low and indicate that a high proportion of the cation exchange is occupied by acid cations (Al^{3+} and H^+).

16.4.4 Biological Properties

Ultic Soils in their natural state have low pH, relatively low nutrient availability, potential for aluminium toxicity, and periods of drought in summer and saturation in winter or early spring, thus they are a relatively difficult environment for plant growth. The natural vegetation comprises slow-growing native forest with plants such as manuka and kanuka, which are tolerant of the poor conditions, dominant in early stages of succession.

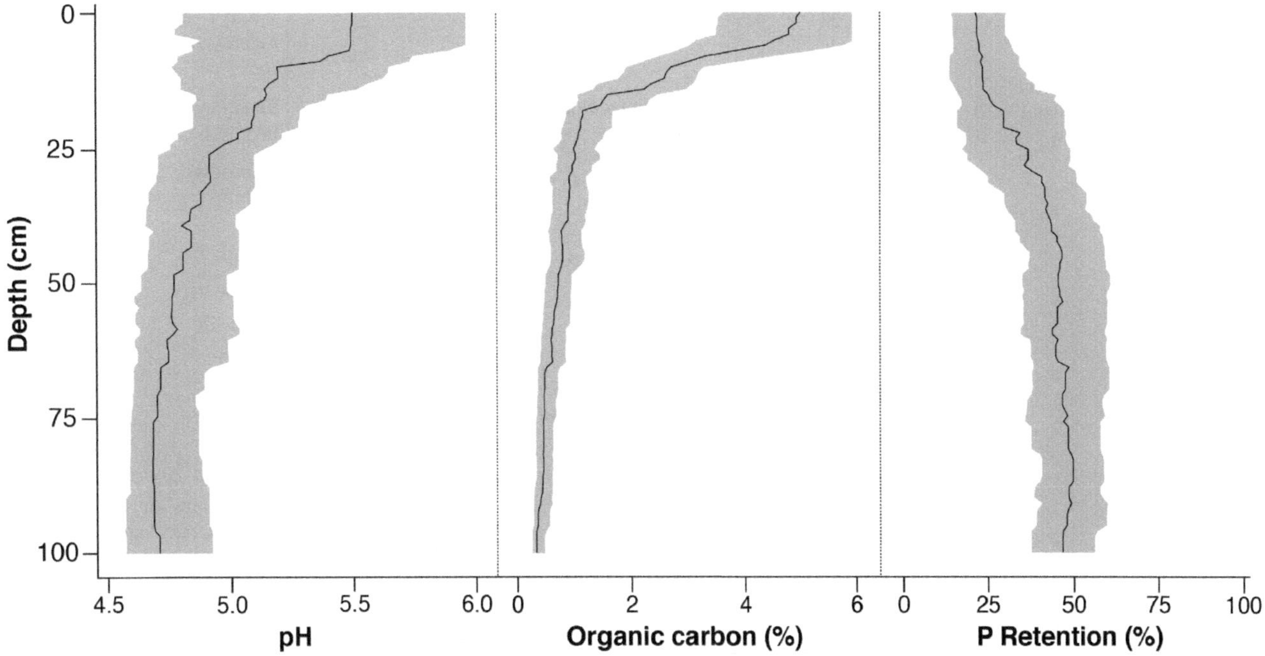

Fig. 16.10 Median, and upper and lower quartiles, of soil pH, organic carbon (%), and P retention (%) for Ultic Soils in the New Zealand Soil Data Repository

Table 16.3 Example of chemical properties of a Mottled Albic Ultic Soil (Mapua series, SB09292)

Horizon	Depth (cm)	pH (in H_2O)	Carbon (%)	Nitrogen (%)	CEC[a] (cmol$^{(+)}$/kg)	Sum bases (cmol$^{(+)}$ kg)	P retention (%)
Ah	0–14	5.5	3.8	0.25	13.4	7.51	29
Bw(f)	14–20	5.2	1.3	0.08	8.4	2.57	28
Bt(f)1	20–37	5.1	0.8	0.05	9	2.39	28
Bt(f)2	37–65	5	0.6	0.04	15.2	3.24	42
Bt(g)	65–104	5	0.5	0.04	18.1	3.24	48
C(g)1	104–120	5.1	0.4	0.01	14.3	2.37	42
C(g)2	120–180	4.9	0.2	0.01	13.7	2.07	44

[a]CEC cation exchange capacity

16.5 Distinguishing Between Ultic Soils and Related Soil Orders

The Ultic Soils are distinguished by the predominantly high clay content (except for a relatively minor class of sandy soils within the order), markedly pale upper subsoils and generally pale yellowish subsoils, which can also be reddish in colour in some situations, and evidence of strong leaching and weathering. The other soil orders with a high clay content are Granular Soils which, in contrast to most of the Ultic Soils, are derived mainly from strongly weathered silicic tephra deposits (especially Hamilton Ash), and Oxidic Soils formed in dark-coloured mafic-rich volcanic deposits including basalt. Both Oxidic and Granular Soils have reddish brown or deep brown colours, in contrast to the dominantly pale, yellowish, upper subsoil colours of Ultic Soils.

16.6 Correlation with Other Classification Systems

The Ultic Soils generally correlate with the Ultisols of *Soil Taxonomy* and the Planosols of the *World Reference Base* (Table 16.4). However, Planosols may not correlate well to most New Zealand Ultic Soils because the New Zealand soils may not meet strict Planosol requirements for textural change or pale (gley) colours. In the New Zealand genetic classification, most Ultic Soils were classified as northern yellow-brown earths, although some were considered gley soils.

Table 16.4 Correlation[a] between Ultic Soils and equivalent classes of *Soil Taxonomy*, *World Reference Base*, and the New Zealand genetic soil classification

New Zealand Soil Classification	*Soil Taxonomy*	*World Reference Base*	NZ genetic soil classification
Densipan Ultic Soils	Albaquults, Epiaquults	Endogleyic Planosols	
Albic Ultic Soils	Albaquults	Planosols (Albic)	Northern yellow-brown earths
Perch-gley Ultic Soils	Kandic Albaquults, Aeric Epiaquults	Epigleyic Planosols	Gley soils
Sandy Ultic Soils	Haplohumults	Planosols (Arenic)	Yellow-brown earths
Yellow Ultic Soils	Kandihumults, Typic Kandiudults, Typic Hapludults, Typic Paleudults, Aeric Paleaquults, Aeric Endoaquults	Haplic Planosols	Northern yellow-brown earths

[a]The correlations given here are a guide only and, for accurate classifications, the relevant soil classification documents should be consulted. The two major international soil classification systems are *Soil Taxonomy*, which was developed initially in the USA, and *World Reference Base*, which was developed primarily in Europe. The NZ genetic soil classification was used in New Zealand prior to 1992

16.7 Use and Management of Ultic Soils

The strongly weathered Ultic Soils have clayey subsoils, slow permeability, low soil fertility, and low water holding capacity, which, along with winter wetness and summer drought, means the soils are usually challenging to use productively. With lime and fertiliser inputs, the soils will support pastural agriculture. Many of the Ultic Soils occur on sloping hill country and therefore are not well suited to more intensive horticultural production or cropping due to erosion risk in ploughed fields. Thus, sheep and cattle grazing are the dominant land uses as well as forestry.

For sheep and beef farming, once soil fertility has been built up, maintenance fertiliser will be needed to replace nutrients removed with the products (mainly wool and meat). Because the soils have slow permeability, they are prone to becoming waterlogged in periods of high rainfall (in winter and early spring). The dispersive tendency of A and E horizons enhances susceptibility to livestock-treading damage and erosion. Thus, when grazing cattle, care needs to be taken to prevent pugging which leads to bare ground at the soil surface with consequent erosion, weed invasion of pasture, and loss of productivity.

Trampling of wet soils can also cause compaction and sealing of the soil surface which promotes runoff and, thus, increased erosion and sediment carried to downslope water bodies. Management options include ensuring stocking rates are low enough to prevent damage, or to graze an area for a part of the day then move stock to stand-off pads where effluent must be appropriately managed. Sheep may be a better option for grazing the steeper hill country as they do not damage the ground surface as much as heavy cattle. Riparian fencing and appropriate riparian vegetation can help trap sediments and nutrient runoff, thus protecting stream water quality.

Ultic Soils can be used for urban or peri-urban development which provides a good alternative for city expansion, avoiding other more productive soil. For urban development, slopes may need some recontouring and people should be encouraged to replace topsoil on the new land surfaces where they want to establish gardens. The 'yellow clay' of Ultic Soil subsoils is well known to developers and Auckland gardeners. It is acidic and hard in summer and very wet in winter. Lime can correct the acidity but needs to be mechanically incorporated to correct subsoil acidity, and compost can be dug-in to help correct the heavy clay-influenced structure.

Care must be taken when establishing expensive urban infrastructure on Ultic Soils because many slopes are prone to mudslides (also called earthflows), and landslides, particularly those on the Waitemata Group sedimentary rocks in the Auckland/Northland region (Fig. 16.11). The high proportions of swelling clays can lead to the formation of contraction cracks during dry periods. The cracks allow rapid infiltration of water into the soil but drainage from the soil profile is very slow. The clayey soil material becomes saturated and, where it overlies a gently dipping rock surface and with low shear strength, failure occurs along a basal shear surface at the soil/bedrock contact. Modelling undertaken by Vicki Moon showed that failure was initiated on slopes of 15–20 degrees. It is also evident that 'slow mudslide activity', a type of soil creep, has occurred in some developed urban areas on Ultic Soils in the Auckland region.

That the landscape is eroding and slowly but surely downwearing is highlighted by the sediment yields from rivers such as the Wairoa River near Dargaville, which carries an average sediment load of 1.1 Mt per year. Severe

Fig. 16.11 Mudslide on Ultic Soils developed on Waitemata Group parent materials at Tawharanui Peninsula in Northland. The two bowl-shaped features on the hillslope are source areas from which the material has flowed downslope into the floor of the valley (photo: V. G. Moon)

storm events (such as that of 17–18 July 2020, with a rainfall of 215 mm in 24 h at Whangarei, and nearly 60 mm at Kaikohe in 1 h) result in widespread flooding, erosion, and land-sliding in parts of Northland.

Ultic Soils are generally not well suited to intensive cropping or horticulture. However, an exception is the Buried-granular Yellow Ultic Soil (Kainui soil) that occurs on flat summits or gently sloping shoulders of low hills in the northern Waikato. The Kainui soils have been used successfully for market gardening in the Hamilton area, mainly for summer crops planted early in spring as soon as the soil becomes sufficiently dry to work. Early planting is needed to avoid commonly occurring late summer droughts during which the soils become cracked and hard unless irrigated (which can generate heavy cropping). Very wet conditions associated with water perching in winter (on the impermeable subsoil paleosol on Hamilton Ash) do not normally allow winter crops to be grown because the sticky, wet soil usually precludes the use of wheeled vehicles. A tea plantation has been developed, partly on Kainui soils, north of Hamilton (Fig. 16.12).

Another example of intensive land use on Ultic Soils is orcharding in the Mapua area near Nelson. Although posing significant problems, the wet, acidic, clayey, and enleached Mottled Albic Ultic Soil (Mapua series, Tables 16.2 and 16.3) is capable of growing high-quality apples.

Plantation forestry (Fig. 16.5) has been utilised successfully on Ultic Soils in Northland. In the context of optimising sustainable forest management, Haydon Jones examined the likely impacts of forest harvesting on the properties of Ultic Soils in the southern Mahurangi Forest. He found that forest harvesting resulted in a decrease in macroporosity, Bray Mg, and pH, and an increase in mean total C, Bray P, and K. Jones attributed these changes to disturbance by harvesting machinery, slash deposition, and the development of grass cover. The results showed that forest harvesting had a negative impact on some soil properties, such as causing soil compaction and acidification, and

Fig. 16.12 Tea growing on Kainui soils at Zealong tea plantation near Hamilton

a positive effect on others, such as increasing some nutrients. Jones suggested the reduction in macroporosity to be of concern with respect to future forest productivity and susceptibility to further compaction.

Further Reading

Bruce JG (1978) Soils of part Raglan County, South Auckland, New Zealand. N Z Soil Bur Bull 41. NZ DSIR, Wellington. Maps + 102p

Edbrooke SW (compiler) (2001) Geology of the Auckland area. Institute of Geological and Nuclear Sciences 1: 250 000 Geological Map 3. 1 sheet + 74 pp, Institute of Geological and Nuclear Sciences, Lower Hutt

Fieldes M (1968) Clay mineralogy. In: Soils of New Zealand part 2. N Z Soil Bur Bull 26(2):22–39

Fieldes M, Weatherhead AV (1968) Mineralogy of sand fractions. In: Soils of New Zealand part 2. N Z Soil Bur Bull 26(2):8–21

Fitzpatrick RW, Taylor RM, Schwertmann U et al (1985) Occurrence and properties of lepidocrocite in some soils of New Zealand, South Africa and Australia. Aust J Soil Res 23:543–567

Gibbs HS (1965) Soils of Northland. Reprinted from National Resources Survey, Part III-Northland Region (compiled by the Town and Country Planning Branch, Ministry of Works). New Zealand Soil Bureau Publication 286

Gibbs HS (1980) New Zealand soils—an introduction. Oxford University Press, Wellington, 115p

Gibbs HS, Cowie JD, Pullar WA (1968) Soils of North Island. In: Soils of New Zealand part 1. N Z Soil Bur Bull 26(1):48–67

Harmsworth GR (1996) Land use capability classification of the Northland region. Landcare Research Science Series No 9. Manaaki Whenua Press, Lincoln, New Zealand, 269p

Hayward BW, Jamieson A, Morley MS (2017) Out of the ocean, into the fire. History of the rocks, fossils and landforms of Auckland, Northland, and Coromandel. Geoscience Society of New Zealand Miscellaneous Publication 146, 336p

Hewitt AE (1993) Methods and rationale of the New Zealand Soil Classification. Landcare Res Sci Ser 2:1–71

Hewitt AE (2007) Soils. Te Ara—the Encyclopaedia of New Zealand. http://www.teAra.govt.nz/en/soils. Accessed 26 Aug 2020

Hewitt AE (2010) New Zealand Soil Classification, 3rd edn. Manaaki Whenua Press, Lincoln

Hicks M, Shankar U, McKerchar et al (2011) Suspended sediment yields from New Zealand rivers. J of Hydrol (NZ) 50:81–142

Isaac MJ, Herzer RH, Brook FJ et al (1994) Cretaceous and Cenozoic sedimentary basins of Northland, New Zealand. Institute of Geological and Nuclear Sciences Monograph 8, 203p + 4 sheets

Jones HS (2004) Impacts of forest harvesting on performance of soil-landscape modelling in a radiata pine forest, northern New Zealand. Unpublished PhD thesis lodged in the Library, University of Waikato, Hamilton

Jones HS, Lowe DJ, Rijkse WC et al (2000) A soil-landscape model for southern Mahurangi Forest, Northland. Mahurangi Forest-Soil Project Report No 1 to Forest Research. Department of Earth Sciences, University of Waikato, 63p

Jones HS, Lowe DJ, McLay CDA et al (2002) Impacts of forest harvesting on soil properties in southern Mahurangi Forest, northern New Zealand. Programme & Abstracts, Golden Jubilee Conference, New Zealand Society of Soil Science, Victoria University of Wellington, p 68

Jones HS, Lowe DJ, Jongmans AG et al (2002) Red-weathering, saprolite permeability, and soil drainage in southern Mahurangi Forest, northern New Zealand. Programme & Abstracts, Golden Jubilee Conference, New Zealand Society of Soil Science, Victoria University of Wellington, p 114

Jones HS, Lowe DJ, McLay CDA (2004) Spatial prediction of soil properties in a radiata pine forest, northern New Zealand: how does harvesting affect the performance of prediction techniques? In: SuperSoil 2004 programme and abstracts, 3rd Australia New Zealand soils conference, 5–9 Dec 2004, University of Sydney, Australia, p 66

Leathwick J, Wilson G, Rutledge D et al (2003) Land environments of New Zealand. Bateman, Auckland, 184pp + 2 maps

Lowe DJ (2019) Using soil stratigraphy and tephrochronology to understand the origin, age, and classification of a unique Late Quaternary tephra-derived Ultisol in Aotearoa New Zealand. Quaternary 2(1):9 (34 p). https://doi.org/10.3390/quat2010009

Lowe DJ (2019b) Origin and classification of the Kainui silt loam: update on the leopard that changed its spots. N Z Soil News 67 (1):75–85

Lynn IH, Basher LR (1994) Principles underlying land systems in resource assessment of hill and mountain lands in New Zealand. In: Webb TH (ed) Soil-landscape modelling in New Zealand. Manaaki Whenua Press, Lincoln, pp 38–51

Molloy L (1998) Soils in the New Zealand landscape, 2nd edn. New Zealand Society of Soil Science, 239 p

Moon VG, de lange A, de Lange WP (2003) Mudslides developed on Waitemata Group rocks, Tawharanui Peninsula, North Auckland. N Z Geogr 59:44–53

Ministry of Works (1964) National resources survey, part III. Northland region. Government Printer, Wellington, 202p

Ministry of Works (1971) National resources survey, part 1: Northland region. Town and Country Planning Division of Ministry of Works. Government Printer, Wellington

Mitchell ND (1991) The derivation of climate surfaces for New Zealand, and their application to the bioclimatic analysis of the distribution of kauri (*Agathis australis*). J R Soc N Z 21:13–24

McLaren RG, Cameron KC (1996) Soil Science sustainable production and environmental protection, 2nd edn. Oxford University Press

Newnham RM (1999) Environmental change in Northland, New Zealand, during the last glacial and Holocene. Quat Int 57(58): 61–70

Newnham RM, Lowe DJ, Green JD et al (2004) A discontinuous ca. 80 ka record of Late Quaternary environmental change from Lake Omapere, Northland, New Zealand. Palaeogeogr Palaeoclimatol Palaeoecol 207:165–198

Newnham RM, McGlone MS, Moar NT et al (2013) The vegetation cover of New Zealand at the Last Glacial Maximum. Quat Sci Rev 74:202–214

Newnham RM, Alloway BV, Holt KA et al (2017) Last Glacial pollen–climate reconstructions from Northland, New Zealand. J Quat Sci 32:685–703

Noyes AM, Balks MR, Moon VG et al (2016) Soil recovery following landsliding at Whatawhata Research Station, Waikato, New Zealand: preliminary results. In: Proceedings of hill country symposium 2016 (New Zealand Grassland Association, New Zealand Society of Animal Production, and New Zealand Grassland Trust, 12–13 Apr 2016, Rotorua), pp 83–88

Orbell GE, Parfitt RL, Furkert RJ (compilers) (1980) Specialist North Auckland. Guidebook for Tour 6. In: Soils with variable charge conference, Palmerston North (Feb 1981), 105p

Palmer RWP (1974) Soil variation on hillslopes at Whatawhata. Unpublished MSc thesis, University of Waikato, Hamilton

Rijkse WC (1996) Soils of Mahurangi Forest, North Island, New Zealand. Landcare Research Contract Report LC9697/003 (prepared for Carter Holt Harvey Forests Ltd, Albany)

Soil Survey Staff (2014) Keys to soil taxonomy, 12th edition. USDA Natural Resources Conservation Service, 360p. http://www.nrcs.usda.gov/wps/PA_NRCSConsumption/download?cid=stelprdb1252094&ext=pdf

Sutherland CF, Wilson AD, Cox JE et al (1985) Soil map of part Whangaparaoa-Auckland area. NZMS 290 sheet Rl0/11, North Island, New Zealand, 1: 100 000. New Zealand Soil Bureau map 221

Taylor NH, Sutherland CF (1942) Soils of North Auckland. Soil Bur Bull 3:25–38

Taylor NH, Sutherland CF (1953) Soils of North Auckland. In: Proceedings of the 15th conference of the New Zealand Grasslands Association, pp 15–38

Tonkin PJ, Lowe DJ, Nelson CS (2015) Obituary—Emeritus Professor Dr John Davidson McCraw (1925–2014) MBE, MSc *NZ*, DSc *Well*, CRSNZ, FNZSSS. N Z Soil News 66:14–35

World Reference Base (2015 update) http://www.fao.org/3/a-i3794e.pdf

Soils in the Ross Sea Region of Antarctica 17

Way down south in Antarctica
is where the icy Cryosols are.
With permafrost and desert pavement
some of the soils are truly ancient.

Abstract

The Ross Sea region of Antarctica contains the largest ice-free areas in Antarctica (about 6700 km^2). Soils in the Ross Sea region have a mean annual temperature of about −15 to −25 °C and so all contain permafrost. Antarctic soils generally comprise: a stony desert pavement from which fine material has been removed, predominantly by wind erosion; an active layer that freezes and thaws each year; and underlying permafrost in which temperatures remain below 0 °C. The McMurdo Dry Valleys, described as a 'cold desert', are Antarctica's largest continuous ice-free area, formed where evaporation exceeds precipitation. In some of the driest areas, there is not enough soil moisture to form ice-cement and the soils are described as having 'dry permafrost'. However, in coastal areas, and areas where there is some input of snow or meltwater, the permafrost is ice-cemented, patterned ground features form at the land surface, and soils may contain lenses or wedges of ice. Soils range from having minimal weathering and soil development on active surfaces, such as wind-blown sand dunes and sites of fluvial activity, through to ancient soils on stable, older, higher altitude surfaces, with highly weathered desert pavements, and salt-rich layers in the soil.

17.1 Important Features of Antarctic Soils

17.1.1 Concept and Key Features of Antarctic Soils

The question of whether the materials on the ice-free surfaces in Antarctica should be regarded as soils or just loose geological materials has been much discussed. In many past definitions of soil, the presence of organic matter and ability to support higher plants were considered integral features of soils. However, the weathered land surface materials in profiles in Antarctica are now accepted as 'soil'. Ongoing

A. E. Hewitt et al., *The Soils of Aotearoa New Zealand*, World Soils Book Series,
https://doi.org/10.1007/978-3-030-64763-6_17

study has shown Antarctic soils to be fascinating and complex with active physical, chemical, and biological processes. The most outstanding feature of soils in the Ross Sea region of Antarctica is the presence of frozen soil materials as a result of the extremely cold climate. Inland areas are extremely dry, giving rise to a 'cold desert' environment. Within the McMurdo Dry Valleys, extensive areas are ice-free as evaporation exceeds precipitation. Similar conditions occur in other smaller areas in the Transantarctic Mountains including the Britannia Range and Darwin Mountains. Thus, where there is no permanent ice or snow cover, the rock materials gradually weather to form soil.

Most of the soils in the Ross Sea region of Antarctica comprise a surface desert pavement that overlies a seasonally thawed 'active layer' with permafrost at depth (Figs. 17.1 and 17.2). Because of the cold climate, and lack of moisture, physical weathering is dominant. Thus, most soils have a gravelly sand texture with many larger boulders and stones within the soil profile. Soils with noticeable clay content are rare, and are generally formed from fine-textured parent materials such as lake sediments.

The desert pavement comprises a surface layer of rocks, gravel, and sand. The active layer is the near-surface soil that thaws each summer, then freezes again in winter. The active layer ranges in depth from about 1 m deep at warmer coastal sites in the northern part of the Ross Sea region to no thawing at all, and thus no active layer, in cold higher altitude sites on the margins of the Polar Plateau. The active layer is thawed only for a few weeks at the height of the Antarctic summer in late December and early January and at other times may be frozen solid. Thus, the only way to accurately determine the depth of the active layer is to install temperature monitoring equipment.

Beneath the active layer is permafrost ('permanently' frozen ground). Permafrost is defined as having a temperature below 0 °C for at least two consecutive years. A 'transition layer' between the active layer and the permafrost is increasingly being recognised where the soil occasionally warms to temperatures >0 °C in a particularly warm summer. Where there is sufficient moisture (>about 10% gravimetric moisture), the permafrost is ice-cemented and is consequently difficult to excavate and study. In some areas solid ice, in the form of ground ice, ice lenses, or ice wedges, may lie below the soil surface.

Ground ice is extensive in some areas where glaciers have gradually ablated (through sublimation, or melting and evaporation, of the ice). As the ice is melted and evaporated a layer of mineral material, which was caught in the ice, is left behind at the surface. Eventually, the mineral materials accumulate to such a thickness that they insulate the underlying ice and prevent further evaporation and melting (Fig. 17.2). Some of the surfaces above ground ice are very old and the mineral material can become highly weathered with a strongly developed soil that, to the casual observer, appears as terra-firma.

On some older high surfaces and inland parts of the McMurdo Dry Valleys the environment is so dry that the soils contain insufficient moisture to form ice-cement. Thus, even where the temperature is well below 0 °C, the permafrost soil material is loose and easily excavated, and is described as 'dry permafrost' (Fig. 17.1).

Antarctic soils are included in this book as New Zealand has a strong interest in the Ross Sea region due to a territorial claim over the 'Ross Dependency', the area of Antarctica directly south of New Zealand. Under the Antarctic Treaty (to which New Zealand was one of the 12 original signatories in 1959), all territorial claims are put aside and the Antarctic continent is managed in the interest of peace and science. There has been a long history of New Zealand research into soils in the Ross Sea region of Antarctica, commencing with the expedition of John McCraw and Graeme Claridge in 1959. Iain Campbell and Graeme Claridge continued investigations into Antarctic soils from the 1960s to 1990s. Their 1987 book is a classic source of information related to Antarctic soils. In addition, various historic exploration-era expeditions, including those of Scott and Shackleton, operated from Christchurch in New Zealand, as do modern research parties including US and Italian Antarctic programmes. Hence, there is a >100-year-long, enduring, connection between New Zealand and Antarctica.

17.1.2 Areas of Occurrence

The Ross Dependency lies directly south of New Zealand and stretches from latitude 60° S to 90° S (the South Pole) and from longitude 160° E through to 150° W (Fig. 17.3) with a total area of 450,000 km^2. While most of the Ross Dependency is sea- or ice-covered, it also contains the largest ice-free area in Antarctica. There are many small ice-free areas scattered along the coast of the Ross Sea, and on isolated outcrops (nunataks) and valleys within the Transantarctic Mountains. The largest continuous area (about 6,700 km^2) of exposed soil is in the McMurdo Dry Valleys. The ice-free areas, on which soils develop, make up less than 2% of the Ross Dependency region.

17.1.3 Variation Within Antarctic Soils

The variation within Antarctic soils is surprisingly large, driven primarily by soil age, parent material, and moisture availability. The *New Zealand Soil Classification* does not extend to Antarctica so people working in Antarctica use international soil classification systems to classify Antarctic

Fig. 17.1 An Antarctic soil and associated landscape, Beacon Heights. At this site, the soil contains insufficient water to form ice-cement and hence the subsoil is referred to as 'dry permafrost'. Two major rock units, the Beacon Sandstone (pale-coloured layers) and Ferrar Dolerite (dark brown upper cliffs), are visible in the mountains in the distance. Snow is visible on the ground surface, shortly after a rare snowfall event, but had all sublimated within four hours. The white material within the profile (at about 25–50 cm depth) is salt

Fig. 17.2 Structure of frozen soils on Ross Island. Left: ice-cored moraine near McMurdo Station in 1990. This site was exposed by active stream erosion. The surface veneer of soil material insulates the underlying ice from further melting. Right: profile exposed in a road cutting illustrating the desert pavement, active layer, and permafrost with ice wedges in patterned ground cracks. In both cases, the ice was only visible for a few days because, following exposure (in summer), the ice melts and ablates until an insulating cover of mineral material accumulates to prevent further ablation

soils. In *Soil Taxonomy*, the soils in the Ross Sea region are classified as Gelisols. In *World Reference Base*, they are called Cryosols.

The Gelisols of the Ross Sea region generally fall within the Turbels where there is evidence of cryoturbation (mixing of the soil as a result of freeze-thaw processes) or Orthels. At the great-group level, the soils are further divided into those that do *not* have ice-cemented permafrost (Anhyorthels or Anhyturbels) and the more typical soils that contain ice-cemented permafrost (Haplorthels or Haploturbels) with

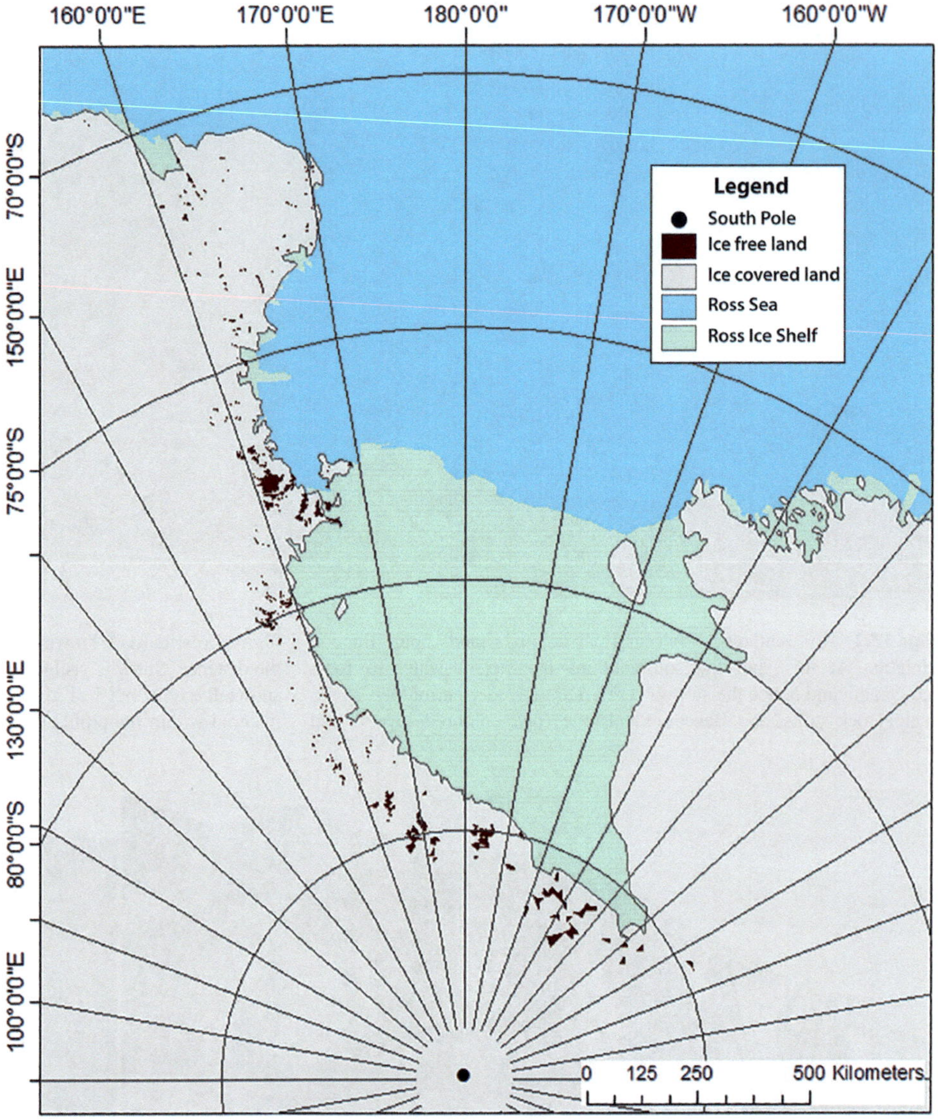

Fig. 17.3 Location of the Ross Sea region (map from Antarctica New Zealand). Soils are developed on the ice-free land areas

a few sandy soils on sand dunes (Psammorthels). The soils which form in penguin colonies with a strong influence from bird activity are recognised as Ornithogenic soils. Erica Hofstee suggested that the ornithogenic soils should be recognised at the suborder level of *Soil Taxonomy* as an 'Ornel' with similar great-groups to those of the other suborders.

If we were to apply the current *New Zealand Soil Classification* to Antarctic soils they would mainly key out as Orthic Raw Soils because of their relatively minimal soil development, although some have features pertaining to Semiarid Soils.

17.1.4 Origin of the Soil Name

In the *World Reference Base*, the soils in the Ross Sea region all classify as Cryosols. The term 'cryo' comes from a Greek word κρύο (cryo), meaning 'cold' or frozen conditions, along with other such terms as cryogenic.

The *Soil Taxonomy* term 'gel' is from the Latin word '*gelare*', to freeze, that expresses the cold frozen nature of the soils. The Gelisol order, defined by the presence of permafrost or gelic materials (mineral or organic soil materials that show evidence of frost churning and/or ice segregation in the active layer and/or the upper part of the permafrost) was first added to the 8th edition of 'Keys to Soil Taxonomy' in 1998. Recognition of Gelisols was partly to encompass soils with cryogenic properties in Alaska (matching the adjacent Canadian classification for soils on permafrost, the Cryosolic Soils) and also to accommodate Antarctic soils. The USDA changed their definition of 'soil' to remove a requirement for organic matter so that the Antarctic Gelisols could be considered 'soil' and therefore included in *Soil Taxonomy*.

17.2 Soil Profile Genesis

Due to the cold dry environment, most soils in the Ross Sea region have limited weathering and so largely retain the inherited properties of the parent materials which are mainly glacial tills. Although climate is the dominant influence on soil genesis, parent material, topography, and time all have recognisable influence. Organisms impact the soils markedly only in a few isolated sites on coastal margins where penguin colonies occur or where summer meltwater supports moss growth.

Soil desert pavement surfaces comprise gravels and pebbles that are left behind as finer materials are eroded, primarily by wind, until a protective layer of coarser material remains at the surface preventing further erosion. Over time, wind-blown sand and salt-weathering abrades and corrodes the exposed rock surfaces. The abrasion, along with a small amount of chemical weathering, particularly oxidation of iron-rich minerals, leads to the formation of polished rock surfaces with a 'desert varnish' coating. Ventifacts (rocks carved and polished to smooth surfaces by the effects of prolonged exposure to wind-blown particles) are evident at many sites.

The older the land surface the more weathered the desert pavement. Some high-altitude surfaces are thought to be extremely (up to about 5 million years) old. Even though weathering processes proceed only slowly in the Antarctic environment, the long exposure time frames mean that desert pavements on older surfaces are highly developed. Older surfaces typically have a closely packed layer of gravel, cobble, and boulder-sized material with ventifacted surfaces that are pitted, and coated with desert varnish (Fig. 17.4).

In younger soils, the colours are often described as 'lithochromic'. In other words, the soils are the colours of the rocks or deposits from which they have formed as there has been minimal chemical weathering. Soils are most often formed in glacial tills that contain a poorly sorted mix of sand, gravel, and larger pebbles and boulders that have been broken up by physical weathering processes.

Ice lenses occur where small areas of ice accumulate in the soil. Moisture tends to condense from the soil atmosphere at the freezing front (the depth at which the temperature in the soil is 0 °C), thus over time an ice-enriched layer may form at the interface at the top of the permafrost.

'Patterned ground' (Fig. 17.5) forms where contraction cracks open in the extreme cold of winter forming polygons, much like those formed by the cracks in a drying mud puddle, but on a much larger scale. Patterned ground polygons are often 5–10 m across. Blowing snow tends to accumulate in the cracks and thus meltwater seeps down, then refreezes. Over time ice wedges up to about 2 metres deep and 50 cm across may form (Fig. 17.2).

The Dry Valley landscape has been considered stable over periods extending to millions of years particularly at higher altitudes which have escaped some of the disturbance of valley floors and walls that occurred during more recent glacial advances and retreats. The older the surface the more strongly weathered the soils. Older soils often contain visible salt accumulation, redder soil colours, more clay, and stronger weathering of surface soils. Exfoliation, cavernous weathering, and ventifaction are often evident.

Salts accumulate in the soils because leaching does not occur at most sites (through lack of water) and any moisture that may dissolve and move salts eventually evaporates, leaving the salts behind. In areas where snow accumulation and melting lead to some surface or subsurface moisture flow, the water may move and concentrate salts in lakes such as Lake Vanda, or ephemeral ponds such as Don Juan Pond, both of which are in the Wright Valley.

17.3 Soil-Landscape Relationships

17.3.1 Overview

The soils in the Ross Sea region of Antarctica have strong relationships to the landscapes in which they form. At the regional scale, the soils can be divided into three groups, the coastal margins, the McMurdo Dry Valleys, and the isolated ice-free areas at high altitudes in the Transantarctic Mountains and on the margins of the Polar Plateau. At local scales, there are distinct relationships between topographic and geomorphic position and the soils that form. The age of landscape surfaces greatly influences the amount of weathering and the degree of soil development. Recently deposited, or mobile, material such as sand dunes, stream and beach deposits, and the youngest glacial deposits, have soils which exhibit minimal soil weathering or development (Typic Haplorthels). Due to the cold dry environment, soils on bedrock are generally shallow and classify as Lithic Haplorthels. On older surfaces, weathering is more advanced with iron oxides imparting dark brown colours to soil

Fig. 17.4 Weathered rocks on a high altitude desert pavement. The small stones that lie in pits or hollows in rocks (photo at right) are likely to be ground around by the actions of wind and frost to gradually enlarge the hollows and slowly wear the rock down. The polished nature of the surface is an effect of wind abrasion and chemical weathering known as 'desert varnish'

Fig. 17.5 Patterned ground. The polygons are generally 5–20 m across, here accentuated by drifting snow caught in the hollows between polygons. The deep hollows indicate ice wedges or underlying ground ice

materials and salt accumulating within the profile and under pebbles and gravel in the desert pavement (e.g. forming Salic Anhyorthels).

The availability of moisture in soil has a major impact on the soil processes and resulting soil profile form. Moisture availability is related to both the wider regional landscape (including distance from sea and influence of katabatic winds) and the local topographic position. Examples are discussed in the following sections.

17.3.2 Antarctic Coastal Margins

Due to the sea acting as both a temperature moderating influence, and a source of moisture, soils in the coastal margins are generally warmer and wetter than soils at sites further inland. The coastal margins receive greater precipitation and, because of the warmer, more humid, environment, more of the snow that falls melts, rather than directly sublimating, as is common further inland. For a few weeks, every summer many coastal soils may have water flowing over or through the soil. The marine ecosystem is the source of the food chain for all the wildlife along the coast and the ice-free coastal areas provide important resting and breeding sites for birds. Thus, the activity of birds transfers some nutrients into the coastal soils from the marine ecosystem.

The major landscape units of coastal areas include raised beaches, areas of glacial till, and exposed bedrock. Because of isostatic rebound, many sites along the coast have a series of raised beach-ridges. The soils on raised beaches comprise weakly weathered, highly rounded and smoothed, pebbles and gravels in a sandy matrix with occasional shells or other marine-derived material.

The glacial drift deposits on coastal margins have generally been exposed by glacial retreat over the last few thousand years so are only weakly weathered. The lithology tends to be dominated by the local geology. For example, soils are marble-dominated at Marble Point, gneiss-dominated at Gneiss Point, and granite-dominated at Granite Harbour. However, a mixture of lithologies, including granites, sandstones, and dolerite occurs at all sites indicating that some materials have been carried some distance. The tills are poorly sorted, ranging from angular to sub-rounded large (> 1 m diameter) boulders, all the way through to fine sands. On Ross Island, the bedrock comprises scoriaceous basalt. Being porous, meltwater penetrates into the scoria and freeze-thaw processes break the clasts so soils comprise rock fragments interspersed with gravels and sands.

In the relatively moist coastal environments, all soils contain enough water to form ice-cemented permafrost. On younger surfaces, weak patterned ground cracks are visible. However, on some of the older surfaces, such as the foot slopes of some hills, there are areas with strong patterned ground development. The availability of moisture can lead to stronger biological (moss, algae, and lichen) growth in the coastal regions than in other areas (Sect. 17.4.4).

17.3.3 The McMurdo Dry Valleys

The McMurdo Dry Valleys are unique (Fig. 17.6) and are considered the closest environment to Mars that we have on Earth. In the dry valleys, evaporation exceeds precipitation; hence, the absence of snow and ice that covers most of the Antarctic continent. Evaporation is enhanced by strong katabatic winds that flow down off the Polar Plateau through the dry valleys. Precipitation is low because the dry valleys are shielded from water sources by the Ross Ice Shelf to the east, the Transantarctic Mountains to the north and south, and the Polar Plateau to the west. Thus, the McMurdo Dry Valleys are a 'cold desert' environment and the soils share many properties with soils of hot deserts including minimal organic matter, accumulation of salts, and sand-dominated soil textures.

Within the McMurdo Dry Valleys, the main landscape drivers that impact the soils are the availability of moisture, the microclimate, the surface age (with older surfaces tending to occur at higher altitudes), and the underlying parent materials.

Soil moisture levels are higher on the margins of the Polar Plateau (due to inputs of blowing snow) and nearer the coast because of higher snow fall. The driest areas tend to be in the dry valley floors and in upland areas that are sheltered from the wind-blown snow associated with polar cold air drainage. Within the valley walls and floors, there are areas that receive intermittent moisture in the summer from the ephemeral streams that carry meltwaters from the high valley walls down into the valley floors. Moist soils are also found on the margins of lakes and streams where moisture is drawn up into the adjacent soil under capillary forces where it evaporates, often leaving a salt margin (Fig. 17.7).

Where there is sufficient moisture, the soils contain ice-cemented permafrost and strong patterned ground features tend to form. Ice-cored moraine (where a mineral soil is formed over remnant glacier ice) occurs in a number of areas —for example at Beacon Valley, where mineral material overlies and insulates underlying glacier ice. The surface soils on ice-cored moraine can be relatively highly weathered, with strongly oxidised and weathered surface pavements, and may contain dry permafrost material that overlies the ice core. A distinctive feature of ice-cored moraine is the presence of large-scale patterned ground that often has patterned ground cracks 1–2 m deep between the polygons. Increased melting, resulting in greater subsidence, occurs in the crack area, and hence generates an extremely hummocky surface which is somewhat difficult and tedious to walk over.

Fig. 17.6 The McMurdo Dry Valleys. Left: the Wright Valley provides an example of the dry valley environment. Right: example of an Anhyorthel, with dry permafrost below about 40 cm depth, developed on an alluvial fan on the floor of the Wright Valley (scale in cm)

Fig. 17.7 Left: soil dampened by moisture that has been drawn out from an ephemeral stream that flows from meltwater from the Goodspeed Glacier in the Wright Valley (Onyx River in the foreground). Right: moist soil and white salt accumulation evident on the margins of the Goodspeed Stream

In the driest parts of the valleys where soil moisture contents are less than 10% by weight, there is insufficient moisture to form ice-cement; thus, the soil within the dry permafrost, where temperatures remain well below 0 °C at all times, generally remains loose and unconsolidated (Fig. 17.6). However, on the older high-altitude surfaces where dry permafrost occurs, soil weathering can lead to weak (salt or iron oxide) cementation of soil particles giving some coherence to the soil profile. In some areas, the dry permafrost is formed in soft fine sand to silty material with very few coarse fragments. However, such soils still have a stony desert pavement and are, therefore, not easily identified just by observing the soil surface.

There is a clear soil topographic pattern across the dry valleys with often over 1000 m of altitude between the high terraces and the valley floors (Fig. 17.8)

The oldest, most weathered soils are on the high terraces, above the valley walls (beyond the edge of the area shown in

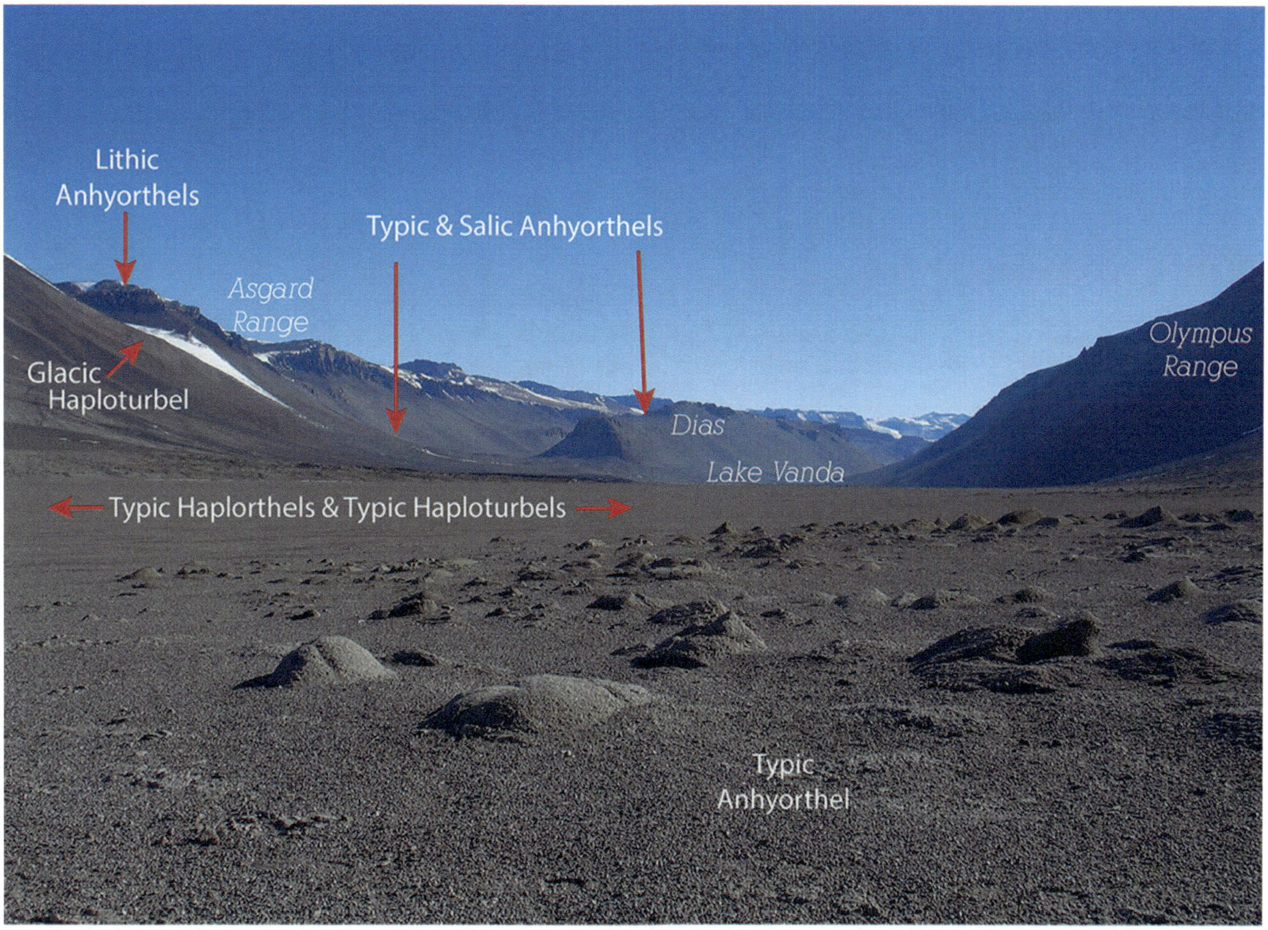

Fig. 17.8 Typical soil-landscape pattern across the Wright Valley, Antarctica. Lithic Anhyorthels occur on rock outcrops, Glacic Haploturbels occur on the lateral moraine of the active present-day glacier, Typic and Salic Anhyorthels occur on older high till or colluvial surfaces, Typic Haplorthels and Typic Haploturbels occur on the modern flood plain of the Onyx River, and Typic Anhyorthels occur on the valley floor above the influence of the Onyx River or other ephemeral streams

Fig. 17.8). The upper valley walls often comprise near-vertical outcrops of rock, with the dark red-brown-coloured Ferrar Dolerite particularly prominent.

On the steeply sloping valley walls, screens and fans form and the soils contain sloping layers (beds) of gravels and sands, reflecting periodic movement of material down slope (Fig. 17.9). Some scree movement has no doubt been activated, since the start of the twentieth century, as explorers have walked up and down the slopes. Periodic bursts of meltwater occur when ephemeral lakes, high on the slope margins, occasionally give way causing a flood of water to flow down the slope. Such events cause gully erosion in upper slopes and deposition of new material in the foot slopes. The effects of one such active fan event were observed in the 2008–09 summer near Lake Vanda where new material had been deposited across an experimental site (Fig. 17.9). Water from snow and glacier meltwater flows down, within the fans of the valley walls, through the summer months, sometimes forming ephemeral streams that feed water to the Onyx River (Fig. 17.7).

17.3.4 Soils on the Margins of the Polar Plateau

South of the McMurdo Dry Valleys multiple large glaciers flow from the Polar Plateau down to the Ross Ice Shelf. These include the 200 km long, and up to 40 km wide, Beardmore Glacier, the fast moving, 136 km long, Byrd Glacier, and the 130 km long Darwin Glacier. The mountain ranges on either side of these glaciers have some ice-free areas where soils have formed. Some of the mountain outcrops, such as the Otway Massif, and the Dominion and Britannia Ranges, have areas of spectacularly ancient soils with strong accumulations of sodium nitrate salts that are considered to be derived predominantly from deposition from the atmosphere over extremely long time periods (millions of years, Fig. 17.10).

Fig. 17.9 Left: the fan here has been a site of periodic activity, most recently, immediately before this photo was taken in 2008. New gully erosion in the upper slopes and deposition of relatively fine material on the lower slopes of the colluvial fan were evident. Right: the layered gravelly sand dominated soil on the lower fan (scale in cm). This soil contained ice-cemented permafrost below about 30 cm depth, with the moisture likely to be from subsurface flow beneath the fan or surface flow during a fan building event, or both

As glaciers, including the Beardmore Glacier and Law Glacier, flowed down from the Polar Plateau along the margin of adjacent ranges they have scraped rock material off the edges of the mountains and deposited it, forming a sequence of lateral moraines along the margins of the range (Fig. 17.11). The glaciers have been slowly retreating from their maximum heights, mainly due to ablation, and so have built up a pattern of soils across the lateral moraines.

There is a weathering sequence across the lateral moraines with soils becoming deeper and more weathered with increasing distance from the glacier edge. Soils range from <10 cm of minimally weathered mineral material accumulated over glacier ice near the glacier margin (Fig. 17.11), through a developing desert pavement with a gradually deepening soil a few kilometres from the glacier edge (Fig. 17.12) to about 70 cm of dry permafrost, with a moderately weathered desert pavement, near the base of the mountains.

All of the lateral moraine sequences are formed over remnant glacier ice, and when viewed up close they have the characteristic very large hummocky patterned ground typical of ice-cored moraine, evident in the landscape pictures in Figs. 17.11, 17.12, 17.13 and 17.14. Even in mid-summer at the high altitudes and latitudes, the air temperature remains at around −15 °C and the soil surface temperatures generally stay below zero. Thus, there is no, or minimal, active layer. However, moisture contents are so low that the soils comprise dry permafrost.

The pattern of shallow soils near the glacier edges, and deeper soils formed with increasing distance inland, is repeated at a number of sites in the region including near Mt Acheron (Fig. 17.13) and the Ong Valley.

Another glacier-mountain system that has substantial ice-free areas is the Britannia Range, which is flanked on one side by the Hatherton Glacier that flows into and joins the Darwin Glacier. James Bockheim and others (of USA), and a University of Waikato group led by Michael Selby, recognised soils on five different drift (moraine) surfaces in the Britannia Range. The drifts and soils increase in age with increasing elevation, the youngest at ~1250 m above sea level (asl) and the oldest at ~1700 m asl, and with increasing distance from the modern ice margin. Ages were

Fig. 17.10 Soil and landscape at Otway Massif at an altitude of about 3500 m on the margin of the Polar Plateau 500 km from the South Pole. This is one of the oldest, and coldest, soils on Earth

Fig. 17.11 Between the edge of the Dominion Range and the Beardmore Glacier is a 4-km-wide stretch of lateral moraines. The Beardmore Glacier has been gradually retreating and within a hundred metres of the glacier edge, there is a weakly formed desert pavement with only a few centimetres of soil material over the slowly sublimating glacier ice

initially estimated by limited radiocarbon dating and correlation with ages on moraines in the Dry Valleys. More recent work, however, has used surface exposure dating (cosmogenic dating) to develop a more robust age framework for the drifts and associated soils (conflicting in part with data from the nearby Darwin Mountains) as follows (in order of increasing age): (i) the Hatherton soil (on Hatherton drift) is between c. 700 and 2000 years old; (ii) the Britannia I soil (on Britannia I drift) is between 5000 and 8000 years old; (iii) the Britannia II soil (on Britannia II drift) is 125 ka in age; (iv) the Danum soil (on Danum drift) is between 0.45 and 0.64 Ma; and (v) the Isca soils (on Isca drift), are aged between 1 and 2 Ma.

The drifts and soils were recognised and mapped using multiple criteria including differences in lithology that showed up as (i) marked differences in colour (Fig. 17.14),

Fig. 17.12 The land surface on lateral moraines about 2 km from the Beardmore Glacier edge is hummocky patterned ground, making it difficult to traverse (left). Here, the soil comprises about 30 cm of mineral material over the underlying glacier ice. Notice how the rocks are concentrated at the soil surface, forming a weak desert pavement and providing protection from wind erosion for the underlying sandy ablation till material

Fig. 17.13 Left: the lateral moraine sequence adjacent to the Law Glacier near Mt Achernar is up to 10 km wide. The colour change in the rocks partway up the steep slope marks the height reached by the last major glacial advance. The distinctive dark red Ferrar Dolerite is exposed in the cliffs. Right: a soil profile from the inland edge of the area contains relatively unweathered, loose, gravelly sand, dry permafrost with a moderately developed desert pavement near the surface. The underlying glacial ice is at a depth of about 70 cm. Permafrost would be at a depth of <5 cm at this site

(ii) drift surface morphology (e.g. thin, loose gravelly sands, pebbles, and boulders on ice *versus* strongly patterned ground on relatively deep profiles), (iii) surface boulder weathering, (iv) soil morphology (depth, texture, colour, pseudomorphs, mineralogy), and (v) other profile features such as salt content and pH. Some of the soils contain buried soils, showing up as oxidised horizons at depth.

17.3.5 Soils and Meteorites

Iron meteorites (16 specimens in total, the largest weighing 138 kg) were discovered on Derrick Peak in the Britannia Range in 1978 (Fig. 17.15). Peter Kamp and David Lowe used the soil/drift ages to help constrain the terrestrial age of the meteorites (i.e. when the meteorites had landed on

Fig. 17.14 View of three different drift (moraine) surfaces and soils in Dubris Valley (occupying the main part of photo) in the Britannia Range showing the distinct boundary between the pale-coloured 0.125 Ma Britannia II deposits at far left (with a low dolerite: sandstone ratio), the dark reddish brown c. 0.5 Ma Danum deposits, with marked patterned ground, dominating centre foreground (high dolerite: sandstone ratio), and c. 1–2 Ma very dark Isca deposits on the elevated areas upper right. In the far background, running left to right, is the narrow Isca Valley flanked at the back by spectacular columnar-jointed Ferrar Dolerite cliffs. Hatherton Glacier is out of view to the left

Earth). They suggested that the wide bench on Derrick Peak, on which many, but not all, of the specimens, were found, was a glacially carved feature dating back to c. 3–3.5 Ma on the basis of old tillite found on the glacially carved Pontes Ridge, up valley from Derrick Peak (Fig. 17.15). However, the additional occurrence of meteorites on the Danum drifts 300 m below the bench provided a minimum age for the fall, assuming the specimens are all in situ (an interpretation supported by subsequent fieldwork by Margaret Bradshaw and others), meaning that the meteorite fall event probably took place after 0.45–0.64 Ma (the age of the Danum drift/soil). The Bradshaw-led party found a further nine meteorites on Derrick Peak in 1988 including one on Danum drifts in nearby Onnum Valley. The Derrick Peak meteorites were dated, by Kunihiko Nishiizumi and others, at 1 ± 0.1 Ma using cosmogenic dating, an age somewhat in conflict with the younger moraine-based age (0.43–0.64 Ma) reported by Kurt Joy and others in 2014. However, if the cosmogenic age on the meteorites is accurate, then stratigraphic principles dictate that the Danum drift soils must be at least 1 Ma old, much closer in age to the Isca than the Britannia drifts.

The 1 ± 0.1 Ma terrestrial age for the meteorites is consistent with novel morphological features on them that attest to a prolonged period of exposure to strong winds ('sand blasting') and corrosion. The meteorites all show well-developed 'thumb-print' depressions formed during entry to the atmosphere, a dark, slightly iridescent patina, and the surfaces include a bronze-coloured, chewed-toffee-like mineral called schreibersite ($(Fe, Ni)_3P$), that protrudes from the dominant iron matrix. Normally, schreibersite (which is not found in crustal rocks near the surface of the Earth) is preferentially ablated during the meteor's descent through the atmosphere. The protruding form in the Derrick Peak specimens means the encasing metal has been significantly corroded and eroded to enlarge the surface depressions, and expose the schreibersite crystals, since the meteorite landed. The meteorites, all of the

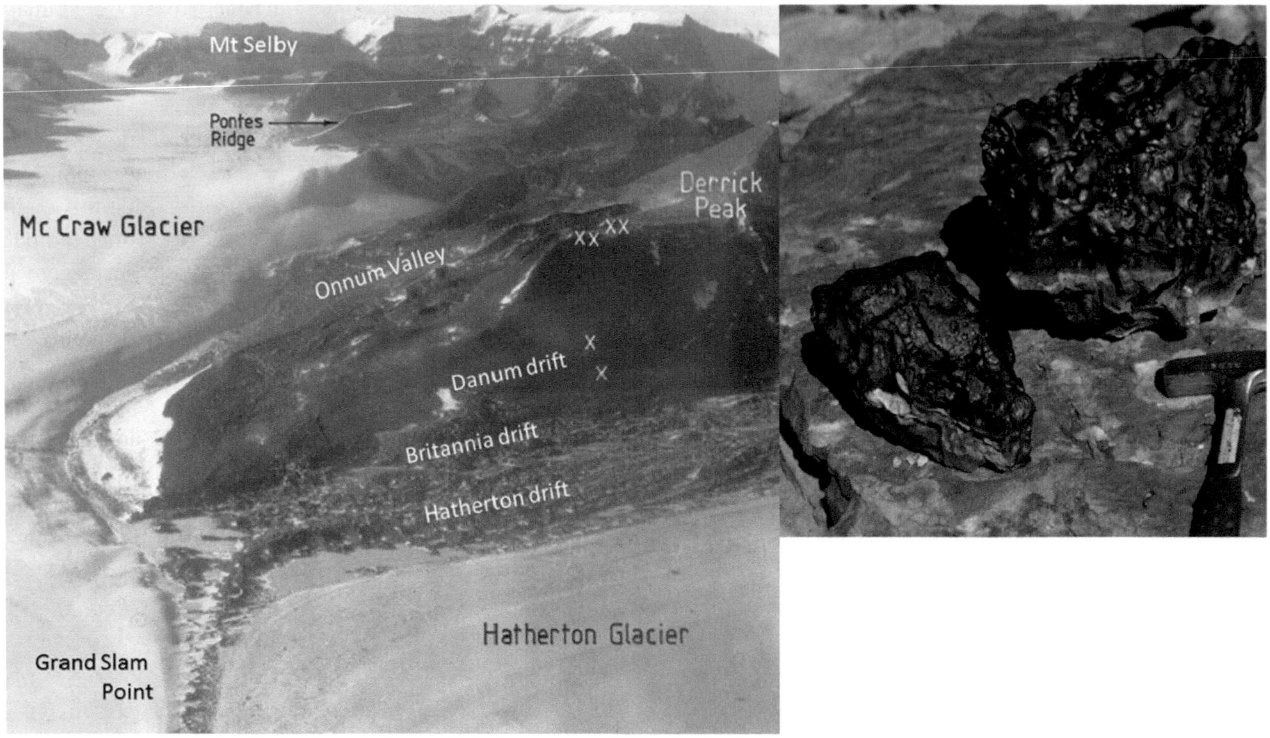

Fig. 17.15 Derrick Peak meteorites. Left: oblique southward view of the Derrick Peak area in the Britannia Range showing the locations (crosses) of iron meteorites discovered in 1978 on the wide bench just below Derrick Peak, at ~1400 m asl, and on Danum drifts 300 m below the bench. From Kamp and Lowe (1982), photo by M. J. Selby. Right: two of the iron meteorites found, DRPA 78013 (10.6 kg), on display in the Canterbury Museum (left), and DRPA 78011 (31.8 kg)

same composition, with the same unique surface morphology, derive from a single, large meteorite (possibly the size of a small car) that fragmented on or before impact. They are made mainly of iron together with nickel (~6 wt%), cobalt (~0.5 wt%), phosphorus (~0.4 wt%), and germanium (~150 ppm), similar to alloys of stainless steel (though without chromium).

17.4 Key Soil Properties

17.4.1 Soil Composition

Because of the predominance of physical weathering, the mineral composition of most soils in the Ross Sea region is largely that of the parent rocks. Many of the soils are formed in glacial tills and tend to contain a mixture of the rock materials that occur in the region. The broad geology comprises a basement of gneiss and granites, with occasional outcrops of marble. The basement is commonly overlain by a widespread quartz-dominated sandstone (the Beacon Sandstone, visible in Fig. 17.1). Both the basement rocks and overlying sandstones are intruded by an extensive system of dolerite sills and dykes (the Ferrar Dolerite). Basaltic volcanoes dominate Ross Island and occur in a few places on the Antarctic mainland. Small pockets of lake bed sediments, carboniferous shales, and other minor rock units, also occur in some places. Thus, the soil materials tend to be dominated by the local rock type, most often a mix of quartzo-feldspathic basement and sandstone rocks, along with Ferrar Dolerite. However, most soils contain small portions of a wide range of rock materials transported and deposited by glacial activity. On Ross Island, the soils are dominated by scoriaceous basalt with only minor inputs from further afield.

Chemical weathering is limited by the cold temperatures and lack of liquid water, thus limited oxidation and clay formation occurs. Clays rarely make up more than 1% of the soil. Illite and hydrous micas are the most common clay minerals with smectites and aluminium cholorites sometimes reported in strongly weathered soils. Halloysite occurs at some sites and is attributed to weathering of tephra from Antarctic volcanoes.

In the cold desert environment, free salts are often visible in the soil where evaporation exceeds precipitation. Calcite and gypsum often occur as encrustations on the undersides of rocks. Salts may also be visible on the soil surface where they are left behind as moisture evaporates. At some sites, particularly on older, more highly weathered soils, at higher altitudes, the salts may be evident in a layer within the soil.

If a soil profile pit is left open sometimes, after a few hours, you will see salts accumulating on the soil profile face—left behind as moisture evaporates from the exposed surface. The salts consist largely of chlorides, nitrates, and sulphates of sodium, potassium, calcium, and magnesium. Graeme Claridge undertook a detailed study of the salt contents of Antarctic soils and concluded that the salts are derived primarily from rock weathering but the effect of atmospheric inputs is evident in the anions with salts at sites nearer the coast dominated by chlorides. Further inland, on the margins of the Polar Plateau where the wind is dominated by katabatic flows off the plateau, the salts are dominated by nitrate which is considered to be derived from the upper atmosphere circulation system where air flows downwards over the South Pole.

17.4.2 Physical Properties

a. Introduction

Soils in the Ross Sea region mainly comprise bouldery, stony, gravelly sands. Some soils, particularly those on ablation tills (the material that has accumulated on top of an ablating glacier) may have a higher proportion of fine sand and silt due to the grinding action of the glacier. Clay contents are often <1% of whole soil. Thus, we infer that most soils have a very low soil moisture holding capacity and are quite porous. Soils that have ice-cemented permafrost effectively have an impermeable layer, due to ice accumulated at the top of the permafrost, that limits downward movement of water which freezes when it reaches the top of the permafrost. The physical properties of most interest are the temperature and moisture content because they impact strongly on all soil and ecosystem processes.

b. Soil temperature

The soil temperature determines the depth of the active layer, the availability of liquid moisture, and the conditions for the survival of soil organisms. As the Ross Sea region is all within the Antarctic Circle, it experiences darkness for three months in winter, and daylight 24/hours per day in summer, with a more usual day/night pattern in the spring and autumn. Thus, temperatures are cold in winter, gradually warming with the return of the sun in spring and summer. For example, the mean annual temperatures range from −18 °C at Marble Point on the coast to −20 °C in the floor of the Wright Valley, at about 50 m altitude, and −24 °C at 1700 m altitude on the slopes of Mt Fleming at the head of the Wright Valley.

Antarctic soils remain at temperatures below 0 °C for much of the year. However, when the summer sun's radiant energy directly warms the dark-coloured soil surface, temperatures may rise to as high as 20 °C even though the air temperature remains close to zero. Heat is conducted downwards, warming and thawing the soil, but often the soil remains frozen just a few centimetres below the soil surface; thus, there are strong thermal gradients within the soil. Where there is ice within the soil it takes a lot of energy to melt the ice (because water has a very high latent heat of fusion). Thus, the temperature within the soil may remain near zero for long periods as the soil moisture absorbs the available energy while ice is melted or sublimated.

There is a lag time between the soil surface reaching maximum temperature and the temperature rising at depth within the soil. Also, the extremes of warmth and cold experienced at the soil surface are gradually dampened with depth. At a depth of about 15–20 m within the soil, there is no appreciable change in temperature throughout the year and this depth is referred to as the depth of zero annual temperature amplitude. The temperature at that depth is a good indication of the mean annual temperature.

The maximum depth of thaw (which determines the depth of the active layer) is generally deeper at lower altitude and lower latitude. For instance, the mean active layer depth at 1600 m altitude on Mt Fleming is about 6 cm, whereas at Scott Base, near sea level (and at a similar latitude), it is about 30 cm. At Cape Hallett, towards the north of the Ross Sea region, the active layer is about 1 m deep on sites away from the effects of shading from nearby hills. The active layer depth also varies from one summer to the next—for example, the active layer depth at Scott Base between 1999 and 2019 ranged between about 20 and 40 cm. The maximum depth of thaw is greater in moist soils, such as on lake margins, because the thermal conductivity of moist soil is greater than that of dry soil. At some sites, meltwater flows through the soil and effectively conducts heat into the soil, leading to a deeper active layer.

When there is snow on the soil surface it insulates the underlying soil from variations in air temperature. In winter, snow-cover, where present, will protect the soil from extremely low temperatures. However, snow cover is relatively rare in the McMurdo Dry Valleys such as the Wright Valley. In winter, cold air tends to pool on the valley floor and both air and soil surface temperatures gradually drop to as low as −40 °C. Occasional winter storms bring strong winds and the air is mixed causing a rapid rise in air temperature, sometimes changing by over 30 °C, in a few hours with the near-surface soil temperatures also recording a sharp rise in temperature.

c. Soil moisture

In the Ross Sea region, a lack of soil moisture is often the major limitation to biological activity. Annual precipitation is generally very low and usually falls as snow. However, occasionally a small amount of rain has been observed in summer when air temperatures are above zero. The precipitation is higher in the coastal areas where usually there is enough snowfall over winter to cover the ground surface, leading to a spring meltwater event. Occasional summer snowfalls may cover the ground near the coast for a week or two before melting and evaporating.

In some coastal areas, runoff from snow or glacier melt occurs for several weeks over summer, saturating the soil and forming ephemeral streams. For example, the soil on the foot slope on the edge of Granite Harbour at, and near, Botany Bay, has water flowing through it for a large portion of the summer. The water conducts heat into the soil providing an unusually deep (>90 cm) active layer. The moisture also supports an area of exceptional moss growth at Botany Bay.

In the McMurdo Dry Valleys, there is minimal snow-fall. Annual precipitation of 45 mm was recorded at Vanda Station in the Wright Valley in the 1970s. On the valley floors, an occasional summer snow shower occurs where snow settles on the ground surface. The snow is usually lost to sublimation within a few hours with very little moisture penetrating the soil surface. However, snow may accumulate on the adjacent higher altitude mountain tops for extended periods over summer. Some of the meltwater from higher altitude areas moves down through the soils of the valley walls to evaporate on the valley floor, sometimes leaving accumulations of salt behind, such as at Don Juan Pond in the Wright Valley. The areas of subsurface flow down valley walls are sometimes called 'water tracks' and have been described as 'solute superhighways' as they move both salt and water downslope relatively rapidly.

In moister coastal areas, and soils with ice-cemented permafrost, the soil moisture content near the soil surface can vary greatly through the summer, ranging from saturation during snow-melt to less than 5% gravimetric moisture content as the surface soil becomes effectively 'air dry'. Ten years of monitoring at Marble Point showed sites generally had three or four moistening (snow-melt) events per summer where water extended to at least 20 cm depth and lasted for up to 2 weeks before drying back to 'air dry'. Moisture accumulates at the top of the permafrost where the temperature remains at or below zero and moisture condenses from the air and freezes, often forming ice lenses (which give a volumetric moisture content of 100%). Within the permafrost, the soil moisture (ice) content can be highly variable, with volumetric moisture contents ranging from less than 10% in soil and rock materials, to 100% in ice wedges and lenses.

Within the dry valleys, where there is dry permafrost, the soils have moisture contents of less than 5% by weight, and often as low as 1%, throughout the soil profile. In 10 years of soil moisture monitoring on a dry permafrost site in the Wright Valley, there were no soil moistening events that extended to 20 cm deep in the soil. Once every few years, there may be surface soil moistening for a few hours if a snow patch forms and some snow melts before evaporating. Some soils adjacent to lakes or ephemeral streams may be moistened for extended periods during summer as capillary rise can draw moisture through the soil for several metres from the lake or stream. The outer margins of such areas are often marked by the accumulation of salts left behind where the water evaporates.

The soil moisture holding capacity is low because of the coarse gravelly sand texture of most Antarctic soils. However, even at low soil moisture contents, much of the water may not be strongly held due to the generally low (<1%) clay and organic matter content, so it may be possible for microbes to operate at lower moisture contents than would generally be expected.

The first Antarctic soil science expedition

Norman Taylor, the Director of New Zealand Soil Bureau from 1952 to 1962, planned to attend the 1960 International Soil Science Society meeting in Madison, Wisconsin, and wanted to impress the world by showing that New Zealand soil survey reached from the equator to the pole. Accordingly, he sent John McCraw and Graeme Claridge to Antarctica during the 1959–60 austral summer to investigate the soils, "if any could be found", and produce a soil map.

McCraw and Claridge set off without any pre-planned programme, armed only with a spade, cameras, and notebooks. Their first venture was to visit the site of the 1907–09 Shackleton expedition hut at Cape Royds, inhabited at that time by Norman Taylor's son who was studying penguins. They travelled to Cape Royds, via sea ice, by Massey Fergusson tractor and sledge. It was early in October, and very cold, but they managed to explore the surroundings and dug their first soil profile pits (albeit very shallow ones as the active layer was not yet thawed, so was ice-cemented). They were somewhat mystified by what they saw.

Claridge and McCraw next set off to cross McMurdo Sound, again by tractor and sledge. They were dropped off in the Taylor Valley with no radio contact and a vague arrangement that they would be picked up by people from Scott Base a month later. To keep warm on the sledge they converted a large packing case into a shelter. At the mouth of the Taylor

Valley the case was set on end so that it became a small hut which was used for cooking and as an office for work on writing and map sketching (Fig. 17.16). From this camp site they wandered over as much of the valley as they could, armed with a heavy, steel-handled, spade, camera, and notebooks.

Eventually the pair were picked up and returned to New Zealand. Published papers included a detailed soil map of the Taylor Valley—the first Antarctic soil map. They were able to demonstrate that the materials they found were true soils, showing horizon development.

17.4.3 Chemical Properties

The chemistry of soils in the Ross Sea region of Antarctica is of interest, though not for the potential to grow plants. The soil chemistry provides an indication of the degree of weathering at different sites and is a means to better understand conditions for microbial activity and the processes related to salt formation and accumulation.

Most of the soils contain free salts. Thus, the soils have a very high pH and electrical conductivity (Table 17.1). Due to the low amounts of clay and organic matter, the cation exchange capacity (CEC) is low to very low and base saturations are usually 100%. The total carbon and nitrogen values reflect carbonate and nitrate salt contents, rather than any measurable organic matter content.

Because of the low clay and organic matter contents, the soils have a low pH buffering capacity. Therefore the salts that accumulate have a strong effect on soil pH. Soil pHs range from about pH 6 to pH 9 (Table 17.1) with a tendency for higher pH closer to the sea (reflecting a marine origin for some of the salts). On older inland sites nitrate is often the dominant cation, derived from the atmosphere, in contrast to chlorides derived from the sea nearer the coast.

17.4.4 Biological Properties

The larger animals (seals, and sea leopards) and birds (predominantly skua and penguins) in the Ross Sea region of Antarctica are dependent on the marine ecosystem for food. However, many seek out coastal ice-free areas to rest, nest, and produce their young. Penguins form nesting colonies of thousands of birds and have a strong impact on the soils in the areas where they congregate.

In the Ross Sea region, the most common penguin species is the Adelie which build nests from small stones to make a platform to keep eggs and chicks above the influence of meltwaters. The preferred stones are about 2–6 cm in diameter (coarse gravel), big enough to be useful in forming a nest and small enough for the penguins to pick up in their beaks. Gradually, the penguins build up deep layers of stones, intermixed with penguin guano and the remains of dead animals forming an ornithogenic soil (Fig. 17.17). At the Adelie penguin colony at Cape Hallett, radiocarbon dates show the penguins have been nesting there for at least

Fig. 17.16 Graeme Claridge (left) and John McCraw with the packing crate office, Taylor Valley, Antarctica, 1960. Photo: J. D. McCraw

Table 17.1 Examples of soil chemical properties of <2 mm fraction of two soils, Marble Point (a coastal Haplorthel) and Mt Fleming (an old-surface, high altitude Haplorthel)

Depth cm	pH in H_2O	Total[a] carbon %	Total nitrogen %	Electrical conductivity dS m^{-1}	CEC[b] $cmol^{(+)}$ kg^{-1}	Acid oxalate extractable P	Ammonium acetate exchangeable bases $cmol^{(+)}$ kg^{-1}			
							Ca	Mg	Na	K
Marble Point										
0–3	8.1	0.88	0.02	67.8	6.5	169	21	3	24	2
3–15	8.5	0.91	0.01	12.2	8.1	180	12	1	7	1
15–32	8.0	0.86	0.01	13.2	7.3	139	24	2	5	1
32–45	7.8	1.45	0.01	12.1	6.4	148	31	2	2	1
45–69	7.8	2.15	0.01	11.11	7.3	138	32	3	1	0.3
69–100	7.6	4.51	0.01	18.4	11.1	217	43	5	2	0.3
Mt Fleming										
0–2	5.8	0.07	0.09	22.8	5.7	118	35	6	6	1
2–8	7.5	0.12	0.20	75.1	7.4	128	41	23	40	2
8–20	8.0	0.98	0.15	38.8	7.0	131	3	10	14	1
20–45	8.3	1.26	0.10	18.9	5.8	100	3	5	7	1
45–70	8.4	0.95	0.08		4.9	109	7	4	7	1

[a]These soils contain free calcium carbonate and nitrate salts so total carbon and total nitrogen do not reflect organic matter content. Analyses by USDA, Lincoln, NE

[b]CEC cation exchange capacity

1000 years, and in that time about 1 m of 'soil' has accumulated. Erica Hofstee and others reported that the guano-derived soil at Cape Hallett (after removing stones) was about 12% phosphorus. It also had high levels of elements such as arsenic and cadmium which are likely to be responsible for the relatively low numbers and diversity of microbes found in these soils.

In the McMurdo Dry Valleys, there is little visible evidence of soil organisms, no vascular plants, and soil organic matter content is negligible. However, despite the hostile environment, soil biological communities exist that include Collembola (springtails), mites, and nematodes. Bacterial numbers of up to 10^9 cells g^{-1} dry soil have been reported by Jackie Aislabie, among others. Recent molecular genetic studies show high levels of microbial diversity. Soil pH, salinity, and available water content, can impact soil biological systems and high spatial variability can result in high variability in species diversity and community structure over microtopographic scales.

There are no higher plants in the Ross Sea region. However, where liquid water occurs at the ground surface, even if it is just for a few hours each day over the short summer (6 weeks or so), mosses, cyanobacteria, and lichens are evident. In some areas, there is sufficient melt-water adjacent to melting snow patches to support life. In other places, there is a regular flow of water from melting of adjacent glaciers to feed streams or surface water flow that supports moss or growth of cyanobacterial mats.

17.5 Use and Management of Antarctic Soils

Antarctic soils fall under the protection of the Protocol on Environmental Protection to the Antarctic Treaty (sometimes referred to as the Madrid Protocol). The Environmental Protocol demands (among other things) that protection of the Antarctic environment as a wilderness with aesthetic and scientific value shall be a 'fundamental consideration' of activities in the area. Antarctic soils are not productive in the agricultural sense and the emphasis on soil management in Antarctica relates to preventing damage as a result of human activities.

The first known landfall in the Ross Sea region of Antarctica was by a whaling ship owned by Carsten Borchgrevik at Cape Adare in 1895. Since Robert Falcon Scott and his crew arrived in the Ross Sea region in 1901, and established a hut (at Hut Point on Ross Island) to store some of his supplies to make more room on board the ship, humans have impacted on the Antarctic terrestrial environment. Scott chose his site well and both New Zealand and the USA have established bases nearby.

The bases represent the greatest impacts on the soil environment. Areas have been recontoured to facilitate building platforms and provide storage space. Nearby areas have been trampled, bulldozed for aggregate and road construction, and rubbish and fuel have been spilled. However, in the twenty-first century, there is a strong mandate to try to

Fig. 17.17 Soil formed from guano in the Adele Penguin colony at Cape Hallett

minimise the impacts of human activity on the environment. Precautions are in place to prevent fuel spills, waste is handled carefully, and new earthworks are planned to keep any 'foot print' as small as possible. Most past impacts have, where possible, been remediated. Notably, the former rubbish dump, and a short-lived nuclear power station and adjacent contaminated soil materials, have been removed from the vicinity of McMurdo Station and the affected areas remediated as much as practicable.

In areas adjacent to bases on Ross Island, and at Marble Point where the surface active layer has been removed, the soil then thaws into what was previously the underlying permafrost. Such thawing results in ice melting and evaporating at the surface. The ground ice near the coast is often moderately salty and when the water evaporates, a layer of salt is left behind at the soil surface (Fig. 17.18). Such accumulations of salt impact survival of soil organisms. Over time, the salt is redistributed through the actions of wind and snow meltwater flow and the soil surface gradually recovers.

At points further removed from the bases, there has been concern about the impacts of the camping and trampling activities of scientific expeditions. In the past, there has been some spillage of fuels and wastewater at some sites. Now, under the current environmental management protocols, all human wastes are carried out and accidental spills are rare with immediate actions taken to remedy any soil contamination.

The number of tourists, based on cruise ships, visiting the region increased greatly between about 1990 and 2020. Site guidelines have been developed to manage visitors at the most heavily visited sites, and onshore visits by large numbers of tourists are generally limited to a few key sites.

While there is concern that footprints will last forever, it is evident, given the large number of people who have walked over most of the McMurdo Dry Valley region, and the rarity with which previous footsteps are observed, that on most surfaces there is enough soil surface movement to naturally recover the visible impacts. Trampling experiments have shown that the desert pavement readily recovers within a few seasons, however larger indents in the surface, such as

Fig. 17.18 Removal of the active layer may result in melting and evaporation of underlying ice-core causing soil slumping and accumulation of salt at the soil surface. Left: hummocky terrain below fuel tanks formed as a result of underlying permafrost melt out following surface soil removal (photo taken in the 1990s—fuel tanks have since been removed). Right: salt accumulated on a bulldozed surface near McMurdo Station as a result of melting and evaporation of ice that was previously protected in the permafrost (picture shows an area approx. 15 cm across)

those of vehicle tracks on soft materials, are still visible over 40 years after establishment. Visible footpaths will form, and remain evident in the landscape, with as few as 20 passes over the same track on softer surface materials, whereas walking on hard rock leaves little visible evidence.

Further Reading

Adams B, Bardgett R, Ayres E et al (2006) Diversity and distribution of Victoria Land biota. Soil Biol Biochem 38:3003–3018

Adams B, Wall D, Gozel U, Dillman C, Hogg I (2007) The southernmost worm, *Scottnema lindsayae* (Nematoda): diversity, dispersal and ecological stability. Polar Biol 30:809–815

Adlam LS, Balks MR, Seybold CA et al (2010) Temporal and spatial variation in active layer depth in the McMurdo Sound region, Antarctica. Antarct Sci 22(1):45–52. https://doi.org/10.1017/S0954102009990460

Aislabie J, Balks M, Astori N et al (1999) Polycyclic aromatic hydrocarbons in fuel-oil contaminated soils, Antarctica. Chemosphere 39(13):2201–2207

Aislabie J, Balks MR, Foght JM et al (2004) Hydrocarbon spills on Antarctic soils: effects and management. Environ Sci Technol 38 (5):1265–1274

Aislabie JM, Chhour KL, Saul DJ et al (2006a) Dominant bacteria in soils of Marble Point and Wright Valley, Victoria Land, Antarctica. Soil Biol Biochem 38:3041–3056

Aislabie JM, Saul DJ, Foght JM (2006b) Bioremediation of hydrocarbon-contaminated polar soils. Extremophiles 10:171–179

Aislabie JM, Duggan S, Barker GM (2008) Relation between soil classification and bacterial diversity in soils of the Ross Sea region, Antarctica. Geoderma 144:9–20

Balks EM, Balks MR (2000) GPS surveying in Antarctica. Surv Q (24):32–34

Balks MR, Lopez-Martinez J, Goryachkin S et al (2013) Windows on Antarctic soil-landscape relations across selected regions of Antarctica. Geol Soci Lond Spec Publ 381:397–410. https://doi.org/10.1144/SP381.9

Balks MR, O'Neill TA (2016) Soil and permafrost in the Ross Sea region of Antarctica: stable or dynamic? Cuadernos de Investigacion Geografica 42(2):415–434. https://doi.org/10.18172/cig.2923

Balks MR, Paetzold R, Kimble J et al (2002) Effects of hydrocarbon spills on the temperature and moisture regimes of Cryosols in the Ross Sea region. Antarct Sci 14(4):319–326

Balks MR, Zabowski D (2016) Celebrating soil: discovering soils and landscapes. Springer, Dordrecht, 243 p

Bockheim JG (ed) (2015) The soils of Antarctica. World soils book series. Springer, Dordrecht. https://doi.org/10.1007/978-3-319-05497-1_2

Bockheim JG, Wilson SC, Fenton GH et al (1989) Late Quaternary ice-surface fluctuations of Hatherton Glacier, Transantarctic Mountains. Quat Res 31:229–254

Brooks ST, Tejedo P, O'Neill TA (2019) Insights on the environmental impacts associated with visible disturbance of ice-free ground in Antarctica. Antarct Sci 31(6):304–314

Campbell IB, Balks MR, Claridge GGC (1994) The effect of human activities on moisture content of soils and underlying permafrost from the McMurdo Sound region, Antarctica. Antarct Sci 6(3):307–314. https://doi.org/10.1017/S0954102094000477

Campbell IB, Balks MR, Claridge GGC (1993) A simple visual technique for estimating the impact of fieldwork on the terrestrial environment in ice-free areas of Antarctica. Polar Rec 29(171):321–328. https://doi.org/10.1017/S0032247400023974

Campbell IB, Claridge GGC (1987) Antarctica: soils, weathering processes and environment. Developments in soil science, vol 16. Elsevier, Oxford, 368p

Campbell IB, Claridge GGC, Campbell DI et al (1998) The soil environment of the McMurdo Dry Valleys, Antarctica. In: Priscu J (ed) Ecosystem dynamics in a polar desert. The McMurdo Dry Valleys, Antarctica. Antarctic research series, vol 72. American Geophysical Union, Washington DC, pp 297–322

Campbell IB, Claridge GGC, Balks MR (1998b) Short- and long-term impacts of human disturbances on snow-free surfaces in Antarctica. Polar Rec 34(188):15–24

Campbell DI, MacCulloch RJL, Campbell IB (1997) Thermal regimes of some soils in the McMurdo Sound region of Antarctica. In: Lyons WB, Howard-Williams C, Hawes I (eds) Ecosystems processes in Antarctic ice-free landscapes. Balkema, Rotterdam, pp 45–55. ISBN 90-5410-925-4

Chinn TJH (1979) Impacts of "extreme" events on the Dry Valleys area. N Z Antarct Rec 2(2):9–13. ISSN 0110-5124

Claridge GGC (1965) The clay mineralogy and chemistry of some soils from the Ross Dependency, Antarctica. N Z J Geol Geophys 8:186–220

Claridge GGC, Campbell IB (1968) Origin of nitrate deposits. Nature 217:428–430

Claridge GGC, Campbell IB (1977) The salts in Antarctic soils. Their distribution and relationship to soil processes. Soil Sci 123:377–384

Claridge GGC, Campbell IB, Balks MR (1999) Movement of salts in Antarctic soils: experiments using lithium chloride. Permafr Periglac Process 10:223–233

Clarke RS Jr (1982) The Derrick Peak, Antarctica, iron meteorites. Meteoritics 17:129–134

Fountain AG, Levy JS, Gooseff MN et al (2014) The McMurdo Dry Valleys: a landscape on the threshold of change. Geomorphology 225:25–35

Freckman DW, Virginia RA (1997) Low diversity Antarctic soil nematode communities: distribution and response to disturbance. Ecology 78:363–369

Goddard HE (2013) Investigation of air and soil climate across the latitudinal and altitudinal gradient of the Ross Sea region. MSc thesis, University of Waikato. 166p

Goddard HE, Balks MR, Seybold CA (2012) Soil temperatures over a latitudinal gradient in the Ross Sea region of Antarctica: preliminary results. In: Proceedings of the tenth international conference on permafrost. Salekard, Yamal-Nenets Automonous District, Russia, 25–29 June 2012, vol 4/1: Extended Abstracts, pp 183–184. ISBN 978-5-91128-050-5

Gooseff MN, Barrett JE, Levy JE (2013) Shallow groundwater systems in a polar desert, McMurdo Dry Valleys, Antarctica. Hydrogeol J 21:171–183. https://doi.org/10.1007/s10040-012-0926-3

Guglielmin M, Balks MR, Adlam LS et al (2011) Permafrost thermal regime from two 30-m deep boreholes in Southern Victoria Land, Antarctica. Permafr Periglac Process 1–11. https://doi.org/10.1002/ppp.715

Hofstee EH, Balks MR, Petchey F et al (2006) Soils of Seabee Hook, Cape Hallett, Northern Victoria Land, Antarctica. Antarct Sci 18 (4):473–486

Hogg I, Green A, Storey B et al (2010) Perceptions and reality: a new understanding of terrestrial life in the Ross Sea region. Antarctic 28 (2):27–29

Ikard SJ, Gooseff MN, Barrett JE et al (2009) Thermal characterisation of active layer across a soil moisture gradient in the McMurdo Dry Valleys, Antarctica. Permafrost Periglac Process 20:27–39. https://doi.org/10.1002/ppp.634

IUSS Working Group WRB (2014) World Reference Base for soil resources 2014. International Soil Classification System for naming soils and creating legends for soil maps. World Soil Resources Reports No 106. FAO, Rome. ISBN 978-92-5-108369-7

Joy K, Fink D, Storey B et al (2014) A 2 million year glacial chronology of the Hatherton Glacier, Antarctica, and implications for the size of the East Antarctic ice sheet at the last glacial maximum. Quat Sci Rev 83:46–57

Kamp PJJ, Lowe DJ (1982) Geology and terrestrial age of the Derrick Peak meteorite occurrence, Antarctica. Meteoritics 17:119–127

Kirkbride M, Bradshaw M, Harmsen F (1991) Further finds of the Derrick Peak meteorite occurrence, Transantarctic Mountains, and implications for terrestrial age. Meteoritics 26:213–216

Kennedy AD (1993) Water as a limiting factor in the Antarctic terrestrial environment—a biogeographical synthesis. Arct Antarct Alp Res 25:308–315

McLeod M, Bockheim JG, Balks MR (2008) Glacial geomorphology, soil development and permafrost features in central-upper Wright Valley, Antarctica. Geoderma 144(1–2):93–103

McLeod M, Bockheim J, Balks M et al (2009) Soils of western Wright Valley, Antarctica. Antarct Sci 21(4):355–365

McLeod M (2012) Soil and permafrost distribution, soil characteristics and soil vulnerability to human foot trampling, Wright Valley, Antarctica. PhD thesis, University of Waikato, 219p

McCraw JD (1967) Soils of Taylor Dry Valley, Victoria Land, Antarctica, with notes on soils from other localities in Victoria Land. N Z J Geol Geophys 10(2):498–539

McGaughran AB, Hogg ID, Stevens MI (2008) Patterns of population genetic structure for springtails and mites in southern Victoria Land, Antarctica. Mol Phylogenet Evol 46:606–618. https://doi.org/10.1016/j.ympev.2007.10.003

Nishiizumi K, Elmore D, Kubik PW (1987) Update on terrestrial ages of Antarctic meteorites. Earth Planet Sci Lett 93:299–313

O'Neill TA (2017) Protecting Antarctic soil environments: a review of the current and future challenges for the environmental protocol. Environ Sci Policy 76:153–164

O'Neill TA, Aislabie J, Balks MR (2015) Human impacts on soils. In: Bockheim JG (ed) The soils of Antarctica. World soils book series, Springer, Dordrecht, pp. 281–303. https://doi.org/10.1007/978-3-319-05497-1_2

O'Neill TA, Balks MR, López-Martínez J (2014) Ross Island recreational walking tracks: relationships between soil physiochemical properties and track usage. Polar Rec. https://doi.org/10.1017/S0032247414000400

O'Neill TA, Balks M, Stevenson B et al (2013a) The short-term effects of surface soil disturbance on soil bacterial community structure at an experimental site near Scott Base, Antarctica. Polar Biol 36 (7):985–996. https://doi.org/10.1007/s00300-013-1322-8

O'Neill TA, Balks MR, Lopez-Martinez J (2013) The effectiveness of environmental impact assessments on visitor activity in the Ross Sea region of Antarctica. In: Lundmark L, Lemelin R, Müller D et al (eds) New issues in polar tourism: communities, environments, politics. Springer Publishers, pp 87–110. https://doi.org/10.1007/978-94-007-5884-1_6

O'Neill TA, Balks MR, Lopez-Martinez J et al (2012) A method for assessing the physical recovery of Antarctic desert pavements following human-induced disturbances: a case study in the Ross Sea region of Antarctica. J Environ Manag 112:415–428. https://doi.org/10.1016/j.jenvman.2012.08.008

Scarrow JW, Balks MR, Almond PC (2014) Three soil chronosequences in recessional glacial deposits near the Polar Plateau, in the central Transantarctic Mountains, Antarctica. Antarct Sci 26:1–11. https://doi.org/10.1017/s0954102014000078

Seybold CA, Balks MR, Harms DS (2010) Characterization of active layer water contents in the McMurdo Sound region, Antarctica. Antarct Sci 22(6):633–645. https://doi.org/10.1017/S0954102010000696

Soil Survey Staff (2014) Keys to Soil Taxonomy. Twelfth edition. United States Department of Agriculture, Natural Resources Conservation Service, 360p

Wall AM, Balks MR, Campbell DI (2004) Soil moisture measurement in the Ross Sea region of Antarctica using Hydra soil moisture probes. In: Super soil 2004: 3rd Australian and New Zealand soils conference, pp 1–9

Conclusion: Global Context, Formation Pathways, Mapping, and Assessment of the Soils of Aotearoa New Zealand

18

The soils of New Zealand can be studied nowhere else but in New Zealand and if we are to play a worthy part in helping to solve our country's problems it is our clear duty to pursue these studies.... To know and understand our soils and their relations with their environment, their distribution and their properties, the processes operating within them, the changes that are occurring or can be induced, and the significance of all these to human activities, these are the goals of soil science.

—Norman H. Taylor 1955.

Abstract

Soils in Aotearoa New Zealand are diverse. Soils of all 12 orders in *Soil Taxonomy* occur in New Zealand, which is unusual for a relatively small land area. Also, all the soil orders of the *New Zealand Soil Classification* can be identified elsewhere in the world. However, some features of New Zealand soils are globally uncommon including the large area of Pumice Soils; the world's oldest Allophanic Soils; Pallic Soils formed in loess with minimal calcium carbonate and often with fragipans; Ultic Soils with podzol-like morphologies including distinctive 'egg-cup' E horizons and densipans; Perch-gley Podzol Soils on the West Coast of South Island; extensive Brown Soils that mainly lack subsoil illuvial clay; relatively high soil carbon contents in topsoils, and yet few soils that have deep (>25 cm) dark topsoils; and the 'dry permafrost' soils of the McMurdo Dry Valleys in Antarctica. Ten methods used in New Zealand to evaluate soils and assist in land management decisions are described, ranging from traditional knowledge, through digital soil mapping, concepts of high-class soils, soil versatility, and soil vulnerability, to complex technical evaluation systems including ecosystem service modelling. Aotearoa's diverse and productive soils are a life-supporting taonga/treasure which all need to treat with kaitiakitanga/care.

18.1 Introduction

In the previous chapters, the 15 New Zealand soil orders, along with Antarctic Gelisols/Cryosols, were reviewed focussing on the features that make the soils of each order distinct. This chapter seeks to take a wider view. The distinctiveness and diversity of New Zealand soils in a global context are discussed. The major genetic pathways that are active in developing the diverse soils that form in New Zealand are reviewed. Finally, the main soil and land evaluation tools that are used to develop and disseminate information about the soils of New Zealand and their use and management are briefly reviewed. A variety of approaches has been developed to evaluate soils and their associated landscapes of which ten are discussed here.

18.2 New Zealand Soils in a Global Context

18.2.1 Introduction

It is informative to compare New Zealand soils with those in other countries to understand our 'pedological place' in the world. The question of the distinctiveness of New Zealand soils is important for the person who wishes to import or export soil-based knowledge and for the application of

A. E. Hewitt et al., *The Soils of Aotearoa New Zealand*, World Soils Book Series,
https://doi.org/10.1007/978-3-030-64763-6_18

international environmental standards. How unique are New Zealand soils? Do our soils have close global relatives, or do they have some truly distinctive character that reflects, in some way, the geographic and environmental isolation of New Zealand? There are a number of ways in which the New Zealand archipelago stands apart from other comparable landmasses including the following:

1. spanning 14° in the mid-latitudes, New Zealand has temperate island (hence oceanic) status, whereas much of the rest of the mid-latitude world is continental with cooler winters, warmer summers, and often drier conditions;
2. other comparable landmasses (e.g. Japan) have adjacent continents that experience contrasting climatic extremes, whereas Australia does not experience the intense cold of the Northern Hemisphere continental masses, and is over 4000 km from New Zealand;
3. lying astride an obliquely convergent plate boundary, consequent active plate tectonics and volcanism have generated a geomorphologically dynamic landscape that includes uplifted mountains, in the path of westerly winds, that are high enough to carry ice caps and glaciers in glacial periods and to cause a marked modern-day rain shadow;
4. native ecosystems comprise unique flora, dominated by evergreen mixed conifer-broadleaf forest, developed in global isolation, and terrestrial animal life that is dominated by birds, apart from sea mammals and two species of bat;
5. a recent history of human settlement that dates from c. 1280 AD, thus human impacts represent only the past c. 750 years, with only c. 250 years since the arrival of James Cook, in 1769, which marked the start of well-recorded changes arising from European exploration and settlement.

Thus, in summary, the New Zealand archipelago has

- widely varying rock types (including extensive, relatively young, unconsolidated deposits);
- a diverse topography with strong relief and mainly youthful landscapes, the result of active and (mainly) uplifting tectonism, volcanic activity, and the effects of the last glaciation, together with the reworking of materials by vigorous rivers, and fast rates of erosion by mass movement;
- markedly varied climate;
- a distinctive flora and fauna;
- and human impacts spanning less than about 750 years.

As a result, New Zealand has a diverse array of soils with broad patterns driven largely by climate and parent material together with surface age (Chap. 1). The soil diversity is related to scale, with New Zealand soils showing distinctive changes at the *land regional scale* (~10 km–~100 km). Using a land systems approach (following a theoretical framework of pedodiversity developed by Juan Jose Ibánez and others) Allan Hewitt showed that much of the regional-scale distinctiveness in New Zealand relates to steep gradients in climate and underlying geological structures and parent materials which leads to strongly contrasting soils. At paddock or farm scales (~10 m–~10 km), diverse soils may form complex spatial relationships, in common with patterns in many other countries.

Apart from the regional-scale distinctiveness, and various differences such as the extensive Pumice Soils, none of the features listed above are globally unique. Thus, where there are similar conditions to those in other parts of the world, there are likely to be similar soils.

18.2.2 Soil Diversity

The diversity of New Zealand soils does not make the question of global distinctiveness easy to answer. However, in considering *Soil Taxonomy* as a global standard of taxonomic diversity then New Zealand may be remarkable as one of the few countries that contains soils representing all of the 12 soil orders defined in *Soil Taxonomy*, albeit some orders are relatively uncommon (Table 18.1). The only *Soil Taxonomy* order that has not yet been formally identified in New Zealand is the Gelisols (soils that contain permafrost or evidence of churning by strong frost action). However, it is thought that Gelisols exist in the proximity of glaciers, at high altitude, in the Southern Alps. Although such soils have not been studied in detail, the likely occurrence of Gelisols is supported by recent topoclimate-based modelling by Katrin

Table 18.1 Representation of *Soil Taxonomy* soil orders in New Zealand. Percentage of NZ land area in brackets (data from Manaaki Whenua – Landcare Research)

Very common	Common	Less common	Rare
Inceptisols (47.4)	Alfisols (9.9)	Mollisols (1.2)	Oxisols (0.2)
Spodosols (13.1)	Entisols (7.4)	Histosols (0.9)	Vertisols (0.1)
Andisols (12.9)	Ultisols (4.2)	Aridisols (0.9)	Gelisols[a] (<0.1)

[a]Small areas of Gelisols are presumed to occur in proximity to glaciers and rock glaciers, above ~2000 m elevation in alpine areas of New Zealand. About 2% of the NZ landmass is not soil, i.e. lakes, urban sealing, etc.

Sattler and others that indicates the presence of contemporary permafrost in debris-mantled slopes above ~2000 m elevation in the central Southern Alps, and above ~2150 m in the northern Kaikoura ranges. These altitudinal limits are considered highly unusual for permafrost globally, and Sattler and others emphasise that such permafrost may relate to reduced ice loss because of maritime New Zealand's moderate summer temperatures. Relict periglacial features, including patterned ground such as stone stripes, occur in upland Central Otago and active solifluction deposits are commonplace on slopes in central and southern New Zealand. Ploughing boulders on the Rock and Pillar Range also indicate that periglacial processes remain active in some areas at elevations of ~1500 m. Gelisols form the main soil order in the Ross Sea region of Antarctica (Chap. 17).

The diversity of soils in the relatively small area of New Zealand provides some unique challenges because high soil diversity has implications for management decision making. For example, when a predictive model is developed at a particular place, and then applied in other areas, confidence in the results is enhanced if it is known that the soils are similar and thus likely to behave in a similar way. Thus, considerable work is needed to provide confidence in the predictive capacity of a model, such as "Overseer", when applied over a heterogeneous range of soils.

Soil diversity at the paddock scale may be managed using precision agricultural techniques. Where the diversity is well mapped, and the soil well characterised, then the techniques of precision agriculture can provide production and environmental benefits. However, in areas where soil investigations show low paddock-scale diversity, the investment in precision agriculture techniques may not be worthwhile.

18.2.3 Global Relatives of New Zealand Soils

The New Zealand soil-landscape can be divided into four broad regions (Fig. 18.1). The relatively tectonically stable Northland area is dominated by Ultic Soils, Granular Soils, and Oxidic Soils. We expect to find related soils in parts of the world that are relatively tectonically stable, have warm, generally wet, climates, and have missed the major disturbances of late Quaternary volcanism and associated mantling of tephra deposits (Table 18.2). Although temperatures were lower during the last glacial period, Northland remained under forest cover, unlike the rest of New Zealand.

Correlatives of the Oxidic Soils are distributed globally in the tropics typically on stable landforms of continental shields, but they also occur in orogenic belts on highly weatherable basic rocks, as is the case in northern New Zealand. Although Oxidic Soils in New Zealand are generally similar to their international correlatives, equivalent soils elsewhere are typically bioturbated by termites. The apparent lack of termites in pre-human times in New Zealand suggests that our soils are much less bioturbated and may differ biologically in some way.

Soils similar to both Granular and Ultic Soils are well represented, on old landscapes, in many parts of the world. However, some of the Ultic Soils in northern New Zealand show strong podzol morphological features, despite being enriched in clay, because of the potency of the kauri tree (*Agathis australis*) as a scattered podzolising agent in the region's pre-human forests. The Densipan Ultic Soils are also globally uncommon.

The widespread tephras in the central North Island are dominated by Pumice Soils and Allophanic Soils. Similar soils form in other parts of the world where tephra from rhyolitic and andesitic volcanoes dominate the soil parent materials (e.g. Japan and Oregon, USA). However, substantial proportions of the Allophanic Soils in North Island are dominated by silica-rich rhyolitic tephras and are consequently less fertile, chemically, than the main andesite- to basalt-dominated equivalent tephra-derived soils elsewhere in the world. Essentially unlimited Al availability means that so-called 'non-allophanic' Andisols (i.e. soils on tephra dominated by Al-humus complexes rather than allophane, as found in parts of Japan, for example) are probably not present in New Zealand. Based on a soil profile about 1 m deep, New Zealand has a much wider range of ages on Andisols than elsewhere. Most of the Pumice Soils (Vitrands in *Soil Taxonomy*) are ~700 or ~1800 years old. The Allophanic Soils (Udands in *Soil Taxonomy*) are formed mostly by developmental upbuilding pedogenesis and date back ~20,000–25,000 years with some as old as ~50,000 years (composite ages for 1.0–1.5-m deep soil profiles). Thus, New Zealand has possibly the oldest Andisols in the world on this basis. The rhyolitic Pumice Soils (Vitrands) are exceptionally siliceous and glassy, and lack macro- and micronutrients (including Cu, Se, Co, B, and I).

Soils similar to the Semiarid Soils are common in low rainfall areas throughout the world. A feature of the New Zealand Semiarid Soils is that they generally have no lime (calcium carbonate) inherited directly from the parent materials. The lime accumulated in the profiles is not from limestone or other calcareous rocks but formed instead as a result of weathering processes in the soil, i.e. the lime is secondary in origin. In mid- and western USA, Mollisols (soils with deep (>25 cm), dark, organic matter-rich topsoils, i.e. mollic epipedons) occur in similar semiarid and marginally moister environments. However, Mollisols are relatively rare in New Zealand. One reason is that their formation may have been suppressed by the lack of natural lime (to help stabilise organic carbon). Other inhibiting factors for Mollisol formation include the overwhelming predominance of podocarp-broadleaf forest vegetation, rather than grasslands, from postglacial (Holocene) times

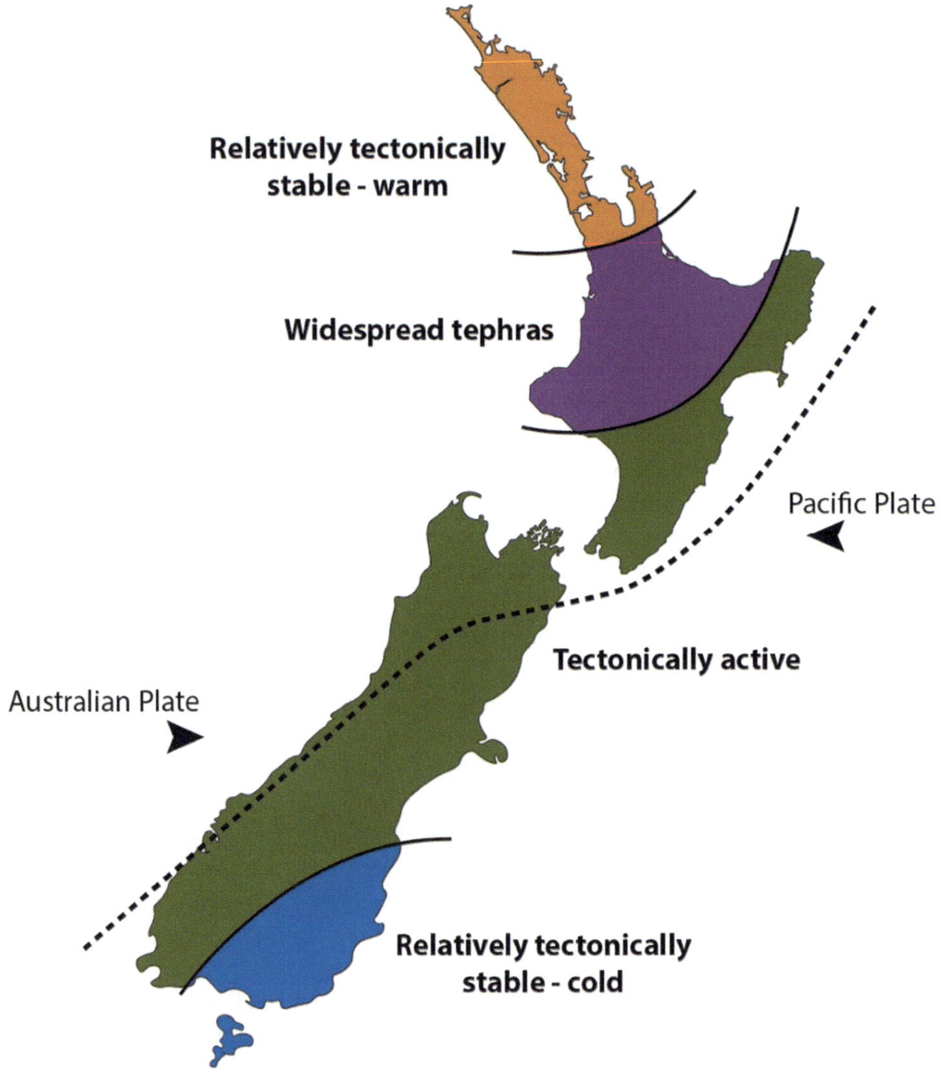

Fig. 18.1 Broad soil regions of New Zealand within a general plate tectonic and climate framework after Hewitt (1997)

until human arrival; the nature of the (relatively restricted) New Zealand native grassland; a lack of soil vertebrates or native burrowing animals; and the relatively mild winters.

The Pallic Soils occur within areas on loess in Europe and North America, but New Zealand Pallic Soils are distinctive due to their low contents of calcium carbonate (which, in contrast, is prevalent in the northern hemisphere loess). The Fragic Pallic Soils, i.e. soils with a non-cemented 'brittle' fragipan, have been described as globally unusual.

Brown Soils are the most extensive New Zealand soil order (43% of New Zealand's land area). Similar soils are also widely recognised throughout the world. However, in New Zealand 'argillic' Brown Soils, with enrichment of subsoil clay, are rare, partly because the climate (high average rainfall) and mostly evergreen forest vegetation promote acidification and podzolisation rather than the clay movement and accumulation (argilliluviation) pathways that tend to be favoured under deciduous forest.

Most of the New Zealand Podzol Soils are well represented globally. However, there are two possible peculiarities. First, experience in testing *Soil Taxonomy* showed that some Podzol Soils in Northland had clay values that were too high to allow them to be recognised as Spodosols (the *Soil Taxonomy* equivalent of a podzol) despite the soils having other podzol morphological features. Consequently, the clay-rich Northland Podzol Soils (Wharaekohe series) were classed as Ultisols in *Soil Taxonomy*, rather than Spodosols. That podzolic features could be developed on a slowly permeable clayey parent material highlighted the potency of podzolising forest vegetation, notably kauri (*Agathis Australis*) and rimu (*Dacrydium cupressinum*) in northern New Zealand. Second, it is likely that some of the Perch-gley Podzols, such as the Okarito series (Chap. 11), are uncommon and possibly globally unique. Few temperate regions can match the extreme leaching provided by the 5–10 m year^{-1} rainfall of the West Coast of South Island, and

Table 18.2 Generalised international prevalence of soils similar to New Zealand soils

NZ soil order	Examples of location of similar soils	Global prevalence
Northern North Island: tectonically relatively stable, moderate to warm, moist, climate		
Oxidic Soils	Brazil, tropics	Common
Granular Soils	Australia	Common
Ultic Soils	Eastern-southern USA	Common
Central North Island including Taranaki: volcanically active—widespread young tephras—temperate climate		
Allophanic Soils	Pacific rim, rift systems	Common
Pumice Soils	Pacific rim, rift systems	Common
Lower North Island and South Island: tectonically active hill and high country, as well as the tectonically stable Otago region—cool climate—mature soils		
Semiarid Soils	Western USA	Common
Pallic Soils	USA, Europe	Common
Brown Soils	Europe	Common
Podzol Soils	Europe	Common
Melanic Soils	Europe	Common
New Zealand wide: immature and wetland soils		
Recent Soils	Global	Ubiquitous
Raw Soils	Global	Ubiquitous
Anthropic Soils	Global	Ubiquitous
Gley Soils	Global	Ubiquitous
Organic Soils	Northern latitudes	Ubiquitous

also provide the incremental accumulation of loess (up-building pedogenic conditions) to generate two-storeyed soils, that comprise a Podzol Soil on alluvium, with effectively a Gley Soil in loess overlying it.

The Raw Soils, Recent Soils, Gley Soils, Organic Soils, and Anthropic Soils, are all globally ubiquitous. However, many of the plant species that form New Zealand's Organic Soils are found nowhere else on Earth, notably, the wire and cane rushes of the Restionaceae (restiad) family. The Ross Dependency in Antarctica includes the cold desert environment of the McMurdo Dry Valleys where a range of globally unique soils occur. Of particular note are the areas of permafrost where there is insufficient moisture to form ice-cement, resulting in the formation of "dry permafrost" soils (Chap. 17).

The international correlations discussed here may not be precise because the data needed are often lacking. Thus, we have based correlations mainly on the similarity of central concepts. Correlations between *Soil Taxonomy*, the *World Reference Base*, and the *New Zealand Soil Classification* (given in the international soil correlation sections of Chaps. 2 through 16), show that all New Zealand soils, except Raw Soils, can be placed within the major international soil classifications systems. Many Raw Soils are not recognised by *Soil Taxonomy* or the *World Reference Base*. Raw Soils are generally not considered to be soil in *World Reference Base* and they are included as Entisols (the equivalent of Recent Soils) in *Soil Taxonomy*. Anthropic Soils are recognised in *World Reference Base*, but in *Soil Taxonomy* they are recognised, but only at lower levels within the Entisols order. The Gley Soils of the *New Zealand Soil Classification* have no equivalent order in *Soil Taxonomy* but are recognised, mostly at the suborder level, through their 'aquic soil moisture' conditions (e.g. Aquepts, Aquents). Similarly, Pumice Soils are ranked as a suborder of Andisols (mainly Vitrands) in *Soil Taxonomy* (See also Appendix B).

18.2.4 Soil Organic Matter

New Zealand soils are relatively high in organic carbon (or soil organic matter) compared to the levels in soils in many other countries. When *Soil Taxonomy* was first tested in New Zealand, it was observed that many mature alluvial terrace soils were assigned to Fluventic classes because they contained more carbon than the designers of *Soil Taxonomy* expected.

When comparing similar soils, the soil organic carbon to 1 m depth in most New Zealand soils is higher than the global mean of the correlated soil. For Allophanic Soils, the New Zealand and world values are similar (although soils with thick melanic horizons in Japan, for example, have exceptionally high organic carbon contents of up to c. 30%). Only in the Pumice Soils is the New Zealand mean organic carbon content less than the global mean for similar soils (Vitrands). The lower carbon content in Pumice Soils may be explained by the relative youth (<2,000 years) and large extent of New Zealand pumice materials (Chap. 12) compared to global soils formed in pumice.

Although most New Zealand soils are relatively high in carbon, few have thick enough topsoils to qualify for the definition of "Black Soils" (as currently being used internationally, by organisations such as the FAO (Food and Agriculture Organisation) in the United Nations). Black Soils include the Mollisols/Chernozems of the Northern Hemisphere that feature deep, organic matter-rich, topsoils, and are currently of interest due to the desire to prevent them from losing carbon to the atmosphere. It is good that carbon is relatively high in New Zealand soils, however, that means that on average, there is little potential for our soils to act as carbon sinks. However, like the 'Black Soils', it is important to ensure that we maintain, or improve, existing soil carbon levels.

Why are New Zealand soils relatively high in carbon compared to world averages? The main factors include the dominant cool, moist, temperate New Zealand climate, and the preponderance of soil acidity, which support plant growth but impede organic matter decomposition. The existence of soil carbon at depth in some soils has been attributed to the relatively short time since native forested lands were converted to pasture. Under mostly evergreen native forest (in contrast to deciduous forests), the soils had slow leaf turnover, low rates of decomposition, and low nutrient availability as a nutrient-conserving mechanism. Lower carbon levels in other countries may also be because the New Zealand soils are predominantly now under pasture, which is more carbon conservative than intensive arable use. Also, New Zealand soils have not had the accumulated impacts of thousands of years of cultivation by ancient civilisations which has likely led to soil carbon loss in many regions.

The Allophanic Soils are a special case (Chap. 2) as they may contain relatively large amounts of organic matter (generally matching levels in many of their counterparts overseas). The high organic carbon content is usually correlated with allophane content and the high specific surface areas associated with tiny allophane spherules. Allophanic Soils thus not only adsorb substantial organic carbon but also protect it from degradation, including in tight nanopores within allophanic micro- and nanoaggregates. Also, Allophanic Soils commonly contain carbon-enriched, buried A or buried AB horizons (within 1-m-deep profiles) so that the overall carbon content for the soils is higher than that associated with mineral soils of other orders that do not contain buried soil horizons.

18.3 Pathways of Soil Formation in New Zealand

What would become of a Raw Soil, on a fully formed parent material with no, or negligible, additions of geological materials to the land surface, given an extensive period of time, a stable site, and continuity of climate, landform, and uninterrupted ecological succession? Under such conditions, the Raw Soil will likely age and develop through 'classical' topdown pedogenesis (soil formation). Topdown pedogenesis comprises soil-forming processes operating mainly from the land surface (driven by the organic and water cycles) that result in the gradual deepening of the profile as a downward moving front on a static parent material. Initially, a well-developed topsoil (A horizon) will form and, subsequently, subsoil horizons develop beneath it driven by the environmental conditions (soil-forming factors). The properties and arrangement of consequent soil horizons provide clues to the processes that have driven the pathways experienced by the maturing soil over hundreds to thousands of years, or more.

However, in many landscapes in New Zealand, ongoing deposition of new materials, such as river-flood alluvium, tephras, or loess, is common. The evolution of soils in such landscapes therefore has an additional complexity because the impact from topdown (pedogenic) processes is modified by the rates at which new materials are added to the land surface via geological processes. The resultant soils are formed by upbuilding pedogenesis. Upbuilding pedogenesis is the ongoing formation of soil via topdown processes, while tephras or sediment (such as loess or alluvium) are intermittently added to the land surface as the result of normal geological processes. The resultant soils may show distinctive layering and buried horizons (sometimes referred to as paleosols), forming multi-layered profiles. The frequency and thickness of the accumulating materials determine how much impact topdown processes have on the ensuing soil profile, and if either *developmental* or *retardant* upbuilding, or both, will take place.

Developmental upbuilding occurs when the rate of addition of tephra or loess, or other geological materials, to the land, is incremental and slow enough to allow topdown soil formation to keep pace as the ground surface gradually rises (a corollary is that each part of the profile has been an A horizon at one time). In New Zealand, the most extensive soils formed by developmental upbuilding pedogenesis include many Allophanic Soils (Chap. 2) formed on tephra and essentially all the Pallic Soils (Chap. 10) that are formed in loess deposits. Other soils, such as the Perch-gley Podzol Soils, are also formed by developmental upbuilding as are many Recent Soils (Chap. 14) and some Brown Soils (Chap. 4).

In contrast, retardant upbuilding occurs when a relatively thick layer of tephra, alluvium, or colluvium is instantaneously added to the surface of a soil which thus becomes a buried soil, cut off and isolated from the new land surface in which topdown pedogenesis then begins anew (hence the development of the original soil is retarded). Retardant upbuilding can also occur where thin geological additions accumulate sufficiently rapidly for the original soil surface to be buried quickly, thereby isolating it from the surface processes. Once a paleosol is isolated, any changes may be regarded as largely diagenetic (low-temperature alteration of geological deposits), not pedogenic (active soil processes), especially when cut off completely from surface processes (i.e. deeply buried). Examples of soils formed by retardant upbuilding pedogenesis were given in Chaps. 2 (Allophanic Soils), 12 (Pumice Soils), and 13 (Recent Soils).

The pedogenic processes most important in New Zealand soil formation include:

- weathering of rock and soil materials and the concomitant synthesis of clay minerals (argillisation), including the formation of allophane and halloysite (Chap. 2);
- leaching of solutes (including nutrients) from the soil;
- formation of pale E horizons and underlying iron and organic matter enriched horizons via leaching under acid conditions (podzolisation, Chap. 11);
- darkening of inorganic matter by admixing with humified organic matter (melanisation, Chaps. 7 and 14);
- acidification (replacement of cations in the soil by hydrogen with the cations lost via leaching);
- soil mixing by mechanisms such as treefall;
- migration of fine clay and humus down through the soil profile mainly in suspension (illimerisation, or illuviation, Chaps. 6 and 16);
- formation of pale-coloured soils due to chemical reduction of iron and manganese compounds under anaerobic (oxygen deficient), or redox (alternating oxidation and reducing) conditions (gleying, Chap. 5);
- initiation of peat formation (paludisation, Chap. 8); and
- accumulation of iron and related oxides to give soils yellow to brown colours (brunification, Chap. 4).

The major soil-forming (genetic) pathways for New Zealand soils (Fig. 18.2) relate soil processes and their interactions with the soil-forming factors. The *New Zealand Soil Classification* can be used to help trace the genesis of New Zealand soils through time. The pathways initially show a fundamental distinction between mineral (inorganic) and organic soil materials. In mineral parent materials, Raw Soils develop into Recent Soils. The nature of the parent material—potentially including ongoing geological additions associated with upbuilding pedogenesis—then strongly

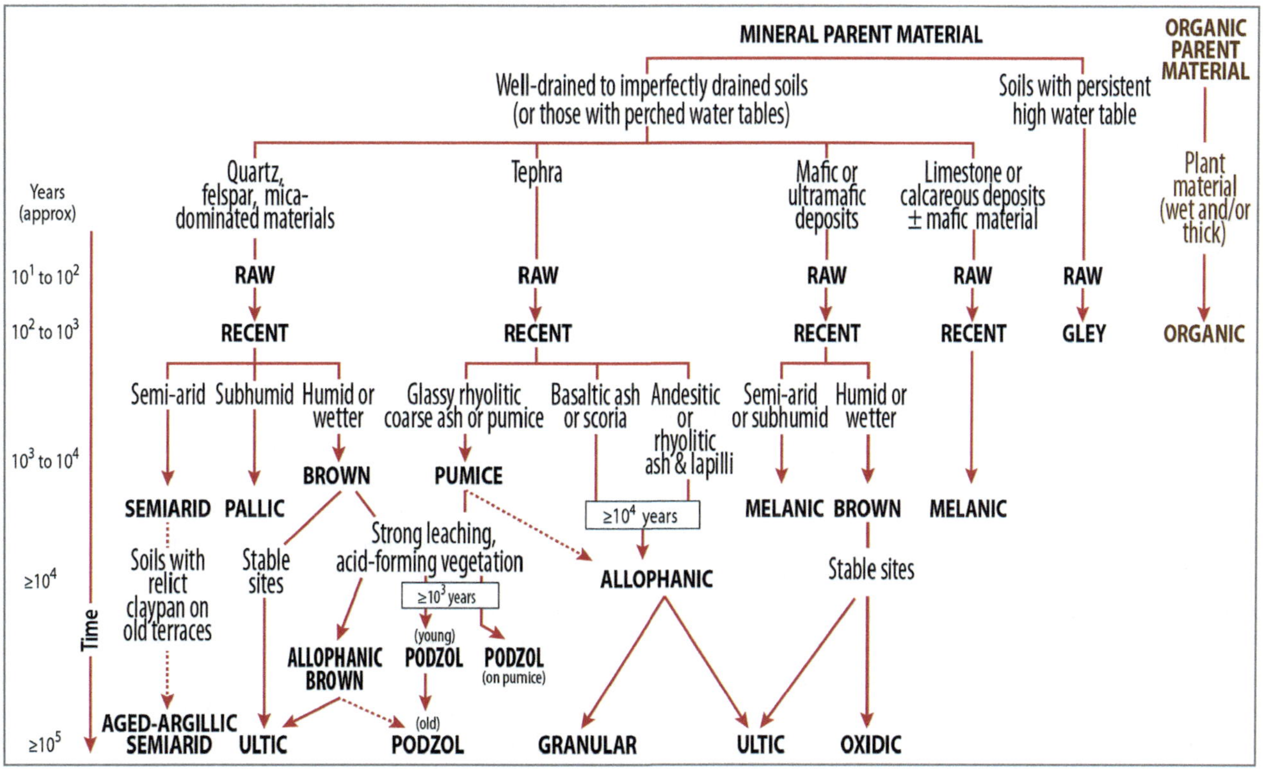

Fig. 18.2 Major genetic pathways in the evolution of New Zealand soils with approximate age ranges for their formation

determines the subsequent soil formation pathway. Later, climate and vegetation, and the passage of time, become the more important controls over the character of the soil that develops from particular parent materials.

Soils will only proceed the whole way down a soil formation pathway on relatively stable sites. Erosion, soil mixing or disturbance (pedoturbation), or the addition of fresh sediment or tephra, will potentially impede, retard, halt, or even reverse, soil progress along a genetic pathway. Many, perhaps most, Raw Soils may never develop into Recent Soils or other soils because they occur in environments with continual cycles of erosion or frequent additions of new materials. Some soil orders may form quickly (in c. 1000–5000 years) when conditions are favourable, such as 'young' Podzol Soils that have formed under high rainfall on free-draining sands dominated by *Nothofagus* (beech) in Westland, or on well drained pumice (Taupo Tephra, c. 1800 years old) under podocarp-dominated forest at >500 m altitude, and moderately high rainfall, on the Mamaku Plateau. At the same time, some Podzol Soils are likely to be some hundreds of thousands of years old as one of the most evolved (mature) soils in New Zealand. The most strongly developed soils, Podzol Soils, Ultic Soils, Granular Soils, and Oxidic Soils, have each passed through earlier phases at an intermediate stage of their evolution (Fig. 18.2). Similarly, the (Weathered) Aged-argillic Semiarid Soils, which may be c. 350,000 years old, represent a soil developed on the oldest (most elevated) of four terrace surfaces in Central Otago. It is therefore inferred to have been through each of the preceding stages of soil development recognised on the three successively younger terrace (and fan) land surfaces (Chap. 15).

Undisturbed Organic Soils are likely to develop from Fibric, through Mesic, to Humic groups, with time, as the organic materials progressively decompose. If no fresh organic material is added, and if the water table drops or becomes more oxygenated, the organic content will decompose to a point that the soil eventually becomes mineral dominated and will then develop down a mineral pathway.

The pathways included in Fig. 18.2 represent mainly those likely to have occurred over extensive areas. For clarity, minor pathways or interlinkages have been omitted, but we have included two exceptions: (i) allophanic material may be formed from non-tephric deposits by the strong leaching of unconsolidated quartzo-feldspathic and micaceous materials, and hence lead to allophanic groups such as Allophanic Brown Soils; and (ii) Ultic Soils may be formed in mafic or tephra-derived parent materials given long time periods in a warm moist climate.

Although the major genetic pathways of soil formation (Fig. 18.2) are essentially global, some features reflect the distinctive parent materials and climatic conditions of New Zealand, in turn, related strongly to the active tectonic setting in an oceanic, mid-latitude location. For example, Andisols (Allophanic Soils, Pumice Soils, and Tephric Recent Soils combined) cover ~13% of New Zealand's land area (the third most common *Soil Taxonomy* order in New Zealand), yet globally Andisols represent only 0.84% of the land area, 12th in the global ranking by area.

The distinctive elements of New Zealand soil-forming processes reflect some combination of low calcium carbonate (lime) content, lack of surface casting native soil fauna, generally low incidence of fire in our forests, and a nutrient conservative flora that cannot drive the soil-plant system to its potential. Thus, rapid acidification in lime-free parent materials has occurred under nutrient conservative native forest vegetation with little bioturbation of surface horizons, apart from tree throw. The accumulation of acid organic matter has led to less movement (illuviation) of clay down the profile, and more brunification and (especially) podzolisation, than would be expected in similar environments internationally.

18.4 Soil and Land Evaluation in New Zealand

18.4.1 Introduction

If we could breed soils what would a champion soil look like? This was a question posed by Eddie Cutler of Lincoln University in the 1960s. His answer would have been conditioned by the prevailing productive pastoral agriculture of the landscape in which he worked. His ideal soil would have had an earthy (crumb) structure (like breadcrumbs), a fine-sandy loam texture, and been well drained. More than 50 years later, however, we would have to answer by saying 'it depends'. To meet the needs of different plants, land management, and environmental requirements, a range of 'ideal' soils that is fit for a specific purpose are recognised. For example, a champion soil to support pasture for dairy grazing requires a high capacity to store water, and resistance to bypass flow so that water stress that would impede pasture growth is ideally avoided, and rapid leaching of nutrients or contaminants to groundwater is minimised. In the same locality, a soil for the production of medal-winning wines may need the opposite capability, with low water storage capacity required to induce drought stress and help the management of sugars and alcohol.

Although we cannot breed soils, they can be modified in ways that may persist for a long time. Modifications include drainage to remove excess water and liming to make soils less acidic and more hospitable to food production. A more severe example is deep ripping or 'flipping', where the entire soil profile is inverted, to improve drainage or water holding capacity (Chap. 3, Anthropic Soils).

Land evaluation thus involves the assessment of soils (or land) to determine the suitability of the soils or land for an intended enterprise. The bigger the contrast between the 'ideal' soil for a given land use, and the soil's present state, the greater the inputs or changes that may be needed to accommodate a particular activity. Thus, the less suitable a land area is for a given purpose the greater the consequent development, maintenance, and environmental mitigation costs are likely to be.

Information about soil and land management ranges from traditional knowledge, through the soil classification (which provides an understanding of the likely soil properties and behavioural characteristics, Chaps. 1–17), to various systems of soil mapping and evaluation as well as soil and land assessment for a range of purposes. Ten soil and land evaluation systems are briefly described in the following sections. Ways to access valuable sets of land information from regional and national databases are also included.

18.4.2 Traditional Knowledge Derived from Local Experience

People who live and work on the land, and depend on it for their food and/or income, observe the soil and associated crop growth on a daily basis and thus build up a depth of understanding of their land, often over many generations. It is arrogant and misguided for scientists to arrive and start to tell people how they should manage their land without first taking the time to listen to, and understand, the local knowledge and management methods. As every landholding is unique, there are often local insights that may not be immediately evident to the new arrival.

With regard to traditional Māori knowledge, tikanga, and kaitiakitanga, some insight may be found in Elsdon Best's book, 'Māori Agriculture'. In the twenty-first century, it is important to consult with iwi to understand local concerns and insights. The Māori language, in common with the languages used by farmers worldwide, points to a universal understanding of key soil properties. Descriptions of soil colour, texture (sandy, clayey, heavy, light), the depth of the soil, and its ability to grow particular plants or crops are universal.

Astute land managers will develop a useful personal set of knowledge of the behaviour of their land, and, with continuing experience, grow this knowledge base over time. Such knowledge has traditionally been passed on through oral traditions and hands-on teaching. However, it is of advantage to others on similar land, or to those who follow, to communicate the understanding in forms that can be shared more widely. Science-based soil classification, mapping, and land evaluation tools provide a means not only to capture and communicate information about the soil and its potential uses but also to extrapolate knowledge gained in one area to other places with similar soils and conditions. Soil mapping and land evaluation thus provide information needed to build the knowledge base of those who lack the intimate knowledge of experienced farmers and gardeners.

18.4.3 Soil Maps and Associated Information

New Zealand has a long and excellent legacy of soil mapping. The first recorded soil 'map' in New Zealand was of the Wairau Plains, published in 1914 by Leonard Wild. The need for soil surveys was increasingly recognised in the early 1920s. (Sir) Theodore Rigg, at the Cawthron Institute in Nelson, initiated soil mapping of Waimea County with a soil map prepared in 1922. Geologist Hartley Ferrar, and others, undertook agro-geological surveys in Northland, Otago, and the King Country. Ferrar recognised the importance of mapping soils and developed soil maps based on the geological parent materials and soil texture. In 1929, he published a report on the soils of the irrigation areas in Central Otago.

In central North Island, two quests drove early mapping. The first was the need to develop an understanding of volcanic hazard, and Leslie Grange began mapping volcanic deposits including 'ash showers' (tephras) in 1926. The second quest was to try to understand the cause of the so-called 'bush sickness' disease that affected sheep and cattle (now known to be a vitamin B12 deficiency caused by a lack of cobalt in soils and herbage, Chap. 12). Grange and Norman Taylor began mapping soils in the bush-sick areas in central North Island in 1930 (directed by Rigg) and showed a correlation between the Pumice Soils and incidence of bush sickness. The close association between soil type (particularly the soils formed on the pumiceous Taupo and Kaharoa tephras) and bush sickness demonstrated the potential power of soil surveys. This led to the formation of the Soil Survey Division as a separate entity within the Department of Scientific and Industrial Research (DSIR) in 1936, with Grange its first director. By 1939, a number of detailed soil surveys, including Westport District and Waipa County, had been published as DSIR bulletins. The Waipa survey, led by Grange and Taylor and carried out between 1933 and 1935, included the novel use of purpose-flown aerial photographs. It established a format and methodology for mapping at reasonably large scale (1: 31,680) that was then applied to the rest of New Zealand. The once close connection established between landowners and the soil surveyors means that the soil series names adopted in the survey remain in use by farmers to this day.

World War II brought increased impetus to the soil mapping effort with a reconnaissance soil map of the whole North Island completed, along with a number of more detailed surveys of soil and erosion throughout the country. Soil erosion had become a topic of growing concern following a series of major erosion and flood events on the east coast of North Island in the 1930s. The findings of early erosion surveys by Harry Gibbs and Grange, critical of farm management in places, were highly contentious. However, in 1941, the Soil Conservation and Rivers Control Act was passed. The act defined catchment districts and associated catchment boards were established to prevent or minimise damage from floods and erosion. By 1945, 11 catchment boards had been formed, and by 1967 the entire New Zealand landscape was covered. The catchment boards eventually became part of regional councils in the wide-ranging reforms under the Resource Management Act (1991).

The Town and Country Planning Act (1953) was another noteworthy piece of legislation that arose in part from recognition by Gibbs, James Raeside, Eddie Cutler, and Grange, that urban spread on to high-class soils (Sect. 18.4.11) should be limited to preserve such soils for food production.

By 1960 when Taylor, the Director of DSIR Soil Bureau, attended the International Soil Science Society meeting in the USA, he took with him a map that demonstrated that New Zealand's soil survey effort extended from the equator to the South Pole. As well as work throughout New Zealand, a number of soil surveys were undertaken in the Pacific Islands, and Graeme Claridge and John McCraw had completed the first expedition to study soils in Antarctica in the summer of 1959–60. Much of this work was published as Soil Bureau bulletins or soil survey reports.

The soil maps have been built upon as the basis for the modern-day digital and geographical information system (GIS)-based soil maps. However, the original bulletins, and legacy maps, are well worth seeking out and reading (many are now available online at the digital library: http://digitallibrary.landcareresearch.co.nz/) because they contain a wealth of information about the regions and soil-landscape relationships that are not always readily discernable in modern maps and databases.

In July 1992, Soil Bureau was merged with some other research organisations to form a new crown research institute that is currently known as Manaaki Whenua – Landcare Research. The focus of soil survey shifted to developing online, geographic information system (GIS)-based databases, resulting in the advent of S-map (https://smap.landcareresearch.co.nz/) and the National Soil Data Repository (NSDR) (https://soils.landcareresearch.co.nz/soil-data/national-soils-data-repository-and-the-national-soils-database/) (Fig. 18.3). Much of the new survey work involved developing an enhanced soil-landscape modelling approach. Soil properties may be quantified via digital soil mapping, and the development of 'functional soil horizons' is used to provide a basis for predicting soil physical properties via 'pedotransfer functions' (modelling based on data from soil profile descriptions).

Digital soil mapping enables the production of maps of soil classes or properties using various environmental data layers (sensor data, including digital elevation models) along with statistical, GIS, and remote sensing, software. Digital soil mapping, developed from the early-mid-2000s, began by adding existing soil information into a GIS framework and then applying geostatistics. A map in a digital survey is linked to a spatially coordinated database that can consist of many layers of information. Digital soil maps may be zoomed in or out and thus viewed at any scale, a function referred to as scale of visualisation. (The scale of visualisation is different from the scale at which the maps were constructed and hence caution must be observed when zooming in regarding the accuracy of detail that becomes evident.)

Digital soil maps can be upgraded with new data, and can accommodate data at any scale. The S-map New Zealand digital soil map programme, which was started in 2004, has the following aims:

- identify every soil in New Zealand within a single database with the same classifying criteria;

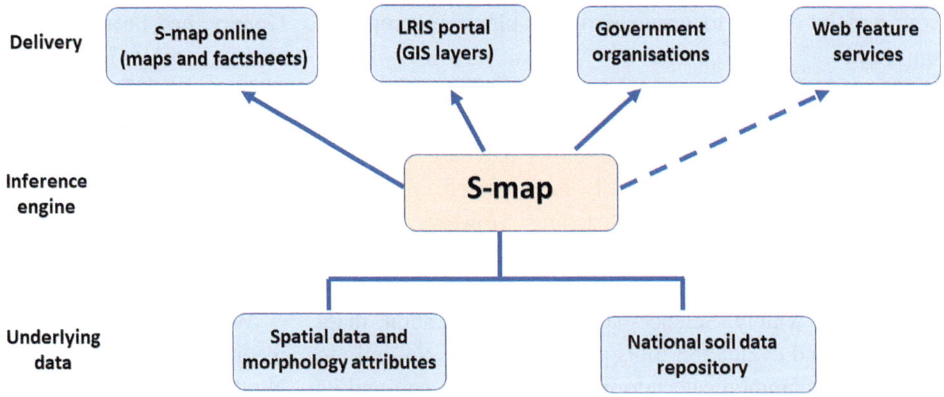

Fig. 18.3 Structure of the New Zealand digital soil data and mapping system that includes S-map, which links to the National Soil Database, that is devloped and managed by Manaaki Whenua – Landcare Research

- realistically depict the spatial variability of soils in the landscape;
- capture most of the data collected in the field from pre-existing mapping programmes along with new data;
- provide indices of uncertainty for each attribute described;
- provide the most accurate and up-to-date soil data 'to the kitchen table' via the Internet; and
- automatically generate data and derivative maps to provide a range of data formats and interpretations, including derivative maps of properties, such as drainage classes, depth to rock, or depth to an impeding layer.

One of the main advantages of digital soil maps over other mapping techniques is that the techniques are transparent, repeatable, and up-dateable (when more information becomes available), and that probability maps are provided with model and validation statistics.

By 2020, S-map coverage of New Zealand had reached about 37%, but coverage of intensively used land, including horticulture, cropping, and intensive pasture, had reached 67%. Currently, S-map is being extended and new tools and applications are being added. A soil database platform (one-stop-shop) has been developed, namely the Land Resource Information System (LRIS, which includes S-map and the National Soil Data Repository http://lris.scinfo.org.nz/#). Already the LRIS portal contains a wealth of readily accessible information about soils in New Zealand and is undergoing continuing development and upgrade.

In addition to the development of digital soil maps, soil profile data are now being acquired using digital tools including (digital) cameras, hand-held portable X-ray fluorescence devices (pXRF) for elemental analysis in the field, and various spectrometers that have been used to predict a wide array of physical and chemical properties (once the results are matched and calibrated against traditional lab measurements of the properties, such as grain sizes, bulk density, and soil organic carbon). The use of these devices to digitally characterise soils is referred to as digital soil morphometrics, defined as the application of instruments and techniques for measuring, mapping, and quantifying soil properties.

18.4.4 New Zealand Land Resource Inventory and Land Use Capability Classifications

The New Zealand Land Resource Inventory (NZLRI) and the derived Land Use Capability (LUC) classifications were adapted for New Zealand from the system first published by the Soil Conservation Service, US Department of Agriculture, in 1961. The Water and Soil Division of the Ministry of Works developed the original database, and the first edition of the LUC Survey Handbook for New Zealand was published in 1971. Although primarily developed to help provide an inventory of erosion and potential mitigation, it has subsequently become much more. In the 3rd edition (published in 2009), LUC classification is defined as 'a systematic arrangement of different kinds of land according to those properties that determine its capacity for long-term sustained production'. Ian Lynn and colleagues at Manaaki Whenua – Landcare Research have continued the update and management of the database.

The NZLRI provides national maps of key land resource information including soil, underlying rock materials, slope, vegetation, and erosion susceptibility. The LUC classifies land into eight 'capability' classes, based on information from the NZLRI (Table 18.3). The LUC classes range from Class 1, the most versatile land with minimal limitations for most land uses, to Class 8 which is land on steep, rocky, often high altitude, areas, that is not suitable for productive use. However, Class 8 land forms some of New Zealand's more spectacular scenery and conservation land. The capability classes can then be interpreted as to the suitability of particular land areas for specific land uses including cropping, pasture, and forestry (Table 18.3). New Zealand coverage of NZLRI and the LUC can be viewed at the 'our environment' website https://ourenvironment.scinfo.org.nz/

Table 18.3 Overview of the land use capability classes in relation to general versatility of land use, from the LUC Survey Handbook (Lynn et al. 2009)

LUC class	Arable cropping suitability	Pastoral grazing suitability	Production forestry suitability	General suitability
1	High to low	High to low	High to low	Multiple use of land
2				
3				
4				
5	Unsuitable			Pastoral or forestry land
6				
7				
8		Unsuitable	Unsuitable	Conservation land

and at https://lris.scinfo.org.nz/layer/48076-nzlri-land-use-capability/.

The data for the NZLRI were assembled from a range of previously published soil and geological maps complemented by aerial photograph interpretation and some field ground-truthing. It has been periodically reviewed and updated. Like all GIS, map-based information care should be exercised in its interpretation. The original target scale for the NZLRI was 1:63,360 (now 1:50,000) and some of it was based on the 1:253,440 (four miles to the inch) scale maps. While it is possible to zoom in to much larger scales, as a visualisation tool, the map units need to be ground-truthed to ensure accuracy at the farm or paddock-scale, i.e. there are fundamental differences between the scale at which soils on a parcel of land were surveyed, the scale at which a map was presented or published, and the scale at which the map can be viewed in a GIS.

As the first complete national inventory of soils, rock type, slope, erosion, and vegetation, the NZLRI was an important accomplishment and widely used in many applications, including some that were beyond those anticipated by the original designers and mappers. The LUC system has been, and remains, an important tool in sustainable land management and is now embedded in legislation (e.g. LUC surveys are integrated into the 'One Plan' of the Horizons [Manawatu-Whanganui] Regional Council, which defines how the natural and physical resources are to be cared for and managed).

There are three levels in the LUC classification (Fig. 18.4). The LUC class is the highest level and provides an indication of the general category of land use that may be best suited to the land, essentially a general 'index of versatility' (Table 18.4). At the second level, the LUC class is subdivided into four subclasses based on the dominant limitation that could render the land unsuitable, or require particular management practices to overcome, for a particular use. The four limitations recognised are **e** erosion; **w** wetness; **s** soil; and **c** climate. The third level is subdivided into land use capability units. Each unit with a particular combination of capabilities and limitations is assigned a number. The units are defined for individual maps and are used to discriminate similarities or differences between parcels of land within farms, catchments, or regions.

Detailed definitions and instructions for undertaking LRI and LUC surveys are included in the Land Use Capability Survey Handbook which is available from https://www.landcareresearch.co.nz/__data/assets/pdf_file/0017/50048/luc_handbook.pdf or hard copies may be purchased from the New Zealand Society of Soil Science.

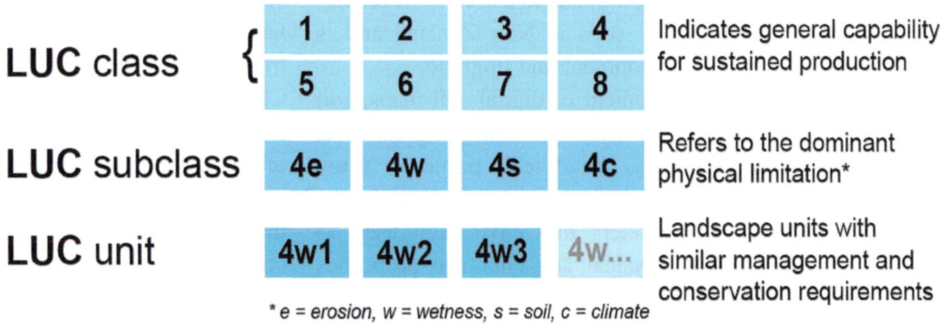

Fig. 18.4 The three levels of the land use capability (LUC) classification (adapted from the Land Use Capability Survey Handbook (see below)

Table 18.4 An example of the matching table for versatility for non-arable land use from the Topoclimate South study. From Carrick (2002)

Land quality code	Land qualities that materially influence non-arable horticulture production potential					
	y	r	a	n	w	d
Non-arable horticulture versatility rating	Potential rooting depth (cm)	Maximum limitation to root penetrability	Maximum limitation to profile aeration capacity	Salinity (EC, mS/cm)	Maximum short-term water-logging conditions	Soil water deficit (mm)
O1	>90	Minimal	Slight	<0.4	Slight	0–100
O2	60–90	Moderate	Moderate	0.4–0.8	Moderate	100–200
O3	25–60	Severe	Severe	0.8–2	Very severe	200–400
O4	<25	Very severe		>2		>400

18.4.5 Land Use Suitability

Land use suitability assessment was described in a 1976 report of the Food and Agriculture Organisation of the UN http://www.fao.org/3/X5310E/x5310e00.htm#Contents. Land use suitability assessment is allied to land use capability assessment as it also seeks to assess the suitability of a tract of land for a range of potential land uses. Within an area of study, a set of nominated land use types (LUT) are delineated. A set of variables, usually known as 'land qualities' or 'soil qualities' that will impact the performance of LUTs across the study area are determined. This terminology should not be confused with that in the 'soil quality' analysis (Sect. 18.4.8). A key step is to determine the land and soil variables that materially affect the performance of the LUTs. For each LUT the relationship between land quality, and the estimated LUT suitability class, is determined. The land qualities can be expressed in either continuous variables or classes.

A project called Topoclimate South, undertaken in Southland from 1998 to 2001, was based on the land use suitability method. The purpose of Topoclimate South was to identify land for further regional investment in agriculture. The project also delivered a sound basis for environmental assessment and mitigation. Sam Carrick and the Topoclimate South soils team upgraded the regional soil classification and soil map of the Southland lowlands and low hill country. The land use suitability analysis in Southland provides an example of the land use suitability assessment process (Table 18.4). The Southland team assessed four land use types: non-arable, arable, pastoral, and forestry. One of the goals was to assess the versatility of the land, where versatility relates to the range of crops that would be assessed as suitable for a given land area.

Applications of land use suitability analysis are wide-ranging and include:

- understanding of the effects of land and soil on the performance of specific land uses;
- determining favourable locations for specific enterprises;
- analysis of land use opportunities and risks;
- identification of indicators of soil resilience;
- determination of the level of versatility of land to support a wide range of enterprises; and
- identification of the soil (class) and land character required to support a particularly specialised enterprise.

18.4.6 Soil Versatility

The concept of soil versatility classification for arable land was introduced in New Zealand by Les Molloy in 1980, and based on two broad parameters: the physical characteristics of the root zone, and climate zones. A highly versatile soil is one that is capable of growing a wide range of crops suited to its particular climate (essentially synonymous with 'high class' soils (Sect. 18.4.11), or LUC class 1 and 2 soils, (Sect. 18.4.4)). Thus, highly versatile soils occur on flat land or very gentle slopes (<7°), have a potential rooting depth of at least 0.9 m, offer little resistance to root penetration, and suffer very few days per year of soil water deficit.

Versatility classifications were developed for arable crop production and orchard crop production by Alistair Wilson and Trevor Webb. Versatility indices provide the basis for defining soil and land characteristics perceived to influence crop production and soil management. Criteria used to assess such characteristics include topography (land slope), soil physical and chemical characteristics, environmental characteristics, and agro-climatic characteristics. For example, agro-climatic characteristics include the duration of frost-free periods (days), growing-degree days (November-April), chill units, and sunshine hours, and these attributes are then arranged into six classes (very high to extremely low) each of which has a numerical rating 1–6 (1 being the optimum rating, or the class with fewest agro-climatic limitations, and 6 the lowest rating).

An example of the derivation and use of soil versatility classification is a study of the Conroy land system of Central Otago undertaken by Allan Hewitt in 1995. The aims of the study were to identify the most versatile land capable of non-arable cropping that was irrigable, the least productive land with little potential for productive use, and 'other' land with limited productive potential but which may support native plants or pasture. Versatility indices were estimated by summing numerical suitability class assignments for irrigation, non-arable, arable, and pasture land uses, with the lowest index being the most versatile soil. The most versatile soils had versatility indices of 9. In contrast, soils with a versatility index of 17 were identified as the least versatile according to the parameters evaluated.

18.4.7 Soil Vulnerability

Soil vulnerability is a risk analysis that identifies possible undesirable outcomes that may occur as a result of specific soil properties. Commonly, the vulnerability is expressed in qualitative terms. Soil process modelling may enable quantitative expression of risk. An example is the model 'CLUES' which assesses nitrate contamination risks and pathways through the New Zealand landscape. The design and implementation of CLUES were led by Sandy Elliot and Graeme McBride. The model provides a spatial assessment of the likelihood of nitrate contamination in groundwater and surface water across the landscape.

Pedological inputs to the CLUES model included

- an assessment for each soil of its potential to leach nitrate and contribute contaminants to ground or surface water bodies; and
- assessment of the potential for nitrate removal from soil water by passage through poorly oxygenated (reduced) soils, particularly Gley Soils, either Perch-gley or groundwater gleys, before the water reaches a groundwater or a river or lake.

The two inputs were combined to produce a generalised national map of the potential for nitrate to reach freshwater bodies.

The nitrate soil vulnerability concept was developed further by extending it to include mapping of other potential contaminants or degradation processes and by expressing the concept in terms of risk analysis. Risk was assessed by identifying a possible hazard, its likelihood of occurrence, and its possible consequence. Likelihood depended on the pressure of a particular management practice, and the vulnerability of the landscape to that pressure. The consequence of the hazard depended on the environmental sensitivity of the response to the hazard, and the effect of the risk on the values of environmental services. The vulnerability classes were based only on soil or land factors. Trevor Webb and others provided an example of mapping of soil vulnerability of nitrate and phosphorus leaching, microbial bypass flow, and soil runoff potential, for two areas of Canterbury in a book published by Canterbury Regional Council. Similar maps are now available through S-map that use a range of 'pedotransfer' (soil modelling) functions to generate derivative maps of vulnerability and other properties.

18.4.8 Soil Quality

Over the last century, agriculture has changed many of our soils that were originally ill-suited for farming. Acidity has been corrected by liming, low fertility by fertilisers, drought by irrigation, wetness by drainage, and subsoil pans by deep ripping. These and other interventions have contributed greatly to national prosperity and to increasing intensity of land use. However, more intensive production, whether by horticulture, cropping, or intensive pastoralism, raises the issue of sustainability. Can the soil functions we rely on to be sustained without loss of soil quality or adverse effects on the wider environment? It is essential that soil quality is monitored so that loss of quality can be identified and managed or mitigated.

Soil quality was defined broadly by Else Bünemann and others in 2018, as 'the capacity of a soil to function within ecosystem and land-use boundaries to sustain biological productivity, maintain environmental quality, and promote plant and animal health'. Thus, soil quality more specifically is about the status of important soil properties that change with the effects of land management (called 'dynamic soil properties' or 'management dependant properties' or simply 'soil quality indicators'). Soil quality indicators may change in response to land management over short periods of time (days, weeks, or years). Soil quality is assessed using soil quality indicators that show levels of soil properties that imply the status of important soil health processes. Indicators are used to identify any factor that may be amiss and requires further attention. If the indicator points to a possible issue then further investigation, using a wider range of soil properties, including more detailed and expensive measurements, may be required. The scientific literature mostly uses the term 'soil quality' rather than 'soil health' because 'health' is encumbered by many shades of meaning. However, general literature on the state of the environment often uses the terms interchangeably.

Systematic soil quality monitoring in New Zealand was initiated at a regional scale by Manaaki Whenua Landcare – Research with support from regional councils. Following the investigation of a wide range of potential indicators, they selected seven soil quality indicators: total carbon, total nitrogen, pH, Olsen phosphorus, mineralisable nitrogen, soil dry bulk density, and macroporosity (Table 18.5). The selected indicators were based on dynamic soil properties that indicated soil biological reserves, fertility reserves, chemical reserves, and physical reserves. Research continues into soil quality indictors and monitoring and there is scope for the development of new and powerful indicators employing new technology.

The SINDI (soil indicators) website [http://sindi.landcareresearch.co.nz/] provides a means for interpreting the indicator values derived from the soil sampling and analysis to aid in developing appropriate management response. The focus for indicator research has been on soil condition and does not include soil erosion or soil contamination. Although the SINDI was designed for regional-scale monitoring it can be applied at any scale given an appropriate sampling strategy.

Soil quality indicators are being incorporated into the national state of the environment monitoring programmes with rapid changes occurring over recent years. The latest information is on the Ministry for the Environment website [https://www.mfe.govt.nz/land/state-of-our-land].

18.4.9 Visual Soil Assessment

Visual soil assessment (VSA) is a relatively simple, quick, method of assessing the state of the soil, in the field, to help inform land management decisions and, if repeated

Table 18.5 Soil quality indicators used by the SINDI (Soil **INDI**cators) soil quality monitoring system in New Zealand

Indicator[a]	Indicator of
Total carbon	Total carbon resource and carbon sequestration
Total nitrogen	Total nitrogen resource, and carbon/nitrogen ratio as an indication of soil saturation with nitrogen
pH	Acidity/alkalinity
Olsen-extractable plant-available phosphate	Plant-available phosphorous; excess or deficiency
Anaerobically mineralisable nitrogen	The pool of soil microbes, and the pool of labile nitrogen
Macroporosity	The percentage of pores capable of storing readily available water, providing drainage, supplying aeration for roots
Dry bulk density	Soil structure and compaction

[a]Measurements are made from samples of the top 10 cm of the soil profile, bulked from cores sampled in 25 m transects. Other indicators (e.g. aggregate stability) are also used for specific crops as needed

regularly, can be used to monitor changes over time. Globally, there are a number of visual soil assessment methods developed for particular situations. In New Zealand Graham Shepherd led the development of a VSA system that is published in four volumes, comprising field guides and soil management guidelines for both flat to rolling land and for hill country. The guides are available at https://www.landcareresearch.co.nz/publications/books/visual-soil-assessment-field-guide/download-field-guide.

In the New Zealand VSA system, a number of soil and plant indicators are rated on a defined three-point scale as either poor, moderate, or good. Each indicator is then multiplied by a weighting factor and the total of all the indicator rankings gives an overall ranking score for the site as good, moderate, or poor. Undertaking the exercise provides the operator with an increased awareness of the state of their soil and an incentive to work to improve the land management to improve the scores if they are low.

In Antarctica, Iain Campbell, Megan Balks, and Graeme Claridge developed a VSA system to assess impacts of human activities on the relatively pristine Antarctic soil environment.

18.4.10 Soil Stocks, Soil Natural Capital, Soil Functions, and Ecosystem Services

Four linked concepts—soil stocks, soil natural capital, soil functions, and soil ecosystem services—are useful for developing an understanding of the relationships between soil and land evaluation, soil processes, and sustainable land management. Such understanding starts with the soil profile, and moves through the links between soils, land, and soil processes, to the larger focus of processes of the atmosphere, hydrosphere, geosphere, and biosphere, together with human society. The four concepts together help form technical linkages and enable multidisciplinary understanding.

a. Soil stocks and soil natural capital

Soil stocks are defined as the soil attributes that materially influence ecosystems. Generally, people are familiar with the idea of soils having carbon stocks which support carbon sequestration. The concept has been extended to include a range of other soil stocks, for example, the stock of soil pores that supports soil water storage capacity, or the stock of available soil phosphorus that supports plant growth. The basic unit is the soil profile or its three-dimensional counterpart, the pedon. The pedon then becomes the fundamental (spatial) soil individual which comprises a 'bundle' of soil stocks. A soil-landscape or soil map unit is an aggregate of similar, contiguous, pedons (referred to as a polypedon, Fig. 18.5).

Most businesses use some form of capital resource to support their enterprise, be it labour, finance, or land, to create goods or services for profit. The term 'natural capital' identifies the natural resources used by businesses and ecosystems. The term '*soil* natural capital' identifies soil as one of the natural capital stocks to be valued for the services they provide, for example, filtering of contaminants, supplying nutrients to plants, storing and releasing water, and drainage.

An objective of introducing the concept of 'soil natural capital' is for soils to be ascribed value that can be factored into economic analyses. Thus the soil may be better appreciated for its roles in supporting the major earth systems including productive agroecosystems and the health of natural ecosystems. Therefore, soils may be considered by the business community, with investment in soils being 'an opportunity to enhance services and mitigate risks'.

A soil natural capital approach may be integrated into most techniques of land evaluation, for example, land capability or land use suitability, by analysing the goodness of fit between land use needs and available soil stocks. The goodness of fit becomes the estimate of capability or

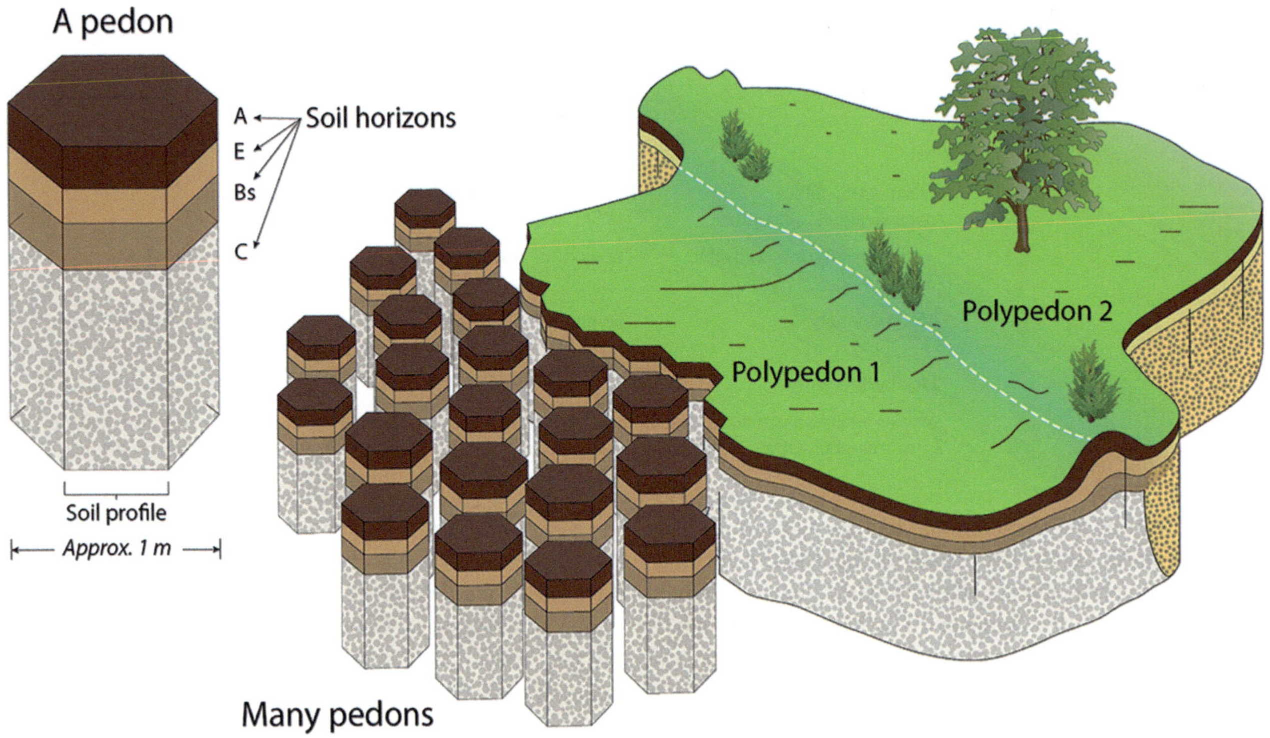

Fig. 18.5 The relationship between a soil profile, a soil pedon, and the wider landscape. Adapted from Blum et al. (2006)

suitability. Where good fits are identified there is the best likelihood of good production and environmental performance. The goodness of fit is, therefore, a non-monetary value that is expressed in terms of the efficient utilisation of a landscape.

b. Soil functions

Soil functions are soil processes that provide something directly useful for people and/or ecosystems. For example, removal of nitrate from soil water is achieved by two important soil functions: soil water storage, and denitrification. Soil water storage depends on the soil pore network (a soil stock). When nitrate-charged water drains through the soil, some is stored in the soil pores. The greater the soil water storage capacity, the more water will remain in storage until both the water and nutrients, including nitrate, are taken up by plants, thus minimising nitrate loss to groundwater while maximising nutrient and water availability to the plant. Denitrification (conversion of nitrate-N to N_2 gas which is released to the atmosphere) occurs in soil horizons saturated by water in a state of biochemical reduction (gleying). Both water storage and denitrification are functions that decrease the amount of nitrate that may escape through the soil into groundwater or streams.

c. Ecosystem services

Ecosystem services are the useful things generated by ecosystems to support life on Earth. Originally understood in reference to biological systems, the concept has been broadened to incorporate the processes involved in land, air, water, soil, and topography as well as ecology. The value in providing environmental services and protecting the environment was estimated, by US researchers Collins and Larry (in 2007) to be approximately 40–70 times greater than the value of agricultural production (although if a food shortage was to ensue then undoubtedly food production would be valued more highly). Ecosystem services are divided into four main categories (Table 18.6).

Soil science has a role to understand soil ecosystem services in the context of; the wider earth systems, natural and agro-ecosystems, and to bridge the gaps between soil science and related sciences that are working in the ecosystem services field. Estelle Dominati, Alec Mackay, and Murry Patterson have defined ecosystem services, and the related concepts, in the context of New Zealand's soils, as well as exploring soil natural capital valuation. In 2016, Dominati and others developed a 'more holistic, integrative modelling approach' which they called 'an ecosystem approach to quantify soil performance for multiple outcomes'. As an

Table 18.6 Four categories of ecosystem services with soil-related examples

Ecosystem service category	General examples	Soil-related examples
Provisioning	Production of food and clean water	Source of nutrients for plants Hosts biodiversity Source of clay for industrial uses
Regulating	Controlling earth systems such as the control of climate or disease	Filtering of contaminants, Carbon sequestration, Regulation of rainfall runoff/infiltration/drainage Repository for dead organisms that prevents spread of disease
Supporting	Nutrient cycles and crop pollination	Building foundation strength Nutrient cycling and storage
Cultural	Spiritual, intellectual, recreational, aesthetic appreciation, and social and heritage values	Preserves archaeological heritage Basis for sports fields Spiritual ties to the land/earth/whenua Source of clay for cultural uses Preserves record of climate and landscape history, e.g. pollen (in wet organic soils), plant phytoliths, and tephra records

example, they used their modelling approach to quantify the ecosystem services provided by dairy farming on two contrasting soils in the Waikato region. Ecosystem services quantification is a means of estimating the differences in the value of services provided by the two contrasting soils (Fig. 18.6).

Overall, Dominati and colleagues reported that, the economic value of services was 35% higher in the Horotiu soil than the Te Kowhai soil. The differences between the two soils in the provision of food production, flood mitigation, and the filtering of nutrients and contaminants, were greater than the differences in the provision of support and recycling of wastes.

d. Quantifying soil natural capital using the stock adequacy method

The stock adequacy method was designed to use soil information stored in soil survey databases to evaluate the adequacy of key soil stocks to support the soil functions required for specific land uses.

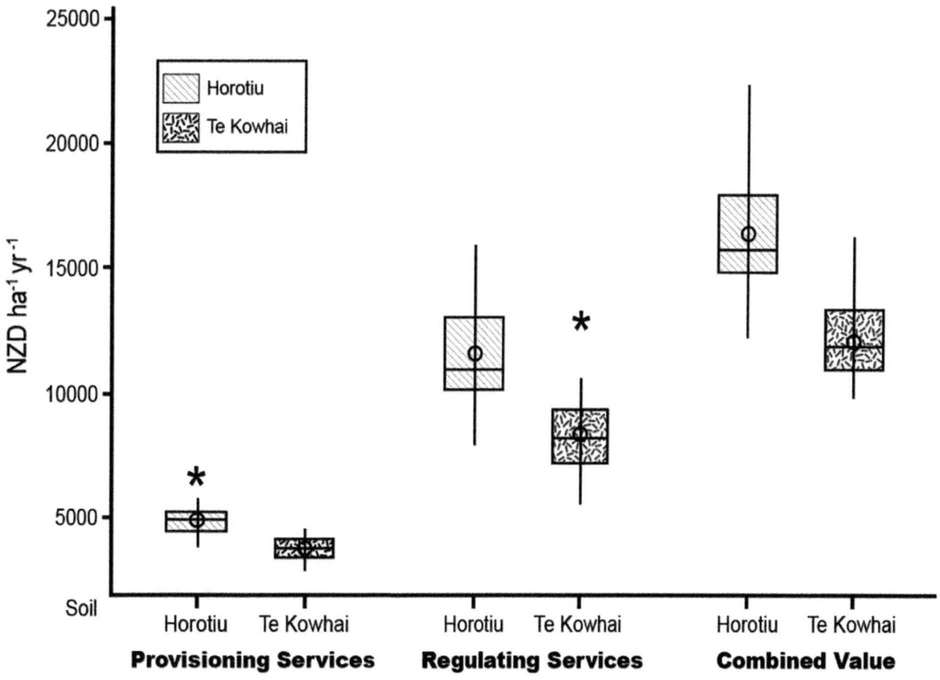

Fig. 18.6 Comparison of the estimated value of provisioning and regulating services for Horotiu (Allophanic Soil) and Te Kowhai (Gley Soil) soils under dairy farming over 35 years (adapted from Dominati et al. 2016). Box plots show median values (line across box), means (circles), and upper and lower quartiles at the ends of the box. Whiskers extend to the highest and lowest values; the asterisks indicate outlier values

The stock adequacy method to quantify soil natural capital follows five key steps:

1. Define the land use type (LUT) of interest (e.g. extensively grazed pastoral sheep farming);
2. Identify the key soil services needed for the specified LUT (e.g. soil water storage capacity to maintain water supply to grassroots to maintain grass production, Olsen P level, and soil pH as indicators of nutrient availability, capacity of soil to hold and prevent nitrate leaching to groundwater, capacity of soil to deliver water to groundwater to maintain summer stream flows, and so on);
3. For each soil service that is identified determine the soil functions that drive the service (e.g. the soil profile data (soil stocks) needed to calculate soil water storage include potential root depth, soil texture, and stoniness, or total, readily available, and field capacity soil moisture retention data, for each soil horizon in the plant root zone);
4. Estimate the adequacy of each stock (e.g. for a given rainfall and crop evapotranspiration, evaluate the proportion of the soil water requirements able to be met by the available soil moisture storage);
5. Collate the adequacy calculations for all the different required soil stocks to give an overall estimate of the soil stock adequacy for the given soil and land use.

It is evident that such calculations, if they are to have any relevance, will quickly become quite complex and depend on a range of assumptions about crop growth, root depths, water and nutrient requirements, and the weather conditions, among other parameters.

18.4.11 Assessment of Soils of High Value for (Potential) Food Production

Prior to introduction of the Resource Management Act 1991 (RMA), the Town and Country Planning Act 1953 controlled urban development and specifically demanded that soils of high value for food production (actual or potential) be protected from urban development. A system of determining what constituted soils of high value for food production was developed. Subdivision of such areas (for instance, much of the mainly flat-lying, tephra-mantled alluvial land, supporting a patchwork of Allophanic Soils, between Hamilton and Cambridge) was limited to a minimum lot size of 10 acres ($\sim$4 ha), which was deemed to be large enough for an economic horticultural production unit.

Under the RMA, there is no specific protection of soils for food production as the act is permissive, rather than prescriptive. The purpose (Section 5) of the RMA includes, among other things, sustainable management of physical resources to enable social, economic, and cultural well-being *while* safeguarding the life-supporting capacity of air, water, soil, and ecosystems for the reasonably foreseeable needs of future generations. The soil is thus considered as one of many 'environmental' variables. See http://www.legislation.govt.nz/act/public/1991/0069/latest/DLM231905.html for the full statement. Hence, under the RMA, any land use is acceptable, in theory, so long as it does not reduce the land's potential to meet the needs of future generations, reduce its life-supporting capacity, or damage the environment (within an agreed framework). A key point is that the activity itself is not assessed but rather the effects of the activity on the environment. Provided environmental effects can be mitigated, land use changes are normally permitted.

Between 1991 and 2019, under the RMA, there has been rapid urban expansion onto land that had previously been protected under the Town and Country Planning Act. The lack of protection of the most versatile soils (Sect. 18.4.6) was the result of the enabling and balancing requirements of the RMA where economic and social considerations have held sway. In 2019, the government started the process to introduce a National Policy Statement (NPS) for Highly Productive Land to protect some of the soils that are of high-value for food production, sometimes referred to as 'versatile' or 'high class' soils. Most regional and district councils use the additional terms 'elite' for LUC class 1 soils and 'prime' for LUC class 2 soils. Thus, the definitions of what constitutes our most valuable land for food production are again up for debate.

One widely accepted measure, given that the data are readily available, is to consider that all land of land use capability (LUC) classes 1 and 2 represent our most productive land that deserves protection from urban development. Classes 1 and 2 comprise flat or gently undulating (<7°) land that can be repeatedly cultivated without danger of excessive water erosion, which is versatile, and suitable for a large range of crops. LUC 1 and 2 soils are most common among the Recent Soils and Allophanic Soils but also occur in some areas of other soil orders. In some regions, such as the Pukekohe area, a specific combination of climate and soils has successfully supported intensive vegetable growing on some rolling (8–15°) land (LUC 3) and, thus, there is a strong case for some areas of LUC 3 land to also be protected from urban development.

The case for protection of LUC 1 and 2 land is a compelling one. In New Zealand, only 0.7% of the land is LUC 1 with a further 4.5% in LUC 2. Thus, only 5.2% of New Zealand comprises our most versatile and productive land (Figs. 18.7 and 18.8). Soils on LUC 1 and 2 land have high versatility (hence have the capability to produce a wide range of crops), high energy-use efficiency, high yields, high pollution absorption capacity, and moderate or better soil resilience. Thus, the further loss of versatile soils from

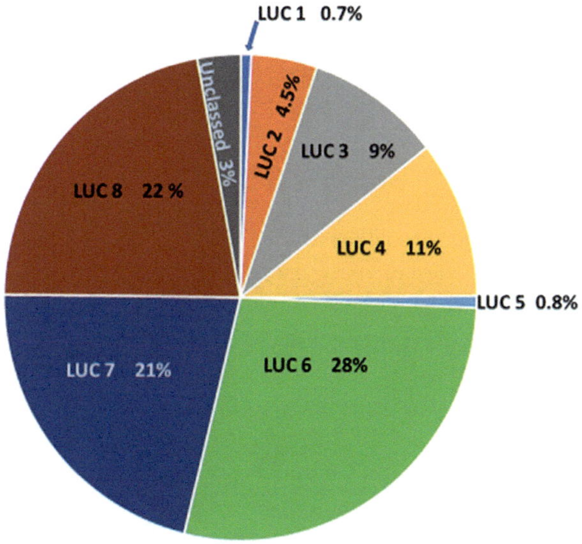

Fig. 18.7 Proportion of NZ land area in each land use capability class. Data from Manaaki Whenua – Landcare Research

potential arable use limits the ability for current and future generations to undertake arable production on land most suitable and sustainable for food production. The loss of the relatively rare LUC 1 land, and the scarce LUC 2 land, also contradicts the desirability to maintain soil and biodiversity and to preserve soil ecosystems. Conserving representatives of all our soils, as taonga, could also be considered in much in the same way as kiwi and kauri are treasured and preserved.

It is no coincidence that much of the most productive land is adjacent to our most rapidly growing cities. Yet there are plenty of less versatile soils available for housing.

Around 36% of New Zealand comprises rolling (7–15°) and hilly (15–25°) lands which are typically suitable for supporting permanent pasture and sheep and cattle grazing, but which are generally not suitable for intensive crop production for the most important foods such as wheat, corn, potatoes, and other fruits and vegetables. Only ~54% of New Zealand's soils can sustain agriculture as the balance is steep, erosion prone LUC 7 and 8 lands which are better utilised for protection forestry (Fig. 18.8).

Thus, if roads, buildings, and other permanent sealed features, are allowed to continually expand over the 5% or so of New Zealand that is ideally suited, for intensive cropping and horticultural production, there will be no suitable alternative areas in New Zealand to support such food production. Already more than 10% of the highly productive LUC class 1 and 2 soils of New Zealand have been converted into lifestyle blocks and urbanisation (Fig. 18.9). The Ministry for the Environment and NZ Stats reported, in 2019, that in the Waikato and Auckland areas about 35% of highly productive land is now lost to lifestyle blocks and urbanisation.

The National Policy Statement that is currently under consideration is likely to lead to all councils having to implement measures to protect soils of high value for food production. However, the planning processes that may lead to such implementation are slow, protections can be 'got around', and in the meantime urban motorway, industrial, and housing areas, are rapidly expanding onto high-class soils in a number of regions.

18.4.12 Soil Security—The Future of Land Evaluation?

Alex McBratney and collaborators have summarised the challenges faced by soil science to 'secure' our soils. Five dimensions of soil security have been articulated: capability, condition, capital, connectivity, and codification. The concept of soil security seeks to combine land evaluation methods with concepts of sustainable land management. The five dimensions are described in more detail below.

1. *Capability* of the soil/land encompasses the well-established LUC assessment (Sect. 18.4.4) along with land use suitability assessment methods (Sect. 18.4.5). Soil vulnerability (Sect. 18.4.7) could also be included in an assessment of capability.
2. *Condition* of the soil/land is encapsulated by identifying a set of soil quality indicators that allow monitoring and assessment of the functioning of soils and ecosystems in sustaining life (Sect. 18.4.8).
3. *Capital* is the value of soil natural capital and ecosystem services (Sect. 18.4.9). A goal is to express soil capital values in monetary terms but technical and acceptability issues need to be overcome. However, alternative, non-monetary, means are available employing indices or soil function output values, and the Ministry of Primary Industries has recognised the role of soil in terms of natural capital within the developing National Policy Statement on high-class soils.
4. *Connectivity* is the linkage of soil information and land management decision making, including monitoring systems. In New Zealand a precursor of what this might become is provided by the linkage of S-map outputs (Sect. 18.4.3) to the Overseer nutrient management model.
5. *Codification* is the adoption of soil data into policies to influence decision making in the choices of how land is managed and in the design of sustainable landscapes. For example, the LUC classification, or an adaptation of it, has potential to guide decision making on which areas of land should be protected from urban development, and the Overseer model has potential to be used to guide decisions about fertiliser application. However, great

Fig. 18.8 High-class (LUC 1 and LUC 2) soils of New Zealand. The high-class soils, in black, cover only 5.2% of the area of New Zealand (adapted from diagrams supplied by Manaaki Whenua – Landcare Research)

caution needs to be exercised in using such arbitrary models which all have limitations due to the assumptions made in their development, the accuracy of input data, and their ability to account for the inherent variability in soils and landscapes. Unforeseen inequities can arise with a blind reliance on such models.

18.5 Concluding Comments: Aotearoa New Zealand Soils—A National Taonga and Our Role in Kaitiakitanga

Les Molloy, in 1988, concluded that '…only rarely have we stood back and celebrated our [New Zealand] soils as something beautiful, and perhaps even mysterious. For what other natural body, worldwide in its distribution, has so many interesting secrets to reveal to the patient observer?'. But how good are New Zealand soils? Given the diversity of soils, and their uses, there is no one answer to the question. European explorers initially thought that New Zealand soils were particularly fertile because of the large size of many native forest trees. However, their hopes were dashed as pioneer farmers discovered the limitations inherent in the soils of many areas. The early misconception arose from the ability of many New Zealand plants to be highly nutrient conservative. On acidic, low fertility, soils the trees utilise what small nutrient resources are available and very efficiently recycle nutrients to support their growth. For example, the kauri, among the largest trees in the world, often grows on low nutrient sites. However, we do have high-performing, versatile, high class-soils but they comprise only about 5% of New Zealand.

Fig. 18.9 Lifestyle blocks on LUC Class 1 and 2 lands near Hamilton. Such subdivisions were generally not allowed prior to the RMA (1991)

The introduction to this book (Chap. 1) likened soils to snowflakes in the sense that no two are alike. The contrasting range of soils described in Chaps. 2–17 suggests this is true. However, the *New Zealand Soil Classification* provides a meaningful pattern and coherent summary which thus enables understanding of the whole population of New Zealand soils. Soil classification enables all soils to be named and the names can be used to communicate the character, and assess the status, of the soil resource. Both the soil classification name, and an evaluation of soil suitability for various purposes, help develop concepts of each soil so we can comprehend the properties of the soil and the associated landscape. Accumulating, and communicating, knowledge about soil resource can help to utilise and manage the land wisely. Thus, it is important that New Zealanders are familiar with our main soils and so an understanding of the soil orders of the *New Zealand Soil Classification* is a useful way for people to become familiar with the range, variety, and key characteristics of our soil resource.

Soil science has provided New Zealand with an excellent understanding of the limitations of our soils, and a range of tools to overcome those limitations, making New Zealand an important producer of food and fibre. For example, Allophanic Soils have excellent soil physical properties but need regular inputs of phosphorus fertiliser to overcome the high P retention that renders phosphorus unavailable to plants. As we accumulate knowledge of the potential of, and risks to, our soils, and if we utilise our science-based knowledge, skills, and resources (including people, technology, infrastructure, education, and governance) appropriately in our production systems, then we can become better at wise, sustainable, stewardship (kaitiakitanga) of our soils and environment.

Our soils, like most New Zealanders, are fairly resilient and if well cared for will continue to support our food, biofuel, and fibre production, and waste disposal, as well as providing for biodiversity, natural ecosystems, and clean water. However, the time is fast approaching when the most versatile and highly productive soils will be a dwindling and irreplaceable resource if the current rates of urbanisation continue unabated. Soil is indeed a precious resource that should never be taken for granted.

The vision is that all New Zealanders appreciate the soil as a taonga that is critical for life on Earth, recycling nutrients, supporting plant growth, and providing us with food. Soils and associated ecosystems support all the unique terrestrial living organisms of Aotearoa. Soil is also our final resting place. Thus understanding and providing kaitiakitanga for the soil, with a sense of awe at soil's infinite variety, beauty, resilience, and the growth that soil supports is an important duty of all people in Aotearoa.

Further Reading

Almond PC, Tonkin PJ (1999) Pedogenesis by upbuilding in an extreme leaching and weathering environment, and slow loess accretion, south Westland, New Zealand. Geoderma 92:1–36

Andrew R, Dymond JR (2013) Expansion of lifestyle blocks and urban areas onto high-class land: an update for planning and policy. J R Soc N Z 43:128–140

Atkinson JD (1976) DSIR's first fifty years. New Zealand Department of Scientific and Industrial Research Information Series 115, 220p

Batjes NH (1996) Total carbon and nitrogen in soils of the world. Eur J Soil Sci 47:151–163

Beare MH, Williams H, Cameron KC (1999) On-farm monitoring of soil quality for sustainable crop production. In: Currie LD, Hedley MJ, Horne DJ, Loganathan P (eds) Best soil management practices for production. Proceedings of the 12th annual fertilizer and Lime Research Centre workshop, Massey University, pp 81–90

Best E (1925) Maori agriculture. Dominion Museum Bulletin No 9, 315 p. Reprinted 1976. Whitcombe & Tombs, Wellington

Blum WEH, Warkentin BR, Frossard EE (2006) Soil, human society and the environment. Geol Soc Lond Spec Publ 266:1–8

Brook MS, Ferrar S (2019) Hartley Travers Ferrar (1879–1932) and his geological legacy in Antarctica, Egypt and New Zealand. Earth Sci Hist 38:43–58

Brooking T, Pawson E (2011) Seeds of empire: the environmental transformation of New Zealand. I. B. Tauris & Co. Ltd, London, 276p

Büneman EK, Bongiorno G, Bai Z et al (2018) Soil quality—a critical review. Soil Biol Biochem 120:105–125

Campbell IB, Balks MR, Claridge GGC (1993) A simple visual technique for estimating the impact of fieldwork on the terrestrial environment in ice-free areas of Antarctica. Polar Rec 29:321–328

Carrick S (2002) Methodology for soil versatility evaluation in the Southland region. Landcare Research Contract Report: LC0203/013. Funded by Crops for Southland and Environment Southland, 45p

Carrick S, Hainsworth S, Lilburne L et al (2014) S-map @ the farm scale? Towards a national protocol for soil mapping for farm nutrient budgets. In: Currie LD, Christensen CL (eds) Nutrient management for the farm, catchment and community. http://flrc.massey.ac.nz/publications.html. Occasional Report No 27. Fertilizer and Lime Research Centre, Massey University, Palmerston North, 11p

Chapin FS (1993) The evolutionary basis of biogeochemical soil development. Geoderma 57:223–227

Collins S, Larry E (2007) Caring for our natural assets: an ecosystem services perspective. USDA, Forest Service, Pacific Northwest Research Station, Portland, OR, 11p

Curran-Cournane F (2018) Land use pressures confronting finite land and soil resources—Pukekohe test case. Report prepared for Ministry for Primary Industries by Auckland Council. Research and Evaluation Unit, Auckland Council, 29p

Curran-Cournane F, Fraser S, Hicks D et al (2013) Changes in soil quality and land use in grazed pasture within rural Auckland. N Z J Agric Res 56:102–116

Curran-Cournane F, Vaughan M, Memon A et al (2014) Trade-offs between high class land and development: recent and future pressures on Auckland's valuable soil resources. Land Use Policy 39:146–154

Curran-Cournane F, Cain T, Greenhalgh S et al (2016) Attitudes of a farming community towards urban growth and rural fragmentation—an Auckland case study. Land Use Policy 58:241–250

Curran-Cournane F, Golubiewski N, Buckthought L (2018) The odds appear stacked against versatile land: can we change them? N Z J Agric Res 61:315–326. https://doi.org/10.1080/00288233.2018.1430590

Cutler EJB (1966) Some pedological factors of importance to the pasture ecosystem in the South Island. Proc N Z Ecol Soc 13:65–68

Davies J (2017) The business case for soil. Nature 543:309–311. https://doi.org/10.1038/543309a

Deloitte (2018) New Zealand's food story: the Pukekohe hub. Quick stats on horticulture in New Zealand and reports. https://news.fuseworksmedia.com/977fca0b03b941e188dc482236d0ce09?utm_source=dlvr.it&utm_medium=twitter

Dominati E (2011) Quantifying and valuing the ecosystem services of pastoral soils under dairy use. Unpublished PhD thesis, Massey University, Palmerston North, New Zealand, 413p

Dominati E, Patterson M, Mackay A (2010) A framework for classifying and quantifying the natural capital and ecosystem services of soils. Ecol Econ 69:1858–1868

Dominati E, Mackay A, Bouma J et al (2016) An ecosystems approach to quantify soil performance for multiple outcomes: the future of land evaluation? Soil Sci Soc Am J 80:438–449

Duchaufour P (1982) Pedology (trans: Paton TR) George Allen & Unwin, London, 448p

Duchaufour P, Souchier B (1978) Roles of iron and clay in genesis of acid soils under a humid, temperate climate. Geoderma 20:15–26

Elliott S, McBride G, Shankar U et al (2008) CLUES spatial DSS: from farm-scale leaching models to regional decision support. In: Sanches-Marre M, Bejar J, Comas J et al (eds) Proceedings of the international modelling congress on environmental modelling and software integrating sciences and information technology for environmental assessment and decision making 4th Biennial meeting. International Environmental Modelling and Software Society (iEMS)

FAO Food and Agriculture Organisation of the United Nations (1976) A framework for land evaluation. FAO Soils Bulletin 32. Rome, Italy. ISBN 92-5-100111-1. http://www.fao.org/3/X5310E/x5310e00.htm#Contents

FAO (1988) Soil map of the world, revised legend. World Soil Resources Report 60. Food and Agriculture Organisation of the United Nations, Rome

Farmer VC, Mew B, Clayden B et al (1984) Hypothesis on the pedogenesis of the 'gley podzols' of the west coast of South Island, New Zealand. J Soil Sci 35:99–102

Fraser S, Jones H, Hill R et al (2014) Soil windows—unravelling soils in the landscape at farm scale. In: Currie LD, Christensen CL (eds) Nutrient management for the farm, catchment and community. Occasional Report No 27. Fertilizer and Lime Research Centre,

Massey University, Palmerston North, New Zealand, 11p. http://flrc.massey.ac.nz/publications.html
Grab SW, Dickinson KJM, Mark AF et al (2008) Ploughing boulders on the Rock and Pillar Range, south-central New Zealand: their geomorphology and alpine plant associations. J R Soc N Z 38:51–70
Gibbs HS, Raeside J, Dixon JK (1945) Soil erosion of the high country of the South Island. N Z DSIR Bull 92, 72 p
Grange LI, Gibbs HS (1947) Soil erosion in New Zealand. Part 1. Southern half of North Island. N Z Soil Bur Bul 1, 28 p
Grange LI, Taylor NH, Sutherland CF (1939) Soils and agriculture of part of Waipa County. DSIR Bulletin No 76. Government Printer, Wellington, 85p
Grealish GJ, Fitzpatrick RV, Hutson JL (2015) Soil survey data rescued by means of user friendly soil identification keys and toposequence models to deliver soil information for improved land management. GeoResJ 6:81–91
Grealish G (2017) New Zealand soil mapping protocols and guidelines. Report for Technical Advisory Group for Soil Mapping Protocols, Landcare Research, Palmerston North, 30p. http://www.envirolink.govt.nz/assets/Envirolink/Tools/R12-4-New-Zealand-soil-mapping-protocols-and-guidelines.pdf
Gregg P (2008) Soil erosion and conservation—surveys and research, Te Ara—the Encyclopedia of New Zealand. http://www.TeAra.govt.nz/en/soil-erosion-and-conservation/page-4. Accessed 30 Dec 2019
Greenhalgh S, Samarasinghe O, Curran-Cournane F et al (2017) Using ecosystem services to underpin cost–benefit analysis: Is it a way to protect finite soil resources? Ecosyst Serv 27(Part A):1–14. https://doi.org/10.1016/j.ecoser.2017.07.005
Hartemink AE, Minasny B (eds) (2016) Digital soil morphometrics. Springer, Dordrecht. 442 p
Hewitt AE (1992) Soil classification in New Zealand: legacy and lessons. Aust J Soil Res 30:843–854
Hewitt AE (1995) Soils of the Conroy land system, Central Otago. Landcare Research Science Series 13, 56p
Hewitt AE (1997) Are NZ soils distinctive? A subterranean view of NZ ecosystems. NZ Soil News 45:7–16
Hewitt AE (2007) Soils, Te Ara. The Encyclopaedia of New Zealand. http://www.teara.govt.nz/TheBush/Landscapes/Soils/1/en
Hewitt AE (2010) New Zealand Soil Classification, 3rd edn. Landcare Research Science Series No 1. Lincoln, Manaaki Whenua Press
Hewitt AE (2013) Survey of New Zealand soil orders. In: Dymond JR (ed) Ecosystem Services in New Zealand: conditions and trends. Section 1.10, pp 121–130. ISBN 978-0-478-34737-1
Hewitt AE, Van Wambeke A (1982) Soil, land and site evaluation methods for landuse planning. In: Proceedings of 1st international symposium: soil, geology and landforms: impact of land-use planning in developing countries, Bangkok, Thailand, pp Gl.l–Gl.43
Hewitt AE, Webb TH, Dominati E, Hedley H (2012) Soil natural capital definition using land evaluation principles. In: Minasny B, Malone BP, McBratney AB (eds) Digital soil assessments and beyond. CRC Press, Boca Raton, FL, pp 15–18
Hewitt A, Dominati E, Webb TH et al (2015) Soil natural capital quantification by the stock adequacy method. Geoderma 241–242:107–114. http://www.sciencedirect.com/science/article/pii/S0016706114004121
Hill RB, Sparling GG, Frampton C, Cuff J (2003) National soil quality review and programme design. Ministry for the Environment report. http://www.mfe.govt.nz/publications/ser/tech-report-74-land-jan03.html
Hogan D, Williamson B (eds) (1999) New Zealand is different: chemical milestones in New Zealand history. Clerestory Press. NZ Institute of Chemistry, 308p
Huang Y-T, Lowe DJ, Churchman GJ et al (2016) DNA adsorption by nanocrystalline allophane spherules and nanoaggregates, and implications for carbon sequestration in Andisols. Appl Clay Sci 120:40–50
Ibanez JJ, De-Alba S, Lobo A et al (1998) Pedodiversity and global soil patterns at coarse scales. Geoderma 83:171–192
Johnson DL, Watson-Stegner D (1987) Evolution model of pedogenesis. Soil Sci 143:349–366
Johnson DL, Keller EA, Rockwell TK (1990) Dynamic pedogenesis: New views on some key soil concepts and a model for interpreting Quaternary soils. Quat Res 33:306–319
Klingbiel AA, Montgomery PH (1961) Land capability classification, agriculture handbook 210, Washington, DC, US Department of Agriculture
Land Monitoring Forum (2009) Land and soil monitoring: a guide for state of the environment and regional council reporting. New Zealand
Lilburne L, Hewitt AE, Sparling G et al (2002) Soil quality in New Zealand: policy and the science response. J Environ Qual 31:1768–1773
Lilburne L, Sparling G, Schipper L (2004) Soil quality monitoring in New Zealand: development of an interpretative framework. Agr Ecosyst Environ 104:535–544
Lilburne LR, Webb TH, Hewitt AE (2013) S-map database manual. Unpublished report. Landcare Research
Lilburne L, Webb T, Palmer DJ et al (2014) Pedotransfer functions from S-map for mapping water-holding capacity, soil-water demand, nutrient leaching vulnerability and soil services. In: Currie LD, Christensen CL (eds) Nutrient management for the farm, catchment and community. Occasional Report No 27. Fertilizer and Lime Research Centre, Massey University, Palmerston North, 7 p. http://flrc.massey.ac.nz/publications.html
Lowe DJ, Tonkin PJ (2010) Unravelling upbuilding pedogenesis in tephra and loess sequences in New Zealand using tephrochronology. In: Proceedings of the 19th world congress of soil science "soil solutions for a changing world", Brisbane, Australia, 1–6 Aug 2010; Symposium 1.3.2 Geochronological Techniques and Soil Formation, pp 34–37. https://iuss.org/19th%20WCSS/Symposium/pdf/0116.pdf
Lowe DJ, Newnham RM, McCraw JD (2000) Quaternary environmental change in New Zealand and its effects on the soil pattern. In: Adams JA, Metherell AK (eds) Soil 2000: new horizons for a new century, proceedings of the Australian & New Zealand 2nd joint soils conference, vol 3, Christchurch, New Zealand, 3–8 Dec 2000. New Zealand Society of Soil Science, Lincoln, New Zealand, pp 117–118
Lynn I, Basher L (1994) Principles underlying land systems in resource assessment of hill and mountain lands in New Zealand. In: Webb TH (ed) Soil-landscape modelling in New Zealand. Landcare Research Science Series, 5:38–51
Lynn IH, Manderson AK, Page MJ et al (2009) Land use capability survey handbook—a New Zealand handbook for the classification of land, 3rd edn. Hamilton, AgResearch; Lincoln, Landcare Research; Lower Hutt, GNS Science
Manderson A, Palmer AS, Mackay A et al (2007) Introductory guide to farm soil mapping. AgResearch, Palmerston North, 64 p
Manderson A, Mackay A, Palmer AS (2007) Environmental whole farm management plans: their character, diversity, and use as agri-environmental indicators in New Zealand. J Environ Mgt 82:319–331
McBratney A, Damien J, Field DJ (2014) The dimensions of soil security. Geoderma 213:203–213
McCraw JD (1959) Periglacial and allied phenomena in western Otago. N Z Geogr 15:61–68

McIntosh PD, Hewitt AE, Giddens K, Taylor M (1997) Benchmark sites for assessing the chemical impacts of pastoral farming on loessial soils in southern New Zealand. Agric Syst Environ 65:267–280
McIntosh P, Laffan M, Hewitt A (2005) The role of fire and nutrient loss in the genesis of the forest soils of Tasmania and southern New Zealand. For Ecol Manag 220:185–215
Ministry for the Environment and Stats NZ (2019) New Zealand's environmental reporting series: environment Aotearoa 2019. www.mfe.govt.nz and www.stats.govt.nz (http://www.mfe.govt.nz/environment-aotearoa-2019)
Ministry for Primary Industries (2019a) Proposed national policy statement for highly productive land. https://www.mpi.govt.nz/news-and-resources/consultations/proposed-national-policy-statement-for-highly-productive-land/
Ministry for Primary Industries (2019b) Valuing highly productive land. A discussion document on a proposed national policy statement for highly productive land. MPI Discussion Paper 2019/05 [co-published by Ministry for the Environment], 70p. https://www.mpi.govt.nz/dmsdocument/36624-discussion-document-on-a-proposed-national-policy-statement-for-highly-productive-land
Newnham RM, Lowe DJ, Williams PW (1999) Quaternary environmental change in New Zealand: a review. Prog Phys Geogr 23:567–610
Parfitt R, Saigusa M, Eden DN (1984) Soil development processes in an Aqualf-Ochrept sequence from loess with admixtures of tephra, New Zealand. J Soil Sci 35:625–640
Pedro G, Jamagne M, Begon JC (1978) Two routes in genesis of strongly differentiated acid soils under humid, cool-temperate conditions. Geoderma 20:173–189
Richardson BF (2021) The price we pay for land: the political economy of Pukekohe's development. J N Z Pac Stud 9(1) (in press)
Roche MM (1994) Land and water: water and soil conservation and central government in New Zealand [1941–1988]. Historical Branch, Department of Internal Affairs, Wellington, 219p
Rossiter DG (1996) A theoretical framework for land evaluation. Geoderma 72:165–190
Rossiter DG, Van Wambeke A (1996) ALES Automated Land Evaluation System software
Rossiter DG, Hewitt AE, Dominati E (2018) Pedometric valuation of the soil resource. In: McBratney A, Minaseny B, Stockman U (eds) Pedometrics. Progress in soil science. Springer, pp 521–546 https://doi.org/10.1007/978-3-319-63439-5_17
Sase T, Hosono M (1996) Vegetation histories of Holocene volcanic ash soils in Japan and New Zealand—relationship between genesis of melanic volcanic ash soils and human impact. Earth Sci (Chikyu Kagaku) 50:466–482
Sattler K, Anderson B, Macintosh A et al (2016) Estimating permafrost distribution in the maritime Southern Alps, New Zealand, based on climatic conditions at rock glacier sites. Front Earth Sci 4:4. https://doi.org/10.3389/feart.2016.00004
Savin IY (2019) Scale problem in modern soil mapping. Bull V V Soil Inst Dokuchaev 97:5–20. https://doi.org/10.19047/0136-1694-2019-97-5-20
Shepherd G (2000) Visual soil assessment volume 1: field guide. Cropping and pastoral grazing on flat to rolling country. Horizons Manawatu and Landcare Research, Palmerston North, 84p. ISBN 1-877221-92-9. https://www.landcareresearch.co.nz/publications/books/visual-soil-assessment-field-guide/download-field-guide. Accessed Mar 2020
Shepherd G, Ross C, Basher L et al (2000) Visual soil assessment volume 2: soil management guidelines. Cropping and pastoral grazing on flat to rolling country. Horizons Manawatu and Landcare Research, Palmerston North, 44p. ISBN 1-877221-93-7. https://www.landcareresearch.co.nz/publications/books/visual-soil-assessment-field-guide/download-field-guide. Accessed Mar 2020
Shepherd G, Janssen H (2000) Visual soil assessment volume 3: field guide for hill country landuses. Horizons Manawatu and Landcare Research, Palmerston North, 48p. ISBN 1-877221-94-5. https://www.landcareresearch.co.nz/publications/books/visual-soil-assessment-field-guide/download-field-guide. Accessed Mar 2020
Shepherd G, Jassen H, Bird L (2000) Visual soil assessment volume 4: soil management guidelines for hill country landuses. Horizons Manawatu and Landcare Research, Palmerston North, 24p. ISBN 1-877221-95-3. https://www.landcareresearch.co.nz/publications/books/visual-soil-assessment-field-guide/download-field-guide. Accessed Mar 2020
Soil-security-dimensions-brief. https://www.soils.org/files/am/pdfs/soil-security-dimensions-brief.pdf
Soil Survey Staff (2014) Keys to Soil Taxonomy, 12th edn. USDA Natural Resources Conservation Service, Washington, DC, 362 p
Sparling GP, Schipper LA (2002) Soil quality at a national scale in New Zealand. J Environ Qual 31:1848–1857
Sparling G, Schipper L (2004) Soil quality monitoring in New Zealand: trends and issues arising from a broad-scale survey. Agr Ecosyst Environ 104:545–552
Sparling G, Parfitt RL, Hewitt AE, Schipper LA (2003) Three approaches to define desired soil organic matter contents. J Environ Qual 32:760–766
Sparling GP, Schipper LA, Bettjeman W, Hill R (2004) Soil quality monitoring in New Zealand: practical lessons from a 6-year trial. Agr Ecosyst Environ 104:523–534
Sparling GP, Lilburne L, Vojvodic-Vukovic M (2008) Provisional targets for soil quality indicators in New Zealand. Landcare Research Science Series No 34. Manaaki Whenua Press, Lincoln
Tate KR, Giltrap DJG, Claydon JJ et al (1997) Organic carbon stocks in NZ terrestrial ecosystems. J R Soc N Z 27(3):315–335. https://doi.org/10.1080/03014223.1997.9517541
Taylor NH (1955) First presidential address. The role of soil science in New Zealand problems. Trans R Soc N Z 82:961–971
Taylor NH, Pohlen IJ (1968) Classification of New Zealand soils. In: Soils of New Zealand. Part 1. Soil Bur Bull 26(1):15–33
Taylor MD (2010) Discussion paper on indicators target ranges for soil quality monitoring. Internal report: environment Waikato, 30p
Taylor MD, Kim ND, Hill RB, Chapman R (2010) A review of soil quality indicators and five key issues after 12 yr soil quality monitoring in the Waikato region. Soil Use Manag 26(3):212–224
Taylor M, Cox N, Littler R, Drewry J (2017) Trends in soil quality monitoring data in the Waikato region 1995–2015. Waikato Regional Council Technical Report 2017/26
Tonkin PJ (2007) A history of soil survey and selected aspects of soil conservation in New Zealand. N Z Soil News 55:59–71 and 102–115
Trangmar BB, Cutler EJB (1983) Soils and erosion of the Sumner region of the Port Hills, Canterbury, New Zealand. NZ Soil Survey Report 70. New Zealand Soil Bureau, DSIR. https://doi.org/10.7931/dl1-ssr-70
Webb TH (ed) (1994) Soil-landscape modelling in New Zealand. Landcare Research Science Series 5, 173p
Webb TH (2003) Identification of functional horizons to predict physical properties for soils from alluvium in Canterbury, New Zealand. Aust J Soil Res 41:1005–1019
Webb T, Hewitt A, Lilburne L, McLeod M et al (2010) Mapping of vulnerability of nitrate and phosphorus leaching, microbial bypass flow, and soil runoff potential for two areas of Canterbury. Environment Canterbury Regional Council Report No R10/125
Webb TH, Wilson AD (1994) A manual of land characteristics for land evaluation of rural land. Landcare Research Science Series 10, 32p

Webb TH, Wilson AD (1995) Classification of land according to its versatility for orchard crop production. Landcare Research Science Series 8, 31p

Webb T, Jessen M, Mcleod M et al (1995) Identification of high class land. Broadsheet, New Zealand Association of Resource Management, Nov 1995, pp 109–114

Wilde RH, Willoughby EJ, Hewitt AE (1999) The New Zealand soil spatial database data manual. Unpublished Landcare Research report

Wilson AD (1984) Soil evaluation and classification system for crop production involving tillage. N.Z. Soil Bureau Soil Resources Report SR 3, 45p

Wilson AD, Giltrap DJ (1984) Soil evaluation and classification system for orchard production. N.Z. Soil Bureau Soil Resources Report SR 6, 54p

Wood J, Wilmshurst J, Newnham RM et al (2017) Evolution and ecological change during the New Zealand Quaternary. In: Shulmeister J (ed) Landscape and quaternary environmental change in New Zealand. Atlantis Press (Springer), pp 235–291

Wood V (2003) Appraising soil fertility in early colonial New Zealand: the 'biometric fallacy' and beyond. Environ Hist 9:393–405

Appendix A
Ratings Tables for New Zealand Soil Properties

See Tables A.1 and A.2.

Table A.1 Ratings for soil chemical properties after L. C. Blakemore, P. L. Searle, and B. K. Daly 1987. Methods for chemical analysis of soils. NZ Soil Bureau Scientific Report 80. 103p. ISSN 03041735. Reproduced with permission of Manaaki Whenua – Landcare Research

	Rating	Very high	High	Medium	Low	Very low
A1: Ratings for soil pH, carbon, nitrogen, and phosphorus						
	pH (1:2.5 soil: water)	>9.0 (extremely alkaline) 8.4–9.0 (strongly alkaline) 7.6–8.3 (moderately alkaline)	7.1–7.5 (slightly alkaline) 6.6–7.0 (near neutral)	6.0–6.5 (slightly acid) 5.3–5.9 (moderately acid)	4.5–5.2 (strongly acid)	<4.5 (extremely acid)
Organic matter	Organic carbon (%)	>20	10–20	4–10	2–4	<2
	Total nitrogen (%)	>1.0	0.6–1.0	0.3–0.6	0.1–0.3	<0.1
	C/N	>24	16–24	12–16	10–12	<10
Phosphorus	Truog (μg/g)	>50	30–50	20–30	10–20	<10
	Olsen (μg/g)	>50	30–50	20–30	10–20	<10
	0.5M H_2SO_4 (cmol$^{(+)}$ kg^{-1})	>40	20–40	10–20	5–10	<5
	Inorganic (cmol$^{(+)}$ kg^{-1})	>50	30–50	20–30	10–20	<10
	Organic (cmol$^{(+)}$ kg^{-1})	>70	50–70	20–50	10–20	<10
	Total (cmol$^{(+)}$ kg^{-1})	>120	80–120	40–80	20–40	<20
	P retention (%)	90–100	60–90	30–60	10–30	0–10
A2: Ratings for cation exchange related properties						
Cation exchange properties (NH_4OAc, pH7)	CEC (cmol$^{(+)}$ kg^{-1})	>40	25–40	12–25	6–12	<6
	Σ Bases (cmol$^{(+)}$ kg^{-1})	>25	15–25	7–15	3–7	<3
	BS (%)	80–100	60–80	40–60	20–40	<20
	Ca (cmol$^{(+)}$ kg^{-1})	>20	10–20	5–10	2–5	<2
	Mg (cmol$^{(+)}$ kg^{-1})	>7	3–7	1–3	0.5–1	<0.5
	K (cmol$^{(+)}$ kg^{-1})	>1.2	0.8–1.2	0.5–0.8	0.3–0.5	<0.3
	Na (cmol$^{(+)}$ kg^{-1})	>2	0.7–2	0.3–0.7	0.1–0.3	<0.1
KCl—extract Al (cmol$^{(+)}$ kg^{-1})		>5	2–5	0.5–2.0	0.1–0.5	<0.1
Exchange Acidity (pH 8.2) (cmol$^{(+)}$ kg^{-1})		>60	30–60	15–30	5–15	<5
Reserve	K_c	>0.5	0.35–0.5	0.20–0.35	0.10–0.20	<0.10
	Mg_r	>30	15–30	7–15	3–7	<3

(continued)

A. E. Hewitt et al., *The Soils of Aotearoa New Zealand*, World Soils Book Series,
https://doi.org/10.1007/978-3-030-64763-6

Table A.1 (continued)

	Rating	Very high	High	Medium	Low	Very low
A3: Ratings for aluminium, iron, silica and sulphur contents, and soluble salts						
Acid oxalate—extractable	Al (%)	>3.0	1.0–3.0	0.5–1.0	0.2–0.5	<0.2
	Fe (%)	>2.0	1.0–2.0	0.5–1.0	0.2–0.5	<0.2
	Si (%)		>0.5	0.15–0.5	0.05–0.15	<0.05
Pyrophosphate—extractable	Al (%)	>2.0	0.8–2.0	0.4–0.8	0.1–0.4	<0.1
	Fe (%)	>1.2	0.6–1.2	0.3–0.6	0.1–0.3	<0.1
Dithionate-citrate—extractable	Al (%)	>2.0	1.0–2.0	0.5–1.0	0.2–0.5	<0.2
	Fe (%)	>4.0	2.0–4.0	1.0–2.0	0.5–1.0	<0.5
	Phosphate—extractable sulphur or sulphate (μg S/g)	>150	50–150	15–50	5–15	<5
Soluble salts	Conductivity (millimho/cm) 1:5 extract at 25 °C	>2	0.8–2	0.4–0.8	0.15–0.4	<0.15
	Salts (%)	>0.7	0.3–0.7	0.15–0.3	0.05–0.15	<0.05

Table A.2 Ratings for soil physical properties

Physical parameter	Very low	Low	Medium	High	Very high
Potential rooting depth[a] (cm)	<20 v. shallow	20–40 shallow	40–80 mod. deep	80–120 deep	>120 v. deep
Soil dry bulk density (g cm^{-3})	<0.6	0.6–0.9	0.9–1.5	1.5–1.8	>1.8
Total porosity (%v/v)	<35	35–45	45–60	65–75	>75
Macroporosity (%v/v)	<5	5–10	10–20	20–25	>25
Penetration resistance[b] (kPa)	<1000	1000–1500	1500–2200	2200–3000	>3000
Degree of packing[c] (kPa)	<500 loose	500–1000 very friable	1000–2200 friable	2200–3000 firm	>3000 very firm
Permability[d] (mm h^{-1})	<0.4 extremely slow	0.4–4 slow	4–18 moderately slow	18–72 moderate	>72 rapid
Profile readily available water[a] (mm) over potential rooting depth or 1.5 m depth	<25	25–50	50–100	100–130	>130
Soil depth[d] (cm)	<20 very shallow	20–45 shallow	45–100 moderately deep		>100 deep

[a]From A. D. Wilson and D. J. Giltrap 1984. Soil evaluation and classification system for orchard crop production. Soil Resources Report SR6. Soil Bureau, DSIR. 54p

[b]Adapted from E. Griffiths 1984. Interpretation of soil morphology for assessing moisture movement and storage. NZ Soil Bureau Scientific Report 74, 20p

[c]Determined using Singleton Blade and 6 mm tip penetrometer after E. Griffiths 1984. Interpretation of soil morphology for assessing moisture movement and storage. NZ Soil Bureau Scientific Report 74, 20p

[d]Adapted from S-map ratings

Appendix B
Correlation of Orders of New Zealand Soil Classification with the Nearest Equivalent Taxa of *Soil Taxonomy* and *World Reference Base*

NZ Soil Classification Soil order	*Soil Taxonomy* Equivalent soil order (or suborder/great group)	*World Reference Base* Equivalent reference soil group
Allophanic Soils	Andisols (except Vitrands)	Andosols
Anthropic Soils	Entisols or unclassified	Anthrosols or Technosols
Brown Soils	Inceptisols (also Entisols)	Cambisols
Gley Soils	Aquic suborders of Inceptisols, Entisols, Oxisols, etc.	Gleysols
Granular Soils	Ultisols	Nitisols
Melanic Soils	Mollisols, Vertisols	Vertisols, Calcisols, Chernozems or Phaeozems
Organic Soils	Histosols	Histosols
Oxidic Soils	Oxisols	Ferralsols
Pallic Soils	Alfisols, Inceptisols	Cambisols, Luvisols, or Planosols
Podzol Soils	Spodosols	Podzols
Pumice Soils	Andisols: mainly Vitrands (also Vitricryands, Vitraquands)	Andosols
Raw Soils	Not-soil or Entisols	Regosols, Arenosols or unclassified
Recent Soils	Entisols (also Inceptisols)	Fluvisols, Arenosols, Leptosols, Cambisols, Regosols, or Umbrisols
Semiarid Soils	Aridisols	Luvisols, Cambisols, or Solonetz
Ultic Soils	Ultisols	Planosols

[a]After Hewitt (2010), Soil Survey Staff (2014) and World Reference Base (2015). The correlations given here are a guide only—for accurate classifications of a specific soil the soil properties and details of the particular classification documents should be consulted

Hewitt AE (2010) New Zealand Soil Classification. 3rd edn. Landcare Research Science Series No. 1. Lincoln, Manaaki Whenua Press 136 p
Soil Survey Staff (2014) Keys to Soil Taxonomy, 12th ed. USDA Natural Resources Conservation Service, Washington, DC 362 p
World Reference Base (2015) http://www.fao.org/3/i3794en/I3794en.pdf accessed on 2 March 2020

A. E. Hewitt et al., *The Soils of Aotearoa New Zealand*, World Soils Book Series,
https://doi.org/10.1007/978-3-030-64763-6

Index

A. E. Hewitt et al., *The Soils of Aotearoa New Zealand*, World Soils Book Series,
https://doi.org/10.1007/978-3-030-64763-6

D

H

I

T

GPSR Compliance

The European Union's (EU) General Product Safety Regulation (GPSR) is a set of rules that requires consumer products to be safe and our obligations to ensure this.

If you have any concerns about our products, you can contact us on ProductSafety@springernature.com

In case Publisher is established outside the EU, the EU authorized representative is:

Springer Nature Customer Service Center GmbH
Europaplatz 3
69115 Heidelberg, Germany

Batch number: 10731528

Printed by Printforce, the Netherlands